D0164337

Exploring
the Planets

Exploring the Planets

Second Edition

ERIC H. CHRISTIANSEN
W. KENNETH HAMBLIN
Brigham Young University

Prentice Hall, Englewood Cliffs, New Jersey 07632

Library of Congress Cataloging-in-Publication Data
Christiansen, Eric H.
    Exploring the planets / Eric H. Christiansen, W. Kenneth Hamblin.
  -- 2nd ed.
      p. cm.
    Hamblin named first on previous ed.
    Includes index.
    ISBN 0-02-322421-5
    1. Planets--Geology. 2. Astrogeology. I. Hamblin, W. Kenneth
(William Kenneth). II. Title.
QB603.G46H36 1995
559.92--dc20   94-18017
                                                        CIP

Editor: Robert A. McConnin
Production supervision: Francesca Drago
Interior design: Angela Foote
Cover design: Singer Design
Manufacturing buyer: Allan Fisher

© 1995, 1990 by Prentice-Hall, Inc.
A Simon & Schuster Company
Englewood Cliffs, New Jersey 07632

All rights reserved. No part of this book may be
reproduced, in any form or by any means,
without permission in writing from the publisher.

The author and publisher of this book have used their best efforts in
preparing this book. These efforts include the development, research,
and testing of the theories and programs to determine their effectiveness.
The author and publisher shall not be liable in any event for incidental
or consequential damages in connection with, or arising out of, the
furnishing, performance, or use of these programs.

Printed in the United States of America
10 9 8 7 6 5 4 3 2 1

ISBN 0-02-322421-5

Prentice-Hall International (UK) Limited, *London*
Prentice-Hall of Australia Pty. Limited, *Sydney*
Prentice-Hall Canada Inc., *Toronto*
Prentice-Hall Hispanoamericana, S A, *Mexico*
Prentice-Hall of India Private Limited, *New Delhi*
Prentice-Hall of Japan, Inc., *Tokyo*
Simon & Schuster Asia Pte. Ltd., *Singapore*
Editora Prentice-Hall do Brasil, Ltda., *Rio de Janeiro*

TABLE OF CONTENTS

PREFACE

Endowed with an irrepressible curiosity, the human species has always explored. Humans first explored their immediate surroundings, then neighboring mountains, valleys, and rivers, then adjacent lands, and eventually the vast oceans and faraway continents. Even the night skies were the object of intense scrutiny and wonderment. With time, people would use telescopes to explore the planet vicariously. Eventually, our desire to explore and the capabilities of our technology broke the bonds of Earth's gravity and took us to visit, sometimes through remote eyes, our planetary neighbors.

Twenty-five years ago, we first extended our reach beyond our home planet with a brief visit to the Moon. The exploration of the solar system that followed was just as revolutionary in its way as the exploration of our own planet was in the centuries that preceded this step. In little more than two decades we collected pieces of the Moon, probed the atmosphere of Venus, mapped the huge volcanoes and canyons of Mars, photographed the surface of Mercury, examined comets and asteroids close up, and explored the frigid surfaces of the moons of Jupiter, Saturn, Uranus, and Neptune. Of all the planets, only Pluto has not been visited by spacecraft. There will be other exciting periods of space exploration to be sure, and many great and significant discoveries are waiting to be made, but no other generation will explore for the first time so many worlds in our solar system. Subsequent missions to the planets, among them *Galileo* (to Jupiter), *Clementine* (to the Moon and an asteroid), and perhaps *Mars 94* (Russian mission to Mars) will continue the flood of data to help us understand the processes that shaped these planetary objects, but the most exciting phase—when we explored the surfaces of these bodies for the first time—is over.

With the completion of the Magellan mission to Venus, it is now an appropriate point to summarize the scientific results of these exploratory missions. We have written this book to share with you some of the knowledge gained from this important period of discovery and to introduce you to the study of geology by considering the fundamentals of how planets originate and evolve.

We hope this book will appeal to all who are curious about planets, moons, and other objects in the solar system. Our approach is nonmathematical, but nonetheless analytical; as such, we hope it will be used by a broad range of interested people, even those with little scientific background. The book emphasizes the surfaces, internal structures, and histories of the planets from a geological point of view. We assume only what is typically obtained in high school science classes and the intuition gained by experience with our surroundings. The metric system of measurement is used throughout the book; distances are expressed in meters and kilometers, masses in kilograms, and temperatures in degrees Kelvin.

Pedagogy

This book could serve as a text for an introductory college science course or as a supplement to college astronomy or geology courses. The chapters are ordered to provide a logical flow of ideas regarding the development of the planets. After considering some fundamental principles needed to understand the planets, we start with a discussion of the smallest and simplest bodies, the asteroids and meteorites, and their role as building blocks of the inner planets. From there we discuss progressively larger terrestrial planets. We proceed from the Moon to larger and more complex planets

Venus and Earth. The outer planets are discussed in the order of their occurrence outward from the Sun. Comets are described as the icy parent bodies of the outer planets. A closing chapter compares the planets by briefly examining the principal processes that have shaped them. Although Chapters 1 and 2 would be most helpful if studied first, the other chapters generally contain enough background material to stand alone or to be read out of sequence.

We have also incorporated several learning aids to help you in your study of the planets. Each chapter begins with a succinct list of the major concepts that should be understood. A preliminary reading of these items will focus your attention on the fundamental principles. Important new terms are printed in bold type and are reiterated in a list at the end of each chapter. Thought-provoking questions should guide your review and help you to extrapolate beyond simple repetition of observations. Additional readings are also listed at the end of each chapter for those who wish to explore further on their own. A glossary provides short definitions of terms that may be unfamiliar. The definitions are in accordance with the latest edition of the Glossary of Geology published by the American Geological Institute.

Supplementary Materials

If you are using this book as a text for a college course, there are several aids that you can obtain from the publisher upon adoption. These include:

1. **Slide Set.** A set of 150 slides to complement lecture or laboratory presentations has been carefully selected and reproduced. Most of the slides are in color, greatly adding to the impact of classroom presentations.
2. *Geodisc*. This full-color video disc contains many slides in the slide set, but it also includes several videos just for planetary science, including *The Movie Series* (Earth, Mars, Miranda, L.A.) produced at the Jet Propulsion Laboratory and U.S. Geological Survey. Video clips of Venus dramatically show its amazing surface. Each video segment is a computer-generated virtual flight across the surface of a planet. In all, the *Macmillan Geodisc* has over 1200 photos, 200 drawings, 50 minutes of full-motion video, and 6 minutes of animation on various Earth-science topics to make your classroom presentations more vivid.
3. **Instructor's Guide.** This guide should help you utilize the text and related material more effec-

tively as an instructor. It contains an outline of each lecture for a typical semester course, as well as suggestions for lecture preparation, discussion material, out-of-class assignments, and visual aids.
4. **Test Bank.** A bank of more than 500 examination questions is available on diskette for IBM or Apple computers. The questions include true/false, multiple choice, short answer, and essay questions that can be modified to suit your needs. We have tested and revised these questions based on over 10 years of teaching this class.

New to This Edition

This book has been significantly revised in an effort to make it more complete and accurate.

1. All of the chapters have been updated with new information and rechecked for accuracy.
2. We have reorganized the book so that it flows more logically from the smallest to the largest terrestrial planets. Asteroids are discussed in Chapter 3, before any other planetary bodies. They are small and have simple histories. The position of this chapter also helps to reiterate the planetary formation and differentiation processes described in the preceding chapter. New photographs of Gaspra and Ida are the centerpieces of this chapter.
3. We have completely rewritten the chapter on Venus, based on the dramatic results of the recently completed Magellan mission. New photos, maps, and diagrams help to illustrate the nature of this, the most Earthlike, planet.
4. To better illustrate the dynamic nature of geologic processes on the planet, we have added new paintings of the planets by Teryl Bodily. These are intended to place the reader in unique positions in space and time where accretionary collisions, eruptions, plumes, and rings can be visualized.

Acknowledgments

We sincerely thank all those who have made the exploration of the planets possible—principally the citizens of the United States and the former Soviet Union. Through their desire to know and understand their surroundings, the people of these countries have appropriated funds to construct rockets, spacecraft, and sophisticated instruments that have endowed us with the remote vision we require to explore the planets. Other nations are now joining this endeavor and we hope that the exploration of space will yet become a collective endeavor, shared by all the peoples of Earth.

We are also grateful to the following colleagues who reviewed the entire text or various chapters and offered many helpful suggestions: C. J. Casella, Northern Illinois University; J. Cain, Florida State University; Ronald Greeley, Arizona State University; James W. Head III, Brown University; R. Craig Kochel, Southern Illinois University; Jeffrey Moore, Arizona State University; Carlton Moore, Arizona State University; Quinn Passey, Exxon Research; Lawrence A. Soderblom, U.S. Geological Survey; Joseph Veverka, Cornell University.

Although many are acknowledged in the illustration credits, many other people also helped to provide the various images of the planets, both photographic and shaded relief maps: Raymond M. Batson, U.S. Geological Survey; Kathy Hoyt, U.S. Geological Survey; Jay L. Inge, U.S. Geological Survey; Barbara Pope, National Space Science Data Center; Patricia Ross, National Space Science Data Center; and Jody Swann, U.S. Geological Survey.

If you have any comments about this book, please contact us at the address below or by E-mail (eric_christiansen@byu.edu).

Eric H. Christiansen
W. Kenneth Hamblin
Dept. of Geology
Brigham Young University
Provo, Utah 84602

TABLE 1.1

Physical Characteristics of the Planets and Their Major Satellites

Planetary Body	Semi-Major Axis (AU for Planets, 10^3km for Satellites)	Orbital Period (Days or Years (y))	Rotation Period (Days)	Density (g/cm^3)	Diameter (km)	Surface Composition	Atmosphere Composition
Mercury	0.387	87.97	58.65	5.44	4,800	basaltic	Na (thin)
Venus	0.723	224.7	243.0 R	5.25	12,104	basaltic	CO_2
Earth	1.000	365.26	1.00	5.52	12,756	basaltic & H_2O	$N_2 + O_2$
Moon	384	27.3	27.3	3.34	3,476	basaltic	None
Mars	1.524	686.98	1.03	3.93	6,787	basaltic	CO_2
Largest Asteroids							
Vesta	2.362	3.63 y	0.22	2.9	520	basaltic	None
Ceres	2.768	4.61 y	0.38	?	932	DCS	None
Pallas	2.773	4.62 y	0.33	?	533	D S	None
Jupiter	5.203	11.86 y	0.41	1.3	143,800		H_2 and He
Io	422	1.77	1.77	3.50	3,640	S compounds	SO_2 (thin)
Europa	671	3.55	3.55	3.03	3,130	water ice	None
Ganymede	1071	7.15	7.15	1.93	5,280	water ice D	None
Callisto	1884	16.69	16.69	1.79	4,840	water ice D	None
Saturn	9.54	29.46 y	0.43	0.69	120,660		H_2 and He
Mimas	186	0.94	0.94	1.12	392	water ice	None
Enceladus	238	1.37	1.37	1.00	500	water ice	None
Tethys	295	1.89	1.89	1.00	1,060	water ice	None
Dione	377	2.74	2.74	1.49	1,120	water ice	None
Rhea	527	4.52	4.52	1.24	1,530	water ice	None
Titan	1222	15.94	15.9	1.88	5,150	water ice C	N_2
Hyperion	1484	21.3	?	?	250	water ice	None
Iapetus	3562	79.33	79.33	1.03	1,436	H_2O ice DCS	None
Phoebe	12930	550.4 R	0.4	?	220	H_2O ice DC?	None
Uranus	19.18	84.01 y	0.72	1.28	51,120		H_2 and He
Miranda	130	1.41	1.41	1.35	470	water ice	None
Ariel	191	2.52	2.52	1.66	1,150	water ice	None
Umbriel	266	4.14	4.14	1.51	1,170	water ice	None
Titania	438	8.70	8.70	1.68	1,580	water ice	None
Oberon	586	13.46	13.46	1.58	1,520	water ice	None
Neptune	30.07	164.79 y	0.73	1.64	49,560		H_2 and He
Triton	355	5.88 R	5.88	2.01	2,700	N_2 and CH_4 ice	N_2, CH_4
Proteus	118	1.12	1.12	?	400	D H_2O ice	None
Nereid	5562	359.9	?	?	340	D H_2O ice	None
Pluto	39.44	247.7	6.4	2.06	2,284	nitrogen ice	N_2
Charon	17	6.39	6.4	2.06	1,192	H_2O ice	None

D = dark materials; silicates, carbonaceous, or methane
C = carbonaceous materials
S = silicates
R = retrograde orbit

CHAPTER 1

The Solar System

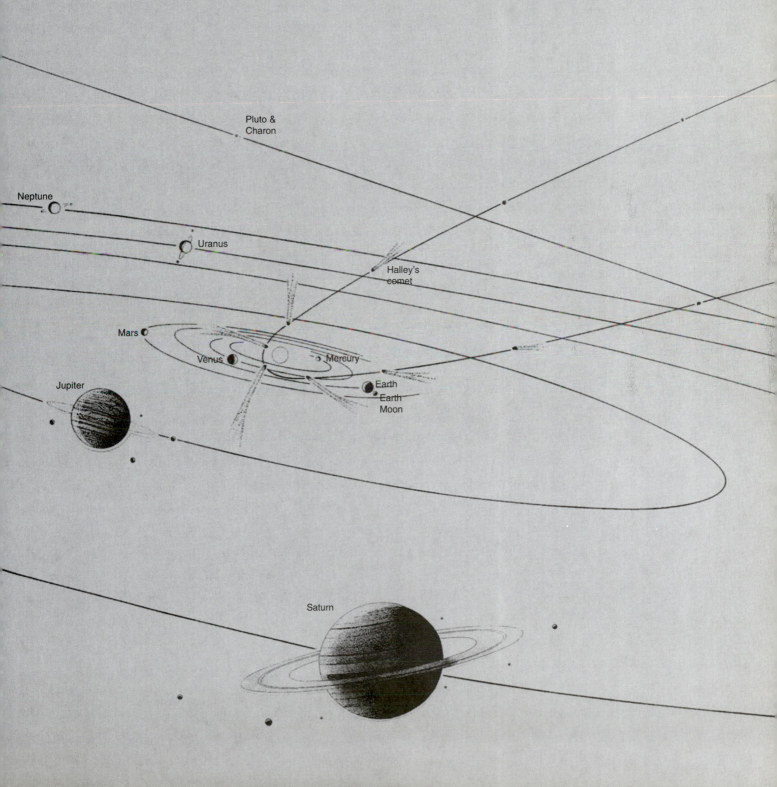

Pluto & Charon

Neptune

Uranus

Halley's comet

Mars

Venus

Mercury

Jupiter

Earth

Earth Moon

Saturn

Introduction

Our solar system consists of one star, a family of nine planets, at least 58 moons, thousands of asteroids, and billions of meteoroids and comets. In terms of mass, however, the solar system consists of very little else than the Sun itself. This ball of gas comprises 99.87 percent of the mass of the solar system. Most of the remaining 0.13 percent resides in Jupiter. Thus, most of the solar system is empty space. From the nearest star, using technology currently available on Earth, nothing would be seen of our solar system except the Sun, and it would appear only as a small yellowish star of a type common in the galaxy. Nonetheless, our new knowledge about the planetary objects that orbit the Sun is extremely important and is changing the way we look at the solar system and Earth itself.

In the annals of history, the second half of the twentieth century may well be remembered more for the exploration of the planets than for any other single achievement of humankind. No other generation has had the opportunity to reach beyond our own world, to see, touch, and hear the forces that shape our universe. Our objective in this chapter is to introduce you to the results of this vast undertaking by taking a brief survey of the nature of the major planetary objects in our solar system.

Major Concepts

1. A planet is a body, not large enough to generate nuclear fusion reactions, that orbits a star. Planets are largely composed of metals, silicates, ices, and gases.

2. The solar system contains nine planets and 24 other planetary bodies (moons and asteroids) with diameters greater than about 350 kilometers. Some icy comets in the far distant reaches of the solar system may also be this large.

3. The small inner planets (Mercury, Venus, Earth, and Mars) are composed mostly of silicate rocks and metals; the outer planets (Jupiter, Saturn, Uranus, and Neptune) are much larger, consist mostly of gaseous hydrogen and helium and ice, and have large systems of icy moons. Pluto, the smallest planet in the solar system, is similar to the moons of the outer planets in size and composition.

4. The asteroids, fragments of once larger silicate and metallic bodies, orbit the Sun and are concentrated between the orbits of Mars and Jupiter.

5. Comets are small icy bodies formed in the outer solar system. If their orbits become changed, they may enter the inner solar system for a short time before they vaporize and break up.

6. The surface features, compositions, internal structures, and other characteristics of planetary bodies are records of the events that shaped them. The distinctions between planets, moons, asteroids, and comets are largely arbitrary. Each world has a story to tell us about its development, adding to our understanding of how matter and energy interact in our solar system.

Planets, Moons, Asteroids, and Comets

By tradition, it has generally been assumed that there is a clear-cut distinction in the hierarchy of planets, moons, asteroids, and interplanetary debris (planets being the larger objects that orbit the Sun, and moons being smaller and orbiting a planet). As a result of the recent exploration of space, we are now obliged to look at the solar system differently. The solar system is really made up of 26 worlds larger than 1000 km in diameter, each distinct from the others (Table 1.1). These worlds record a variety of events in planetary evolution and shed spectacular new light on the origin and history of our own planet, Earth. The traditional practice of classifying objects in the solar system as planets, moons, asteroids, comets, or meteorites establishes artificial categories that may blur important similarities.

Planets are objects that orbit (or revolve) around a central luminous star (Figure 1.1). Our Sun is a medium-sized star, which like other stars generates energy by nuclear fusion at high temperatures. It is only one of about 100 billion stars that collectively form a slowly rotating spiral galaxy (Figure 1.2). Planets are commonly solid, but liquids and gases are also important constituents. Planets are too small for nuclear fusion reactions to have initiated within their interiors. The principal planets in order outward from the Sun are Mercury, Venus, Earth, Mars, Jupiter, Saturn, Uranus, Neptune, and Pluto. A **planetary body** is a general term referring to any body orbiting a star. Planetary bodies include planets and their natural satellites or moons, as well as smaller objects such as asteroids and comets. They range in size from the small asteroids to the largest of the planets.

The relative sizes of the major planetary bodies in the solar system are shown in Figure 1.3. Descriptive information about the planetary objects discussed in this book is listed in the table at the

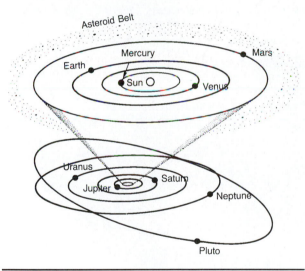

Figure 1.1
The orbits of the principal planets of the solar system are centered on the Sun—a medium-sized star. All of the planets, except Pluto, revolve about the Sun near the equatorial plane of the solar system. Pluto's elliptical orbit is tilted by over 17 degrees. Viewed from above the solar system, all the planets move in normal counterclockwise directions. The necessity of enlarging the inner solar system emphasizes the scale problem encountered in such a display. If Pluto's path were about the size of a bicycle tire, the orbits of the four inner planets would fit inside a circle smaller than a quarter. In addition to the planets, there are at least 44 moons, thousands of small rocky asteroids, and billions of icy comets. The asteroid belt, where most asteroids occur, is shown between the orbits of Jupiter and Mars.

Figure 1.2
A galaxy, like the one shown here, is a grouping of billions of stars. The Sun and the planets are part of the Milky Way galaxy, which is probably similar to this spiral galaxy. Located in one of the arms, our solar system revolves about the center of the galaxy at a speed of over 200 km per second. Yet the galaxy is so vast that it takes about 250 million years to complete one revolution. The Sun has yet to complete its nineteenth orbit about the galactic center.

beginning of this chapter; it is most useful for comparative purposes. The **inner planets** (Mercury, Venus, Earth, and Mars) are also called the **terrestrial** planets (meaning that they are Earth-like). They are relatively small worlds composed of silicate (silicon-oxygen compounds) rock surrounding metallic cores. Farther from the Sun lie the four gas giants, with deep atmospheres that thicken downward into hot liquid, probably surrounding small, solid rocky or icy cores. These planets have no solid surfaces. Jupiter, Saturn, Uranus, and Neptune all have rings of small particles that encircle them. The outermost and smallest planet is Pluto. Natural satellites, or moons, orbit every planet except Mercury and Venus. The satellites of Earth and Mars are composed of silicate

rock. Jupiter has a miniature planetary system comprised of four large moons and at least 12 smaller ones. The larger moons have rocky cores, which are covered with thick mantles of water ice. Saturn is also the center of a miniature planetary system, involving more than 17 planetary bodies. The moons of Saturn are relatively small icy bodies (except for Titan, which is larger than Mercury) with possible small rocky cores. Uranus and Neptune also have many small icy satellites. Little is known about the moon of Pluto, but it probably resembles the icy satellites of Uranus and Neptune.

Most of the information summarized in this chapter and throughout the rest of this book is the result of the space programs of the United States and the former Soviet Union. Table 1.2 lists the

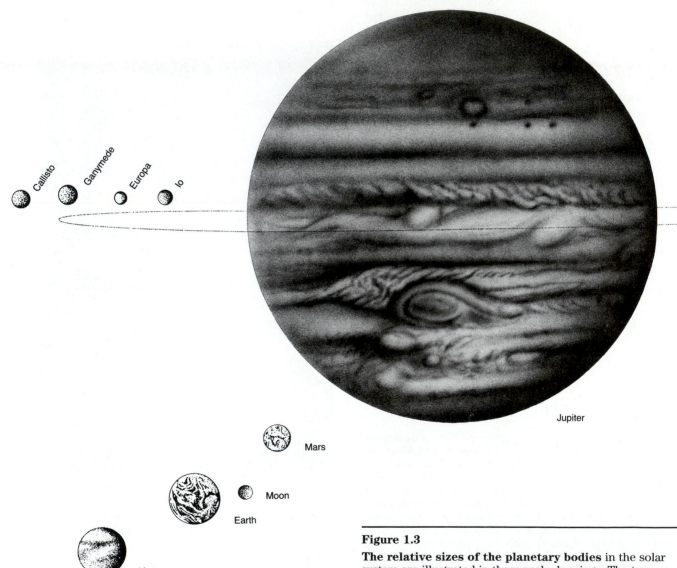

Figure 1.3

The relative sizes of the planetary bodies in the solar system are illustrated in these scale drawings. The terrestrial planets along with the asteroids and Jupiter's satellite Io are much smaller and composed mostly of rocky silicate materials. The **giant planets** (Jupiter, Saturn, Uranus, and Neptune) have deep atmospheres of hydro-

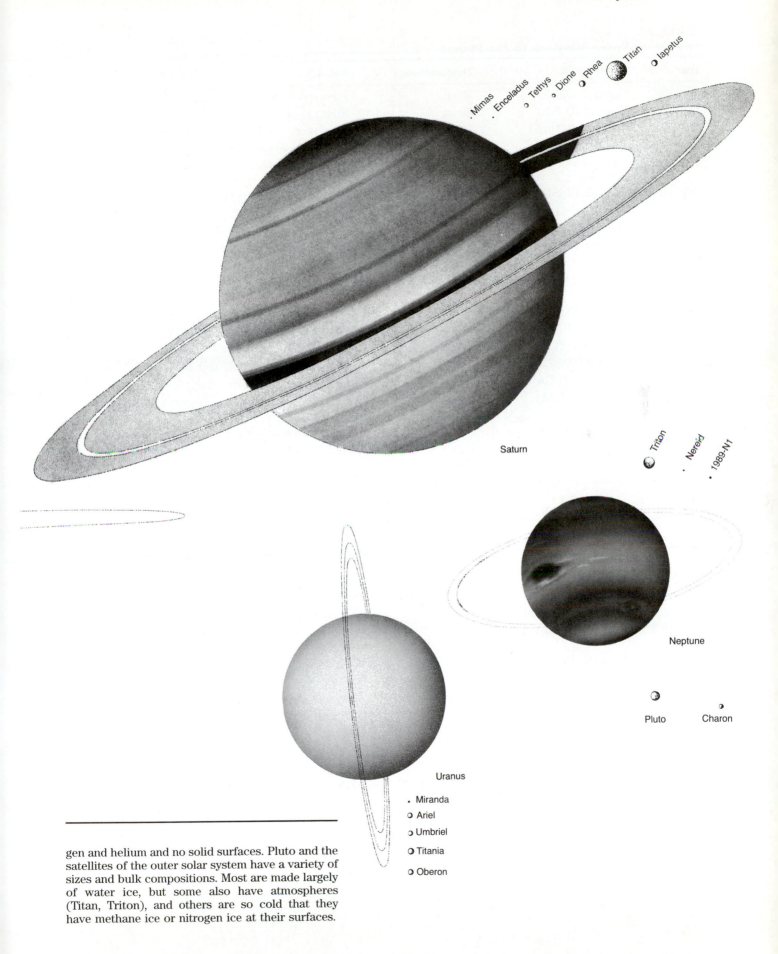

Mimas · Enceladus · Tethys ○ Dione ○ Rhea Titan · Iapetus

Saturn

Triton · Nereid · 1989-N1

Neptune

Uranus

· Miranda
○ Ariel
○ Umbriel
○ Titania
○ Oberon

Pluto   Charon

gen and helium and no solid surfaces. Pluto and the satellites of the outer solar system have a variety of sizes and bulk compositions. Most are made largely of water ice, but some also have atmospheres (Titan, Triton), and others are so cold that they have methane ice or nitrogen ice at their surfaces.

TABLE 1.2
Important Space Missions to the Planets

Year	Spacecraft	Destination	Remarks
1959	Luna 2 (USSR)	Moon	First space vehicle to impact on the Moon
1959	Luna 3 (USSR)	Moon	First photos of farside of Moon
1962	Mariner 2 (USA)	Venus	First reports of high surface temperature
1964	Mariner 4 (USA)	Mars	First pictures of Mars from spacecraft
1964	Ranger 7 (USA)	Moon	Took many closeup photos of Moon before it impacted
1966	Luna 9 (USSR)	Moon	First soft-landing on Moon
1966	Surveyor 1 (USA)	Moon	Soft-landed on Moon
1966	Lunar Orbiter 1 (USA)	Moon	Orbited Moon, took 21 pictures
1967	Venera 4 (USSR)	Venus	Atmosphere examined by capsule
1969	Mariner 6 (USA)	Mars	Photographic flyby
1969	Apollo 11 (USA)	Moon	First manned landing on Moon, rocks and soils returned
1969	Apollo 12 (USA)	Moon	Manned landing
1970	Venera 7 (USSR)	Venus	First data sent from surface of Venus
1970	Luna 16 (USSR)	Moon	Returned lunar soil to Earth for analysis
1970	Luna 17 (USSR)	Moon	Returned lunar soil to Earth for analysis
1971	Mariner 9 (USA)	Mars	First spacecraft to orbit Mars, 7,300 pictures
1971	Apollo 14 (USA)	Moon	Manned landing, returned largest amount of rock and soil
1971	Apollo 15 (USA)	Moon	Manned landing with rover
1972	Apollo 16 (USA)	Moon	Manned landing with rover, 100 kg of rocks returned
1972	Luna 20 (USSR)	Moon	Unmanned: returned rocks to Earth
1972	Pioneer 10 (USA)	Jupiter	Passed through asteroid belt and photographed Jupiter
1972	Venera 8 (USSR)	Venus	First chemical analysis of Venus's surface
1972	Apollo 17 (USSR)	Moon	Manned landing with rover, returned 100 kg of rocks
1973	Luna 21 (USSR)	Moon	Robot car collected samples and returned them to Earth
1973	Pioneer 11 (USA)	Jupiter and Saturn	First flyby mission to Saturn
1974	Mariner 10 (USA)	Mercury, Venus	Only spacecraft to photograph Mercury
1975	Venera 9, 10 (USSR)	Venus	Orbiter and landing craft
1975–1982	Viking 1, 2 (USA)	Mars	Landers examined soil and weather Orbiters photographed details of planet
1976	Luna 24 (USSR)	Moon	Returned soil for analysis on Earth
1977	Voyager 1, 2 (USA)	Outer planets	Launch date for flyby mission
1979			Voyager 1 and 2 reached Jupiter
1980			Voyager 1 reached Saturn
1981			Voyager 2 reached Saturn
1986			Voyager 1 and 2 reached Uranus
1989			Voyager 2 reached Neptune
1979	Pioneer Venus (USA)	Venus	Probed surface with radar, capsule through atmosphere
1978	Venera 11, 12 (USSR)	Venus	Surface and atmospheric studies
1982	Venera 13, 14 (USSR)	Venus	Landed on Venus and analyzed soil and atmosphere
1983	Venera 15, 16 (USSR)	Venus	Radar mapping of surface

TABLE 1.2
Important Space Missions to the Planets—Cont'd

Year	Spacecraft	Destination	Remarks
1986	Vega 1, 2 (USSR)	Halley's comet	Took pictures and studied composition, Venus balloon
1986	Suisei, Sakigake (Japan)	Halley's comet	Studied physical properties of comet
1986	Giotto (European SA)	Halley's comet	Studied atmosphere and magnetic field
1989–1994	Magellan (USA)	Venus	Radar images of Venus
1989–	Galileo (USA)	Jupiter, Asteroids	To arrive in 1995, imaged Gaspra and Ida
1990–	Hubble (USA)	Earth Orbit	Space telescope, first clear images of Pluto and Charon
1994–1995	Clementine (USA)	Moon	First USA mission to Moon since Apollo

dates and goals of the most important missions. Most of these missions were accomplished by unmanned robotic explorers, but the work of the Apollo astronauts in the late 1960s and early 1970s on the Moon may represent the pinnacle of human exploration in space.

Mercury

Mercury is the innermost planet (Table 1.1). It orbits the Sun at only 40 percent of the distance that Earth lies from the Sun. Spacecraft observations of Mercury were made on three occasions in 1974 and 1975, when Mariner 10 photographed approximately half of the planet. With no atmosphere to moderate its environment, the temperature range at the surface is extreme, from 90 K on the surface turned away from the Sun to about 740 K on the surface facing the Sun. No clouds hide Mercury's surface, which is dominated by circular craters formed when large meteorites struck it (Figure 1.4). Many of the craters are extremely ancient, between 3 and 4 billion years old. The largest impact crater shown here is almost 1300 km across. Broad patches of smooth plains, deformed by a system of wrinkles, occur between the craters. The plains may have been produced when the interior warmed and partially melted to produce lavas, which were then erupted onto the surface. Mercury appears to be composed of rocky materials similar to those found in the other terrestrial planets. A magnetic field and a high bulk density (mass/volume) for Mercury hint at the presence of an iron-rich interior that may still be molten.

Venus

Nearly 2.5 times larger than Mercury and farther from the Sun, **Venus** stands in striking visual contrast to Mercury (Figure 1.4). Its orbit is closer to a perfect circle than that of any of the planets, yet it moves in odd ways. Its rotation is retrograde or backward and its spin is so slow that eight Earth months go by during the time Venus spins once on its axis. In addition, Venus orbits the Sun in less than the time it takes for one rotation on its axis. The surface of Venus is totally obscured by a thick atmosphere composed mostly of carbon dioxide with sulfuric acid clouds (Figure 1.4). Nonetheless, Soviet and American spacecraft using radar instruments have revealed the details of the planet's surface features. Impact craters, formed when large meteorites slammed into the surface, are scattered sparsely across the landscape. Mountain belts, volcanoes, and highlands rise several kilometers above vast rolling plains, forming a surface similar in some ways to that of Earth with its continents and ocean basins. However, Venus has no liquid water and temperatures at its surface (almost 750 K) are higher than on Mercury.

Earth

The third planet from the Sun, **Earth**, is dominated by the liquid water of its ocean and the white swirling patterns of clouds, underlining the importance of the atmosphere and of water for the development of the surface features of Earth. Several complete cyclonic storms, spiraling over hundreds of square kilometers, are illustrated in Figure 1.4. They pump huge quantities of water from the ocean to the atmosphere. Much of this water is precipitated on the continents and erodes the land as it flows back to the sea. Earth's poles are marked by ice caps—the Antarctic continent, shown here, is covered by a polar ice cap that is 3 km deep in places. Huge oceans of liquid water cover 70 percent of the planet.

Rising above the oceans, Earth's continental highlands are etched by delicate systems of river valleys. In this view, large parts of the continental highlands of Africa and Antarctica are also visible above the level of the sea. In striking contrast to some of the other inner planets, no impact craters

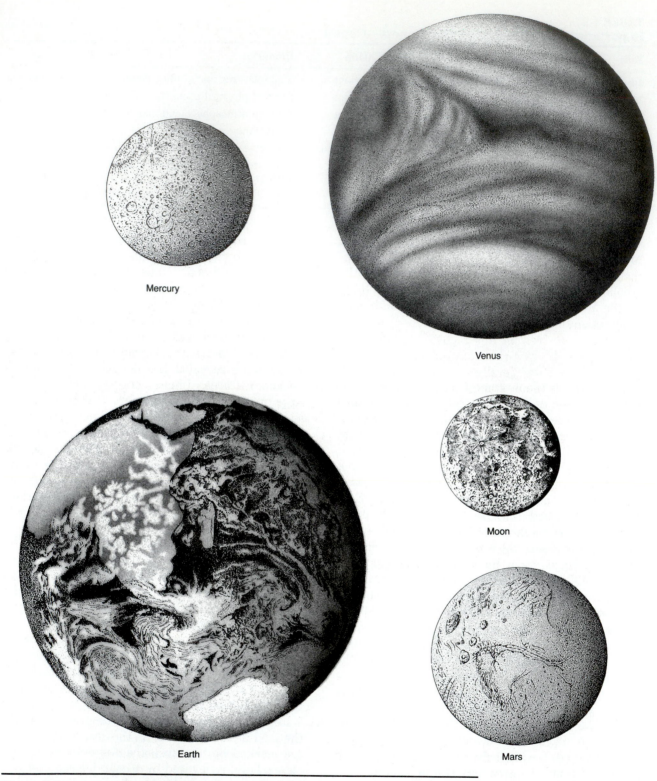

Mercury

Venus

Moon

Earth

Mars

Figure 1.4

The surfaces of the inner planets (Mercury, Venus, Earth, the Moon, and Mars) drawn to a common scale reveal the diversity found in the solar system. Mercury, the Moon, and Mars have many craters formed when meteorites collided with their surfaces billions of years ago. Venus and Earth have conspicuous atmospheres and surfaces that are relatively young.

are visible from this distance. Indeed, such structures are rare on Earth, and most of its rocks are less than 2.5 billion years old. Its landforms are extremely young, shaped by the relentless flow of water and wind. Of particular interest in this view is the rift system of the Red Sea, a large growing fracture in the African continent that separates Arabia from Africa. This fracture attests to the mobility of Earth's **lithosphere**, its outer solid layer. The lithosphere is fragmented, and each segment moves slowly about the planet. Active volcanoes and earthquakes still produce dramatic changes at the surface.

Self-replicating molecules of carbon, hydrogen, nitrogen, and oxygen—life—developed early in Earth's history; they have substantially modified the planet's surface, blanketing huge parts of the continents with greenery. Tropical jungles create the dark band across equatorial Africa, producing dramatic evidence of the unique chemistry of Earth. Life thrives on this planet, which has an oxygen- and nitrogen-rich atmosphere and moderate temperatures (generally above 275 K). Perhaps if nothing more, our studies of the diversity of compositions and conditions of solar system bodies should remind us of the delicate balance of energy, environment, and evolution that allows us to exist at all.

The Moon. With the developments of the space program, the Moon, Earth's natural satellite, has become one of the best-understood planetary bodies in the solar system. As curious as it may seem, we probably understand the earliest history of the Moon better than that of Earth. This is because Earth lacks a rock record of its first 800 million years of history. The surface of the Moon (Figure 1.4) shows two contrasting types of landforms, reflecting two major periods in its history. The bright, densely cratered highlands are similar to the surface of Mercury. Most of the large impact craters had been formed before about 4 billion years ago. The dark, smooth areas are called *maria* and most occupy low regions, such as the circular interiors of impact basins. We know from rock samples brought back by the Apollo astronauts that the maria resulted from great floods of lava, which filled many large craters and spread out over the surrounding area. The volcanic activity therefore occurred after the formation of the densely cratered terrain. Radiometric dates on samples brought back from the Moon indicate that most of the lavas are over 3 billion years old. However, some young impact craters with bright rays formed after these eruptions. Like Mercury, the Moon is a dry, airless world. The lunar surface has not been modified by wind, water, or glaciers, but its surface features record major events early in the history of the solar system, when impact of meteorites was the major geologic process throughout the solar system.

Mars

Mars, the red planet, is smaller and less dense than Earth. For years it was considered to be a planet of mystery because telescopic observations revealed a thin atmosphere, polar ice caps, and shifting markings, which often darkened during the martian spring. Some thought that Mars was populated and that life forms had evolved to a civilized state. Streaks were believed to be canals or vegetated land alongside canals. As it turns out, these fanciful theories were all wrong, but in a different way the real Mars is just as fascinating. Unlike the Moon, many features of Mars indicate that its surface has been modified by atmospheric processes, running water, recent volcanic activity, and lithospheric deformation. Its surface is divided into two distinctly different hemispheres (Figure 1.4). The northern hemisphere has few meteorite impact craters and consists of vast, relatively smooth plains. The ancient southern hemisphere, in contrast, is higher and moderately to heavily cratered. A broad fractured swell capped by several great volcanoes overlaps the boundary between the hemispheres. Huge channels, apparently carved by running water, course across the surface of Mars. Liquid water cannot exist at its surface today; yet Mars has had abundant liquid water on its surface in the past. What happened? The question will be debated for years but one thing seems certain: Mars has experienced significant changes during its history—changes recorded on its surface and in its landscapes, changes that are still occurring—as evidenced by planetwide dust storms, wind-blown sand, tenuous mists, and clouds.

The Asteroid Belt

Occupying the vast tract between Mars and Jupiter is the **asteroid belt** (Figure 1.5). It consists of thousands of small bodies ranging from less than 1 to just over 1000 km in diameter. The asteroids mark the transition from the rocky terrestrial planets to the gas- and ice-rich outer planets. Some asteroids appear to be rocky, some seem covered with lavas, others seem to be metallic, and yet others may have water ice.

Many **meteorites** are derived from the asteroids. Meteorites are small planetary bodies composed of iron or silicate rock that reach Earth from

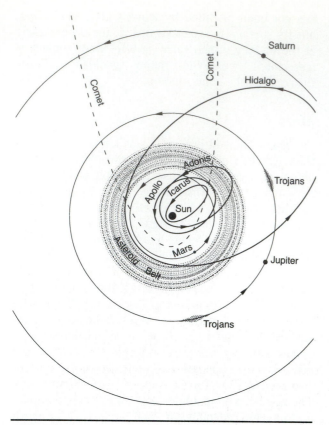

Figure 1.5
The asteroid belt lies between the orbits of Mars and Jupiter and is occupied by thousands of small rocky bodies in orbit about the Sun. The largest asteroids are Ceres, 1000 km in diameter, and Vesta, 550 km in diameter. Vesta may be covered by lava flows. These bodies are indeed small planets. The orbits of many asteroids cross Earth's orbit and these may be the sources of some of the meteorites which fall to Earth.

space. Called meteoroids before they hit a planet, they are actually fragments of asteroids or comets (and more rarely of moons or planets). These pieces of planetary material are extremely valuable because they provide scientists with tangible samples of planetary bodies that can be analyzed in great detail. Moreover, expensive spacecraft are not necessary for obtaining them.

Jupiter

Jupiter is the largest planet in the solar system; in fact, most of the mass of the solar system outside of the Sun is in Jupiter. Jupiter has no solid surface. The spots and colorful bands (Figure 1.6) that parallel its equator are the turbulent manifestations of motion in a thick, cloudy atmosphere of hydrogen and helium. Jupiter and the other outer planets have compositions dramatically different from those of the rocky inner planets. In addition,

Jupiter is the center of a miniature planetary system and a narrow ring. It has four large moons, called the **Galilean satellites** because they were discovered by Galileo. Each of these moons is larger than Pluto and each presents diverse surface features resulting from meteorite impact, volcanism, and surface fracturing. Like the principal planets, these satellites show significant compositional changes outward from their primary (the body they orbit).

Io. The innermost Galilean satellite appears to be one of the most bizarre worlds in the solar system. The density of Io is about the same as Earth's Moon, which suggests that Io has a rocky, rather than an icy, composition. Io is only slightly larger than the Moon, but it has no impact craters. Instead, its surface is dominated by mottled patches in colors of yellow, red, and white, with black pockmarks (Figure 1.7). In 1979, eight active volcanoes were seen on Io by the Voyager 1 spacecraft; four months later, when Voyager 2 flew by, at least six of these volcanoes were still erupting and two new ones had started. Close-up photos reveal that most of the black spots on Io are volcanic craters or calderas. Io must have abundant thermal energy to drive its active volcanoes, but it is the same size as the cold, dead Moon. Much of the surface coloration is caused by various forms of sulfur. Estimates of the thickness of sulfur at the surface range from a thin frosting a few millimeters thick to a globe-encircling layer several kilometers thick.

Europa. The surface of Europa is composed of bright water ice and is distinctive in that it is almost perfectly smooth. Local relief is only a few hundred meters. The major surface features are sets of tan steaks or bands (Figure 1.7). Indeed, it has been said that Europa looks like an icy billiard ball with lines drawn on it with a felt-tip pen. What are the bands? Most appear to be shallow valleys up to 75 km across that extend as far as 3000 km across the surface. They appear to be similar to fractured sea ice in the polar regions of Earth. Internal heat apparently formed a watery slushy "lava," which erupted through cracks and fissures in the crust and coated the surface with fresh new ice. The near-absence of impact craters on Europa suggests that the surface is very young, formed after the early periods of heavy meteorite bombardment that scarred the ancient surfaces of Mercury and the Moon. Resurfacing by eruption of watery lavas continued until relatively recent times.

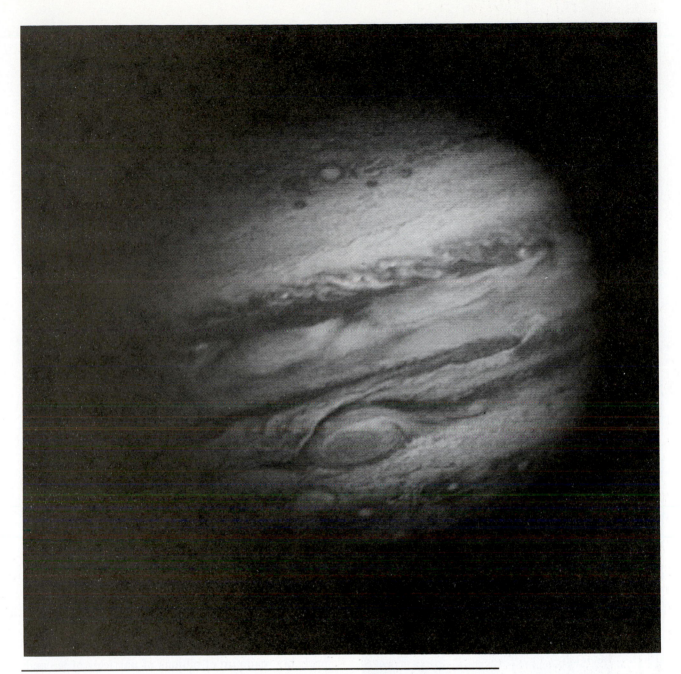

Figure 1.6

Jupiter is the largest and most massive planetary body in the solar system. Its banded appearance is the result of circulation in an extremely thick atmosphere of hydrogen and helium, and details of the bands are constantly changing. The clouds come in a variety of subtle hues of red, yellow, brown, and white. The large oval structure, the Giant Red Spot, in Jupiter's southern hemisphere is a cyclonic storm larger in diameter than Earth itself.

Ganymede. Ganymede is the largest satellite in the solar system. Its surface contains a baffling array of structural and volcanic features unlike anything else in the solar system. Many features appear to result from breaking and lateral movement of lithospheric fragments, so it was with great interest that geologists studied the details shown on the Voyager photographs. The view shown in Figure 1.7 reveals that on a global scale Ganymede has two distinct terrain types. The older is dark and nearly saturated with craters. This is believed to be an old terrain composed of "dirty ice" containing dust and particles from space. This older crust on Ganymede has been fractured and split apart, and many of the fragments have shifted about. The brighter, younger terrain is crossed by a

spectacular series of narrow grooves and stripes, features that result from deformation and cracking of an icy crust.

Callisto.   Callisto is the outermost Galilean satellite; like Ganymede, it is believed to consist of a rocky core surrounded by a thick mantle of ice. The images sent back from the Voyager spacecraft show that Callisto, in contrast to the other Galilean satellites, is saturated with craters, somewhat like the highlands of the Moon (Figure 1.7). The general surface of Callisto is dark, dirty ice, but many craters have bright rays and ejecta blankets. The bright material is probably clean melted ice ejected from impact. The surface of Callisto is believed to be very old, recording events that took place during the early history of the solar system. Why did Callisto escape subsequent modification when the other Galilean satellites record more recent events? How do the events recorded on Callisto in the outer solar system compare with those preserved on the Moon and Mercury in the inner solar system? These and other questions will be considered as we explore the details of the geologic histories of the fascinating satellites of Jupiter.

Saturn

Saturn is similar to Jupiter in many ways. It is a gigantic ball of mostly hydrogen and helium. Its atmosphere is not as colorful as Jupiter's but is marked by dark bands alternating with lighter zones. The rings of Saturn are its most dramatic feature; they have intrigued astronomers for over 300 years (Figure 1.8). Now that we have seen them close up, they are even more astonishing. They extend over a distance of 40,000 km and yet are only a few kilometers thick. The rings are probably made up of billions of particles of ice and ice-covered rock ranging from a few microns to a meter or more in diameter. Rarely are particles in the rings more than a kilometer wide in diameter. Each particle moves in its independent orbit around Saturn, producing an extraordinarily complex ring structure.

Like Jupiter, Saturn is the center of a planetary system, with an elaborate family of satellites. Except for Titan, which is in a class by itself, the moons of Saturn are small bodies of ice admixed with rocky silicate material. One might think that small icy moons would be of little interest to the geologist because of their cold origins, primitive compositional character, and apparent lack of an internal source of heat for any geologic activity. As

Figure 1.7

The Galilean satellites of Jupiter show remarkable differences in their surface features and compositions. Io, the innermost of the satellites, is rocky, has many active volcanoes, and lacks impact craters. Europa has a fractured icy outer shell. Icy Ganymede has large expanses of cratered terrain cut by younger brighter terrains formed by fracturing. Outermost Callisto is heavily cratered with an ancient icy surface.

Figure 1.8

Saturn appears suspended in its delicate, graceful system of rings. Although it is less colorful than Jupiter, Saturn also has a banded cloudy atmosphere composed mostly of hydrogen and helium. Of all the planets, Saturn has the lowest density—only 70 percent that of liquid water. The rings consist of myriads of small chunks of ice in orbit about Saturn which reflect light to give a disklike appearance. Orbiting outside of the rings is a family of at least 17 tiny, icy moons.

it turns out, however, the moons of Saturn experienced many events, recorded in their surface features, which help us better understand the geology of the entire solar system.

The larger satellites (except for Titan) fall into convenient pairs that match their order from Saturn outward: Mimas and Enceladus (between 400 and 500 km in diameter), Tethys and Dione (between 1050 and 1120 km in diameter), and Rhea and Iapetus (about 1500 km in diameter).

Mimas. Tiny Mimas, only 400 km across, is an icy body saturated with impact craters. Its most prominent feature is a huge crater, whose diameter is one-fourth that of the satellite itself (Figure 1.9). The impact that formed the crater must have come close to breaking Mimas into fragments. The opposite hemisphere contains large fractures, which

may have been produced as shock waves from this impact traveled through the satellite.

Enceladus. Enceladus is only slightly larger than Mimas but bears little resemblance to its nearest neighbor. Enceladus has very few impact craters and a complex grooved terrain, similar in some ways to that on Ganymede (Figure 1.9). It is unique among the Saturnian satellites in that it has large expanses of smooth icy plains on the flanks of a planetwide system of fractures. The plains may have been produced by "lavas" of icy slush that erupted through these fissures, covering many older impact craters. This terrain may be a result of heat generated by the variable tidal pull of Saturn and Dione. Such heat would soften the interior of Enceladus and could cause a type of volcanic activity that would resurface the satellite,

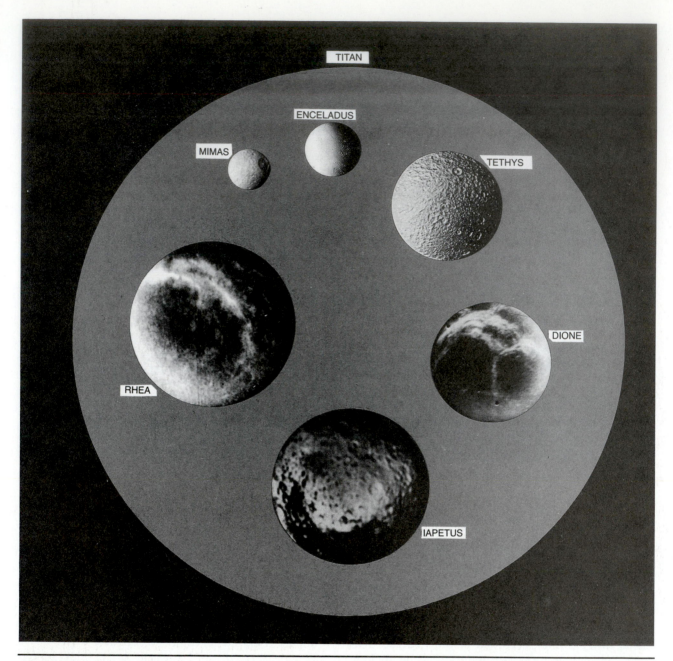

Figure 1.9

The icy moons of Saturn are a geologically diverse group of mostly small satellites, shown here drawn to the same scale. They show an amazing variety of young and old surfaces, impact craters, evidence of icy volcanism, and global fractures. Titan, the largest satellite of Saturn, has a hazy nitrogen-rich atmosphere that obscures its surface features. Some speculate that it has an ocean of liquid hydrocarbons.

creating some of the smooth plains and destroying older craters.

Tethys. The most spectacular feature on the surface of Tethys is an enormous impact basin almost a third the diameter of Tethys itself (Figure 1.9). Heavily cratered Tethys is also scarred with a gigantic fracture system that extends as a continu-

ous feature almost three-quarters of the way around the satellite. What events produced such gigantic features on such a small icy satellite? It has been suggested that Tethys has been fragmented by comet impact and reassembled under its own gravity into a fractured ball of icy debris, or perhaps it simply cracked as a result of global expansion of its icy interior. The density of Tethys, like that of several other Saturnian moons, is very nearly that of pure water ice.

Dione. With its surface saturated with craters and its large, smooth areas that resemble the dark lunar plains, Dione looks much like Earth's Moon (Figure 1.9). Only slightly larger than Tethys, Dione has a sinuous fracture or trough that extends across a major portion of the moon's surface. Dark areas are crisscrossed with light streaks, which may be extrusions of material from its interior, forming a region called wispy terrain.

Rhea. The surface of Rhea, photographed in finer detail than any of the other Saturnian satellites, has been found to be saturated with craters, much like the surfaces of Mercury and Earth's Moon (Figure 1.9). Very large craters, however, which are normally present in the crater populations on other planetary bodies in the solar system, are lacking on Rhea's surface—a fact that may alter our interpretation of the early history of cratering. Two periods of meteorite bombardment may have occurred throughout the solar system during its early history, but evidence of only the latter one is preserved on Rhea. An older cratered surface may have been smoothed over by ice flows from within the satellite. Rhea also has bright wispy bands superimposed on a dark background. The wisps have a cloudlike appearance, but Rhea has no atmosphere.

Iapetus. Iapetus circles Saturn in a lonely orbit nearly three times as far away as Titan or Hyperion, the next closest satellites. Iapetus is slightly smaller than Rhea, with a diameter of 1460 km. However, unlike Rhea, one hemisphere of Iapetus is covered with an exceptionally dark material that reflects so little light that it is similar to coal or soot (Figure 1.9). The boundaries of these unique dark areas are sharp but complex in detail. The dark material is even deposited on the floors of some large craters. Perhaps the dark material was derived from the interior of Iapetus in an exotic form of volcanism. Much of the rest of Iapetus appears to be heavily cratered.

Titan. Titan, the second largest satellite in the solar system, is the only one of Saturn's moons with an atmosphere (Figure 1.9). Haze in the atmosphere totally obscures Titan's surface and is responsible for its bland appearance. It is nearly the same size as airless Ganymede and Callisto, moons of Jupiter. Preliminary data indicate that its atmospheric pressure is 1.5 times that of the surface of Earth and that its temperature is about 100 K. To the surprise of all, Titan's atmosphere is dominated by nitrogen, like Earth's, with only about 1 percent methane (CH_4), along with such hydrocarbons as propane, ethylene, ethane, and acetylene. Remarkably, an ocean composed of hydrocarbons, instead of water, may exist on the surface of Titan!

Uranus

Uranus is a gas- and ice-rich planet encircled by a narrow ring system (Figure 1.10). The rotation axis of Uranus has been flipped on its side, apparently by some early catastrophic collision with another body. In addition, Uranus has five major satellites larger than about 300 km in diameter; it also has many smaller satellites. All of the satellites of Uranus (Figure 1.11) appear to be icy, like those of Saturn.

About the size of Saturn's tiny moon Mimas, **Miranda** is not a simple crater-dominated sphere. Large ovoidal patches disrupt its surface and appear to be unique in the solar system. **Oberon** and **Umbriel**, respectively about three and two times the size of Miranda, have heavily cratered surfaces that appear to have been little modified by volcanism or fracturing. In contrast, the surfaces of **Titania** and **Ariel** reveal that some sort of icy volcanism and fracturing has shaped their landforms.

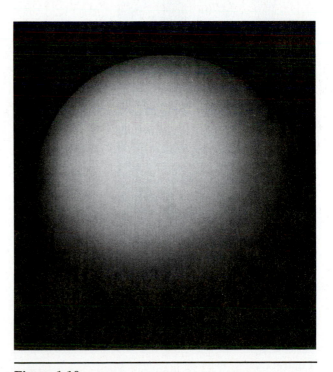

Figure 1.10

Uranus is a gas- and ice-rich planet with a bland hydrogen-rich atmosphere and a narrow ring. The rotational axis is tipped on its side.

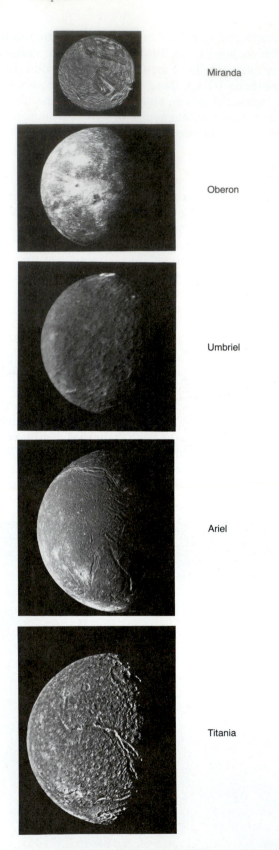

Miranda

Oberon

Umbriel

Ariel

Titania

Figure 1.11

The major satellites of Uranus (Titania, Ariel, Umbriel, Oberon, and Miranda) are icy bodies about the same size as the satellites of Saturn and display similar surface features as well.

Neptune

Neptune, with its beautiful blue atmosphere, was the last of the planets visited by the Voyager spacecraft. Neptune (Figure 1.12) is only slightly smaller than Uranus, and it appears to be similar to its neighbor in its composition. The two planets, called the twins of the outer solar system, are thought to have large cores of water ice and rock surrounded by thick atmospheres of hydrogen, helium, and minor methane.

Only two moons were known to orbit Neptune before the Voyager spacecraft passed it in August of 1989. Six more dark moons were discovered during Voyager's brief flyby. **Triton** (Figure 1.13) is slightly smaller than Jupiter's Europa and has an atmosphere of nitrogen and methane. Its exotic surface is covered by ices of those gases. Impact craters are not common on its surface, unlike most of the other small satellites of the outer solar system. Active geysers fountaining gas and dark dust were photographed. **Nereid** is much smaller, perhaps the size of Uranus's Miranda, and has an extremely eccentric (noncircular) orbit. One of the newly discovered moons, **Proteus**, is larger than Nereid (Figure 1.13).

Pluto and Charon

Tiny **Pluto** was discovered in the remote margins of the solar system in 1930. It is truly a planet

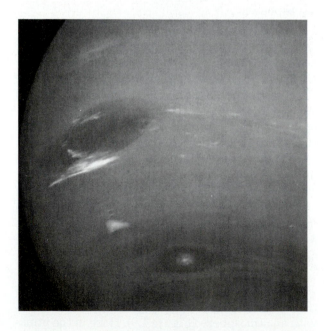

Figure 1.12

Neptune is similar in size, density, and composition to Uranus, but its beautiful banded atmosphere contrasts sharply with the blandness of Uranus. Neptune has a system of narrow rings like the other giant planets.

planet (Figure 1.14). Pluto is not a gas planet like its neighbors (Neptune, Uranus, Saturn, and Jupiter). It is a solid icy body more like Triton than anything else we have seen in the solar system. Its surface is covered by frozen nitrogen, with traces of methane and carbon monoxide ice. Rock and water ice probably comprise its interior. Even at the very low temperatures found on Pluto (probably about 35 K), some of the nitrogen should sublimate (vaporize) to produce a thin atmosphere. Because Pluto has never been visited by a spacecraft, the nature of its surface features are unknown, but from our studies of the icy moons of Uranus and Neptune, the frozen worlds of Pluto and Charon will probably someday reveal an exciting geology.

Comets

At even greater distances from the Sun, **comets** are the characteristic members of the solar family (Figure 1.15). Comets may preserve in their icy interiors some of the original materials from which the outer planets formed, so samples of comets are eagerly sought. Some comets have orbital periods of hundreds of thousands of years.

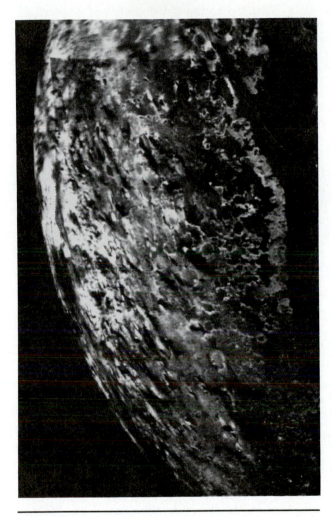

Figure 1.13

Triton and Proteus are the largest satellites of Neptune. Triton has a transparent atmosphere and its surface is covered by ices of methane and nitrogen. The Voyager spacecraft showed that Neptune has at least eight moons.

of extremes—the smallest, the darkest, the coldest, and certainly the most distant. Its orbit surrounds Earth's like an elliptical hula hoop encircling a wedding ring. For years little was known about this planet, but recently enough data have been obtained to establish some of its physical properties. As it turns out, Pluto is hardly larger than the Moon. Yet it was recently discovered that Pluto has a large moon of its own, called **Charon**, and some planetary scientists think of the pair as a double

Pluto

Charon

Moon

Figure 1.14

Pluto, the smallest planet, and its satellite, Charon, form a double-planet system on the extreme outer edge of the solar system. Charon may be as large as one-half the diameter of Pluto. Both bodies have considerable amounts of methane ice at their surfaces and are more like the icy moons of the outer planets than they are like the gas giants.

Figure 1.15

Comets are occasional visitors to the inner solar system, where they develop long tails and bright heads. The tail is composed of gas and dust stripped from a small icy nucleus by the warmth of the Sun. Most comets have orbits that take them far past Pluto and belong to what is called the Oort Cloud.

The group of comets that occupies the outermost limits of the solar system is called the **Oort cloud**. These small balls of ice are only occasional visitors to the inner solar system, where they develop diffuse heads and long streaming tails (Figure 1.15) as the ices sublimate from the warmth of the Sun. Like other planetary bodies, comets glow as a result of reflection of light from the Sun; they are not self-luminous.

An Overview of Important Questions

The illustrations and brief descriptions in this chapter pose many interesting questions about the origins and histories of planetary bodies and, not the least among them, about Earth itself. A few of these questions have already been alluded to; others are listed here.

- How old is the solar system?
- How did the planets form?
- How do the surfaces of planetary bodies tell us their histories?
- Why are so many of the planets battered by meteorite impact scars? How and when were these craters excavated?
- Why are other planets, like Earth and some satellites of the outer planets, deficient in these craters?
- What led to the development of a scorching, arid desert at the surface of Venus but allowed Earth, the planet next door, to be relatively water-rich and temperate in climate?
- Why do the planets have different compositions, exemplified by the marked differences between the rocky inner and icy satellites of the outer planets or more subtly by the densities and atmospheric compositions of the inner planets?
- Why is there a belt of small bodies instead of a single larger planet between Mars and Jupiter?
- What made the Galilean satellites so different from one another? Why is Io rocky and volcanically active whereas its neighbors are icy? Why does Callisto show evidence of only ancient meteorite bombardment?
- What are the sources of energy that shape the surfaces and interiors of planets?
- Why do some planets have rings of orbiting particles?
- What can we learn about the planets from the small bodies of the solar system (asteroids, comets, and meteorites)?
- Finally, what are the basic geologic controls on the evolution of a planet and its surface history?

Our approach to obtaining answers to these questions is to describe the planets as geologic entities and proceed from observations to deductions about their origins and developmental histories. As we explore the planets of the solar system, we will try to reveal the fundamentals of planetary evolution and discover how energy causes planets to change. The characteristics of all planets—their compositions, internal structures, orbital paths, shapes, sizes, and especially their surface features—are reflections of the events and processes that shaped them. Thus, our objective is to show how planets originate and evolve by carefully examining them as individuals and comparing them with one another. Ultimately, we will understand some of the reasons for their differences. In this process, we hope to gain a better understanding of our own Earth and how it functions in its unique place in the solar system.

Review Questions

1. Briefly outline the fundamental properties of a planet. Is there a fundamental distinction between planets that revolve around the Sun and the satellites that revolve around planets?
2. What are the important sources for meteorites? Why are they so important in planetary studies?

3. Describe some of the important differences between the inner planets, as a group, and the outer planets.
4. How are the satellites of the large outer planets similar to one another? What sets them apart from the inner planets?

Key Terms

Asteroids

Asteroid Belt

Inner Planets

Lithosphere

Meteorite

Oort Cloud

Outer Planets

Planet

Planetary Body

Satellite

Solar System

Terrestrial Planets

TABLE 2.1

Important Planetary Materials

Material	Typical Formula	Freezing Temperature (K)	Vaporization Temperature (K)	Condensation Temperature (K)	Density g/cm³
Gases					
Hydrogen	H_2	14	20		—
Helium	He	1	4		—
Carbon Dioxide	CO_2	216	195		—
Nitrogen	N_2	63	77		—
Methane	CH_4	91	109		—
Oxygen	O_2	45	91		—
Liquids					
Water	H_2O	273	373		1.00
Methane	CH_4	91	109		0.42
Ethane	C_2H_6	184	89		0.57
Solids					
Ices					
Water	H_2O	273	373	185	0.92
Carbon Dioxide	CO_2	216	195	95	1.56
Ammonia	NH_3	195	240	110	0.82
Methane	CH_4	91	109	45	0.53
Nitrogen	N_2	63	77	32	0.88
Silicates					
Olivine	Mg_2SiO_4	2183	—	1380	3.21
Pyroxene	$MgSiO_3$	1830	—	1370	3.19
Feldspar	$CaAl_2Si_2O_8$	1824	—	1200	2.77
Metals					
Iron-nickel alloy	$Fe_{91}Ni_9$	1890	—	1390	7.9
Troilite	FeS		—	700	4.6

Densities are given for 1 bar pressure and the temperature of the freezing point for liquids and ices and at 273 K for silicates and metals.

Freezing and vaporization temperatures are for 1 bar pressure.

Condensation temperature is for 10^{-4} bars of pressure to simulate the nebula.

CHAPTER 2

Fundamentals of Planetary Science

Supernova in Distant Galaxy

© TB 93

Introduction

Earth may seem rather insignificant when viewed in the context of the solar system, especially if one considers the vast expanses of empty space surrounding the planet. Nonetheless, our solar system and Earth take on great importance to our understanding of the universe because they were produced by the same processes that formed other stars and planets. The planet we live on is an accessible product of the evolution of a star; by studying it and our neighbors, the Sun and planets, we can learn much about the mechanics of planet formation and evolution.

At the outset, it is important to note that we assume that the physical and chemical laws that govern nature are constant. For example, we use observations about how chemical reactions occur today, such as the combination of oxygen and hydrogen at specific temperatures and pressures to produce water, and infer that similar conditions produced the same results in the past. This is the basic assumption of all sciences. Moreover, much of what we "know" about the planets, as in all science, is a mixture of observation and theory—a mixture that is always subject to change. Scientific knowledge is pieced together slowly by observation, experiment, and inference. The account of the origin and differentiation of planets we present is such a theory or model; it explains our current understanding of facts and observations. It will certainly be revised as we continue to explore the solar system and beyond, but the basic elements of the theory are firmly established.

Major Concepts

1. The elements, other than hydrogen and helium, were produced by nuclear processes that generally occur within stars. Massive explosions of these stars recharge interstellar space with newly formed elements.

2. Our solar system was probably formed by the gravitational collapse of a nebular cloud composed of gas and dust. The outer planets consist of volatile compounds that condensed far from the Sun, where the nebula was cool. The inner planets are poor in these constituents and rich in refractory silicates and metals that crystallized at high temperatures.

3. As these particles accreted in orbit around the forming Sun, the planets probably became hot and internally differentiated to different degrees.

4. A planet's lithosphere is its rigid outer shell; it consists of many rock types that preserve important clues about their diverse origins.

5. Atmospheres have been formed around some planets by gravitational capture of the gaseous solar nebula or by volcanic outgassing from the planets' warm interiors. Hydrospheres (or cryospheres) form if water is released to planet surfaces under appropriate conditions of temperature and pressure.

6. The thermal history of a planet determines its level of volcanic and tectonic activity. In general, planets appear to have evolved from early periods of enhanced volcanic activity and lithospheric mobility to later periods of declining or inactive volcanism. This is tied to the cooling of the planet and its thickening lithosphere. The evolution of the lithosphere is influenced by planet size, composition, and the nature of internal and external heat sources.

Origin of the Elements

Before we consider the origin and evolution of the planets, we must first discuss the nature and origin of the basic material of which they are composed—the elements. An **element** is composed of a single type of **atom**, which contains a unique number of **protons** (positively charged particles that have a mass of one atomic mass unit, or *amu)*. The number of protons helps to determine the distinctive chemical and physical properties of the atom (Figure 2.1). The protons reside in a central nucleus surrounded by orbiting **electrons** (small, negatively charged particles with almost negligible masses). Most atoms also contain a third particle, called a **neutron**, which has a mass of about 1 amu but has no electrical charge. Neutrons reside with protons in the nucleus but do not affect the chemical identity of an element; instead, they produce **isotopes** of an element. These have different atomic weights and may have different nuclear characteristics (e.g., some may undergo spontaneous radioactive decay to isotopes of other elements).

Hydrogen is the most abundant element in the universe. It is also the simplest, usually consisting only of one proton and one electron. Consequently, the atomic weight of hydrogen is one. Other isotopes of hydrogen also contain these two particles but may also have one or more neutrons, creating isotopes of hydrogen with a variety of atomic masses. The relative abundance and simplicity of hydrogen atoms suggest that they could be the building blocks of other elements. That is, if various

particles of hydrogen atoms (protons, electrons, and neutrons) could be combined in different proportions, all of the chemical elements could be formed from primordial hydrogen (Table 2.1). We know that a variety of nuclear transmutations can do just that, but high temperatures and pressures are required—conditions that are found in the interiors of some stars. Once new elements have been generated, they are often ejected into interstellar space by gigantic stellar explosions (novas, supernovas, or planetary nebulas). The dispersed elements may become reconcentrated and recycled to form new stars or planets. Thus, the Sun, the planets, and even our bodies are all composed of elemental "star dust" that saw its origin billions of years ago in another star system. Like the legendary Phoenix, our solar system rose from ashes left on the funeral pyres of an older generation of stars and planets.

The stellar "alchemy" by which new elements are produced may take several forms but commonly involves the **fusion** of light atomic nuclei to form heavier nuclei. As a by-product of these reactions, energy is released (the life-sustaining heat and light we receive from the Sun) and various other particles are produced (gamma rays, electrons, neutrons, and even hydrogen or helium nuclei). Hydrogen (H) may "burn" to produce helium (He, with two protons) but only at temperatures that exceed 10 million K (Figure 2.2). Helium burning can produce carbon (C, atomic weight 12), which can combine with other helium nuclei to produce oxygen (O, atomic weight 16). Similarly these "ashes" may react during carbon burning to produce oxygen, neon (Ne), sodium (Na), and magnesium (Mg); during neon burning to produce oxygen and magnesium; during oxygen burning to produce the element magnesium through sulfur (S); and during silicon (Si) burning to produce elements up to iron (Fe, atomic weight 56).

These latter transmutations take place at progressively higher temperatures and pressures, which are produced only in massive stars. For example, the moderate-sized star that forms the center of our solar system is not large enough to sustain the conditions necessary for carbon burning (Figure 2.3). A star more than four times as massive as the Sun may evolve through a series of stages in which the hydrogen-burning reaction moves progressively outward from the core (Figure 2.3). In the wake of this expanding shell, burning of He, C, Ne, O, and Si may be initiated in discrete shells whose outward expansion is controlled by temperature and the production of fuel from other burning reactions. Once this process reaches a stage where the star has a central core of

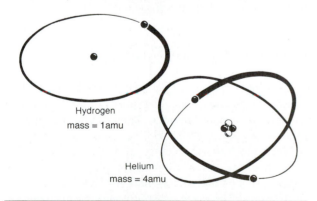

Hydrogen
mass = 1amu

Helium
mass = 4amu

Figure 2.1

The atomic structures of hydrogen and helium illustrate the major particles of atoms. Hydrogen has one positively charged proton (center) and one orbiting, negatively charged electron. Helium has two protons (black), two neutral particles called neutrons (white), and two orbiting electrons. Neutrons and protons contain most of each atom's mass and reside within a central nucleus.

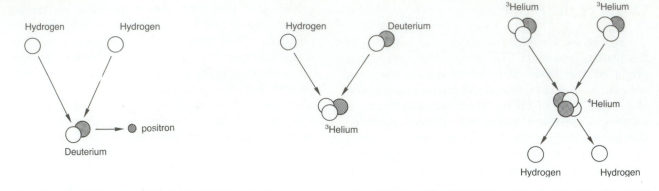

Figure 2.2

Hydrogen burning, the proton-proton chain, is an important energy-producing nuclear reaction that takes places inside stars like our Sun. In this process two hydrogen (H) nuclei combine to form a deuterium (D, hydrogen with an atomic weight of 2) nucleus and a positron (electron with a positive charge). Neutrons are shaded; protons are white. Reaction with an additional hydrogen nucleus produces a helium nucleus with a mass of three (^{3}He). The fusion of two such nuclei results in the production of stable helium (^{4}He, which has two protons) and the ejection of two hydrogen nuclei, which can be consumed in other reactions of this sort. Energy is released by these reactions as a small amount of matter is converted into energy in accordance with Einstein's equation, E (energy) $= m$ (mass converted to energy) c^2 (the speed of light squared).

iron (produced by silicon burning), the stage is set for one of nature's most dramatic events, a supernova explosion.

Iron does not "burn" to create even heavier elements. Consequently, the dynamic balance between the outward-directed pressure caused by heat release and the inward-directed gravitational pressure is lost. The star begins to contract; as a result, the pressure on the core may become so high that normally unreactive protons and electrons combine to form neutrons. The inner support of the star is effectively removed. Consequently, the star immediately collapses and unburned nuclear fuels in the outer shells suddenly react as they collapse toward the hot interior of the star. The reactions proceed with such violence that the star explodes and produces a temporary beacon in our skies, which we call a **nova** or **supernova** (Figure 2.4) depending on its brightness. These explosions are responsible for dispersing many heavy elements into other parts of the galaxy. Our galaxy has seen at least seven supernovas in the last 1000 years; thus, they are not particularly unusual events.

Elements not mentioned in the fusion model described above are formed by a variety of processes in stars. Elements heavier than iron are generally thought to be formed as neutrons are captured by other nuclei. These reactions may take place during the He-burning stages of a star if the proper seed nuclei are present. Rarely, reactions involving the capture of protons may result in the

production of some of the same heavy elements during supernova explosions.

Yet another process is required to produce the isotopes of the light elements lithium (Li), beryllium (Be), and boron (B). These elements are unstable in the deep interiors of stars but may be produced as energetic protons collide with and fragment atoms of carbon, nitrogen (N), or oxygen. These proton–atom collisions may occur during nova explosions or during the formative stages of stars, when many protons are produced. Chance cosmic ray collisions with the proper atoms can also produce these elements in interstellar space.

Origin of the Solar System

It is believed that the Sun and all of the planetary bodies within the solar system originated from the collapse of a solar nebula. These events occurred billions of years ago; therefore, many details of the process are not completely understood. Nonetheless, observations of other developing star systems and studies of meteorites and the systematic compositional differences between the planets provide fundamental constraints on theories of the formation and evolution of the solar system.

When Did the Solar System Form?

Thought to be leftover raw materials from which the planets formed, meteorites that have

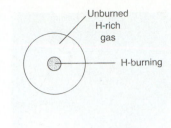

(A) **The interior of a small star** (less than about 4 times the Sun's mass) changes as it evolves from a small hydrogen-burning star to a large hydrogen- and helium-burning star.

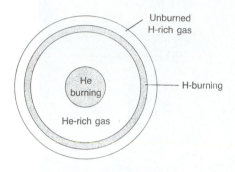

(B) **Small stars burning hydrogen and helium** become cooler at their surfaces and redder and consequently are called red giants. These giants may be 10 to 20 times the diameter of their precursor. Note how the hydrogen-burning shell (shaded) has expanded outward, leaving in its wake a helium-rich shell; eventually hydrogen-burning may extend to the surface causing the disruption of the star's surface and produce a **planetary nebula.**

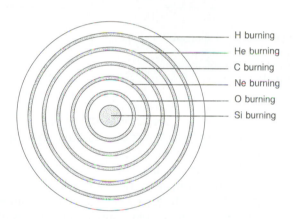

(C) **The internal structure of a massive star** which has evolved past a helium-burning stage. Concentrically arranged shells where burning takes place (shaded) at progressively higher temperatures are separated by unreactive shells (light) where the material is depleted in the fuel being burned in the outer shell and is too cool to participate in the burning reaction of the next inner shell. The "death" of such a massive star is marked by the production of a nova or supernova.

Figure 2.3
The internal structures of stars change with their age and size or mass.

fallen to Earth provide a way to look back through time and estimate when the planets formed. Thus, if we can determine when the meteorites formed, we should have a reliable estimate of the date at which the solar system itself formed. **Radiometric ages** provide a powerful method to determine an absolute age for meteorites and many other types of rocks. A radiometric age can be obtained for a rock if it contains **radioactive isotopes**, which decay to form daughter products at a rate that is well understood. The principal elements used to date planetary materials are isotopes of uranium (which decays to lead), potassium (which decays to argon), and rubidium (which decays to strontium). The concentrations of both the parent and daugh-

ter isotopes are measured with sensitive instruments, and by using an appropriate decay constant, we can calculate the time at which the rock crystallized. The process of using radioactive decay products to determine the age of a rock is much like that of estimating the amount of time elapsed by using an hourglass filled with sand—just substitute the sand in the upper chamber for the radioactive parent isotope, the sand grains in the lower chamber for the daughter isotopes, and the rate at which sand slips from one chamber to the next for the decay rate. By counting the grains of sand in the lower chamber we could obtain an accurate estimate of the time elapsed since the hourglass was turned over. In planets, melting or recrystallization

Figure 2.4

The Crab nebula is the remnant of a supernova explosion of a massive star. This chaotic mass of expanding gas and dust is correlated with the description of a supernova seen by Chinese astronomers in A.D. 1054. Such explosions are an important method of injecting newly formed elements into interstellar space, where they may eventually be recycled to form other generations of stars and planets.

of a rock effectively resets its radiometric clocks to zero. Since most rocks suitable for dating have several different isotopes decaying at once, multiple clocks are ticking away in each rock. Using these dating techniques on meteorites and samples of lunar rocks provides ample evidence that the solar system originated about 4.6 billion years ago.

Nebular Hypothesis for the Formation of the Solar System

Although we perceive "outer space" to be completely without elemental material, it is not a perfect vacuum. Throughout the galaxy, gases are thinly dispersed. For each 10 cm^3 there may be only one atom. (Near Earth's surface, the atmosphere contains about 10^{20} atoms in the same volume.) The most important of these interstellar gases consist of the most abundant elements—hydrogen, helium, carbon, nitrogen, and oxygen. There are also a few metallic elements and dust grains composed of metals and silicates. We have seen how the violent deaths of some stars provide a recycling mechanism to charge interstellar space with these materials.

Occasionally, large concentrations of gas and dust, which may have approximately 1000 atoms/ cm^3, accumulate. Such dense, dusty clouds are called **nebulas** (Figure 2.5) and have been detected in several places in the galaxy. Several nebulas contain young stars; in part, this is why they are thought to be the birthgrounds of stars. Simple gravitational attraction and contraction can occur when the density of the gas is as low as 20 atoms/ cm^3. The **gravitational collapse** of a nebular cloud may be fairly rapid (Figure 2.6). Small density differences and gas turbulence can produce several subregions from a large nebula. Each concentration may eventually collapse independently and become a star. Open star clusters and some multiple star systems are thought to result from the fragmentation of contracting nebulas.

The exact nature of the process that induces the collapse of diffuse dust and gas to form a nebula remains a mystery. A possible clue comes from studies of the most primitive class of meteorites that have fallen to Earth. These meteorites retain chemical traces of the explosion of a massive star, which injected material into the developing (but still gaseous) solar system no more than a few million years before the meteorites solidified. Some scientists have suggested that the nebula from which the planets formed may have been concentrated by such a supernova "trigger," which swept

Figure 2.5

Dusty, gaseous nebulas, such as this one, are thought to form stars. Bright young stars occurring in clusters are obvious in the hazy region of the galaxy.

(A) **A slowly rotating portion of a large nebula** becomes a distinct globule as a mostly gaseous cloud collapses by gravitational attraction.

(B) **Rotation of the cloud prevents collapse of the equatorial disk** while a dense central mass forms.

(C) **A protostar "ignites"** and warms the inner part of the nebula, possibly vaporizing preexisting dust. As the nebula cools, condensation produces solid grains that settle to the central plane of the nebula.

(D) **The dusty nebula clears** either by dust aggregation into larger particles (planets or planetesimals) or by ejection during a T-Tauri stage of the star's evolution. A star energized by fusion and a system of cold bodies remains. Gravitational accretion of these small bodies eventually leads to the development of a small number of major planets.

Figure 2.6

The evolution of a dusty nebula with a surrounding system of orbiting planets is shown in this schematic diagram.

dispersed atoms closer together. It may be that the supernova that preceded the consolidation of the meteorites marked the death of a massive, short-lived companion of the embryonic Sun that later formed from the same nebula.

The contraction of the cloud of gas and dust guarantees more collisions among the atoms within it, producing heat. Some of this heat can be dissipated by infrared radiation into space; the rest is retained and elevates the temperature of the nebula. Because the interior of the nebula gradually warms, increased gas pressure causes the collapsing of the cloud to slow down. When temperatures exceed 10,000 K a **protostar**, probably located near the center of the nebula, may begin to radiate the light produced by the release of gravitational and thermal energy (nuclear reactions begin at a later stage). As a result, the inner portions of the nebula become much warmer than its outer reaches.

During this early contraction, the gas cloud begins to rotate, and as it collapses it rotates even faster, like a figure skater who draws in his or her arms during a spin. Such rapid rotation prevents a flattened disk of material in the equatorial plane from moving inward toward the protostar. The

planets eventually formed from the materials in this type of disk. In fact, disks of dust have been discovered around young nearby stars. Heat loss by **convection** (**radiation** is hindered by dust) allows further contraction of the protostar. Rapid and irregular outbursts of light and strong magnetic fields are associated with this relatively slow gravitational contraction. It is thought that during this so-called **T-Tauri phase**, large amounts of matter are ejected from the nebula in a type of "wind" that sweeps much of the uncondensed gas and even some light dust from the inner part of the evolving nebular disk. This occurs as the star settles onto what is known as the main sequence, a stage in the life of a star when it is relatively stable and long-lived.

As the star continues to contract, critical temperatures (around 8,000,000 K) and pressures are reached at which thermonuclear fusion of hydrogen can be initiated. When the nuclear fires are ignited, the temperature rises farther and essentially halts further contraction. Stars the size of the Sun may maintain this equilibrium state for billions of years, as they gradually consume their budget of hydrogen to form helium. The evolution from stellar nebula to hydrogen-burning star may only take 100,000 years—a short part of the solar system's 4.5-billion-year history.

Differentiation and Condensation of the Solar Nebula

Now let us go back a short time in this grand scenario and try to construct a plausible scheme for the development of the planets in our solar system. To do this, most contemporary theories call on the gravitational coagulation of solid particles that condensed from a dusty nebula such as the one just described.

One important process that occurred in the evolving solar nebula was its separation or **differentiation** into several physically and chemically unlike products. The gross differences in the sizes and compositions of the inner and outer planets demand that differentiation occurred within what must have been a relatively homogeneous nebula. An important part of this differentiation was the **condensation** or crystallization of solid particles from the gaseous nebula.

To determine the type of solids that might condense, we must first know the chemical composition and temperature of the nebula. Most planetary scientists assume that the composition of the nebular disk was about the same as that of the present-day Sun. There is some evidence in the primitive class of meteorites called **carbonaceous chondrites** that temperatures approaching 1800 K may have been approached locally (dust grains may have been vaporized). Using these assumptions regarding temperature and composition and another for pressure within the disk, we can calculate the sequence and composition of the solids condensing from the cooling gas. Figure 2.7 shows the generalized results of such calculations.

The first solids to condense from the nebula were probably small quantities of highly **refrac-**

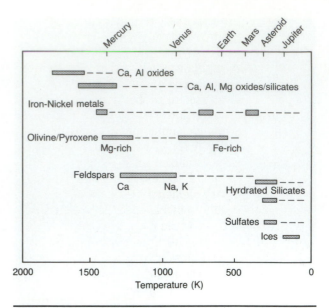

Figure 2.7

The sequence of condensation from a gas of solar composition as it cools from about 1700 K is shown in this diagram. Shaded bars indicate the interval over which condensation probably could occur; dashed lines indicate the persistence of these materials to lower temperatures in the absence of condensation. The upper axis indicates the possible condensation temperatures of components that produced the planets Mercury, Venus, Earth, Mars, the asteroids, and Jupiter.

tory elements such as tungsten, osmium, and zirconium. (Materials that form solids at very high temperatures are called refractory while those that condense or solidify at very low temperatures are called **volatile**.) Indeed, these elements may never have been vaporized in the nebula. The first compounds to form in significant amounts, however, were crystals of calcium and aluminum oxides, which probably condensed at temperatures of around 1700 K. Metallic iron–nickel compounds precipitated directly from the gas at about 1470 K. With continued cooling to about 1450 K, the oxide **minerals** reacted with the gas to form silicate minerals (those with silicon and oxygen) of calcium, aluminum, and magnesium. Magnesium-rich **olivine** and **pyroxene** are examples of these condensates. Alkali [sodium (Na), potassium (K), and rubidium (Rb)] silicates condensed at around 1000 K, forming **feldspars** at the expense of some minerals formed earlier. At about 700 K, previously condensed metallic iron reacted with sulfur in the gas to form troilite (FeS), an important mineral in some meteorites. At lower temperatures, some iron combined with oxygen and participated in reactions with magnesium to form silicates (like iron-rich olivine) or oxides (like magnetite). Eventually, below about 400 K, sulfates, carbonates, and hydrated silicates formed by reaction of early formed minerals with the gas.

Where the nebula was cooler than 300 K, sticky carbonaceous compounds precipitated, and at about 185 K water ice formed, probably in a blizzard of snowflakes. At even lower temperatures, volatile substances such as ammonia (NH_3), methane (CH_4), and nitrogen (N_2), which we normally regard as gases, crystallized as icy solids. As their constituent elements were much more abundant than the refractory elements, the condensation process quite literally snowballed at this point. It is unlikely that more volatile materials such as hydrogen (H_2) or helium (He) ever condensed, even in the cold outer reaches of the nebula. Thus, the low-temperature product of condensation from a solar gas consisted of a mixture of carbonaceous materials, hydrated silicates, sulfates, ices (of water, methane, and ammonia), and uncondensed gases—mostly hydrogen and helium. Realizing that the silicate, or rocky, component of the condensed materials would only be a small proportion of the nebular mass, a condensate that formed in the coldest regions of the nebula would be little more than a dirty snowball.

We have already noted that the nebular disk around the Sun must have possessed a strong thermal gradient. It was initially very hot near the proto-Sun and it must have been much cooler in its outer parts. At any one instant, the composition of the solids in the nebula would be dependent on their distance from the Sun (Figure 2.8). Assume, for example, that the nebula's temperature near the orbit of Mercury, the innermost planet, was about 1400 K. The condensates that would exist at that temperature have been predicted to consist of metallic grains of iron and nickel as well as silicate

minerals rich in calcium, aluminum, and magnesium (Figure 2.7). The density of this assemblage would be quite high. Making some reasonable assumptions, we can calculate the temperature at the orbit of Jupiter to be about 140 K at the very same moment. At this temperature, a greater proportion of the nebula would be condensed; the solids would consist of hydrated and oxidized silicates and carbon-rich materials and also a large amount of low-density ice. Thus, if the solids were somehow isolated from further reaction with the gas and instead formed planets, the composition and size of these planets would be dramatically dependent on their distance from the Sun.

This theoretical model, even if it is simple, explains many of the gross features of the planets. It predicts that Mercury should be rich in metallic iron and consequently dense; that Venus, Earth, and Mars should be less dense and contain more silicon, sodium, and potassium; and that the outer planets should be large, volatile-rich planets, which, because of their masses, have thick atmospheres that are gravitationally trapped remnants of the nebula. The most obvious compositional differences between the inner and outer planets can be explained by this sort of model of nebular condensation.

The differentiation of the nebula was driven by the temperature-dependent condensation of various chemical species. A requirement of this theory is that the solids and gases became physically separated. We do not know by what process this occurred, but we think that the nebula was swept clean of its uncondensed gases by an appropriately timed T-Tauri phase in the development of the early Sun. This housecleaning event may have left refractory and metal-rich dust close to the Sun, hydrated lower-density dust farther from the Sun,

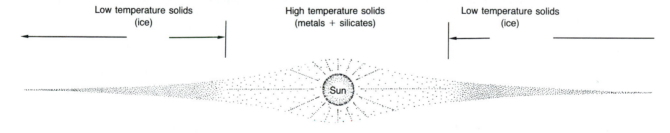

Figure 2.8

A cross section of a hypothetical nebula shows a star forming in its center. Condensation of solids from a solar nebula with a temperature gradient may have given rise to compositional differences in the condensates. At one instant, the condensates in the inner part of the developing solar system would consist of high temperature materials such as silicates, while at the same instant, but farther from the Sun, the nebula may have been cool enough to allow ices to be fully condensed as well. Since water was relatively abundant in the nebular gases, more solid matter formed in the cooler outer part of the nebula.

and low-density ices farthest from the Sun. Alternatively, the solids may have become incorporated in planets and hence grew incapable of communicating or reacting with the gas.

The reasons for the size differences between the inner and the outer planets are suggested by the bulk composition of the nebula itself. Even if the entire refractory or rocky component condensed, it would represent less than 0.5 percent of the total mass available in a nebula of solar composition. Since all of the inner planets are predominantly rocky, over 99.5 percent of their potential mass is missing—it must have been too volatile to condense in the warm inner region of the developing solar system. The outer planets (Jupiter, Saturn, Uranus, Neptune, and Pluto) formed from material that condensed at low temperatures (about 200 to 50 K), where ices of water, ammonia, and methane could form in addition to the rocky component. These ices account for about 1.5 percent of the mass of the nebula. Substantial portions of the outer planets, especially of Uranus and Neptune, are postulated to consist of elements from these ices.

Accretion of the Planets

Such differentiation processes may account for the chemical composition of the planets, but it is unlikely that the planets simply crystallized grain by grain and layer by layer from a dusty nebula. What probably did happen is more complex. Once the condensing grains had settled by the force of gravity into the central plane of the nebula (a process that may have taken only a few thousand years), it appears that **planetesimals** with diameters of a few kilometers formed. Some scientists suggest that this was accomplished by gravitational grain-by-grain **accretion** to produce streams or clusters of small bodies that moved in nearly coincident circular orbits. Low-velocity collisions within or between the planetesimal swarms eventually led to the accumulation (not destruction) of even larger planetesimals. Some of these planets in embryo may have been as large as the Moon or even Mars, but collisions also produced small fragments that were later accreted. Some collisions may have resulted in total disruption of a body, followed by its re-accretion. For a given distance from the Sun, one body eventually became gravitationally dominant and swept up most of the material near its orbit. This process of accumulation of the planets from smaller bodies is called **collisional accretion** and was probably complete within a million years. Some terrestrial planets, especially the Moon and Mercury, retain dramatic

evidence of the last phase of accretion. Their intensely cratered surfaces were produced by the last infalls of material that lingered in their paths even after they had assumed solid spherical shapes. In short, the elemental material of our solar system evolved from gas to dust to clots in co-orbiting streams that eventually accreted to form planets.

Jupiter (and the other outer planets) grew larger and perhaps faster than the inner planets because of the abundance of icy condensates in the cooler outer nebula. Small nebular disks also formed around the larger planets. As these miniature nebulas were probably very similar to the larger solar nebula, condensation and collisional accretion probably produced the large systems of natural satellites that encircle Jupiter and Saturn. Simultaneously, large quantities of the uncondensable nebular gases surrounding the growing icy planets became hydrodynamically unstable and collapsed onto the planets' cores to form thick, colorful atmospheres. In contrast, the present atmospheres of Venus, Earth, and Mars are largely secondary and were most likely expelled from the interiors of their respective planets rather than inherited from the nebula.

Internal Differentiation of the Planets

We have seen how differentiation of the solar nebula led to important differences among the planets. Another type of differentiation occurred within the interiors of the planets and produced a variety of layered internal structures, depending on variables such as size, density, and composition of the planetary body (Figure 2.9). **Internal differentiation** of the planets is a very important process in their evolution, although it goes on by a different mechanism and occurs on a much smaller scale than the differentiation processes that occurred in the nebula. The continuing internal differentiation of planets like Venus and Earth drives their dynamic geologic systems and produces, among other things, volcanoes and earthquakes.

Differentiation within planets occurs because elements have distinctive physical properties (mostly density) and chemical affinities, which allow them to separate from one another. For example, many elements have an affinity to and behave like metallic iron and are consequently called **metals**. This group includes relatively dense elements such as iron, nickel, and cobalt. Other elements have affinities for oxygen and silicon and form rocky materials called **silicates**. These elements include aluminum, calcium, sodium, and potassium; they form solids that are less dense than

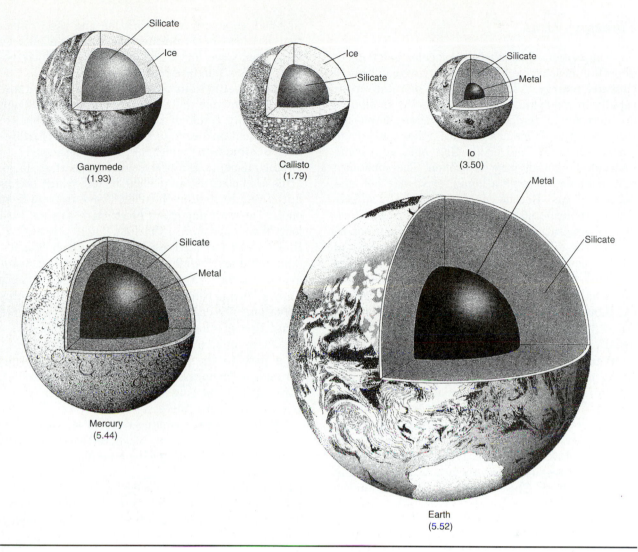

Figure 2.9

The interiors of five planets are compared in this diagram, which illustrates the relative size of various internal components. The densities of the bodies are also given in g/cm^3. Although some scientists believe these layered structures are the result of layer by layer accretion from the nebula, there is good evidence to suggest that the planets were originally relatively homogeneous and that the layered internal structures are the result of planetary differentiation. Note how the proportions of silicate, ice, and metal change from bodies in the inner solar system to those in the outer solar system.

metallic iron. Materials that form low density solids only at low temperature are called **ices**. Of course, the most common is water (H_2O), but methane (CH_4), and nitrogen (N_2) are other important ices. Other elements, like hydrogen, helium, neon, carbon, and oxygen, combine to form stable gases and accumulate in the atmosphere of a planet. These are the **atmophile** elements. Several elements fall in more than one group, depending on external conditions. Distinctive features of the metals include their high densities. Apparently, metals have accumulated in the cores of several planets by gravitational sinking. Churning motions in a molten core of metal may produce a magnetic field that envelopes the planet. If the planets were originally

homogeneous, core formation requires considerable redistribution of elements. This mobility of planetary materials may seem at odds with our experience with the "solid" Earth, but under high temperature and pressure, rocks become weak and behave much like fluids (although they may not be molten). This plasticity allows the transport of materials during planetary differentiation but is largely dependent on a critical temperature within the planet. Therefore, planetary differentiation is intimately tied to the **thermal history** of a planet (a description of temperature variations with time). The fundamental questions may be: From where does a planetary body derive its heat? How is this energy transported from one part of a body to another?

Planetary Heat

The motion of atoms and molecules in an object gives it a measurable form of energy we call heat or **thermal energy**. As atoms vibrate or rotate more rapidly in a substance, we perceive the motion as an increase in temperature. As thermal energy increases, the atoms are forced farther away from one another, and the bonds that hold atoms and molecules together may be broken. The substance may melt and become a liquid, or it may boil and become a gas. Because heat dramatically affects the mechanical and chemical properties of materials, its generation and movement are very important to the differentiation and geologic activity of all the planets. Heat is transferred through three principal ways: (1) conduction, (2) convection, and (3) radiation. **Conduction** is the process by which the vibrational energy of an atom is transferred to adjacent atoms and is the method by which heat is usually transmitted through rigid, opaque solids. **Convection** occurs in fluids and nonrigid solids as warm material expands and moves upward, displacing cooler and denser material downward. Convection within planets transfers thermal energy much more efficiently than conduction. Convection transfers matter and may cause planetary differentiation. **Radiation** involves the emission of electromagnetic waves from the surface of a hot body to its surroundings. The radiation of energy from the Sun to the planets is an obvious means of energy transfer in the solar system.

The major sources of heat involved in planetary dynamics are shown in Figure 2.10. **Accretionary heating** results from the collisional accretion of two or more bodies. It is likely to have influenced the planets during their formation because some of the kinetic energy (the energy of motion) of falling planetesimals was converted to thermal energy during impacts. If the planets grew rapidly enough, some of this energy may have melted a large part of their exteriors, allowing chemical and gravitational differentiation to occur.

Heat produced by loss of gravitational potential energy may have been, in a sense, self-perpetuating by **core formation**. If the iron-rich cores thought to exist inside some planets were formed by accumulation and gravitational segregation from a once-homogeneous planet, then a tremendous amount of heat must have been released. The loss of **gravitational potential energy** of each iron-rich "droplet" would be accompanied by an increase in planetary temperature. The released heat decreased the strength of surrounding materials and allowed core segregation to go on at a more rapid rate. This feedback cycle may have led to rapid, runaway core formation and the production of a heat pulse during this stage of a planet's formation. Core formation on Earth may have required a few hundred million years.

Radiogenic heat is produced by the spontaneous breakdown of the nucleus of an atom. Although we normally think of the planets as cold bodies, they all contain a certain small proportion of **radioactive elements**, which decay spontaneously to lighter elements and release energy as they do so. (This is in contrast to nuclear fusion, which occurs in the Sun to produce heavier elements from light ones.) The most important radioactive elements are long-lived—uranium, thorium, and potassium. **Short-lived** radioactive isotopes of other elements (aluminum and iodine) are not found in significant amounts today but may have been important heat sources during the early stages of planetary differentiation. The nuclear decay of these elements heats the interiors of planets that contain them; rocks may even melt, become less dense, and start on a path to the surface where they may erupt through volcanoes. Each planet once possessed such an internal heat engine, and some may still possess this heat source.

Solar energy, produced from a variety of nuclear fusion reactions, when radiated to Earth maintains the planet's biologic activity and drives the system of circulating water and gas at its surface. Even so, this energy is not sufficient to raise the temperature of planets enough to allow any internal differentiation. A much smaller amount of solar energy reaches the most distant planets.

A less familiar type of solar energy may have heated the planets or their precursors during a hypothetical T-Tauri phase of the Sun. The strong solar winds associated with this phase may have so distorted and enlarged the solar magnetic field that they induced strong electrical currents in the planets. The outer parts of the innermost planets or planetesimals may have been heated, perhaps even melted, as a result. The outer planets probably would not have interacted as strongly with the magnetic field and would be little affected. This theory of electromagnetic heating is controversial but has been used to explain the heat developed in relatively small bodies that were parents to the meteorites that fall to Earth. If this type of heating occurred, it was restricted to the earliest evolution of the inner planets and would have most strongly affected their surface layers.

Tidal heat, another even less obvious source of heat, is important for the evolution of some planetary bodies. The gravitational attraction of a planet on its satellites may, in special cases, lead to

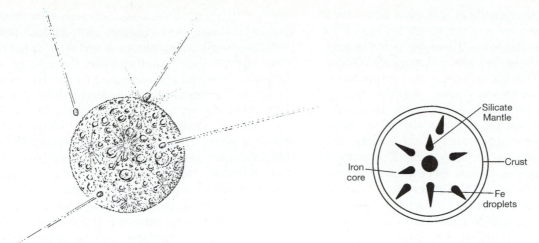

(A) **Accretionary** heat comes from the conversion of kinetic energy to heat. This heat is trapped in the planet if accretion is rapid.

(B) **Core formation** converts gravitational potential energy into heat as molten iron drops to the center of a planet.

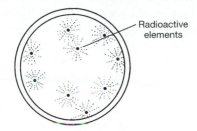

(C) **Radiogenic heat** is caused by the decay of radioactive atoms dispersed in the interior of a planet.

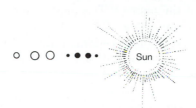

(D) **Solar energy** is produced by nuclear fusion reactions in a star and is transmitted to planets in electromagnetic waves (light).

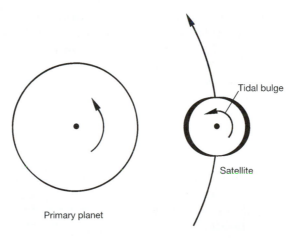

(E) **Tidal heating** results when a satellite is repeatedly flexed by the gravitational attraction of its primary.

Figure 2.10
The variety of planetary heat sources.

higher interior temperatures and even large-scale melting. On Earth, tides in the oceans are raised by the attraction of the Moon. However, it is not only the oceans that bulge but also the more rigid outer layers of rock. Dissipation of the energy produced during these movements may slightly heat a planet, just as an elastic band heats up during repeated stretching. Imagine the size of the tides that might arise on the satellites of Jupiter, the largest planet in the solar system. The mutual gravitational tugs of its four largest moons cause predictable variations in the distance between the satellites and their parent planet that result in tidal variations large enough to generate sufficient heat to melt the interior of at least the innermost satellite, Io. This tidal heat, produced by a sort of mechanical heat pump, appears to have played an important role in the differentiation and evolution of the small moons of Jupiter and Saturn.

In summary, there are several different forms of energy and types of processes involved in the thermal evolution of the planets. The most important heat source for the original differentiation of the planets may have been accretionary heating. Short-lived radioactive isotopes of aluminum and iron may have substantially heated small planetesimals, but it seems unlikely that these elements persisted much past the planetesimal stage. Likewise, electromagnetic heating associated with a T-Tauri phase probably occurred before the planets accreted. Long-lived radioactivity, although important in sustaining the differentiation and geologic activity of the planets, probably contributed as little as 300 K to the temperature of the primitive undifferentiated planets. In contrast, thousands of degrees of heat may have been generated by collisional accretion. It is estimated that the average temperature of Earth would have been about 30,000 K if all of this heat had been retained (1300 K for the Moon, 4000 K for Mercury, 6000 K for Mars, and 25,000 K for Venus). Of course, much of this heat was quickly radiated away into space, but since iron- and magnesium-rich silicates melt at temperatures of about 1400 to 4000 K (increasing with pressure) and ices melt at temperatures below 300 K, retention of even a small fraction of the accretionary heat could lead to melting or even vaporization of planetary materials. Once initiated by accretionary heating, the process of differentiation may have been invigorated by core formation.

Thermal History

Planetary scientists attempt to reconstruct the thermal history of a planet from facts (and assumptions) about its surface features, its chemical com-

position (the abundance of radioactive elements and metals, ices, and silicates), its original temperature distribution, and its accretionary history. These factors help to determine the change of temperature with depth and give us an idea about the internal temperature at different times during the planet's history. Indeed, the geologic history of a planet is a reflection of its thermal history. The size of a planet is an extremely important factor that influences how rapidly a planet can lose the heat released from its interior. All the heat generated within a planetary body must eventually be transported to the surface by conduction or convection and then radiated away. Thus, small planets with large **surface area to mass ratios** cool faster as heat is readily radiated away into space. Large planets lose their heat much more slowly. For example, a large, shallow pan of water placed in a refrigerator cools more rapidly than the same amount of water in a glass. This is a result of the larger surface area to mass ratio of the water body in the pan. The planets Mercury, Mars, and Earth, given the same initial temperature and composition, would not cool at the same rate. Mercury, being the smallest, would cool the fastest (Figure 2.11).

As a planet's temperature varies, either by heating during accretion or by subsequent cooling, substantial changes occur in the nature of the geologic processes that shape its interior and surface. A short scenario of a possible thermal history for a terrestrial type of planet shows some of these principles (Figure 2.12). Shortly after (or during) the formation of a planet, its outer shell may become entirely molten as the result of the heat

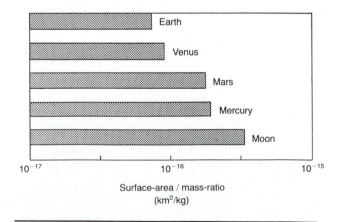

Figure 2.11

The cooling rate of a planet depends on the surface area to mass ratio. A body with a large value for this ratio would cool relatively rapidly because of the large area through which heat may be lost to space relative to its mass.

generated by many impacts of planetesimals during accretion. As this **magma ocean** cools, a crust (or more precisely a **lithosphere**—the solid, rigid outer shell of a planet) composed of light silicate minerals rich in silicon, aluminum, and oxygen may form at the surface. It is easy to imagine that the lithosphere would be very mobile at this stage and could slide across the surface as a silicate scum. Numerous small **plates**, slabs, or "rockbergs" would form as the melt cooled. Collisions and rifts, driven by vigorous convection, must have been frequent during this period of crustal accumulation and high surface temperature. With continued cooling, a dense **mantle** composed of silicate minerals rich in iron and magnesium probably formed beneath the chemically distinct crust, adding to the thickness of the mechanically distinct lithosphere. Extensive melting of at least the outer portions of a planet may have allowed the separation and gravitational accumulation of metallic iron and other dense metals into distinct droplets. Once blobs about 100 km in diameter formed, they could penetrate the lower, previously undifferentiated, part of the planet and form its **core**. Core formation

probably accompanied the accretion of the planets. For example, many planetesimals that collided to form Earth may have previously developed cores. Differentiation caused a tremendous redistribution of mass and released more heat. Further differentiation of the planet could occur when silica-rich magmas were formed and extruded onto the surface to form **volcanoes**, or cooled beneath the surface to form **plutons**. During early differentiation some of the volatile, atmophile elements would be released from the rocky materials as fluids or gases and would accumulate at the surface to form atmospheres or **hydrospheres**. Once core formation is completed, internal geologic activity is maintained by heat released from long-lived radioactive decay. However, in most cases the heat generated by this process is exceeded by heat lost to space, causing the lithosphere to become progressively thicker. Solidification of silicate minerals may begin simultaneously at the core–mantle boundary, producing an ever diminishing volume of partially molten or solid-but-weak material called the **asthenosphere**. This convecting plastic zone allows the more rigid lithosphere to slide and shift about and is a ready source of magma for volcanoes.

Most planets lose their internal heat not only by conduction but also by convection of their interiors. Cylindrical plumes of material rise or fall depending on their temperatures and densities. Hot rising plumes, which bring heat to the surface where it can be radiated away, may cause the lithosphere of a planet to bulge, and cold sinking plumes may depress the surface of a planet. On some planets with thin lithospheres, a different style of shallow convection may develop wherein large lithospheric slabs or plates are continually created and consumed as they shift about, driven by gravity. Linear belts of volcanoes, folded mountains, and earthquakes result from collisions or rifts of these plates. This is called **plate tectonics** and is characteristic of Earth's present stage of development. Tectonics denotes the large-scale deformation of the lithosphere. Earth's plate-tectonic system, which produces continental drift, is a reflection of the continued differentiation of Earth. Eventually, as a planet continues to cool, the lithosphere may become so thick that the lateral motion of separate sections of lithosphere is no longer possible. Separate plates coalesce to form a single, planet-encircling, solid, rigid lithosphere. Vertical movements and lithospheric flexures related to plumes may be the only expressions of the convective flow of mantle materials at this stage. Even later, as the lithosphere continues to thicken, hot magmas cannot penetrate through the lithosphere before they cool. The geologic evolution of a planet

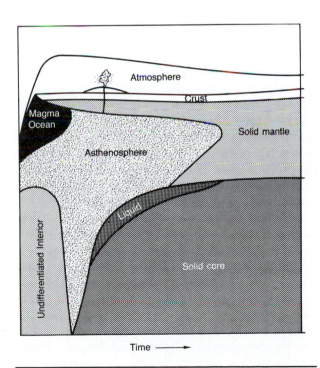

Figure 2.12

The thermal evolution of a terrestrial planet shows the changing temperature inside the planet. The time scale is relative. (Compare this diagram to those in chapters 4, 5, 6, and 7, which include absolute time estimates.) The occurrences, timing, and relative importance of these processes are unique to each planet and are determined by the planet's composition, mass, heat budget, and other characteristics.

is nearly complete by this stage, as its internal heat source (generally natural radioactivity) gradually dies and internal activity ceases to be expressed at its surface. This inexorable cooling may lead to slow planetary contraction. Chemical differentiation slows and eventually ceases, as there is insufficient heat to cause, or even to allow, matter to migrate or segregate any longer.

Planetary bodies whose evolution and differentiation are not limited by their budget of radioactive elements may continue to be dynamic long after their radioactive heat is dissipated. Tidally flexed Io, for example, is only slightly larger than Earth's Moon; yet Io continues to be volcanically active while the Moon ceased to be active almost 3 billion years ago. Nonetheless, the extent, duration, and style of planetary differentiation in most planets are complex functions of their sizes and compositions. For example, small bodies lose heat so rapidly that they quickly move through all these stages in a billion years or so (e.g., the Moon). Very small bodies, such as the moons of the outer planets, may not have become warm enough to differentiate at all and may preserve their primitive compositions to the present day. Larger planets, which cool slowly, evolve at a more leisurely pace, allowing much time for extensive chemical separations to occur (e.g., Earth).

The central theme of this brief summary of the origin and differentiation of planets is that each has a life cycle related to its thermal history. The internal differentiation of planets is a result of this interaction of heat with planetary matter. William Kaula, paraphrasing a famous passage from Shakespeare, has summarized this grand evolutionary scheme:

Our system is a stage,
And both the Sun and planets merely players.
They had their birth and'll have their fiery end.
A planet in its time plays many parts,
Its acts being seven ages.
The first of these is condensation: dust grains
    drifting to
The nebula plane in chondrite clods. And then
The planetesimals: breaking sometimes, but
Most growing, though the Sun's hot breath blows
    gas
Away. And then formation: sweeping up
The bodies in its way, in fierce infalls
To bring them full convective vigor, too hot
For crust to form, though iron may sink and seas
Outgas, by radioactive energy driven.
And then comes plate tectonics: cooling leads
To lithosphere, with many marginal breaks.
Convective thrusts a crust create in belts

Complex. But heating slows; the sixth age shifts
Into final volcanism: no
More lithospheric spreading, only vents
For magma, Nix Olympica* or mare**
To surface, ending fractionation. Last scene
That ends this history is quiescence: time
Sans melt, sans plate, sans almost everything.

Lithospheres, Hydrospheres, and Atmospheres: Products of Planetary Differentiation

The part of a planet that is visible and that may be directly sampled consists entirely of its outermost layers (its atmosphere, hydrosphere, or lithosphere—all products of planetary differentiation). Only the surface, the very skin of the lithosphere, is visible from space, and even this is sometimes obscured from our view by thick or opaque atmospheres (e.g., Venus and Titan), whereas others have no atmosphere at all (e.g., Moon and Ganymede). Before we delve into a study of each planet, let us turn to examine these accessible parts of the planets generally.

Planetary Lithospheres

A planetary lithosphere is commonly conceived of as a mechanically strong, rigid outer layer that overlays a plastic or ductile asthenosphere. Planetary asthenospheres are commonly partially molten. From this point of view, then, the lithosphere is not a chemically distinct unit; instead, it is usually composed of a planet's crust and the upper part of its mantle (Figure 2.13). Obviously, the temperature within the interior determines the thickness of the rigid lithosphere and hence the depth to the top of the asthenosphere. Thus, the origin and evolution of a lithosphere is strongly tied to the thermal history of a planet. If a planet formed hot, an early lithosphere would be thin and mobile and fragments would slide about the slippery asthenosphere below. But as a planet cools, its lithosphere thickens and slowly becomes immobile. The link between the thermal evolution of a planet and its surface geologic features is, therefore, the lithosphere.

The lithosphere of a differentiated planet consists of its solid outer shell and is composed of a variety of solid materials that we call **rocks**. The nature of these rocks provides important informa-

---

*A large volcano on Mars, usually called Olympus Mons.
**Vast volcanic plains on the Moon.

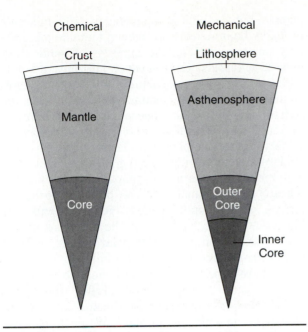

Figure 2.13

The terminology used to describe the outer layers of a planet depends on one's point of view as shown in this cross section of a planet. **Chemically** a planet is differentiated into a crust and mantle of distinct compositions. **Mechanically** the crust and upper mantle of a planet may behave as a single "rigid" unit called the lithosphere. Underlying the lithosphere is the asthenosphere, a semiplastic zone that yields to flow much more easily than does the lithosphere. In some cases the asthenosphere may extend to the base of the crust.

some information about its past. For example, the temperature and pressure of a volcanic lava can be deduced from careful chemical analyses of its now-solid products. Even the internal structures of a sand dune, stream channel, or lava flow are commonly preserved in resultant rocks to inform geologists of the environment in which the rock body was originally formed. Rocks are records of past events. In their composition, structure, and sequence relations to other rock bodies are records of their formation and evolution.

To put it less abstractly, rocks are simply aggregates of smaller entities called minerals (Figure 2.14). By definition, these naturally occurring solids contain a distinctive set of elements arranged in a well-defined internal structure. As a result, each mineral has definite physical and chemical properties by which it can be identified. Differences between minerals arise from the kinds of atoms they contain and the way in which they are arranged (Figure 2.15). Many minerals grow when atoms from a surrounding liquid or gas are added to its crystal structure. Mineral grains may also grow from other solids in the absence of either gases or liquids by recrystallization in the solid state.

tion about the evolution of a planet because their composition and physical character change in response to external physical conditions (pressure, temperature, or composition of the surroundings). A fundamental law of nature holds that all materials attempt to achieve a balance with the chemical and physical forces exerted upon them and arrive at a state of **equilibrium**. This is also the state of lowest total energy. This effort results in progressive changes when planetary materials are exposed to environments different from that in which they formed. Although this equilibrium state is the preferred state of all systems, there are many intermediate or **metastable** states. All geologic materials experience various metastable states, but all changes tend toward achieving a physical or chemical state that is more stable in the present environment.

Because rocks may retain metastable characteristics in their physical and chemical nature, they record information about their mode of formation. Therefore, the composition of a rock preserves

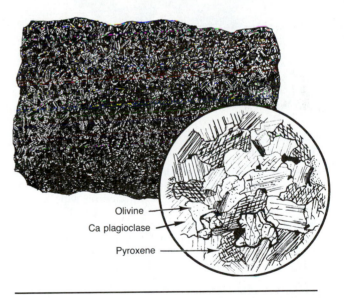

Figure 2.14

A hand specimen (shown full size) of a dark iron- and magnesium-rich rock called gabbro, composed mostly of the silicate minerals plagioclase, pyroxene, and olivine. The individual mineral grains are discernible but are relatively small; the aggregate nature of the rock is clearly demonstrated in the inset, which shows a magnified view of a thin slice of such a rock. Discrete mineral grains of different compositions interlock in this typical igneous fabric, which developed as crystals formed from molten rock.

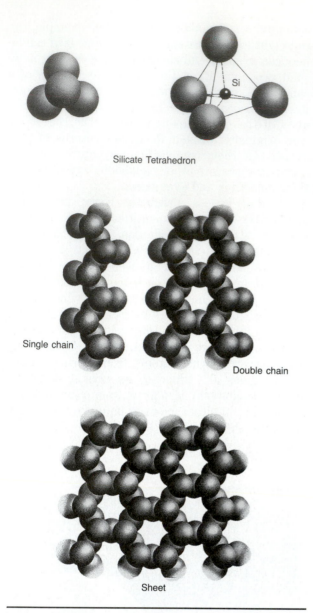

Figure 2.15

The silicon-oxygen tetrahedron is the basic building block of silicate minerals (top). Four large oxygen atoms are arranged in the form of a tetrahedron with a small silicon atom in the small central space. This basic tetrahedron can be combined with others in a variety of ways to create the complex internal atomic structures of silicate minerals. Chains or sheets are created by sharing oxygen ions between two tetrahedral units.

It is useful to divide the major planetary materials into three major groups—metals, silicates, and ices. (With the addition of gases, these are the major constituents we mentioned in the discussion of the solar nebula.) Silicate minerals and ice probably comprise over 95 percent of all planetary lithospheres. Dense **metallic minerals** (principally iron compounds) are rare (but economically important resources) in the lithospheres of differentiated

planets, but the bulk of a planet's metallic minerals probably lie in its core. However, metallic minerals may be important at the surfaces of some fragmented asteroids whose deeper interiors are now exposed. The most important **silicates** are compounds of iron, magnesium, aluminum, calcium, sodium, and potassium. These elements are linked to tetrahedra of silicon and oxygen to form a variety of crystal patterns. These minerals have moderate densities (2.5 to 3.0 g/cm^3) and relatively high melting temperatures (about 1000 to 1300 K). Silicate minerals make up the bulk of the lithospheres of the inner planets. **Icy minerals** are important at the surfaces of the satellites of the outer planets and in the cold polar regions of Mars and Earth. (Some investigators balk at the use of lithosphere—rocky sphere—to describe an icy planet's outer layer; they prefer the term *cryosphere*—ice sphere.) The principal ices consist of water, ammonia, and methane. Water is the most abundant of these ices, but on many icy planetary bodies it is probably mixed with ammonia. Such a mixture may allow liquids to persist to temperatures below the freezing point of water (273 K at 1 bar). Methane ice forms at 109 K (at 1 bar) and is only stable at the surfaces of planets in the far outer reaches of the solar system (e.g., on Pluto and on Neptune's satellite, Triton). These disparate compounds are grouped together as ices because of their low freezing temperatures (300 to 100 K) and uniformly low densities (less than 1.0 g/cm^3). Depending on differentiation histories, other types of minerals may also be important on specific planets (e.g., carbonate minerals are an important but small part of Earth's lithosphere, and various sulfur-based minerals are important at the surface of Io).

Besides these compositional groups, three broad groups of rocks can be distinguished by their mechanism of formation: (1) Igneous rocks form by cooling of a melt; (2) sedimentary rocks form by erosion, transportation, and deposition of rock particles; and (3) metamorphic rocks form by changes resulting from heat, pressure, or the introduction of new elements without the intervention of a melt.

Igneous rocks form from molten rock material called **magma**. A magma may include crystals (solid), melt (liquid), and volatiles (gases). Several types of magmas are important on the planets. The lithospheres of the terrestrial planets have been built by silicate magmas. Figure 2.16 shows the broad range of silicate magma compositions. They may range from **basalt** or **gabbro** (typically hot— about 1500 K—highly fluid, containing up to about 50 percent silica [SiO_2]) to **rhyolite** or **granite** (typically cooler—about 1000 K—and viscous, containing up to about 77 percent silica). On the satellites of the outer planets, **water magmas** were important. Although on Earth we do not generally think of water as a magma, the processes of melting at temperatures above normal surface temperature, mobilization, and cooling of watery liquids lead us to regard water as a magma on cold bodies, whose surface temperatures are only about 50 to 100 K. (Temperatures as high as 275 K are required for the existence of pure water melts on these bodies.) Ammonia and other chemicals may also be dissolved in these melts and subsequent solidification products. **Sulfur magmas** may be important on Io, the innermost satellite of Jupiter, where extensive internal differentiation has led to the accumulation of an outer sulfur-rich shell. Melting of this material would produce volcanic eruptions of fluid sulfurous lavas with temperatures of about 400 K.

The classification of igneous rocks is based upon texture (mostly the size of the constituent mineral crystals) and composition (Figure 2.16). Most **volcanic** or **extrusive rocks**, igneous rocks that cooled at or very near the surface of a planet, are fine-grained because they cooled rapidly. In contrast, many **plutonic** or **intrusive rocks** have larger grains because their magmas cooled slowly beneath the ground, allowing extensive growth of crystals. High-silica (**silicic**) magmas produce rocks of the rhyolite family, with the minerals quartz, potassium feldspar, and sodium plagioclase (Figure 2.16). Low-silica magmas, rich in iron and magnesium (**mafic**), produce rocks of the basalt family, with the minerals calcium plagioclase, pyroxene, and olivine. Magmas with an intermediate

composition produce rocks of the andesite family with compositions between that of rhyolite and basalt. Along with **anorthosites** (plagioclase feldspar rocks), these igneous rocks are the most important constituents of the crusts of the terrestrial planets. In contrast, the mantles of these planets are thought to be formed from dense olivine-rich rocks such as **peridotite**.

Eruptions of magma may occur because the hot molten magma is less dense than the surrounding solids. The magma thus tends to rise and reaches the surface to erupt through a crack or through a pipelike conduit. The eruption style of a magma depends on many complex variables, including its composition, viscosity, gas content, and volume. For example, basaltic magmas are fluid and are commonly extruded in quiet eruptions issuing from fissures in the lithosphere. They produce successions of thin flows that cover broad areas. Basalt eruptions commonly form small **cinder cones** and large and small **shield volcanoes** (Figure 2.17). Water magmas might erupt in a fashion similar to geysers along a fissure, but because they are so fluid the flows would be very thin and topographic volcanoes would be difficult to detect. The fluid eruptions of basaltic or icy lavas may resurface a very large part of the planet, burying the older landforms beneath it. In contrast, andesitic and rhyolitic magmas are extremely viscous and, for silicate magmas, contain large amounts of dissolved gas. They therefore explode violently and extrude lava flows in thick, pasty masses or turbulent rapidly moving ash flows and sheetlike ash falls. These eruptions may produce lava **domes**, **composite volcanoes**, or large **calderas** with **ash-flow**

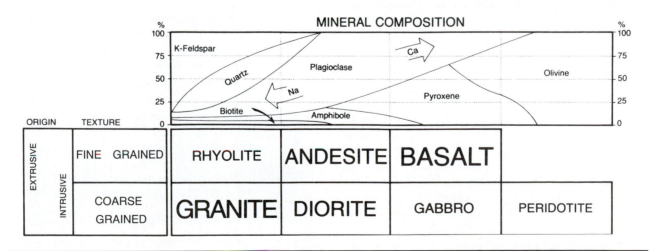

Figure 2.16

The classification of igneous rocks is based on composition and texture. Granite is the most abundant intrusive rock, whereas basalt is the most abundant extrusive rock on Earth.

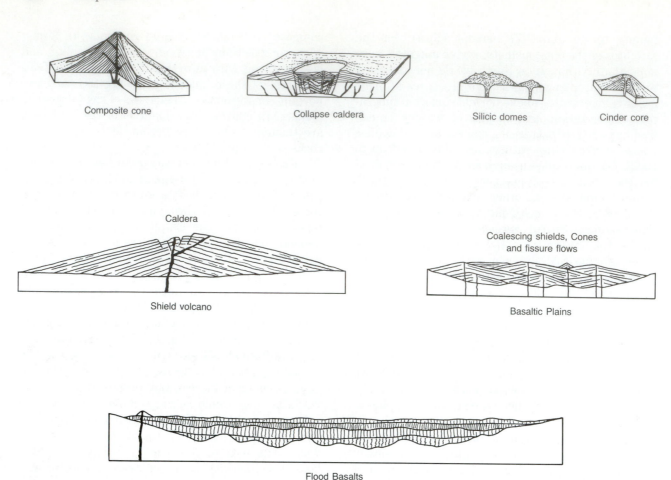

Composite cone   Collapse caldera   Silicic domes   Cinder core

Caldera

Shield volcano

Coalescing shields, Cones
and fissure flows

Basaltic Plains

Flood Basalts

Figure 2.17

A large variety of volcanic landforms are found on the planets; their appearance depends upon the composition of the magma, its eruption rate, and its eruption style. Basaltic lavas are fluid; they are commonly extruded along fissures and flood the surrounding area. Magmas rich in silica are more viscous and form large volcanic mountains or erupt to form vast sheets of volcanic ash.

shields (Figure 2.17). Masses of igneous rock formed by the cooling of magma beneath the surface are called plutons or intrusions. Although volcanic rocks and landforms are much more obvious, plutonic rocks comprise the bulk of many planetary lithospheres. They may be especially common on icy planets because of water's unique properties—its liquid state is more dense that its solid state—that could allow large **magma chambers** to collect in the lithosphere without rising and erupting at the surface. Some of the variety of intrusive rock bodies are shown in Figure 2.18. Their shapes and sizes are strongly dependent on the stress regime (compression or extension) in the lithosphere.

Magma originates by partial melting of the interior of planets. Magmas that reach the surface were probably generated in the lithosphere or asthenosphere of a planet. It is extremely unlikely that planetary magmas arise from their metallic

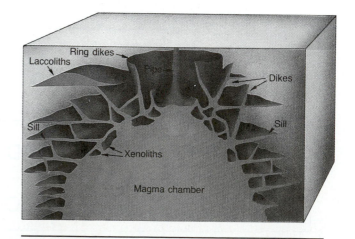

Figure 2.18

Intrusive rock bodies may assume a variety of shapes. Large bodies are elliptical; smaller bodies are formed as magma is squeezed into cracks and zones of weakness in the surrounding rocks.

cores. The heat necessary for lithospheric melting is commonly thought to be transported by "primary" magmas, derived in an asthenosphere, into a lithosphere. Lithospheric and especially crustal rocks may be brought to their melting temperatures by the heat added in this way. Since the first fraction of melt formed usually has a composition very different from its parent material, the processes of partial melting and magma migration promote the internal chemical differentiation of the planets.

Sedimentary rocks originate from fragments of other rocks, from minerals precipitated chemically, and, at least on Earth, from accumulation of organic matter. Although the natures of these materials contrast sharply with one another in many regards, all accumulate at or near the surfaces of planets under the temperatures and pressures characteristic of the planet. Sedimentary rocks are volumetrically minor components of planetary lithospheres but are important surface materials. On Earth, less than 5 percent of the lithosphere consists of sedimentary rocks, and yet they cover 75 percent of the exposed land masses.

Four major processes are involved in the genesis of sedimentary rocks. **Disintegration** of rocks can occur because both mechanical and chemical processes break down rock material at a planet's surface. **Transportation** moves these particles and allows them to accumulate elsewhere during **deposition** processes. Finally, **compaction** and **cementation** convert loose particles into solid rock. It is easy to visualize these processes—rock debris is produced on a volcano's slopes, is carried off by streams that drain the mountain, is subsequently collected as particles of sand or clay in a **delta** at the mouth of a river, and is lithified, as cements form between grains. The same general processes also occur to produce deposits of **impact breccia** associated with impact craters. (Breccias are rocks composed of angular fragments of older rocks.) In this case, disintegration is caused by the transfer of a meteor's kinetic energy to the planet's surface. Fragments are made if the strength of mineral crystals is exceeded. The debris is transported along arching ballistic pathways far from the site of origin and is deposited in a sheetlike accumulation around the newly created crater. Subsequent burial or other processes help to cement the rock fragments together, creating what might be called a sedimentary rock. Another class of sedimentary rocks, however, have much in common with igneous rocks. Called **evaporites**, these rocks form as minerals precipitate from briny water. These include deposits of salt (NaCl) and gypsum ($CaSO_4 \cdot 2H_2O$). In principle, there is little difference between the crystallization of a brine and that of a magma; both are governed by the same chemical laws. Evaporites may be important types of rocks or intergranular cements on Mars, as they are in terrestrial deserts. Other chemical precipitates are also important. Limestone and other carbonate rocks and minerals precipitate from terrestrial water, sometimes aided by biologic processes. Deposits of terrestrial coal and petroleum originate as organic matter accumulates, as plants and animals die.

Reflecting their nature as thin units of surface rock, sedimentary rocks typically occur in layers. In a sequence of sedimentary layers, the youngest is at the top and the oldest is at the bottom, reflecting the **law of superposition** (Figure 2.19). Sedimentary rocks cover large parts of some planetary surfaces; if impact breccias are included, such would be the case for most planetary bodies. Other important structures visible in sedimentary rocks (such as crossbedding, graded bedding, ripple marks, and mud cracks) reveal much about their formation. All of these items tell about the environment of deposition (Figure 2.20); for example, marine limestones prove that oceans once inundated large parts of Earth's continents millions of years ago. Unfortunately, detecting these features requires examination of samples and outcrops—information not usually available to a planetary geologist.

The third major group of rocks includes those produced by **metamorphic** changes that occur deep within a planetary interior or at its surface as a result of impact. Metamorphism results mainly from changes in temperature and pressure or from changes in the chemical environment of a mineral, and excludes those changes that result in melting.

These changes cause new minerals to grow and also cause changes in textures and structural elements of the rock. The diagnostic features of the original sedimentary and igneous rocks are greatly modified or completely obliterated by metamorphism. The deeper parts of most planetary lithospheres probably consist of metamorphic rocks derived from original igneous rocks. As in other rock groups, metamorphic rocks are classified on the basis of texture and composition (Figure 2.21).

Metamorphism can result from a variety of processes, typically by deep burial or from heat released from igneous intrusions. Most metamorphic rocks of regional extent develop in the deeper parts of the lithosphere. This type of metamorphic terrain is only exposed at the surfaces of planets where lithospheric deformation is sufficient to expose such deep levels to erosional stripping. For example, these types of rocks are probably rare at the surface of Mars, but large areas of Earth's

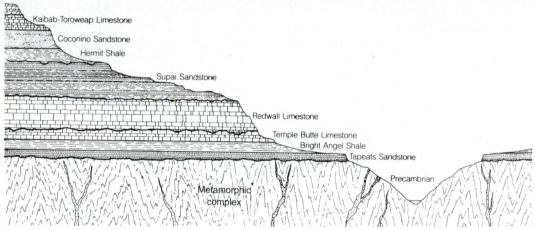

Figure 2.19

Sequences of sedimentary rock in the Grand Canyon show the characteristic stratification or layering of this rock type (A). The cross section in (B) emphasizes the layering and shows the difference between the sedimentary rocks and the older igneous and deformed metamorphic rocks near the floor of the canyon.

continents are underlain by complexly deformed metamorphic rocks that formed in the roots of mountain belts and became exposed by extensive erosion (Figure 2.22). Metamorphism also accompanies meteorite impact. **Shock metamorphism** changes the internal structure of many mineral species. The metastable persistence of these minerals at normal surface pressure is one of the strongest evidences for meteorite impact and must produce the most abundant type of metamorphic rock on the surfaces of cratered planets like the Moon and Mercury.

The evolution of planetary lithospheres is a function of their composition and the physical conditions (pressure, temperature, and state of stress) that are imposed upon them. Rocks result from the flow of energy in this system which drives planetary differentiation. Figure 2.23 shows how rock types change in response to a planet's physical and chemical environment. This cycle can be envisioned as beginning with the crystallization of a magma and proceeding, through interaction with the atmosphere and hydrosphere, to the production of sediments, which lithify to become rocks. Under

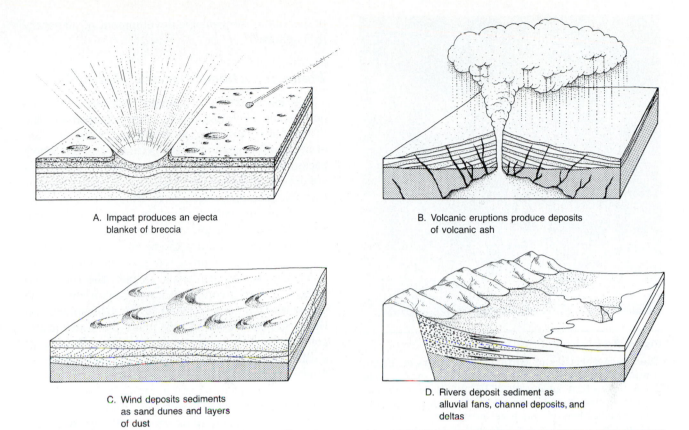

A. Impact produces an ejecta
blanket of breccia

B. Volcanic eruptions produce deposits
of volcanic ash

C. Wind deposits sediments
as sand dunes and layers
of dust

D. Rivers deposit sediment as
alluvial fans, channel deposits, and
deltas

Figure 2.20

The major sedimentary environments on the planets are represented in these idealized diagrams. Sedimentary environments can be divided into four types: (A) Impact produces an ejecta blanket of breccia. (B) volcanic eruptions produce deposits of volcanic ash that are carried by the wind on planets with atmospheres. (C) Wind transports and deposits sediments in sand dunes and layers of dust, and (D) fluvial and marine.

conditions of higher pressure and temperature, sedimentary rocks may become metamorphic rocks, or, with the introduction of large amounts of heat, magma may again be generated, to start the cycle again. Obviously, this is a highly schematic version of the rock-generating processes in planetary lithospheres. Sediments may be derived from other sedimentary rocks or from metamorphic rocks, and not all metamorphic rocks were originally sediments. The atmosphere of a planet is itself the result of processes similar to those that give rise to igneous rocks. These observations are expressed by appropriate lines on the diagram. Placed centrally within this scheme is the energy for the modification of planetary lithospheres. Here we have included not only mantle sources for energy but also the energy provided by accreting planetary material. Accretion (which is continuing even today) was probably the ultimate source of all planetary material, but, more important in this context, it helped form large volumes of shock metamorphic rocks and may have initiated magma generation on the forming planets, important

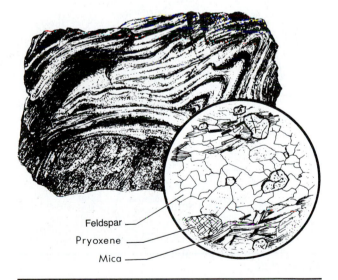

Feldspar
Pryoxene
Mica

Figure 2.21

Metamorphic rocks form by the recrystallization of other rocks, commonly in the presence of deforming stresses. This sample of gneiss, composed of feldspar, pyroxene, and micas, shows deformed layers—evidence of internal deformation that occurred deep within a planet's crust.

Figure 2.22

The characteristics of metamorphic rocks on Earth's continents are shown in this photograph of the Canadian Shield, taken from an altitude of 15 km. These rocks have been compressed and deformed to such an extent that many original features such as horizontal bedding have been obliterated. Metamorphism and folding occurred at great depths. The metamorphic rocks were then intruded by granite plutons (light tones) and cut by fractures. The area was than eroded to expose the complex rock sequence.

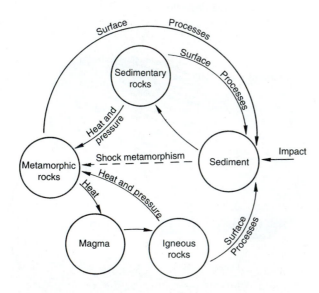

Figure 2.23

The possible changes of one rock type to another are shown in this diagram. By interaction with the atmosphere and hydrosphere of a planet (surface processes), igneous rocks can be broken down into sediments. These may then be compacted and cemented to form sedimentary rock. Igneous and sedimentary rocks may be metamorphosed. Metamorphic rocks may be melted to form magma giving rise to igneous rocks.

processes in the geologic development of planetary lithospheres.

Planetary Atmospheres and Hydrospheres

A geologist is principally interested in the nature and evolution of the various features of the lithosphere. However, an important class of planetary landforms and materials (mostly sedimentary rocks) is the result of the interaction of the atmosphere or hydrosphere with the lithosphere. For example, the ubiquitous river valleys of Earth, the vast sheets of sand and dust on Mars, the chemical precipitates of Titan, and the decomposition of rocks on Venus—all are the result of fluids decomposing, abrading, transporting, or combining with planetary surface materials. In fact, the rocks and oceans of Earth contain a large proportion of the elements that once resided in its atmosphere. In the larger scheme of things, many planetary atmospheres are only the outer layer (albeit gaseous) of a differentiated planet and naturally fall under the purview of the planetary geologist.

We have already pointed out that there are two types of planetary atmospheres (Figure 2.24). The first type, called a **primary** or **inherited atmosphere**, consists of gases gravitationally trapped from the solar nebula. The outer planets appear to have atmospheres of this sort. It is likely that these planets attained large sizes before the gaseous nebula was dispersed. Once the growing planets reached a critical size, the nebular gases became hydrodynamically unstable in their dispersed condition and collapsed around the planets to form dense and often colorful atmospheres. The composition of these atmospheres should be almost the same as that of the nebula, rich in the light elements. Jupiter and Saturn consist predominantly (more than two-thirds by mass) of hydrogen and helium, which appear to have been added to the planets in this way. Uranus and Neptune have smaller proportions of these light elements (less than one-fourth). Apparently the original rock-and-ice cores of Uranus and Neptune did not grow large enough to trap as much of the surrounding nebular gas.

It seems reasonable that the inner planets once had similar "inherited" atmospheres. However, careful measurements of the composition of these atmospheres demonstrate that they retain little or none of this primitive component. Instead, strong solar winds may have stripped away the primitive atmospheres of the already accreted inner planets. This may have been part of the T-Tauri phase of the early Sun, which also cleared away the nebular gases and terminated condensation. An alternative hypothesis is that the nebula was swept away

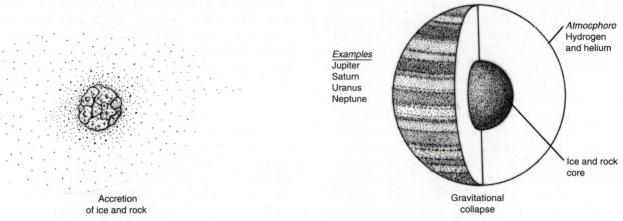

Accretion
of ice and rock

Examples
Jupiter
Saturn
Uranus
Neptune

Atmocphoro
Hydrogen
and helium

Ice and rock
core

Gravitational
collapse

A) **Inherited atmospheres formed** as nebular gases were gravitationally trapped around a growing planetary nucleus of icy materials (left). Eventually the gases became gravitationally unstable and collapsed to form a thick atmosphere around the core (right).

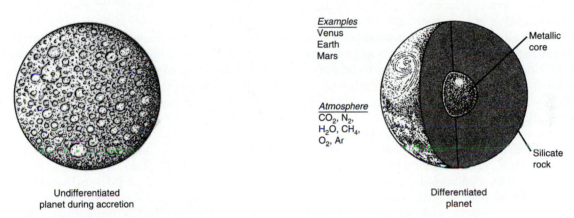

Examples
Venus
Earth
Mars

Atmosphere
CO_2, N_2,
H_2O, CH_4,
O_2, Ar

Metallic
core

Silicate
rock

Undifferentiated
planet during accretion

Differentiated
planet

(B) **Secondary atmospheres** developed on some of the terrestrial planets that retained no nebular gases (left). Magmatism and thermal metamorphism released gases to the surface of the planet where they accumulated (and eventually escaped into space if the planet was small). Secondary atmospheres (right) are thin and compositionally different than inherited atmospheres. Activity in the biosphere and lithosphere may continue to modify the composition of the atmosphere.

Figure 2.24

The development of the two basic types of planetary atmospheres are shown in these diagrams.

before the inner planets (but after the outer planets) had accreted from parental materials. We have little evidence to help us establish which theory is correct, but it seems clear that the gases that now surround the inner planets were formed after the accretion of the planet and are thus **secondary atmospheres**. Many investigators believe that they were derived from the interior of the planets, by a process commonly called **outgassing**.

The most important constituents of secondary planetary atmospheres are various combinations of the elements carbon, hydrogen, and nitrogen, which were incorporated into the planets during accretion. As pointed out earlier, the condensation of the solar nebula led to the formation of hydrated

silicates at temperatures of about 400 K. Carbonaceous compounds formed at slightly lower temperatures. Thus some of the elements (hydrogen, oxygen, and carbon) that eventually formed atmospheres probably became chemically incorporated into solid materials. At these temperatures, small amounts of other unreactive atmophile elements (such as nitrogen, argon, and neon), though not chemically bound in the solids, may have become trapped into the structures of minerals condensing from the nebular gases. Alternatively, ices with some of these elements may have formed at temperatures approaching 100 K and become incorporated into the planets by accretion. Thus, the elements that eventually formed the atmospheres may have resided in nebular condensates in different ways: (1) as chemical constituents of rocky minerals, (2) as occluded elements in other solids, and (3) as ices.

Earlier in this chapter we arrived at the preliminary conclusion that Earth accreted from materials that condensed at temperatures of about 600 K, well above the temperature at which water could be incorporated into solids condensing from the solar nebula (Figure 2.7), and yet over 70 percent of the surface of Earth is now composed of water. Likewise, Venus possesses a dense, carbon dioxide-rich atmosphere, but it should consist of materials that condensed at temperatures above the stability point of carbon-containing solids. Finally Mars, which must have accreted from materials that formed at lower temperatures than the materials that formed Earth, has only a tenuous carbon dioxide-rich atmosphere and small quantities of polar ice (water and carbon dioxide). These inconsistencies are the result of many factors (which will be discussed in subsequent chapters), but certainly they suggest that the planets probably accreted from a variety of materials condensed at different temperatures. Following this interpretation, Earth may have formed from a mix of planetesimals, some that formed beyond the orbit of Mars and, hence, were hydrated, and drier planetesimals that condensed closer to Venus. Thus, each planet may contain components that represent a relatively broad range of nebular equilibration temperatures, but the dominant material probably came from near the planet's present orbit. Some scientists have suggested that the accretion of atmophile substances was a late-stage event. They envision a cloak of volatile-rich planetesimals and comets, formed in the outer solar system, being added to the already formed planets. The influence of Jupiter's gravity field disrupted the original orbits of these planetesimals and led them to eventual collision with the inner planets.

Whatever the mechanism by which these atmophile elements became incorporated into the planets, their segregation and concentration into a gaseous envelope surrounding the planet resulted from its internal differentiation. At the temperatures of planetary differentiation, some volatile elements are released from the rocky materials that bind them. This may be accomplished by melting, or simply by metamorphism at elevated temperature (which may, for example, drive water out of hydrated minerals to produce anhydrous solids and a fluid). Because the resultant fluids or gases have relatively low densities, they rise through fractures and along grain boundaries to the surface and accumulate there. In contrast to primary atmospheres, those produced by outgassing are relatively poor in helium and most hydrogen is combined with oxygen or carbon to form water vapor (H_2O) or methane (CH_4). These gases are accompanied by carbon-oxygen gases such as carbon monoxide (CO) or carbon dioxide (CO_2) and nitrogen (N_2). Argon, other noble gases, and sulfurous compounds are also released during the outgassing of nebular condensates. Free oxygen (O_2) was probably only a very minor part of the gases vented to the surface of the differentiating planets. Ices, in planets that also incorporated them, also melted during differentiation. The satellites of the outer planets, for example, contain outer icy shells many kilometers thick, composed principally of water and ammonia (NH_3), which were probably formed from accumulations of melted ices. Nonetheless, it seems unlikely that the differentiation temperatures for these small bodies were high enough to expel water bound in hydrous silicate minerals; these materials probably constitute their cores.

The heat provided by accretion may have been sufficient to release some of these gases, which, because of their very low densities, then accumulated at the growing planet's surface. Core formation may have been the impetus for rapid outgassing of some planetary bodies (Figure 2.24). In either case, it seems that the secondary atmospheres (or for that matter oceans and icy mantles) of many planetary bodies formed very early in the planets' histories. In addition, Earth and some other planets continue to expel (and perhaps recycle) gases from their interiors to their surface as temperatures are sustained by long-lived heat sources.

Just as there are **sources** for planetary volatiles, there are also **sinks** or traps where these atmophile elements may be temporarily or permanently removed from their respective atmospheres (Figure 2.25). The light elements, particularly hydrogen and helium, are subject to permanent loss to space as they gradually leak out of the top of most planetary atmospheres. One reason for this is that a small proportion of all atmospheric atoms have velocities that exceed a particular planet's mass-dependent **escape velocity**. For example, hydrogen (as H_2) has an average lifetime on Earth of only 1 million years. The process of hydrogen loss is aided by the decomposition of water vapor and other hydrogen-bearing compounds by the action of sunlight in the upper atmospheres of the planets. More dramatic losses of atmospheric gases may have occurred when a hot magma ocean warmed and accelerated the overlying gas to escape velocity. This wind or blowoff may have dissi-

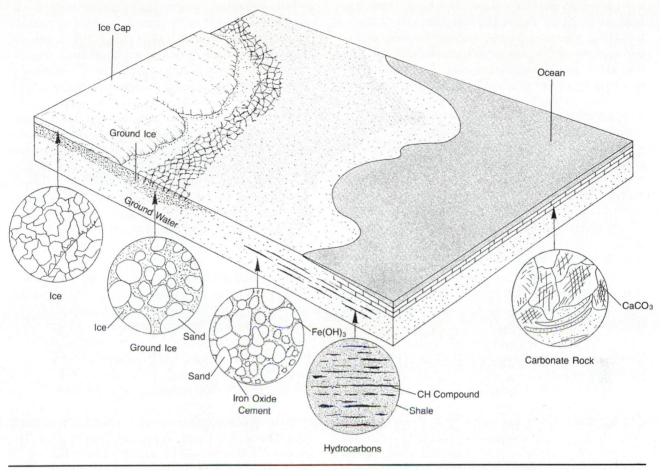

Figure 2.25

Temporary or even permanent sinks where volatiles may be trapped are abundant on the planets and include atmospheres, lakes, oceans, rivers, ice caps and glaciers, reactions with other materials to form rocks (carbonates, red beds), and trapping in soils.

pated much of the gas around the primordial Earth. In addition, the impact of large meteorites may also eject atmospheric gases from a planet.

Other types of volatile traps involve precipitation of gases into solids or liquids, which accumulate on the surfaces of planets. For example, not all of the water vapor released to Earth's atmosphere is now present there. Some water is tied up as ice, some is in rocks, and some has decomposed, allowing hydrogen to escape to space. Carbonate minerals like calcite also form at the surfaces of some planets, removing carbon dioxide and locking it away to form solid rocks. This type of volatile trap may be particularly important for Earth, where thick layers of carbonate rock have formed in its oceans. The atmosphere of Mars, also, may have lost carbon dioxide in this way.

As temperatures declined in the early terrestrial atmosphere, water condensed and showered onto the surface of the planet and accumulated in craters and basins. This condensed portion of a planet's fluid envelope is called its **hydrosphere.** If retained in a liquid state, it can substantially modify a planet's surface by eroding some features away and redepositing the transported material elsewhere to form new sedimentary rock bodies. The solid icy shells of the outer satellites (and the polar caps of Earth and Mars) can be considered as volatile traps, which are similar in origin to planetary hydrospheres. In fact, on some planetary bodies liquid water may be sandwiched beneath a solid outer rind in a fashion similar to that which produces asthenospheres of partially molten silicates on the larger inner planets.

Water was not the only icy substance to have become incorporated in the planets. Nitrogen (N_2) and methane (CH_4) should also be important in bodies formed in the cool outer solar system. It has been postulated that Titan, the largest satellite of Saturn, contains a nitrogen-rich atmosphere with methane clouds from which rain or snow falls onto the surface, producing what may be the solar system's only methane seas. In short, there are important sinks for many volatile substances—including water, methane, carbon dioxide, sulfur, and oxygen. We will return to these later; for now, it is important to realize that even though gases are released by differentiation to an atmosphere, they may escape completely—by loss into space or by incorporation into a variety of surface materials—thereby considerably reducing the mass of the atmosphere. Consequently, the density and composition of atmospheres may change radically as the result of climate change or even as a result of biologic processes. We note finally that the oxygen-rich atmosphere of Earth appears to be the result of the accumulation of oxygen produced through billions of years of photosynthesis by living organisms in Earth's distinctive **biosphere**.

Just as the interiors of planets evolve by differentiation, so do the atmospheres of planets, reflecting the unique compositions, sizes, and physical settings of each planet.

Conclusions

The planets and the Sun are made of elements, which, other than hydrogen and helium, were formed by nuclear processes in earlier generations of stars. Accumulations of gas and dust, called nebulas, occasionally form within interstellar space. Gravitational contraction of part a nebula may have led to the formation of our solar system. As the nebula collapsed by gravitational contraction, the proto-Sun formed at its center; because of rotation the nebula maintained a thin outer disk of gas and dust. New grains of dust condensed from the gas and settled toward the central plane of the nebular disk. As a result of a strong temperature gradient, the compositions of the grains were controlled by their distance from the still-forming Sun. Metals and silicates were important condensates in the inner solar system; ices of water, ammonia, and methane were important in the outer solar system. Eventually, the orbiting dust accreted by collision, resulting in a system of planetary bodies whose size and composition were determined by their distance from the center of the nebula. (A T-Tauri phase of the early Sun may have dispersed the gaseous part of the nebula at this time or a bit earlier.) The accretionary heating of the planets caused internal differentiation, which produced their layered internal structures. Crusts and rigid lithospheres composed of rocks with widely varying compositions formed the outer solid shells of most planets. Atmospheres composed of gaseous elements were trapped from the nebula by the gravity of the giant outer planets. Smaller planets developed atmospheres through outgassing, by which volatile elements were released from their interiors during accretion and differentiation. The geologic histories of the planets are reflected by features preserved on their surfaces and in their atmospheres and are largely the result of their thermal evolution.

Review Questions

1. How and where were the elements formed?
2. Is there any reason to believe that the elements that Earth and other planets are made of once resided in stars?
3. Describe the solar nebula from which the planets are thought to have formed.
4. Contrast the behavior of volatile and refractory materials. Give an example of each.
5. When did the solar system originate? How can we date an event that happened so long ago?
6. Why did planets with compositions as different as those of Earth and Jupiter form from a solar nebula that may have had a uniform chemical composition?
7. What prevented Mars from obtaining an ice-rich outer shell encasing a rocky interior as the temperature of the nebula continued to drop?
8. Contrast the origin and evolution of an inherited atmosphere with that of a secondary atmosphere and give real examples of each type.
9. Why doesn't Earth have a thick atmosphere of hydrogen and helium if these are the two most abundant elements in the universe and in the solar nebula from which Earth is presumed to have formed?
10. How is the nature of a planet's interior reflected in the appearance of its surface?
11. Can solid rock flow?
12. Describe the differences across the crust–mantle boundary in a planet like Earth and contrast those differences with those found at the asthenosphere–lithosphere boundary.
13. Explain why the rocks found at the surface of a planet, say the Moon, are so different in elemental composition from the meteoritic material from which the planet formed.
14. Are all heavy elements concentrated in the cores of differentiated planets? (Hint: Consider uranium, the heaviest natural element.)

15. How do we know that the number of meteorites striking Earth in a year has varied tremendously over the last 4.5 billion years? Is this consistent with our ideas about how the planets formed?

16. Explain the important characteristics of igneous, sedimentary, and metamorphic rocks.

17. Why does magma tend to rise upward toward the surface of a planet?

18. Compare the surfaces of two hypothetical planetary bodies with diameters of 5000 km; the first is found in the inner solar system; the second is in the outer solar system.

19. What are the most important controls on the thermal (heating or cooling) history of a planet?

Key Terms

Accretion
Accretionary Heating
Ash Flow Shield
Asthenosphere
Atmophile
Atom
Basalt
Biosphere
Carbonaceous chondrite
Cementation
Cinder Cone
Collisional Accretion
Compaction
Composite Volcano
Condensation
Conduction
Convection
Deposition
Differentiation
Disintegration
Electron

Element
Equilibrium
Escape Velocity
Evaporite
Extrusive
Feldspar
Fusion
Gabbro
Granite
Gravitational Collapse
Hydrosphere
Ice
Igneous
Inherited Atmosphere
Intrusive
Isotope
Lava Dome
Lithosphere
Mafic
Magma
Magma Chamber

Magma Ocean
Mantle
Metal
Metamorphic
Metastable
Minerals
Nebula
Neutron
Nova
Olivine
Outgassing
Planetary Nebula
Planetesimal
Plate Tectonics
Pluton
Primary Atmosphere
Proton
Protostar
Pyroxene
Radiation
Radioactivity

Radiogenic Heat
Radiometric Age
Refractory
Rhyolite
Rock
Secondary Atmosphere
Sedimentary Rock
Shield Volcano
Shock Metamorphism
Silicate
Silicic
Solar Energy
Supernova
Superposition
T-Tauri Phase
Tidal Heat
Thermal Energy
Thermal History
Transportation
Volatile
Volcano

Additional Reading

Allegre, C. H. 1992. *From Stone to Star*. Cambridge, MA: Harvard University Press.

Berman, L., and J. C. Evans. 1986. *Exploring the Cosmos*. Boston: Little, Brown and Company.

Beatty, J. K., B. O'Leary, and A. Chaiken (eds). 1990. *The New Solar System*. London: Cambridge University Press.

Hartmann, W. K. 1993. *Moons and Planets*. 3rd edition. Belmont, CA: Wadsworth Publishing Company.

Morrison, D., and T. C. Owen. 1987. *The Planetary System*. Reading, MA: Addison-Wesley Publishing Company.

Taylor, S. R. 1992. *Solar System Evolution: A New Perspective*. Cambridge, England: Cambridge University Press.

Wood, J. A. 1979. *The Solar System*. Englewood Cliffs, NJ: Prentice-Hall, Inc.

Zeilik, M., and J. Gaustad. 1983. *Astronomy: The Cosmic Perspective*. New York: Harper and Row.

TABLE 3.1

Selected Asteroids Compared to the Moons of Mars

Object	Distance from Sun (AU) Center	Rotation Period (Hours)	Diameter (km)	Density (g/cm³)
Ceres	2.77	9.1	932	2.3
Vesta	2.36	5.3	520	2.9
Pallas	2.77	7.9	533	—
Hygeia	3.15	18	443	—
Davida	3.18	5.2	341	—
Juno	2.67	7.2	247	—
Bamberga	2.69	8.0	246	—
Eros	1.46	5.3	30	—
Gaspra	2.21	7.0	18 × 10	
Ida	2.4		56 × 24	
Moons of Mars				
Phobos	1.52	7.6	27×19	2.2
Deimos	1.52	30.2	15×11	1.7

Gaspra

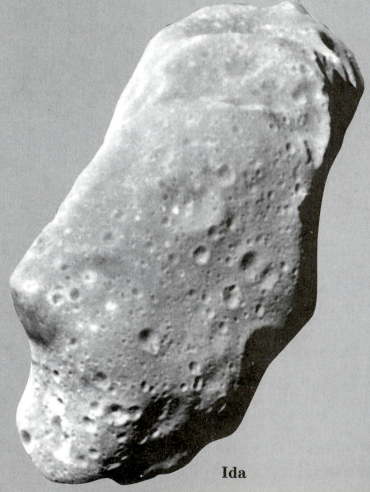

Ida

CHAPTER 3

Meteorites and Asteroids: Small Bodies of the Inner Solar System

Introduction

Although the solar system contains only nine major planets, hosts of smaller, minor bodies revolve about the Sun. Among these objects are asteroids. Because the asteroids are small bodies they have had relatively simple histories. They provide a logical starting point for our discussions of the origin of the larger planets. In addition, meteorites provide us with a set of actual pieces of these bodies—pieces that appear to have come from the surface veneer of impact breccias and volcanic rocks, to metamorphic mantles, to deep metallic cores. The largest asteroid is only 1000 km in diameter; the smallest are only particles of dust. Together this collection of bodies constitutes a small but significant portion of the total mass of the solar system excluding the Sun. What is more important, they have preserved evidence of geologic processes that occurred long ago— evidence that was erased from the inner planets. For example, asteroids and meteorites reveal the probable sizes and types of early planetesimals that accreted to form the rocky inner planets. Moreover, meteorites, as samples of asteroids and other solar system bodies, hold key information about when and how the planets formed and about the composition and physical conditions of the solar nebula.

Major Concepts

1. Meteorites are rocky or metallic bodies that reach Earth from space. Many originated on asteroids. Some come from comets; fewer still come from the Moon, and a handful may come from Mars.

2. Stony, iron, and stony-iron meteorites have distinct histories and reveal the ages, compositions, internal structures, and sizes of the small bodies from which they originated. In particular, chondritic meteorites have changed very little since solids condensed from the solar nebula 4.6 billion years ago. Other types of meteorites show that their parents differentiated by igneous processes in asteroids that became extensively melted.

3. Asteroids are small planetary bodies. Most have orbits that carry them between Mars and Jupiter. These minor planets are the fragmented remains of planetesimals that did not accrete to form a sizable planet. Today, they are a major source for meteorites impacting Earth and the other terrestrial planets.

4. The asteroid belt marks the transition from the rocky inner planets to the ice- and gas-rich outer planets. Volatile-poor asteroids are concentrated in the inner part of the asteroid belt, and volatile-rich asteroids are concentrated in the outer part of the belt. This pattern shows that a temperature gradient existed when these planetesimals formed in the ancient solar nebula.

Meteorites

In ancient times, meteorites were regarded as objects of devotion and veneration, as messengers from the heavens. With the dawn of the Enlightenment, these stones that fell from the sky were regarded with skepticism, and many thought them to be products of volcanic explosions on Earth, pieces of the Moon, or rocks picked up by winds or the "attraction" of clouds and then dropped again. It was not until the late eighteenth century that the extraterrestrial origin of meteorites was generally accepted.

Meteorites are inexpensive samples of a variety of bodies in the inner solar system. They yield detailed chronologic, physical, and chemical information about these planetary bodies, information that we lack about most of the major planets. Except for the lunar rocks returned by the Apollo astronauts and unmanned Russian spacecraft, they represent the only pieces of other planetary bodies that can be studied directly. Many meteorites have been relatively unaltered since the earliest stages of the solar system, and they preserve the chemical and mineralogic characteristics of such ancient materials and of the processes that shaped them. Moreover, because they are tangible samples, they can be analyzed to determine their radiometric ages. Therefore, meteorites reveal much about the conditions in, and the composition of, the dusty nebula from which the inner planets formed, the time of planet formation, and the nature of any subsequent differentiation that may have occurred in small bodies after their condensation and accretion.

Most meteoroids (meteorites before they land on Earth) spend much of their lifetimes in space as dust-sized grains or as small boulders up to several meters across. Most were broken from larger objects 10 to 500 million years ago. When they enter a planet's atmosphere, most are moving at 5 to 25 km/s, fast enough to produce brilliant streaks of vaporized material as their surfaces heat by friction with the air. Many meteors burn up entirely before they reach the surface. Planets with thin atmospheres, like Mars, or none at all, like the Moon, are struck by a much larger proportion of the small meteoroids that pass them in space. Impact of a sufficiently large meteorite produces an impact crater. Judging from the age of the densely cratered lunar highlands, meteoritic impact was much more important around 4 billion years ago than today, but, amazingly, enough meteoritic material still reaches Earth to add about 50 million tons of new matter to its surface and atmosphere each year. Although balanced somewhat by the loss of gases to space, Earth's accretion has slowed but not stopped.

Meteorites do not come from outside the solar system, but from rocks in solar orbits that cross Earth's own. Collisions within the asteroid belt or gravitational perturbations from the much larger planets may gradually move small bodies from the asteroid belt to Earth. Comets, occasional icy visitors to the inner solar system, may also serve as meteorite sources (see Chapter 14). Among the meteorites already collected, pieces of the Moon have been positively identified, and there is an exciting possibility that some meteorites are from Mars. Earth receives only a minute sample of these other bodies, but meteorites are actual fragments of a large and diverse group of solar system objects.

Meteorites are quite varied and can be divided into three broad types: iron, stony-iron, and stony meteorites (Table 3.2). Careful examination of meteorites reveals compositional clues to their origins and tells us much about geologic processes on and in small planetary bodies.

TABLE 3.2
Major Meteorite Types

Type	Abundance (percent)	Composition
Stony Meteorites	94	
Chondrites	86	
Ordinary	82	Metamorphosed chondrites
Carbonaceous	4	Carbon- and volatile-rich, undifferentiated
Achondrites	8	Igneous textures, differentiated
Stony-Iron Meteorites	1	Silicate-metal mixtures, differentiated
Iron Meteorites	5	Iron metal, differentiated

Abundances are percentages of each type of meteorite among all meteorites seen to fall on Earth.

Stony Meteorites

Over 95 percent of the meteorites seen to fall to Earth have compositions similar to terrestrial rocks. They are composed of iron and magnesium silicates, helping to establish the fact that these are the most abundant minerals in the inner planets. Because of rapid weathering and their superficial similarity to terrestrial rocks, stony meteorites are difficult to identify unless they are seen falling from the sky.

There are basically two types of stony meteorites: **chondrites** and **achondrites**. Chondritic meteorites usually contain small spherules of silicate minerals approximately 1 mm in diameter. Many of these globules, known as *chondrules*, are composed of olivine and pyroxene, but other minerals and glass may also be found (Figure 3.1). Most chondrules display intriguing internal structures when

cut and examined under a microscope. These textures are thought to have developed by rapid cooling of molten droplets of rock. Metal grains are also common in the chondrites. Extremely small quantities of grains that come from outside our own solar system have also been identified. These truly exotic materials include diamonds, graphite, aluminum oxide, silicon carbide, and titanium carbide. As many as three different stars may be represented by this material.

Detailed study of chondritic meteorites shows that they have relatively simple, straightforward histories compared to most rocks on the major planets, which have experienced several distinct melting and deformation events. In fact, the bulk composition of one class of chondrites, called **carbonaceous chondrites** (because of an abundance of black carbon compounds) is very similar to the composition of the Sun (minus its hydrogen and other volatile gases) as determined by spectrographic studies (Figure 3.2). These meteorites are probably very similar to the primitive nebular material from which the planets formed. The fine

Figure 3.1

Carbonaceous chondritic meteorites are composed mainly of hydrated silicate minerals, but their most characteristic features are the small spherical grains shown in this slice of a meteorite. These spherical blobs are called chondrules and appear to have formed at higher temperatures than the mineral grains in the surrounding, fine-grained matrix. Chondritic meteorites are the best samples we have of unaltered condensates from the solar nebula and are about 4.6 billion years old.

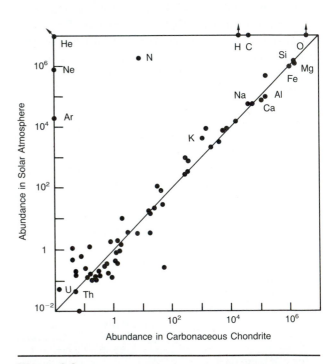

Figure 3.2

The elemental composition of carbonaceous chondrites is similar to the composition of the Sun, except for its extremely volatile elements, such as hydrogen, helium, and carbon. The straight line shows the focus of points where the two materials would be exactly alike. The close correspondence demonstrates that carbonaceous chondrites are primitive, undifferentiated materials that have survived with few changes since the formation of the solar system. The abundances plotted are relative to one million silicon atoms.

grains in these meteorites are composed of hydrated silicate minerals and may contain up to 20 percent water. They are also rich in other volatile elements, but they also have flecks of metallic iron and other inclusions that formed at very high temperatures. Carbonaceous meteorites have low densities, about 2.3 g/cm³. Such meteorites, and other materials with approximately the same composition, are **undifferentiated**. They have not undergone any process, like melting, that segregates elements because of their chemical affinities or properties. All other meteorites (irons, stony-irons, and achondrites) are **differentiated**. They depart from this primitive composition in one way or another. The carbonaceous chondrites have survived for billions of years with little change since their formation from the gas and dust of the solar nebula. They are thus a type of genesis rock. The black matrix material in the chondrites can be described as a sticky agglomeration of this condensed dust. Various dating techniques show that the chondrites formed 4.6 billion years ago, establishing the age of the solar system. Many models of planetary formation assume that the inner planets and the rocky portions of the outer planets accreted from materials with approximately the same chemical composition as the primitive chondrites.

Not all chondrites have primordial compositions rich in volatile elements. Some show differences that are evidence that they formed in warmer parts of the nebula, and some have textures and minerals produced by secondary alteration inside small planets. Many of the latter have experienced varying degrees of thermal metamorphism but have never melted and are often called *ordinary* chondrites because they are more abundant than carbonaceous and other rare types. Recrystallization at elevated temperatures enhances the growth of larger crystals, partially destroying the original chondrule shapes and matrix textures. From studies of the included metallic grains, it appears that these metamorphosed chondrites cooled slowly after they were heated while tens to hundreds of kilometers deep within planetoids. The outer layers of these bodies acted as insulating jackets, protecting them from cold space, and allowed them to cool slowly.

The chondrites, then, preserve evidence bearing on the time of nebular condensation, the chemical compositions and sizes of the bodies created, the manner of condensation, and subsequent heating and metamorphism.

The other main group of stony meteorites, the *achondrites*, has no chondrules. But more than this, the compositions and textures of many are very similar to basalt, the most common kind of lava

0   2 cm

Scale

Figure 3.3

Achondrites, stony meteorites that lack chondrules, have textures and minerals similar to those found in terrestrial basaltic lavas. Some achondrites were probably erupted as lavas on the surface of differentiated asteroids. Vesta, one of the larger asteroids, may have a surface dominated by such lavas now brecciated by repeated impact.

found on Earth and the Moon (Figure 3.3). They also lack the metal grains found in chondrites. Achondrites are igneous rocks. They were either formed by the extrusion of *igneous magmas* on the surfaces of planetoids or as **impact melts** produced during flash heating events when two bodies collide. Their geochemical characteristics show that an igneous origin is likely for many. The parent magmas probably formed by **partial melting** within small bodies that had an overall chondritic composition. As the material heated up and melted, droplets of dense metallic iron formed and moved to the center of the planetesimal. A fraction of the low-melting temperature materials, predominantly silicates, formed magma that rose to the surface because of its low density. Achondrites may be thought of as part of a light scum, or crust, of silicate minerals surrounding the differentiated interior of a small planetoid. Fragmentation of a parent body may have sent small pieces of this crust toward Earth. Because of this more complex history, achondrites are not similar to the bulk composition of the solar nebula. Chemical differentiation of elements from one another occurred during partial melting inside a planetesimal.

If achondrites were produced by melting inside an asteroid, we can gain a general idea of the size of the parent body. Melting could only have occurred in larger bodies, with diameters of hundreds of kilometers. Such bodies could retain enough heat, no matter how it was produced, long enough to melt portions of their interiors. Smaller bodies rapidly radiate away any heat produced before internal melting occurs. The asteroid Vesta, 538 km in diameter, is thought to have an outer layer of achondritic rocks. Achondrites are rare. Apparently, few were produced, or few survived subsequent obliteration. Of the achondrites recovered, many are brecciated and must be **regolith** (fragmental soils) produced by impact on their parent bodies.

Radiometric dating of achondrites shows that most are 4.5 to 4.6 billion years old, but a few distinctive specimens are as young as 1.3 billion years old. These crystallized 3 billion years after most meteorites formed and at least 1 billion years after the last volcanic rocks erupted on the Moon. These rare achondrites (shergottites, nakhlites, chassignites or **SNC meteorites**) have other distinctive features that suggest to some investigators that they were generated on Mars. Over a dozen such meteorites are known, and their numbers are growing as new examples are discovered. The possibility that SNC meteorites are from Mars makes them among the most treasured of all rocks on Earth. Perhaps grazing meteorite impacts on Mars ejected fragments at velocities high enough to allow them to escape from the planet. This exciting possibility is highlighted by the discovery of several small meteorites that originated on the Moon. Discovered in Antarctica, they have all of the characteristics of the lunar rocks returned by the astronauts (see Chapter 4). Careful searches through the large and growing meteorite collections may reveal SNC and lunar meteorites, giving us samples of large planetary bodies that are otherwise unobtainable.

Stony-Iron Meteorites

Meteorites of the second major group are composed of approximately equal proportions of metal and silicate minerals and are called stony-iron meteorites (Figure 3.4). They are quite rare; only about 70 are known to exist. Some are brecciated

Figure 3.4
Stony-iron meteorites consist of silicate minerals (dark grains) coexisting with a large proportion of metal (bright). These rare meteorites are thought to represent brecciated fragments of the silicate mantle—metallic core transition zone from asteroids or small planetesimals. These meteorites provide striking evidence of differentiation processes in small planetary bodies.

and contain inclusions of other meteorite types. The unique makeup of this class requires internal differentiation by igneous processes of bodies near the size of large asteroids (hundreds of kilometers in diameter). As these bodies—which originally may have had chondritic compositions—heated and melted, heavy iron and nickel separated from the molten rock and sank to the center, making a liquid core. At the same time, the lighter silicate minerals or magmas rose to the surface. Although some stony-iron meteorites are impact-produced mixtures, many appear to represent fragments of a transition zone between metallic core and rocky mantle. They may have formed surrounding a massive central core or on the margins of several smaller pockets of molten metal dispersed through an asteroid like raisins in a pudding.

Iron Meteorites

The most familiar, although not the most numerous, meteorites are largely composed of iron and nickel metal (Figure 3.5). Because these meteorites are very durable during their atmospheric passage and resist weathering on the surface of Earth, a disproportionate share of meteorite finds (not falls) is of this type. Many impressive meteorites on display in museums are large irons. They develop their typi-

Figure 3.5

Iron meteorites consist almost entirely of metallic mineral grains, which, when etched with an acid solution, display this dramatic pattern, called a Widmanstatten structure. Iron meteorites crystallized from the once-molten cores of relatively small meteorite parent bodies. During differentiation, the high density and the chemical affinity of molten iron probably forced droplets of iron to drain to the body's interior, displacing silicates outward.

cal melted and pitted appearance as they heat up during atmospheric passage. Before any meteorite reaches the surface, some of its original mass has either fragmented or ablated away.

When cut, polished, and etched with an acid solution, most iron meteorites display distinctive crosshatch patterns, called Widmanstatten figures, that show the internal arrangement of the metallic alloys in the meteorite. From studies of these structures, scientists have shown that these meteorites were originally molten. Cooling rates have been determined from the composition and size of the iron-nickel grains. Most cooled less than 10 K per million years. A small metallic body 1 km in diameter would cool very quickly. In fact it could cool from the temperature of molten iron in about 1000 years if left uninsulated in space. Thus, larger parent bodies, 20 to 300 km (but probably smaller than 1000 km) in diameter that insulated these cooling bodies of iron and nickel and slowed the cooling rate to observed values must have existed. This evidence indicates that iron meteorites originated as the cores of small planetesimals about the size of asteroids. An asteroid core was produced when heavy metals collected toward the centers of hot, fluid, differentiating bodies. The accumulations

of metal may represent central cores in completely differentiated parents or only small scattered pods in a partially differentiated planetoid.

From this simple overview, it can be seen that meteorites have actually recorded and preserved for billions of years the products of some of the major events in the early history of planetary bodies, both small and large. Ages determined from isotopic analysis of the different groups cluster around 4.6 billion years ago. We use these ages to establish the age of the solar system—the point at which solid material began to accumulate into sizable bodies. The undifferentiated carbonaceous chondrites have compositions that are similar to what we think was the composition of the solar nebula. Thermal metamorphism of other meteorites was caused by early planetary heating. Brecciation resulted from accretion or from later impacts that fragmented the bodies. Differentiated meteorites inform us about the nature and extent of the internal reworking that occurred even in small bodies.

Investigations of meteorites are consistent with a simple developmental model. The various chondritic meteorites are products of direct nebular condensation; later, some were mildly metamorphosed. Melting in larger bodies with chondritic compositions produced differentiated planetoids. Differentiated meteorites are crust-to-core samples of several bodies—achondrites are igneous crustal rocks; stony-irons are apparently fragments from mantle-core transition zones; and iron meteorites were once parts of cores formed deep within small bodies.

Asteroids

In addition to the nine major planets that circle the Sun, thousands of smaller planetoids are also part of the solar system (Figure 3.6). These minor planets are called **asteroids**; each of them is as different from the next as the major planets. There are over 10,000 known asteroids, but many others are far too small to be observed from Earth even through the best telescopes. The larger ones only leave blurs on photographic plates, to the distress of many astronomers looking at objects far from our tiny solar system.

By the late eighteenth century, it was recognized that the planets occur at what seem to be predictable intervals from the Sun. Uranus was discovered near the distance predicted, but no planet existed in the gap between Mars and Jupiter. The first asteroid, Ceres (1000 km in diameter),

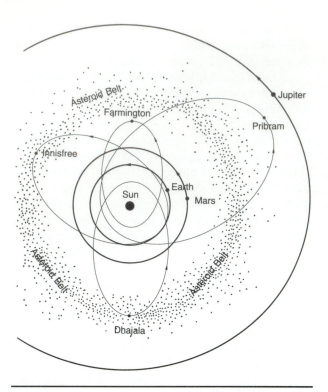

Figure 3.6

Most asteroids have circular orbits that take them between the orbits of Mars and Jupiter in an area known as the asteroid belt. Many other asteroids orbit outside of the main belt—some across Earth's orbit. Asteroids were probably the source of the bodies that impacted Earth and other planets during their early histories.

was discovered in 1801 and was thought for a time to be the missing planet. But Ceres was disappointingly small. Soon, other asteroids were discovered in the same region of the solar system and, when this fact was combined with the recent acceptance of meteorites as fragments from small bodies in space, some scientists postulated that a much larger planet had once existed there but had exploded and sent debris throughout the inner solar system. Subsequent studies of meteorites and asteroids have established that there never was a large planet between Jupiter and Mars, but the emerging story is just as intriguing. In the light shed by meteorite studies, we now know much about the present state and evolution of the asteroids.

Orbital Groupings of Asteroids

Asteroids travel in closely associated groups or families with anywhere from 10 to 60 in a group. Those groups that orbit between Jupiter and Mars are called the **Main Belt asteroids**, of which there are, quite literally, millions (Figure 3.6). The Main Belt is 280 million km across, and all of the largest

asteroids are confined to this group. Almost 5000 Main Belt asteroids have been numbered and have well-known orbits. Members of another, smaller group travel in orbits that cross Earth's path; they are called the **Apollo asteroids**. They have highly elliptical orbits and may be the source of much of the meteoritic material that reaches Earth's surface. At present, it is estimated that about 1000 bodies larger than 1 km in diameter are members of the Apollo group. Their numbers may once have been much larger, but many of them impacted Earth, Moon, Mercury, or Venus. Alternatively, gravitational perturbations and collisions in the Main Belt may serve to replenish the numbers of Apollo asteroids. Decaying or disrupted comets may also contribute to the number of Apollo asteroids in the solar system (see Chapter 14). Certainly some of these asteroids collided with the inner planets to form the impact craters prominent on several of the planets.

Asteroid Sizes

Although most of the asteroids are small, several are large enough to be considered minor planets (Figure 3.7). Ceres is nearly 1000 km in diameter, a little less than one-third the size of the Moon. Two, Pallas and Vesta, are over 500 km in diameter, and about 3000 exceed 20 km in diameter (Figure 3.7). For every asteroid, there are ten others one-third its size. This size distribution, with high numbers of small sizes, is probably produced by collisional fragmentation within the asteroid belt. Meteorites recovered on Earth bear evidence of impact disruption in their brecciated natures and small sizes. Within the asteroid belt, relative speeds between bodies often approach 5 km/s. A collision of two bodies at this speed could produce thousands of smaller fragments or even disrupt an entire body, exposing its deep interior. Nonetheless, in spite of repeated collisions in the asteroid belt, there does not appear to be a superabundance of very fine particles. Meteoroid-impact detectors on board the Pioneer 10 and 11 spacecraft discerned no change in the impact rate when crossing the belt on their way to Jupiter. These fine particles are fairly evenly distributed between the planets. Apparently, many are swept into the Sun.

Geologic Processes on Asteroids

Few of the asteroids are massive enough to have taken on spherical shapes like the larger planets and moons. Just as surface tension in a drop of water forms a spherical shape by deforming the original shape of the drop, so do planetary bodies

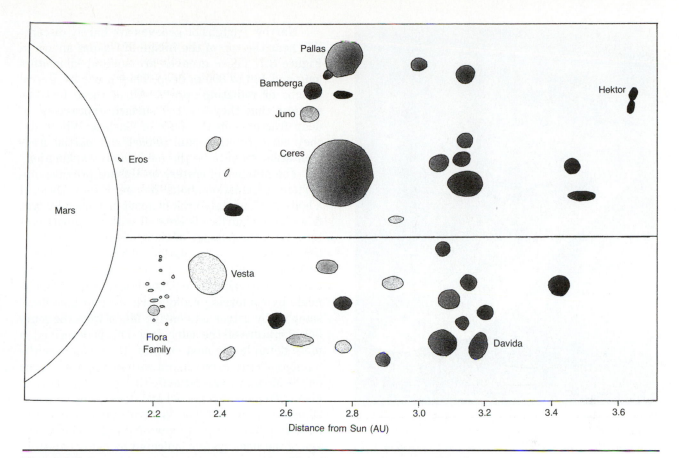

Distance from Sun (AU)

Figure 3.7

Asteroids come in a variety of sizes, shapes, and colors. This diagram illustrates these properties along with relative distances from the Sun and orbital inclinations for all 33 asteroids larger than 200 km in diameter. To give an indication of the size of smaller asteroids, all members of the Flora family are also shown. Thousands of others have comparable sizes. The limb of Mars is shown for comparison. Asteroids near the center of the diagram revolve in near circular, noninclined orbits; those near the top or bottom of the diagram have elliptical and/or inclined orbits. Asteroids become detectably darker with distance from the Sun. This is thought to correlate with the proportion of dark carbonaceous materials at the asteroids' surfaces.

attempt to take on spherical shapes by deformation resulting from gravitational forces. A sphere is a low-energy configuration, with the smallest possible surface area to mass ratio of any geometric shape. However, the small mass of, and consequently low gravitational force generated by most asteroids is not sufficient to alter their shapes radically (Figure 3.8). The materials the asteroids are made of are stronger than the gravitational force that they establish. Mechanical processes, impact, and fragmentation are responsible for their irregular shapes, as is the case for the only two asteroids photographed by passing spacecraft, Ida and Gaspra. Other asteroids probably include the tiny martian moons, Phobos and Deimos. The martian moons, along with a few satellites of the outer planets, are probably captured asteroids. Much can be learned about the geologic processes

of asteroids by close examination of Gaspra, Ida, and the moons of Mars, Phobos, and Deimos.

Gaspra. The first pictures of an asteroid to show surface details were taken in 1991 by the Galileo spacecraft on its circuitous journey to Jupiter (Figure 3.8). Gaspra orbits near the inner edge of the Main Belt, in the Flora region (Figure 3.7). The Galileo images show that Gaspra is small and the most irregularly shaped (18 by 10 by 9 km) planetary body yet observed. The asteroid is crudely cone-shaped, but a giant depression about 8 km across scallops the top, and another the bottom of the asteroid. Its reddish coloration and brightness as observed from telescopes on Earth show that it is similar to most asteroids in this region. This type of asteroid is thought to be composed of iron metal mixed with pyroxene and olivine—

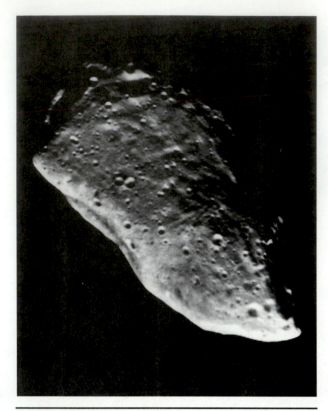

Figure 3.8

Gaspra, the first asteroid to be photographed by a passing spacecraft, shows many of the characteristics we expect for many other asteroids.

two common iron-magnesium silicates. As Galileo swept past Gaspra, it also detected a magnetic field. This field, comparable in strength to Earth's, is probably not caused by convection of molten metal inside Gaspra. It is inconceivable that Gaspra is warm enough to support internal convection. Rather, the magnetic field is probably caused by permanent magnetism of the metal in Gaspra. The magnetism measured in iron meteorites is consistent with this hypothesis.

The surface of Gaspra is composed of smooth facets separated by prominent ridges, spines, and smooth lumps related to the scallops. Impact craters are the most prominent landforms on these facets. As expected, cratering by small objects must be an important process in this part of the asteroid belt. About 60 impact craters have been identified. The largest is only 1.6 km across, less than twice as big as Meteor Crater in Arizona. The largest craters, seen in some detail, have bright, raised rims but no obvious ejecta blankets surrounding the crater. Gaspra has an extremely small gravitational field, and most fragments thrown up by impact should escape to space and never fall back to its surface.

Narrow troughs or grooves are barely discernible near the top of the image of Gaspra shown in Figure 3.7. These grooves are several kilometers long and 100 to 300 m deep. They occur in several groups or radiating sprays. All of these features suggest that they are the surface expressions of deep fractures in the bulk of Gaspra. Where did they come from? Again, impact and partial fragmentation seem to be the most likely explanation.

The number of craters on Gaspra provides important information about the age of its surface. If a body is completely broken apart or buried by lava flows, a new surface is created with no impact craters. With the passage of time, more and more craters accumulate on the surface. Thus, the number of craters in a given area is proportional to its age. An approximate calibration of the impact record can be made by comparing radiometric ages of lava flows sampled by astronauts on the Moon with the number of craters on the same flow. This is described in more detail in the next chapter. If we compare the number of craters on Gaspra with a comparable area on the Moon, we can estimate the age to be less than about 300 million years old and perhaps as young as 20 million years old. This is a truly remarkable conclusion, particularly in view of the 4.6 billion year age of the solar system indicated by meteorite studies. Remember: this age is not an estimate of the time that the rocky material in Gaspra solidified, but rather an estimate of the time that has passed since the surface of Gaspra was created. If Gaspra is metal-rich and strong, its surface age could be older. On the other hand, if the number of projectiles in the asteroid belt is higher than in the Moon's vicinity, its surface age could be even younger. What is important to note here is that the surface of Gaspra is only lightly cratered as compared to the Moon and most other planetary satellites.

Why would the surface of Gaspra be so young? Hints come by considering its geologic setting, irregular shape, and grooves. Gaspra is only one of many asteroids that orbit near one another. When small bodies hit Gaspra, small impact craters formed. But it is not unreasonable to expect that occasionally larger asteroids collide with one another in catastrophic events. The irregular shape is a testimony to Gaspra's tumultuous past. The conclusion is that Gaspra is probably a fragment of the most recent catastrophic collision between a larger precursor Gaspra and another asteroid. The depressions and ridges could be giant spall zones caused by the collision that created Gaspra. Perhaps, thousands of smaller bodies were produced in the same event. Some may still orbit near by, but no evidence of orbiting debris as small as 50 m in diameter has been found.

Ida.   More clues to the nature of the asteroids were revealed by Galileo's pictures of Ida taken in 1993 (Figure 3.9). Ida is more than twice as big as Gaspra, but it also has an irregular shape, measuring 56 by 24 by 21 km. Its color and reflectance are similar to Gaspra's but it may have less metal and more silicate minerals. Galileo showed that Ida has a magnetic field like Gaspra's.

The most dramatic difference between Gaspra and Ida lies in the great abundance of impact craters on Ida's surface. Craters of every size are visible. Huge, incomplete crater walls define Ida's irregular outline. Using the same logic discussed above, crater counts for Ida imply that its surface may be 2 billion years old, roughly 10 times as old

Figure 3.9

Ida has an irregular shape like Gaspra, but its surface is more heavily cratered. A tiny cratered moon, only 1.5 km across, accompanies Gaspra.

as Gaspra's. Somehow it appears that Ida has escaped recent fragmentation.

A small moon of Ida was discovered lurking in the background of the photos taken by Galileo. This is the first asteroid moon to be discovered, but such moons may be quite common in the asteroid belt. The moon may be a fragment of Ida blasted away in an ancient collision.

Phobos.   Phobos is the larger of the two small moons of Mars (Figure 3.10). Its overall shape is irregular (some call it potato-shaped) and is the result of cratering and fragmentation. Its longest dimension is only 27 km. Impact craters are the most common landform and occur uniformly over the surface. Some of the crater rims have blocks that may be impact ejecta returned to the surface in spite of the low escape velocity. But most crater rims lack boulders or any evidence of surrounding ejecta blankets. The escape velocity on Phobos is just a little over 10 m/s; an object tossed from the surface would go into orbit around Mars. Stickney, the largest crater, is 10 km across and covers a large fraction of the entire planetoid.

The most dramatic features on Phobos are a series of linear grooves (Figure 3.10). The grooves are similar to but more spectacular than those found on Gaspra. Most are between 100 and 200 m wide. The grooves appear to be related to Stickney, and they probably formed as Phobos fractured during the impact of the meteor that created Stickney. Some of the grooves are beaded and appear to be a series of pits. This may be caused by slumping of the regolith into the fractures, partially filling them in. The thickness of the regolith layer on Phobos may be up to 100 m deep and must consist of a veneer of finely pulverized material, resulting from repeated impact fragmentation. Layering is visible in some crater walls. Smooth intercrater areas may consist of a thicker blanket of this regolith.

Both Phobos and Deimos have low densities, around 2.2 (+/−0.2) g/cm^3 for Phobos and 1.7 (+/−0.2) g/cm^3 for Deimos. It has been postulated by some that they consist of easily fractured, volatile-rich carbonaceous chondrite material. The size, shape, and orbital properties of Phobos suggest that it started out as an asteroid but was scattered from the outermost asteroid belt by Jupiter. The gravity field of Mars later captured it, making it a small moonlet.

Deimos.   The smaller martian moon, Deimos, is also irregular in shape and has a diameter varying from 11 to 15 km. Photographs of Deimos taken from the Viking orbiters show that it is quite

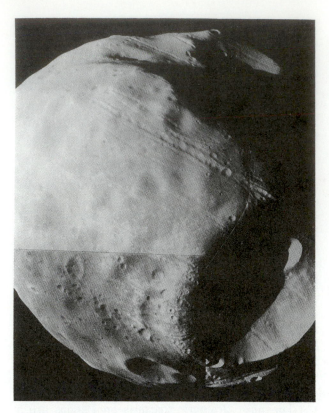

(A) The frontside of Phobos always faces Mars and is dominated by the crater Stickney, which is 10 km across. Linear grooves, less than 200 m wide, with shallow pits emanate from Stickney's rim and converge on a point directly opposite the crater. Phobos was nearly split apart by the impact that created Stickney.

(B) The details of craters and grooves on Phobos are shown here. Most craters lack blocky rims or any sign of an ejecta blanket because of the low escape velocity (10 m/s) for this small moon.

Figure 3.10

Phobos, a tiny satellite of Mars, is only 27 km across; its crater-scarred surface may be similar to many asteroids.

different from its companion satellite (Figure 3.11). Although Deimos, too, is densely cratered, its surface appears much smoother than the pockmarked surface of Phobos. It seems to be mantled with a thick layer of dust or debris that buries many of the craters and ridges. Apparently, Deimos developed a much thicker regolith layer than did Phobos. Small boulders are scattered around a few of the craters. Because of the very weak gravitational fields, these are the only components of crater ejecta that can be retained on the surface.

The images of Gaspra, Ida, Phobos, and Deimos radioed back to Earth are important because they show details of the smallest objects in the solar system yet observed in detail. Phobos and Deimos are very similar to the asteroids in the Main Belt—from which they may indeed have migrated. Judging from what we see here, the asteroids are probably heavily cratered, battered fragments of once larger bodies.

Collisions between asteroids are probably the most important events presently taking place in the vast, sparsely populated regions between the planets. Impact processes have probably determined the size and shape as well as most of the surface features of the asteroids. One of the most striking differences between the appearance of craters on the asteroids and those on the Moon is the absence of ejecta blankets and rays. Escape velocities are very low; material ejected from craters can escape the gravitational attraction of many asteroids, and does not fall back to the surface to form ejecta blankets or bright rays characteristic of young lunar craters; only the largest blocks fall back on the surface. Some of the ejected debris may later hit other asteroids, creating many small craters. At even higher impact energy, as when these small worlds collide, an asteroid may be completely fragmented or a large chunk may be broken off, exposing the deep interior. For example, the formation of Stickney almost disrupted Phobos. Gaspra's young scalloped surface was created by collisional breakup of a once larger body. Dramatic evidence of asteroid collision and frag-

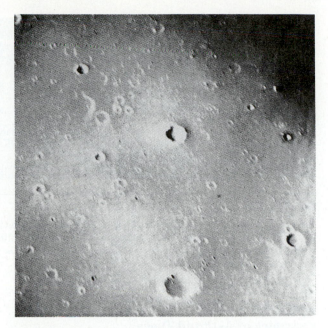

(A) The surface of Deimos appears to be much smoother than that of Phobos. The softer appearance may be a reflection of a thick layer of fine-grained regolith. The bright streaks that extend away from the edges or facets on Deimos probably reflect downslope movement of materials driven by the low gravitational force.

(B) Only the bright rims of many craters are visible on Deimos. Craters are nearly filled by a regolith layer that must be at least 5 to 10 m thick. Why Deimos has such a thick layer of regolith remains a mystery. The blocks, which are tens of meters across, are probably impact ejecta that did not escape from Deimos.

Figure 3.11

Deimos is the smaller of the two irregularly shaped martian moons, with a maximum dimension of only 15 km.

mentation is revealed by images of asteroid Toutatis constructed by bouncing radar off its surface as it passed close to Earth in 1992. It is made of two irregular pieces, probably formed when the body broke up after a massive impact. The two segments are now joined to form a *contact binary* (Figure 3.12). The two fragments measure only 4 and 2.5 km across, but are themselves heavily cratered.

Asteroid Types—Spectral Studies of Asteroids

Asteroids have recently become objects of intense interest for planetary astronomers. They occupy the transition zone between the dense, volatile-poor terrestrial planets and the ice- and gas-rich outer planets and satellites. Obviously, a thorough understanding of their nature and origin is very important. New instruments and techniques have allowed scientists to gather data that help to determine the surface composition of these bodies. The brightness of the asteroids can be measured, using telescopes on Earth, in different colors or wavelengths of light, employing a technique called **spectroscopy**. All minerals absorb and

reflect specific wavelengths (or colors) of light, which are diagnostic of the chemical composition and structure of the mineral. When this spectral information is combined with measurements of the brightness of each asteroid, a good idea can be obtained of its mineralogic composition. Over 600 asteroids have been "fingerprinted" in this way.

These studies have revealed that several distinct spectral types of asteroids exist. Let us examine a few of the interpretations of this new data (Table 3.3).

Asteroid 324, Bamberga, is a dim object, difficult to trace from Earth, in spite of its relatively large size. It has a diameter of 228 km. The asteroid has a black color imparted by what appear to be carbon compounds. Comparisons with the light spectrum of carbonaceous chondrite meteorites indicate that Bamberga may be of this composition. Even though these bodies are dark and difficult to see, the most numerous objects in the asteroid belt are of this type and appear to be volatile-rich carbonaceous material. The distribution of sizes within this asteroid class is dominated by very small bodies, indicating that the individuals are pieces of once-larger bodies fragmented by repeated collisions. Carbonaceous chondrite

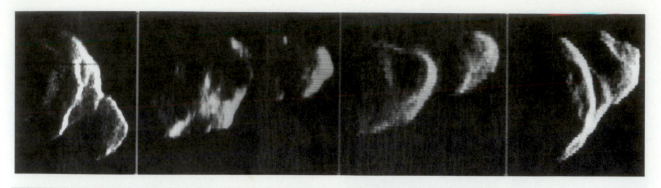

Figure 3.12

Large impacts may catastrophically disrupt an asteroid, as indicated in these radar images of asteroid 4179 Toutatis. Two fragments of a massive collision are now joined to form a contact binary. The four views were taken on different days and show different sides of the pair because it rotates.

TABLE 3.3
Examples of Asteroid Types

Asteroid	Diameter	Composition/Comments
Bamberga	228 km	Carbonaceous chondrite **Low-temperature condensate** **Abundant in outer asteroid belt** **Most asteroids of this type**
Eros	30 km	Ordinary chondrite **Moderate temperature condensate** **Common in inner asteroid belt**
Juno	267 km	Stony-iron meteorite **Differentiated interior of fragmented asteroid**
Vesta	520 km	Basaltic achondrite **Covered with lava formed from partial melting of asteroid interior** **Rare in asteroid belt**

meteorites may be pieces of bodies like Bamberga delivered to Earth.

The spectral properties of the small, irregularly shaped asteroid Eros (maximum diameter of 30 km) indicate that it has quite a different surface composition. It appears to consist of iron and magnesium silicates and appears to lack the carbonaceous coating of Bamberga, a mineralogic makeup that resembles metamorphosed chondritic meteorites. These chondritic meteorites are poor in volatiles, suggesting that they condensed at a relatively high temperature or that they lost their volatiles shortly after condensation, during metamorphism.

The surface of Juno, 267 km in diameter, appears to be comprised of a mixture of silicate minerals (olivine and pyroxene) and metal, (probably iron and nickel) that resembles no terrestrial rock but is very similar to certain stony-iron meteorites. Such a mixture is strong evidence that Juno is a chemically differentiated body that experienced a strong thermal event during its history. If its differentiated interior is now exposed at the surface, Juno may be a fragment of a once larger body.

The three largest asteroids (Ceres, Vesta, and Pallas) are unique in many ways, but some inferences about their natures can be made. Ceres, the largest asteroid, is just under 1000 km in diameter (almost the same size as Saturn's Dione and Tethys). Its spectral properties suggest that, like Bamberga, it is most probably composed of carbonaceous materials. Hydrous silicates, also found in carbonaceous chondrites, have been identified on its surface. In accord with this suggestion is its relatively low density (2.3 g/cm³). Water ice or frost has also been detected on Ceres.

The second largest asteroid (Vesta, over 500 km in diameter) is unique among the asteroids so far studied. Its surface layer is comprised of pyroxene-rich rocks, like achondrite meteorites, which, as we discussed earlier, were most likely lavas erupted on the surface of small planetoids. Vesta was apparently one such object and has preserved much of this volcanic shell. Temperatures must have been very high to allow melting. Vesta also displays variations in the light reflected from its surface as it rotates. These variations may be evidence of other rock units on its surface, perhaps exposed by impact or alternatively not covered by achondrite lavas.

Pallas is only slightly larger (533 km diameter) than Vesta but appears to be more like Eros in its composition, that is, rich in iron and magnesium silicates.

Zonation of the Asteroid Belt

Although many different spectral types and attendant mineral compositions are recognized in the asteroid belt, several generalizations can be made from the interpretations of the spectral data. Almost three-fourths of the asteroids appear to be similar to carbonaceous chondrites. Like Bamberga, they consist of volatile-rich material that probably condensed directly from the planetary nebula at low temperatures and are found mostly in the *outer belt*, far from the Sun (Figure 3.13). Even water ice has been detected in the outer part of the asteroid belt. There is little indication of any post-accretionary thermal events in these asteroids or their meteoritic equivalents. Another 15 percent of the asteroids seem to be similar to Eros, composed of iron and magnesium-rich silicate minerals, olivine and pyroxene, with little dark carbonaceous material. Judging from what we think are their meteoritic equivalents, these asteroids may never have been melted, but metamorphic or condensation temperatures must have reached 1100 K within relatively shallow layers. These asteroids are apparently not as rich in volatile compounds as the first group described and are most common in the *inner asteroid belt*, that part closest to the Sun. Other, relatively rare, asteroids appear to be composed of iron metal or iron-nickel alloys with a silicate component, similar to Juno. The separation of the metallic phases probably occurred deep within small planetesimals when they became nearly molten; temperatures probably exceeded 1700 K. Most of the metallic asteroids are 100 to 200 km in diameter and do not appear to be fragmented like the other types of asteroids. These highly reflective asteroids are concentrated in the *central to inner part of the asteroid belt*.

Evolution of Asteroids

From this information about the zonation of the asteroid belt, it appears likely that the silicate-dominated asteroids closer to the Sun condensed at higher temperatures than those that are near Jupiter. Some of the larger bodies appear to have experienced a degree of internal chemical differentiation. Zones composed of an inner metallic core, a metamorphic silicate mantle, and possibly a volcanic veneer formed on these parent bodies. Subsequent collisions stripped away and fragmented the rocky mantle and crusts, exposing iron-rich cores (Figure 3.14). These metallic cores were stronger and less likely to fragment than the rocky outer layers and therefore preserved much of their original mass, explaining their relatively large, uniform sizes. Fragments of asteroids such as these that

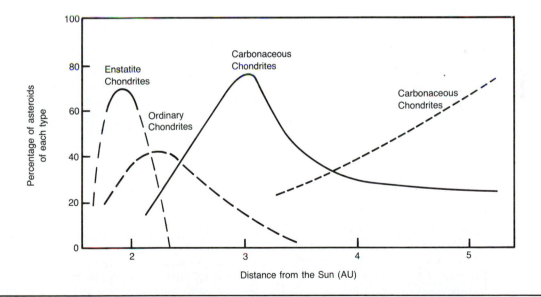

Figure 3.13

A compositional zonation in the asteroid belt has been discovered by careful telescopic studies of the color and brightness of asteroids. Letters indicate names of asteroid subdivisions applied by specialists. The meteorite labels are liberal interpretations of the probable nature of the asteroids in each region of the diagram. Note especially the change in the relative abundance of ordinary versus carbonaceous chondrites as one moves through the asteroid belt. These changes may correspond to different nebular condensation temperatures for these materials. Along the bottom of the diagram, the names and relative positions of several planetary bodies are given.

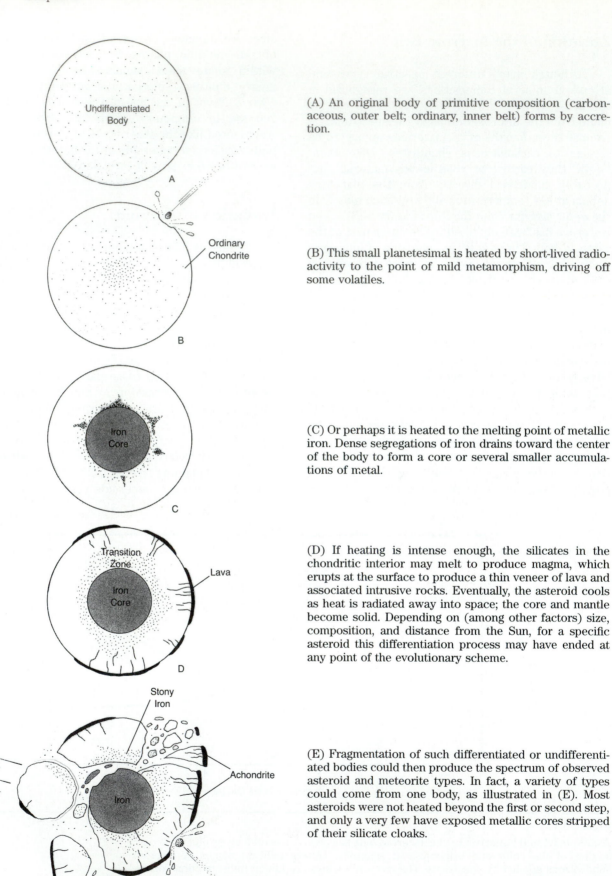

(A) An original body of primitive composition (carbonaceous, outer belt; ordinary, inner belt) forms by accretion.

(B) This small planetesimal is heated by short-lived radioactivity to the point of mild metamorphism, driving off some volatiles.

(C) Or perhaps it is heated to the melting point of metallic iron. Dense segregations of iron drains toward the center of the body to form a core or several smaller accumulations of metal.

(D) If heating is intense enough, the silicates in the chondritic interior may melt to produce magma, which erupts at the surface to produce a thin veneer of lava and associated intrusive rocks. Eventually, the asteroid cools as heat is radiated away into space; the core and mantle become solid. Depending on (among other factors) size, composition, and distance from the Sun, for a specific asteroid this differentiation process may have ended at any point of the evolutionary scheme.

(E) Fragmentation of such differentiated or undifferentiated bodies could then produce the spectrum of observed asteroid and meteorite types. In fact, a variety of types could come from one body, as illustrated in (E). Most asteroids were not heated beyond the first or second step, and only a very few have exposed metallic cores stripped of their silicate cloaks.

Figure 3.14
Stages in the evolution of a meteorite parent body.

reached Earth would be iron meteorites (from the core), stony-irons (core-mantle interface), metamorphosed chondrites (mantle) and achondrites (volcanic veneer). The eventual delivery of such fragments to Earth and other inner planets results from gravitational or impact perturbations in space. An approximate time scale for the important events in the evolution of a meteorite from the asteroid belt is summarized in Figure 3.15.

Other asteroids farther from the Sun accreted from low-temperature condensates and did not melt; some may have reached temperatures high enough to produce mild metamorphism (Figure 3.14). These asteroids appear to be the largest and occupy the outer part of the asteroid belt. When delivered to Earth, fragments of these bodies are called chondrites and carbonaceous chondrites.

Conclusions

From the vast store of information collected by scientists on Earth and with robotic spacecraft, we can answer many important questions about the origins of meteorites and their relationships to asteroids. Many types of meteorites have counterparts in the asteroid belt; a few come to us from the Moon and perhaps Mars. Still other meteorites appear to have been delivered to Earth by comets.

Nonetheless, many questions about meteorites and their small parent bodies remain. The source of the heat that tremendously altered some meteorites and their asteroidal parents still remains a problem. Were the asteroids heated by now-extinct radioactive elements or did early solar activity heat some to melting? Geochemists have found evidence that short-lived isotopes such as iron (^{60}Fe) and aluminum (^{26}Al) were present in early meteorite parents. Rapid decay of these isotopes inside asteroids may have provided a short thermal spike of sufficient intensity to melt and differentiate bodies as small as a few hundred kilometers across. However, it is difficult to reconcile radioactive heating with the steep temperature gradient across the asteroid belt. Consequently, other scientists point to the early Sun as an important heat source. Strong radiation and variable magnetic fields from the young Sun while it was going through a T Tauri stage (Chapter 2) may have heated asteroids to high temperatures by induction heating—a process akin to microwave cooking. In this case the outside of the bodies would be more intensely metamorphosed than the deep interiors. The exteriors of those asteroids nearest the Sun (within 2 times the Earth-Sun distance) may have melted completely, and those farther away (about 3 times the Earth-Sun distance) may not have been melted at all. Even farther away, water ice may have remained stable. This model does not require a special distribution of radioactive aluminum or iron. The pattern of spectral types in the asteroid belt indicates that both the distance from the Sun and the size (which determines the planetoid's ability to retain heat and the amount of radioactive elements) are important. In short, the source of heat to melt the meteorite parent bodies remains uncertain.

Why did thermal activity, manifest by the rare, young SNC meteorites, continue so long after radioactive or solar heat sources should have been ineffective? Is it possible that these young meteorites are really derived from a much larger body with a longer thermal history like Mars? The implications of this radical conclusion for our understanding of Mars are enormous and are being pursued vigorously.

Another problem yet to be totally solved lies in the numbers of individual bodies that exist in the asteroid belt. Why did these bodies not accrete to form a planet? Was there a much larger population of asteroids early in the solar system's history that was depleted as the inner solar system was peppered with impacting bodies? Certainly Jupiter's tremendous gravitational influence affected this turn of events. Possibly Jupiter accelerated the early planetesimals in this region just enough to produce net fragmentation rather than accretion, and then gradually perturbed their orbits so that asteroidal fragments entered the inner solar system, colliding with the surfaces of the planets.

Of course the answers are not all in yet, but further study of the physical nature of meteorites and asteroids and the evolution of their orbits and thermal histories will reveal more about the very early development of the solar system. Preserved in some of these small bodies are quantities of material left relatively undisturbed since the dawn of the planets. In others we can study the igneous and metamorphic changes that occurred deep inside differentiating planets. Moreover, these are the same sorts of planetesimals that, 4.6 billion years ago, accreted to form Earth and its neighbors in space.

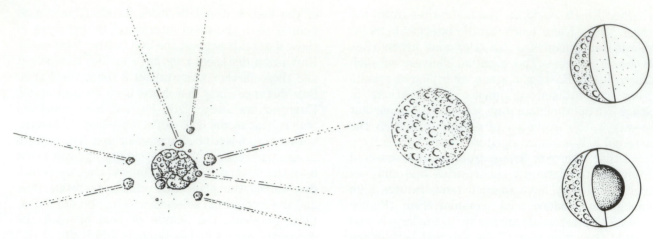

(A) Accretion of a planetesimal 4.6 billion years ago.

(B) Differentiation of interior to form iron core or meta-morphosed rock 4.4 to 4.6 billion years ago.

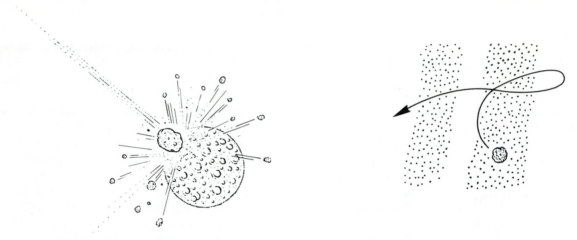

(C) Collision and fragmentation 1 to 0.1 billion years ago.

(D) Deflection into inner solar system 1.0 to 0.1 billion years ago.

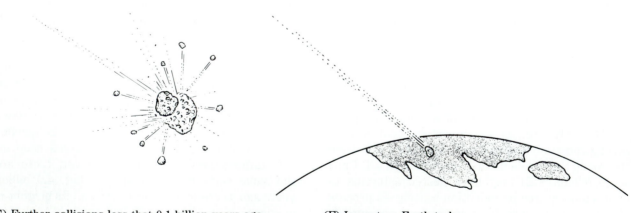

(E) Further collisions less that 0.1 billion years ago.

(F) Impact on Earth today.

Figure 3.15

The evolution of asteroids as meteorite parent bodies is summarized in this diagram. The events that led to the delivery of fragments of the asteroids to Earth and other inner planets are emphasized.

Review Questions

1. Is it possible that any of the meteoritic material that falls to Earth each year comes from asteroids? From other planets? Cite any evidence.
2. What is the evidence that even small asteroids may have melted during their early histories?
3. Contrast the origins of the major classes of meteorites. Could a single asteroidal body give rise to all of these various types? Why?
4. What are the possible explanations for the zonation seen in the asteroid belt?
5. How and why did the parent bodies for meteorites differentiate? What does their differentiation tell us about the early history of the inner planets?
6. What sources of heat may have been important in the thermal evolution of asteroids?
7. What class of meteorites is most like the core of Earth?

Key Terms

Achondrite

Apollo Asteroids

Asteroid

Asteroid Belt

Carbonaceous Chondrite

Chondrite

Differentiated

Impact Melting

Iron Meteorite

Main Belt Asteroids

Meteorite

Partial Melting

Regoliths

SNC Meteorites

Spectroscopy

Stony-Iron Meteorite

Undifferentiated

Additional Reading

Chapman, C. P. 1990. "Asteroids," in J. K. Beatty, B. O'Leary, A. Chaikin, *The New Solar System*. Cambridge, MA: Sky Publishing Co., pp. 97–104.

Dodd, R. T. 1986. *Thunderstones and shooting stars.* Cambridge, MA: Harvard University Press.

McSween, H. Y. 1987. *Meteorites and their parent planets.* London: Cambridge University Press.

Veverka, J. 1977. Phobos and Deimos. *Scientific American.* Vol. 236, No. 2, pp. 30–37.

Lunar Impact

TABLE 4.1

**Physical and Orbital Characteristics
of the Moon**

Mean Distance from Earth	384,400 km
Period of Revolution about Earth	27.3 d
Period of Rotation	27.3 d
Inclination of Axis	6° 41″
Equatorial Diameter	3,476 km
Mass (Earth = 1)	0.0123
Volume (Earth = 1)	0.02
Density	3.34 g/cm^3
Atmosphere	none
Surface Temperature	100 to 400 K
Surface Pressure	0
Surface Gravity (Earth = 1)	0.165
Magnetic Field (Earth = 1)	0
Surface Area/Mass	52×10^{-11} m^2/kg

CHAPTER 4

The Moon

Moon

Earth

Introduction

In many ways the Moon is a geologic Rosetta stone: an airless, waterless body untouched by erosion. It contains clues to events that occurred in the early years of the solar system that have revealed some of the details regarding its origin and have provided new insight about the evolution of Earth. Although they also posed new questions, the thousands of satellite photographs brought back from the Moon have permitted us to map its surface with greater accuracy than Earth could be mapped a few decades ago. We now have over 380 kg of rocks from nine places on the Moon, rocks that have been analyzed by hundreds of scientists from many different countries. Data from a variety of experiments have revealed much about the Moon's deep interior. As it turns out, the Moon is truly a whole new world, with rocks and surface features that provide a record of events that occurred during the first billion years of the solar system. This record is not preserved on Earth because all rocks formed during the first 800 million years of Earth's history were recycled back into the interior.

The importance of the Moon in studying the principles of geology is that it provides an insight into the basic mechanics of planetary evolution and events that occurred early in the solar system. Much of the knowledge we have of how planets are born and of the events that transpired during the early part of their histories has been gained from studies of the Moon.

Major Concepts

1. The surface of the Moon can be divided into two major regions: (a) the relatively low, smooth, dark areas called maria (seas) and (b) the densely cratered, rugged highlands, originally called terrae (land).

2. Most of the craters of the Moon resulted from the impact of meteorites, a process fundamental in planetary development.

3. The geologic time scale for the Moon has been established using the principles of superposition and cross-cutting relations. Radiometric dating of rocks returned from the Moon has provided an absolute time scale.

4. The lunar maria are vast plains of basaltic lava, extruded about 4.0 to 2.5 billion years ago. Other volcanic features on the Moon include sinuous rilles and low-shield volcanoes.

5. The major tectonic features on the Moon, mare ridges and linear rilles, are products of minor vertical movements.

6. Lunar rocks are of igneous and impact origin. The major types include: (a) anorthosite, (b) basalt, (c) breccia, and (d) glass.

7. The Moon is a differentiated planetary body with a crust about 70 km thick. The lithosphere is approximately 1000 km thick. The deeper interior may consist of a partially molten asthenosphere and a small metallic core.

8. The tectonic and thermal evolution of the Moon was very rapid and terminated more than two billion years ago. The Moon has no surface fluids, so that little surface modification has occurred since the termination of its tectonic activity.

9. The major events in the Moon's history were: (a) accretion of material ejected from Earth after a massive collision with a Mars-sized object, (b) differentiation with the formation of the lunar crust by crystallization of a magma ocean, (c) continued intense meteorite bombardment, (d) extrusion of the mare lavas, and (e) light bombardment.

The Moon as a Planet

In July 1969, a human stood for the first time on the surface of another planet, seeing landscape features that were truly alien and returning with a priceless burden of Moon rocks and other information obtainable in no other way. Nonetheless, many of the facts listed in the table of its characteristics at the front of this chapter were known long before we began to explore space; they represent years of diligent study. For example, it was discovered centuries ago that the Moon revolves about Earth and not the Sun and is thus a natural satellite (the largest in the inner solar system). Long ago, the distance from Earth to the Moon was measured and the diameter of the Moon determined. Early astronomers realized that the Moon's rotation period and its period of revolution are the same; thus, it keeps one hemisphere facing Earth at all times. Moreover, many of the Moon's surface features have become well known, especially since the days of Galileo, the first to study the Moon through a telescope. Even the density and gravitational field of the Moon had been determined long before our generation. But not until the 1960s—and the inception of space travel with its sophisticated satellites and probes and the eventual Moon landing—did we begin to appreciate the significance of the Moon as a planet. In spite of its small size and forbidding surface, the Moon has revealed secrets that pertain to the ultimate creation of our planet, Earth, and our neighbors beyond.

Major Geologic Provinces

When Galileo first observed the Moon through a telescope, he discovered that its dark areas are fairly smooth and its bright areas are rugged and densely pockmarked with craters. He called the dark areas **maria,** the Latin word for seas, and the bright areas **terrae** (lands). These terms are still used today, although we know the maria are not seas of water and the terrae are not geologically similar to Earth's continents. The maria and terrae do, however, represent major provinces of the lunar surface, each with different structures, landforms, compositions, and histories.

The maria and terrae can even be distinguished from Earth by the naked eye. As shown in Figure 4.1, the maria on the near side of the Moon appear to be dark and smooth, with only a few large craters. Some maria occur within the walls of large circular basins such as Crisium, Serenitatis, and Imbrium, whereas others such as Oceanus Procellarum occupy much larger, irregular depressions. We know from lunar rock specimens and surface features that the maria are vast layers of thin basaltic lava, which flowed into depressions and flooded large parts of the lunar surface.

The terrae, or highlands, constitute about two-thirds of the near side of the Moon and exhibit a wide range of topographic relief. This is the highest and most rugged topography on the Moon, where local relief in many areas is up to 5000 meters. An important characteristic of the lunar highlands is that they contain abundant craters, many of which range from 50 to 1000 km in diameter. For example, craters larger than 10 km are about 50 times more abundant on the highlands than on the maria.

The far side of the Moon was totally unknown until photographs were first taken by a Soviet spacecraft in 1959. It was a total surprise to learn that although the details were poorly defined, the far side of the Moon was composed almost entirely of densely cratered highlands. Later, orbiting satellites launched by the United States completely photographed the far side of the Moon with definition sharp enough to map the surface in considerable detail (Figure 4.2). These photographs confirmed that the far side of the Moon is densely cratered; only a few large craters contain mare lavas. Why the maria are largely restricted to the near side of the Moon remains a fundamental problem of lunar geology.

Results from several Apollo experiments demonstrate other fundamental differences between the highlands and the maria. Remote sensing measurements of the composition of the lunar surface indicate that the maria and the highlands are composed of distinctly different rock types. Rocks collected by Apollo astronauts show the highlands to be mostly composed of anorthosite and other feldspar-rich rocks, in contrast to the basaltic maria. Tracking Moon-orbiting spacecraft from Earth has confirmed the elevation differences. The highlands may be as much as 5 km above the mean radius of the Moon, whereas the maria lie almost 5 km below the mean radius. (As the Moon has no water, its mean radius can be used as a reference level from which relative elevations or depressions can be measured.) Studies of the velocities of seismic waves have shown that the highland crust is also much thicker, in some cases up to 100 km thick, while the crust under some of the near side maria is only around 40 km thick.

The maria and highlands not only represent different types of terrain, but they broadly represent two different periods in the history of the Moon. The highlands, which occupy about 80 percent of the entire lunar surface, are composed of

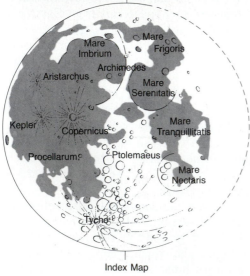

Index Map

Figure 4.1

The Moon is geologically different from Earth. The most striking differences are apparent in this photograph, taken through a telescope on Earth. The absence of swirling clouds, oceans, and an atmosphere reveal the water-poor nature of the Moon. Also seen in this photograph are the fundamental differences between the dark lava-covered maria and the lighter highlands, or terrae, which are intensely cratered. This is the familiar hemisphere of the Moon, for the Moon always has the same face turned toward Earth.

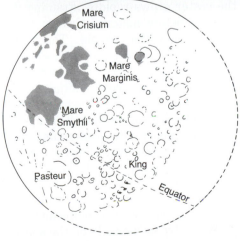

Figure 4.2

The far side of the Moon, photographed by orbiting satellites, reveals that most of the Moon's surface consists of the heavily cratered highlands. This part of the eastern limb of the Moon was never seen until the space age. The irregular, dark, smooth area in the lower left is Mare Marginis with Mare Smythii below it; the rest of this vast region lacks large accumulations of lava. Why there are so few maria on the far side of the Moon remains a topic of discussion.

rocks that formed very early in the Moon's history. The entire outer portion of the Moon is thought to have been molten at the time the highlands crust began to form 4.6 billion years ago. As light silicate minerals accumulated at the top of this "magma ocean," they formed a crust, which soon became densely cratered by a contemporaneous intense bombardment of meteorites. This initial high cratering rate declined rapidly, but the Moon's surface, one of the oldest in the solar system, became covered with craters.

The maria were formed by the extrusion of vast amounts of lava that accumulated in the lowlands of large craters or basins and, in places, overflowed and spread over parts of the lunar highlands. The maria are thus relatively young features of the lunar surface, even though they began to form four billion years ago.

Impact Craters

Although impact cratering is a rare event on Earth today, it has been a fundamental and universal process in planetary development. The Moon is pockmarked with literally billions of craters, which range in size from microscopic pits on the surfaces of rock specimens to huge, circular basins hundreds of kilometers in diameter. The same is true for the surfaces of Mercury, Mars, the asteroids, and most of the satellites of the outer planets as well. Indeed, impact cratering was undoubtedly the dominant geologic process on Earth during the early stages of its evolution. Earth was once heavily cratered like the Moon today. Evaluation of the impact process can provide an important interpretive tool for understanding planets and their development.

The Mechanism of Crater Formation

Although it is difficult to imagine the magnitude of these enormous impact events, which excavated millions of cubic meters of material in seconds, the results of the photographic missions to the planets and studies of experimental craters produced in laboratories have greatly increased our understanding of what happens when a meteorite strikes a planet.

The process of crater formation, like many geologic processes, can be viewed as a transfer of energy of one form to another form. In this case, the kinetic energy of the meteor, which may be moving at velocities of 5 to 50 km/sec (5 km/sec is 18,000 km/hr), is imparted to the planetary surface and changed into other types of energy. It has been calculated that a meteorite with a diameter of only 20 m could excavate a crater 500 m across and would be as energetic as the largest volcanic explosions that have occurred on Earth during historic times. However, not all of the energy of an impact goes into excavating a crater. Much of it is converted to heat. If the meteorite is large enough, a small volume of rock may be completely melted and transformed into an igneous **impact melt** that may come to rest on the floor of the crater or be ejected over its rim during the seconds involved in crater formation. Some materials vaporize completely, and other rocks recrystallize as metamorphic rocks. Some of the energy is used in fracturing and pulverizing the target materials and the meteorite itself. Some energy is dispersed as sound or seismic (earthquake) waves.

The formation of a crater is generally divided into three stages: compression, excavation, and modification (Figure 4.3). During the **compression stage**, the meteor's kinetic energy is transferred through the ground by a shock wave that expands in a spherical pattern away from the point of impact. The pressures experienced by the target materials are so high that the rock behaves like a fluid for a short time. During this stage, only a small amount of material is removed by vapor jets around the side of the meteorite; the main mass of the material remains to be excavated.

After the passage of the shock wave, the rocks relax and a **rarefaction wave** is established that travels upward and allows the rocks to expand explosively as the target attempts to return to the same low pressure as its surroundings. The rarefaction wave is in effect a reflection of the compressional wave because it forces material to flow upward toward the ground surface. The result is similar to an explosion located a few meteorite diameters below the ground. This causes the rock to be lifted upward and bent backward and material to be ejected from the surface and thrown out along **ballistic** trajectories. During this **excavation stage**, the crater grows rapidly, attaining its final diameter even before material thrown from the impact site reaches the ground. The fragmented rock that is blasted away is called **ejecta**. The path of each particle is affected by the strength of the planet's gravitational field. Initially, the ejecta forms a nearly continuous cone-shaped curtain that appears to enlarge at its base. This ejecta curtain quickly tears apart into filamentous elements, which fall to the ground to form the **rays** characteristic of young craters. Near the edge of the crater, layers of rock may be overturned in a flap that forms part of the high crater rim. The excavation is accomplished very rapidly and the duration of the event depends mainly on the size of

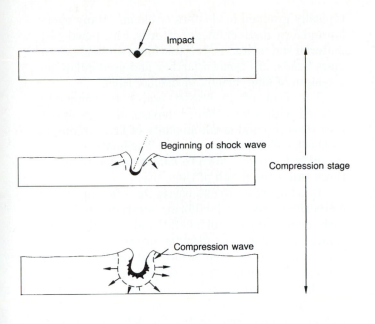

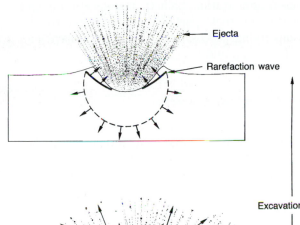

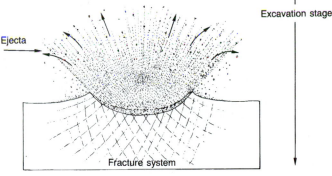

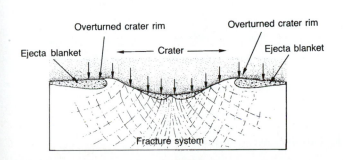

the meteorite and its velocity. For example, Earth's Meteor Crater (Figure 4.4) is thought to have formed in about 10 seconds.

During the final stage, after much ejecta has settled to the surface, the crater becomes modified from its initial, excavated form. The floor of the crater may rapidly rebound upward to compensate for the material removed during excavation. Slices of rock from the steep walls may immediately slip back into the crater, forming a series of terraces that partially fill the depression. Terraces form a nearly concentric natural staircase to the crater floor. The **modification stage** probably continues over a long period as the crust re-establishes a stable configuration in the planet's gravity field (Figure 4.5).

Features of Impact Craters

Impact craters are typically circular, in contrast to many volcanic craters, which are frequently asymmetrical or elongate. When an impacting body strikes the surface at an angle greater than 15°, the crater will nonetheless be circular because the shock waves spread out with equal velocity in all directions from the point of impact and because the rarefaction waves move back toward that point.

As can be seen in Figure 4.5, the rim of the crater is built of deformed bedrock that has been heaved upward and outward, bent back, and even overturned. The material thrown out of the crater accumulates around the crater rim as an **ejecta**

Figure 4.3

Stages in the formation of a hypothetical meteorite impact crater are depicted in this series of diagrams. During the compression stage, the kinetic energy of the meteorite is almost instantly transferred to the ground as a shock wave, which moves outward, compressing the rock. At the point of impact, the rock is intensely fractured, melted, and partly vaporized by shock metamorphism. After the shock wave passes, the rocks return to normal pressures and expand explosively as a "reflected" rarefaction wave, which throws out large amounts of fragmental debris as ejecta—the excavation stage. Near the edge of the growing crater the solid bedrock is forced upward to form an elevated rim; in some cases an overturned flap may develop. The ejecta are propelled along ballistic pathways away from the crater. Ejecta that fall far from the crater have high velocities and may form chains of secondary craters, which radiate away from the point of impact. These secondary impacts may result in extensive mixing of primary crater ejecta with local surface materials. A large amount of fragmental material also falls back into the crater. During a subsequent modification stage, material may slump or slide off the steep crater walls to form terraces. It is common for some type of uplift to form on the floors of larger craters as well.

Figure 4.4

Meteor Crater, Arizona, is a recent impact crater similar to those found on the Moon. It is one of the youngest impact craters found on Earth. This small crater formed about 25,000 years ago when an iron meteorite struck Earth. The bowl-shaped crater is a little over 1 km in diameter and is 200 m deep. Its floor is covered by younger lake sediments. An extensive blanket of ejecta surrounds the crater. The polygonal outline of the crater is probably the result of a fracture system that predates the excavation of the crater.

blanket. This is thickest near the crater and thins outward, becoming discontinuous and patchy at a distance of about 1.5 crater diameters from the rim. Larger blocks thrown out from the main crater may impact with enough speed to form secondary craters, which tend to be irregular in shape and are

typically grouped in clusters or chains. Many overlap and form distinct linear patterns. Fine powdered material and melted material, which resolidifies into glass beads, are thrown farther and accumulate as a system of long, bright, splashlike rays.

The floor of an impact crater is commonly covered with a lens-shaped deposit of fragmented rock (breccia) and small amounts of lava produced by shock melting during impact. The bedrock below is highly fractured down to a depth equal to about three times the depth of the crater.

It is important to emphasize that the impacts of meteorites produce landforms (craters) and new rock bodies (ejecta blankets) and are therefore similar to other rock-forming processes (such as volcanism and sedimentation) that operate at the surface of a planet. In each process, energy is transferred, material is transported and deposited to form a new rock body, and a new landform (crater, volcanic cone or flow, or delta) is created. The rock-forming processes associated with impact include fragmentation, ballistic transportation, and deposition of rock particles, in addition to the rock-modifying processes of shock metamorphism, partial melting, and vaporization.

Types of Impact Craters

The surface of the Moon (Figure 4.6) has been described as being covered by a "forest of craters," and at first glance all craters may look the same. After a few moments of thoughtful study, however,

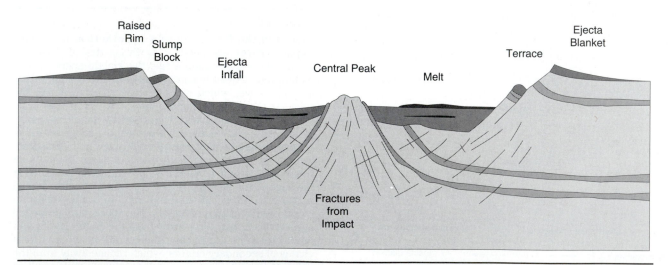

Figure 4.5

The internal structure of a hypothetical impact crater is shown in this cross section. Note particularly the raised rim of the crater, the lens of impact breccia (and impact melt) that floors the crater, the ejecta blanket, which adds to the height of the crater rim, and the deformation of preexisting rocks. Rocks in the target are both folded and extensively fractured. Terraces form as slices of rock drop along steep faults into the interior of the crater.

Figure 4.6
A shaded relief map of the Moon shows the topography and major provinces on the
Moon.

one is able to identify various "trees in the forest" and recognize various types of craters by certain characteristics of their size and form. Detailed study of lunar craters shows that many morphologic differences are associated with size, so it is useful to classify craters into groups that are similar in size and shape and that probably originated from the impact of similar sizes of meteorites.

Craters Less Than 20 Km in Diameter (Bowl-Shaped Craters). Small lunar craters, generally less than about 20 km in diameter, are almost perfectly circular and are typically bowl shaped (Figure 4.7). The deepest measure is about 2 km from the crest of their rims to their floors. Such a crater usually has a well-defined, raised rim created by ejected material and by the expansion of the materials in a ring very close to the crater rim. The ejecta blanket extends from the rim to a distance approximately equal to the diameter of the crater. The surface of the ejecta blanket consists of a series of hummocky ridges that superficially resemble sand dunes. The crater in Figure 4.8 has some large boulders near its rim that were thrown from the crater as it formed.

Another type of fairly small crater possesses a flat floor and is not a simple bowl-shaped hole. The floor of the crater accounts for about 50 percent of the crater diameter. The flat floor may be caused in part by the collection of materials that roll or slump

Figure 4.7

Lunar craters smaller than 20 km in diameter are almost perfectly circular and are typically bowl shaped. In this example from southern Mare Serenitatis, the raised crater rim is well defined and is surrounded by a prominent ejecta blanket composed of light-colored material. The rays formed by secondary cratering extend many crater diameters beyond the rim of the crater.

Figure 4.8

Most craters have blocky rims like this crater (about 14 km across) from the highlands of the lunar far side. Long tongues of lavalike material also extend downslope from the crater rim. These flows may consist of fragmented rock debris, but it is perhaps more likely that they are impact-produced melts ejected at low velocities during the late stages of the formation of the impact crater.

off the walls of the crater and partially fill it. Some prominent flat-floored craters are the result of excavation through distinct layers in the surface materials of a planet (Figure 4.9). The response of materials to the passage of the shock wave depends on their physical properties; if a sharp discontinuity exists in a layered planet, the wave may be reflected by it, creating the flat floors of these small craters. There is substantial evidence that the Moon and several of the other planets are covered by a layer of loose, fragmental material called **regolith** to a depth of many meters. The regolith of the Moon was probably produced by billions of years of meteoritic bombardment, during which the surface material was constantly fragmented, churned, and mixed to produce a brecciated soil. The depth at which the regolith gives way to more solid rock certainly divides two materials with different strengths and other physical characteristics. It seems quite plausible that the lower, less brecciated bedrock would be much stronger than the regolith and hence would not be as easily ejected, creating a sharp break in the slope of the crater wall and a flat floor. Apparently, some fresh bowl-shaped craters were not produced by events sufficiently energetic to encounter this discontinuity.

Nonetheless, virtually all craters over 30 km in diameter have flat floors, and many smaller ones have a distinctive swirllike texture on their floors that results from the slumping of material from the crater wall in thin sheets toward the floor (Figure 4.10).

Craters 20 to 200 Km in Diameter (Terraced Craters with Central Peaks).

Impact craters on the Moon, with diameters of 20 to 200 km, typically have terraces on their inner

Figure 4.10

The accumulation of material on the floor of a crater may result from slumping from its walls commonly forming distinctive surface features. The scarp of the right side of the crater is especially high and creates the crater's irregular outline. The ejecta deposit around the crater rim contains dunelike features formed during its deposition.

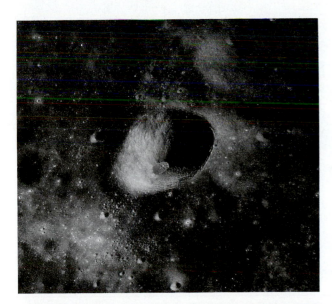

Figure 4.9

A variety of small generally bowl-shaped craters are present in this view of the lunar highlands. The small crater near the center of the picture is about 1 km across. Its small, flat floor and blocky rim contrast sharply with the older and more degraded craters that dominate the surface. Nonetheless, this crater lacks rays and is probably older than the craters shown in Figures 4-7 and 4-8.

walls that become well developed with increasing size. The terraces form as slump blocks when thin slices of rock become unstable on the crater walls and collapse into the depression. All of these craters have more or less flat floors, but some have been arched up, presumably as a result of adjustment during the modification stage. Central peaks that rise abruptly from the floor form prominent features of many terraced craters (Figure 4.11). Almost all lunar craters larger than 50 km in diameter have central peaks.

Although there are several theories to explain the formation of central peaks, the most likely explanation relates them to a type of elastic rebound. During excavation, material at depth is forcefully pushed toward the center of the crater, then upward. Material pathways mimic this as a result of relaxation after passage of a hemispherical shockwave (Figure 4.3).

Craters 200 to 300 Km in Diameter (Peak-Ring Basins).   Large craters are transitional in morphology to still larger, multiring basins. The terraced walls are retained, but the central peaks change progressively from single promontories, through clusters of peaks, to a ring of peaks halfway between the center and the rim of the crater (Figure 4.12). These impact features are called **peak-ring basins.**

Basins Larger Than 300 Km in Diameter (Multiring Basins).   The largest impact features on the planets, called **multiring basins,** are fringed by a series of concentric ridges

Figure 4.11

The lunar crater King illustrates many of the features of impact craters larger than 20 km in diameter. King, about 75 km in diameter and about 4 km deep, is on the far side of the Moon, west of Mendeleev. Among its interesting features are its atypical horseshoe-shaped central peak. Central peaks are common in craters of this size. King crater also has terraced walls that descend to a relatively broad, flat floor. A smooth puddle of solidified impact melt lies above the peak; another large accumulation of impact melt lies outside the crater on its right rim. These pools were fed by lobes of melt that drained into local depressions.

Figure 4.12

Schroedinger Basin lies in the south polar region of the Moon's far side. It is an excellent example of a peak-ring basin, typified by a ring of peaks on its floor in the place of a central peak. The long, narrow valley that extends away from the basin's rim is a chain of closely spaced secondary craters. Schroedinger is 320 km in diameter. (The horizontal lines on the picture are artifacts of the way in which the image was constructed and are characteristic of data collected by the Lunar Orbiter spacecraft.)

and depressions. The youngest and best-preserved multiring basin on the Moon is Orientale (Figure 4.13). Although it is largely hidden from telescopic view, satellites orbiting the Moon have photographed it in detail. The basin resembles a gigantic bull's-eye with three concentric ridges or scarps and intervening lowlands. The diameter of the outer ring of mountains (the Cordillera) is 900 km; the entire state of Colorado could fit within the second ring. The spacing of the rings increases outward, a feature that is common to many multiring basins. Beyond the outer ring, the most prominent topographic features are low ridges oriented radially to the basin (Figure 4.14). A steep scarp, 4 km high in places, forms the outer ring. The texture of the surface just within the outermost ring is distinctly different, consisting mainly of small domes or hummocks. Another steep cliff forms the second ring, but the inner ring is not a well-defined

scarp; rather, it consists of a group of peaks that surrounds the plains in the middle of the basin (Figure 4.15). This ring of peaks is probably very similar to the peak rings formed in smaller craters. The central part of the crater is very dark and smooth and is covered by volcanic lava flows that filled the depression after its formation.

The multiring structure of Nectaris Basin, located near the eastern margin of the near side of the Moon just south of the equator, is somewhat more subtle (Figure 4.1). A definite series of arcuate ridges and intervening lowlands can be seen. This is obviously a much older basin than Orientale because the concentric rings and ejecta blanket are considerably modified by subsequent impact craters. Like Orientale, only the central part of Nectaris is flooded with basalt.

Another basin useful in studying multiring structures is the Imbrium Basin, which dominates

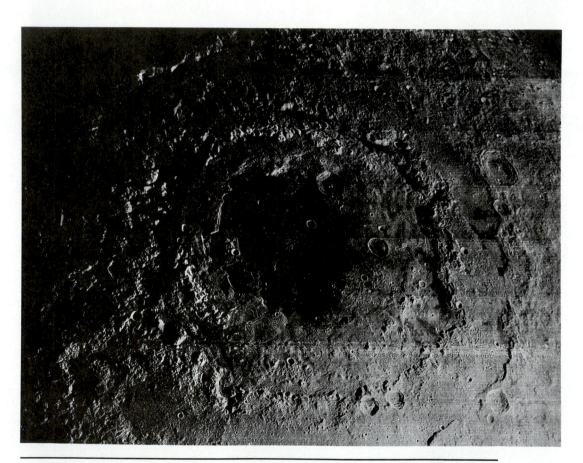

Figure 4.13

Impact craters larger than 300 km in diameter take on the appearance of giant bull's-eyes such as that displayed by Orientale Basin. It is the youngest multiring basin on the Moon. The diameter of the outer ring is 900 km. Two well-defined rings lie within the outer ring. The deposition of ejecta has tremendously modified the original nature of the surrounding terrain; radial textures are prominent. The location of the original rim of the impact excavation is controversial. Some argue that the outer ring is a fault scarp that marks the outer edge of a huge terrace; others suggest that the outer ring is itself the rim of the crater. A relatively small patch of much younger mare lava lies within the inner ring.

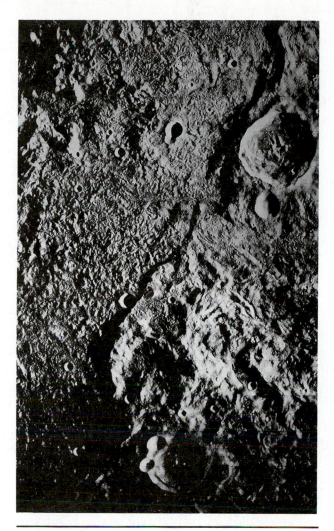 

Figure 4.14

The southeastern edge of the Orientale Basin is marked by a scarp that in places rises 4 km above the adjacent plain. The radially textured ejecta blanket lies to the right of the scarp; to the left is a terrain marked by numerous small domes. In areas where the scarp is low the radial texture of the ejecta crosses the scarp, suggesting that the outer ring is a terrace blanketed by ejecta rather than the crater's original rim. The lack of evidence of debris piled up along the scarp suggests that it formed after the outward movement of ejecta.

Figure 4.15

The two inner rings of Orientale Basin are much less distinct than the outer ring. Locally the middle ring is bounded by a low scarp, but in general it consists of an annulus of large mounds or peaks that are morphologically similar to those in smaller peak-ring basins. The innermost ring is composed of larger domes. The fractured plains that lie to the north may consist of impact-melt-rich rocks, while the dark, smooth plains in the upper part of the photo are mare basalts.

the near side of the Moon (Figure 4.1). It is almost completely flooded with younger lavas, but the concentric structure can still be identified. The innermost ring, exposed only as chains of islands, is 675 km in diameter. The second ring is exposed mainly as Montes Alpes and the rugged terrain near Archimedes. The third ring is the largest and most conspicuous; it corresponds to Montes Carpatus, Montes Apenninus, and Montes Caucasus. This ring marks the beginning of the ejecta blanket characterized by rugged linear ridges best developed at the southeast margin of the basin. Suggestions of still another ring, part of which bounds Sinus Medii on the south, can be seen at a considerable distance southeast of Montes Apenninus.

There are several ideas about how the rings are formed. Two of the more popular theories are illustrated in Figure 4.16. The **megaterrace model** for multiring basin formation postulates that the outer ring(s) of a multiring basin are the margins of huge terraces that slumped downward along steep faults. One of the inner rings marks the rim crest, and the innermost ring is formed by the same process that forms the peak rings in smaller craters. According to this model, the second or middle

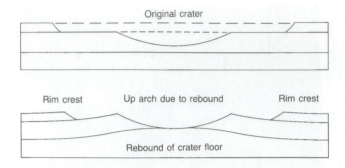

(A) **Megaterraces,** the areas between the outer rings of a basin such as Orientale, are interpreted as large fault-bounded terraces, formed during the modification stage of a basin's evolution. The inner rings are uplifts similar to those in peak-ring basins.

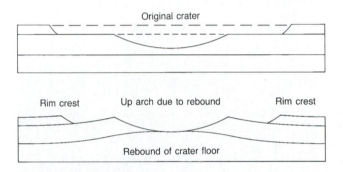

(B) **Nested-crater model** explains the multiring structure of large basins as structural discontinuities in the target. Rebound of the deep crater lifts these discontinuities to the surface, where they are expressed as rings. The crater of excavation is much deeper than that inferred for the megaterrace model. Terraces are considered to be minor modifications.

Figure 4.16

Two models for the formation of multiring basins are contrasted in these vertical cross sections of the outer part of the Moon.

ring of Orientale is a remnant of the rim of the crater of excavation. The **nested-crater model** proposes that the outer ring of the crater is the actual rim crest or limit of excavation for the crater. The inner rings of the basins are produced as shock waves and encounter changes in the layered structure of the planet's surface in much the same way as was discussed earlier for the production of small flat-floored craters. Readjustments after excavation allow the deep crater to rebound to the surface, creating the final subdued topography of the basins. Minor slumping of material off the rings may further modify the final configuration.

It is obvious that the nested-crater model calls for a much deeper original crater and for the excavation of larger amounts of material. It is possible, if this is the way multiring basins form, that chunks of the Moon's mantle have been excavated and strewn across the surface. However, no material similar to the mantle has been unequivocally identified from the samples brought back to Earth by the astronauts. In the megaterrace model, the formation of the outer ring is seen as a part of the modification stage of crater formation, whereas in the nested-crater model, it is principally a result of the excavation of the crater.

The ejecta deposits from such large impacts differ considerably from the hummocky rim deposits and rays of smaller craters. Surrounding the Moon's Imbrium Basin, (as well as many others) is a distinctive ridged and furrowed terrain (Figure 4.17). This sculpted pattern is believed to result from the "flowage" of the rapidly moving ejected debris and from pervasive secondary cratering. Even small craters have chains of secondary craters radiating away from them, produced when blocks of ejecta plow into the surface after being thrown from the primary crater. Apparently, the large basin-related events ejected vast amounts of material that eventually fell back to the planets' surfaces and created the furrowed texture. The surface upon which these fragments fell became churned up and intimately mixed with ejecta. Figure 4.18 shows the lunar crater Ptolemaeus, located in the middle of the near side of the Moon, just south of the equator. It is partially filled by gently undulating, nearly smooth plains. Numerous pools of similar material fill depressions on the crater rim. Patches of light-colored plains like this were originally thought to be volcanic deposits. Careful inspection of the photos reveals that the topography beneath this sheetlike deposit has been preserved. Filling the craters with lava or ash flows would probably not preserve the old rims. Inspection of the rock samples returned from the Moon has shown the fragmental and brecciated nature of this type of deposit, and it is now believed that these smooth plains are part of the ejecta from a distant basin.

Crater Degradation

The descriptions of craters in the previous section generally referred to their original shapes produced during the three stages of the impact process. Once a crater is formed, it may then be modified by a number of processes that gradually change the appearance of the crater until it may

Figure 4.17

The formation of a large multiring basin dramatically alters the nature of the landscape for many kilometers outside of the basin itself. The ridges and grooves seen in this photo lie 650 km from Imbrium Basin but were produced by massive secondary cratering related to the emplacement of Imbrium's ejecta. This terrain is called Imbrium sculpture, and similar terrains surround other large basins on the Moon and other planets. The smooth plains within the highly degraded craters may also have been produced by the emplacement of Imbrium ejecta. Herschel (40 km in diameter) is the fresh crater on the right side of the photo. Ptolemaeus is the large crater partly visible in the lower left of the photo. The dark plains to the north are the basaltic lavas of Sinus Medii.

Figure 4.18

The crater Ptolemaeus (150 km in diameter) is filled to about half of its original depth by a younger deposit with a gently undulating, nearly smooth surface. The outlines of several craters on the floor of the crater are preserved beneath this blanketing deposit. Originally, these light plains were thought to be volcanic deposits, but the study of samples returned by the Apollo astronauts has revealed that such plains are blankets of fragmental material ejected from large basins (in this case from Imbrium Basin).

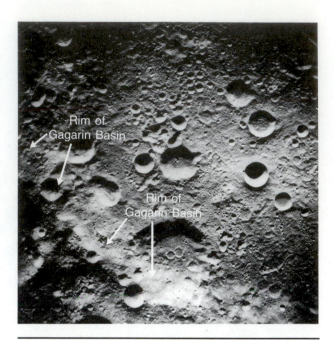

Figure 4.19

The effect of repeated meteorite impact on the lunar surface is to modify gradually or degrade the original shape of a crater. This fact is helpful in determining the relative ages of craters and of other lunar events. This view of the lunar highlands shows a portion of Gagarin Basin (265 km in diameter), the topographic rim of which is discernible along the left and lower margins of this photo. Innumerable impacts after the formation of Gagarin have removed all signs of its ejecta blanket and destroyed features on its floor. Both large and small craters have been important in this degradational process. Young craters, such as the bowl-shaped, sharp-rimmed crater in the center of the image, are surrounded by a halo of light-colored "fresh" ejecta and secondary craters.

be totally unrecognizable or obliterated. The changes that occur are collectively called **crater degradation.**

The effects of crater degradation can be seen in Figure 4.19, a view of the lunar crater Gagarin. The rim of this large crater (265 km in diameter) is barely discernible in this heavily cratered area. Gagarin is shallow, the walls lack terraces and are irregular, and no ejecta patterns can be detected past its rim. Battering by countless impact events has produced these changes. It is obvious from this photo that the appearance of a crater changes with time. Older craters are usually more degraded than younger ones. For example, compare the sharp-rimmed craters that occur on the floor and rim of Gagarin (and are therefore younger) with the irregular, subtle nature of Gagarin itself. Indeed, the rims of the smaller craters superposed on Gagarin show a range from crisp, sharp, unmodified rims to ragged, degraded rims altered by subsequent impact. This concept is important because changes in the morphology of craters make it possible to determine the relative age of specific craters and to

unravel the sequence of some events in a planet's history.

Crater degradation may proceed in several ways. (1) Later impacts may partly or completely destroy the older crater. (2) The crater may be covered with ejecta from a younger crater. (3) The crater may be partially or completely buried by lava flows or sedimentary deposits. (4) Geologic activity of the atmosphere and hydrosphere of the planet (if it has one) may erode the crater. (5) The crater may undergo tectonic modification by faulting or folding. (6) Some tectonic modification may be driven as the crust slowly adjusts to the unloading caused by the excavation of the crater. This gradual flow, called **isostatic adjustment**, is driven by gravity. The sequence of photos in Figure 4.20 shows some of these effects on lunar craters.

(A) **Slumping of crater walls** causes partial filling and the formation of terraces, and isostatic adjustment results in uplift of the crater floor.

(B) **Crater rims** are partly obliterated by subsequent impact.

(C) **The ejecta from the larger crater** at the top partly covers older craters.

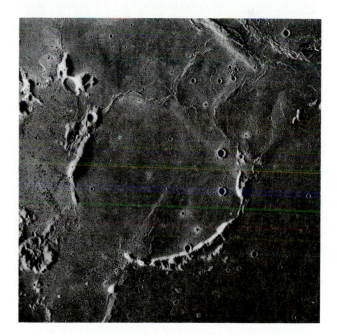

(D) **Some craters are partly covered** by lava flows.

Figure 4.20

Crater degradation and modification can occur by means of slumping, isostatic adjustment, subsequent impact, and burial by ejecta or lava. Examples of the modification of craters are shown in these photographs. Studies of crater degradation and modification are important in reconstructing the sequence of events in lunar history.

Developing a Lunar Time Scale

The deposits of ejecta from craters, together with lava flows and other volcanic deposits, form a complex sequence of overlapping strata that cover most of the lunar surface. The individual deposits can be recognized by their distinctive topographic characteristics and by their physical properties—such as color, brightness, and thermal and electrical properties determined from measurements made with optical and radio telescopes.

In 1962, geologists from the U.S. Geological Survey developed a geologic time scale for the Moon so that major geologic events could be arranged in their proper chronology. The basic principles used to interpret lunar history are essentially the same as those used to study the history of terrestrial events, the most important of which are the laws of superposition and cross-cutting relations. These principles of determining relative ages are, of course, as valid on the Moon or any other planetary body as they are on Earth. In addition, other methods of determining the relative ages of lunar features were developed based on the abundance of craters (crater frequencies) and crater degradation.

The development of a lunar geologic time scale was a major advancement in the study of planetary geology. For the first time, the sequence of events in the history of another planet was firmly established.

Determination of Relative Ages

Although it was recognized long ago that craters and other lunar surface features showed evidence of having been formed at different times, prior to the space program most observers studied features without relating them to their surroundings or their relative ages. Craters were classified according to their dimensions, and statistics were calculated on crater frequency, but little effort was made to establish a sequence of events in lunar history.

The first lunar chronology was developed in 1962 by Shoemaker and Hackman, who interpreted the sequence of events in the vicinity of the crater Copernicus (Figure 4.21). They recognized that ejecta from the craters Copernicus and Eratosthenes was superposed upon mare basalt whereas the basalt was superposed upon the craters of the lunar highland. From this kind of reasoning, they established a lunar geologic column. In many ways, what Shoemaker and Hackman did in providing a rationale for interpreting the Moon's history is comparable to what Smith, Lyell, and

their contemporaries did in establishing the geologic time scale on Earth during the early 1800s. The planet and nomenclature are different, but the logic remains the same.

To understand the basis for establishing the lunar time scale and the meaning of the major events in lunar history, let us carefully consider the ejecta from the major craters studied by Shoemaker and Hackman and their reasons for recognizing the sequence of events these ejecta represent. As you read the following discussion, study the physiographic map of the Moon (Figure 4.6) and the indicated illustrations, for it is only by recognizing the physical relationships between the features discussed that an appreciation for the relative time involved in their formation can be gained. This discussion serves as an illustration of the method that is used to establish the sequence of events that shaped various planetary surfaces.

Copernicus. An outstanding feature of the near side of the Moon is the crater Copernicus, with its spectacular system of bright rays that extend outward in all directions, in some cases for a distance of more than 300 km. Within the rays, predominantly near the crater, are elongate secondary craters. The rays and ejecta blanket surrounding Copernicus are superposed on essentially every feature in their path (Figure 4.21). During full Moon, when the rays are best observed, it is found that rays extend not only across Mare Procellarum and Mare Imbrium but also up the rim and across the floor of the crater Eratosthenes, 190 km to the northeast. From this superposition, it is clear that Copernicus and its associated system of ejecta and ray material are younger than the mare basalts and are younger than the rayless craters such as Eratosthenes. Other systems of rayed craters similar to Copernicus are centered on slightly smaller craters to the west, such as Kepler and Aristarchus. Their ejecta and ray material overlie everything they come in contact with. In the southern hemisphere, material from similar craters, such as Tycho, likewise overlaps all adjacent features. Although the rayed craters may vary in age, as a group they are younger than all other features on the Moon. The period of time during which rayed craters and their associated rim deposits were formed has been called the **Copernican Period**. Rocks and landforms of this age are the youngest on the Moon.

Eratosthenes. About half of the craters larger than 10 km in diameter that occur on the maria are rayed craters. Most other craters of this size range found on the maria are similar, but their

Figure 4.21

Relative ages of lunar features in the vicinity of the crater Copernicus are indicated by superposition of ejecta and mare basalts. Ejecta from Copernicus are superposed on all other features and are therefore the youngest materials. Ejecta from Eratosthenes (northeast of Copernicus) rest on the mare basalts and are younger than the mare material but older than Copernicus. Ejecta from the Imbrium Basin are covered partly by mare lavas and rest on the densely cratered highlands. Thus the basin and the lava flows are younger than the complex ejecta deposits of the highlands.

ejecta blankets are dark, and they lack rays. Consider, for example, the crater Eratosthenes just northeast of Copernicus (Figure 4.21). It has terraced walls, a roughly circular floor, a central peak, a hummocky rim, and a distinctive pattern of secondary craters similar to that of Copernicus. However, unlike Copernicus, it does not have a visible ray system, and the secondary craters are noticeably more subdued than those around Copernicus. Eratosthenes and similar craters, together with their ejecta, are superposed on the maria and are, therefore, younger than the mare lava on which they are formed. However, they are older than the rayed craters, as shown by the fact that the ray material, secondary craters, and ejecta deposits of rayed craters are superposed on the dark rim of Eratosthenes. The period of time when deposits of these dark-rimmed craters formed is referred to as the **Eratosthenian Period.** Deposits of Copernican and Eratosthenian craters are easily recognized on the maria, but it is difficult, in some cases, to discriminate between Eratosthenian and older crater deposits on the lunar highlands where crater densities are much higher. In these instances, relative ages may be established using the principles of crater degradation discussed earlier.

Imbrium Basin. In the northwest part of the near side of the Moon is an enormous multiring basin, now largely filled with lava flows, called Imbrium Basin. We have seen in previous discussions that Imbrium Basin is the largest multiring basin on the Moon and, like other craters, it was formed by impact. Imbrium Basin is surrounded by ejecta deposits similar to those formed by smaller craters—the best exposures being the Montes Apenninus, which extend outward from the southwest rim. The ejecta deposits are called by various names, depending on their topography and location. The most important of these is the Fra Mauro Formation, which can be traced as far as 400 km from the mountains surrounding the basin.

The ejecta from Imbrium Basin are partly covered with lava, as is most of the interior of the basin (Mare Imbrium). The time during which the lava and the ejecta deposits were formed is called the **Imbrian Period**. It is apparent that some impacts occurred after Imbrium Basin and its ejecta were formed, but before the extrusion of the lava flows. Some of these deposits around larger craters such as Archimedes, Plato, and Sinus Iridum are very extensive. It is calculated that about four times as many craters larger than 10 km in diameter were formed on the Moon during the Imbrian Period as during all the time since the last lava flows in Mare Imbrium were extruded.

The Ancient Terrae. The Imbrium ejecta partly overlie a complex sequence of craters and ejecta found in the lunar highlands. Ejecta from the upper part of this densely cratered terrain formed during what is termed the **Nectarian Period**, named after a large subdued basin on the eastern side of the Moon. The ejecta deposit from the Nectaris Basin can almost be traced to the far side of the Moon and serves as an important marker for this part of the Moon. The time before Nectaris Basin formed is called **Pre-Nectarian;** rocks of this age constitute the oldest materials on the Moon's surface. The geologic structure of these deposits is very complex. Large craters are closely spaced and modified by impact. Apparently the surface of the terrae was churned by repeated formation of large craters early in lunar history. Humorum and Serenitatis Basins and their ejecta deposits may all be of Pre-Nectarian age. Of course, the lavas that fill these depressions are much younger, mostly formed during the Imbrian Period. In many places in the highlands, Nectarian and Pre-Nectarian materials are not distinguished from one another and are called **Pre-Imbrian** deposits.

In light of these observations, geologists have mapped most of the Moon and outlined some of the major events in lunar history. These are summarized in the time chart (Figure 4.22).

Crater Frequencies

As we have seen in previous sections, cratering is the major geologic process that has operated on the Moon; the form and number of craters provide a wealth of information concerning the history of the planet and the relative ages of surface features. Cratering can be used to determine the relative age of a terrain unit or reference area because of the simple fact that a greater number of craters will have developed on older terrains than on younger ones. This relationship holds true regardless of whether the rate of cratering is constant, steadily decreasing, or erratic. Comparison of the number of craters in different regions enables estimates to be made of the relative ages of the cratered surfaces. A simple example illustrates this point. Assume that it has been snowing for several days and that the snow is 1 m deep on undisturbed lawns throughout the neighborhood. If the snow is 1 m deep on the sidewalks of a house, you can conclude that nobody has shoveled the snow since the storm began. If the snow is about half a meter deep in front of another house, it would be obvious that the walk had been shoveled, possibly midway through the storm. If another walk has only a few centimeters of snow, it would be obvious that it had

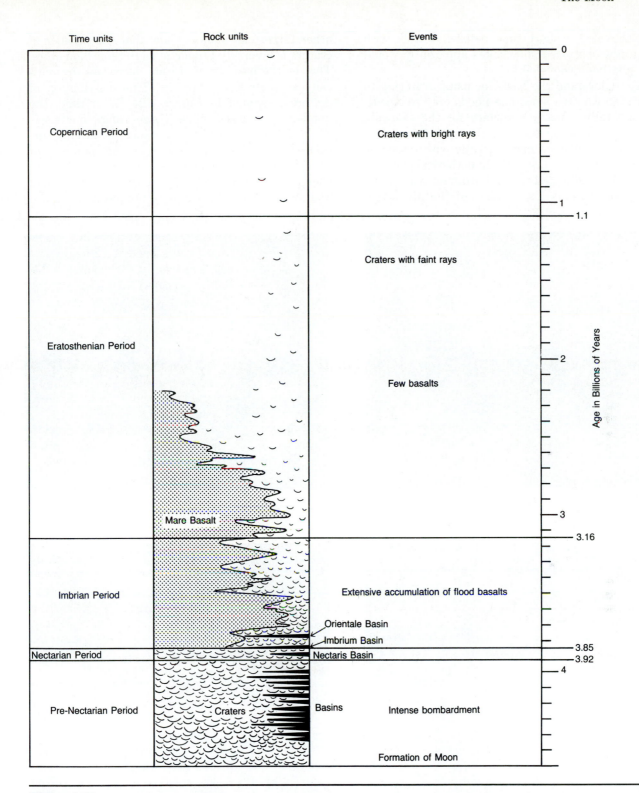

Figure 4.22
The geologic time scale of the Moon is shown in this diagram, along with some of the events that occurred during these periods.

been shoveled a short time before. A walk with only a few scattered snowflakes apparently would have just been shoveled.

To understand the basis for using cratering to determine relative ages, one needs only to substitute mentally a planet's surface for the sidewalk and cratering projectiles for the snowflakes. The idea can be explained more directly with reference to Figures 4.23 and 4.24. On a hypothetical planet, a new surface subjected to bombardment will have a variety of crater sizes. Studies of the Moon and other terrestrial planets show that the number of craters is inversely proportional to their diameters. That is, the number of craters decreases dramatically as their size increases. This distribution of craters is shown in Figure 4.24. With time, the number of craters in each size range increases. There is a limit, however, because eventually further impact does not increase the number of craters—it simply makes new craters out of old ones. This "steady state" occurs first for small craters because they are more numerous.

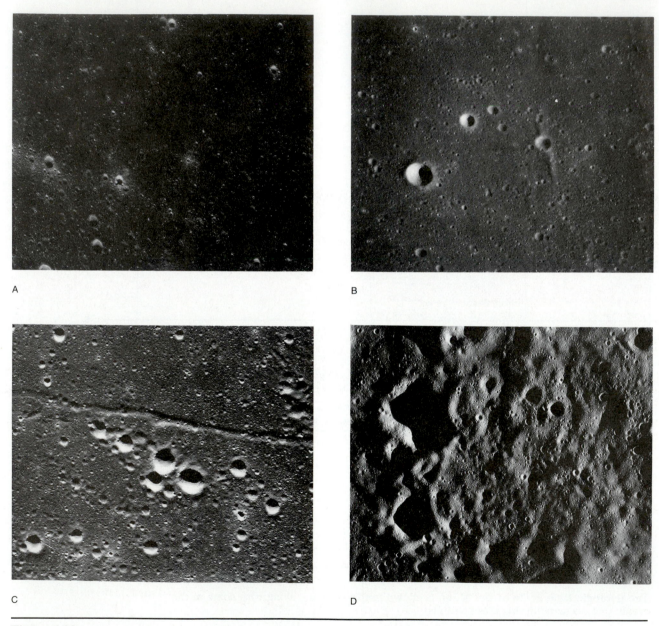

A

B

C

D

Figure 4.23

Crater frequency can be used to determine relative ages of lunar surfaces because an older surface has more craters than a younger one does. In the photographs. (A) has the fewest craters and is therefore the youngest surface. (B) and (C) are progressively older, and (D) is the oldest.

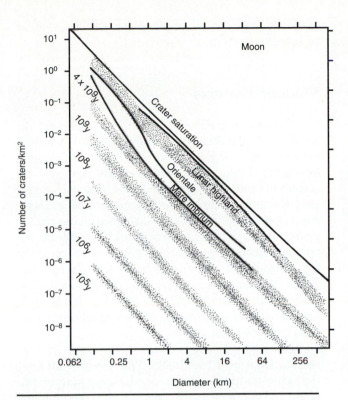

Figure 4.24
The number of craters accumulated on a planetary surface is a good indication of its age. If these crater densities can be related to rock units that have radiometric ages, reliable absolute ages can be assigned to geologic units exposed at the surface—a concept illustrated in this diagram, where the number of craters in a specific size range is plotted against the crater diameter. Lines, or isochrons (equal-time lines), show the number of craters accumulated on surfaces of several different ages. Note that the lunar highlands plot is above and to the right of the isochron for the lunar maria because of the marked difference in the number of superposed craters on the two surfaces. The curve labeled saturation indicates the point at which the total number of craters cannot increase. This steady state is reached when the number of craters formed by these impacts equals the number of craters obliterated.

An example of determining relative age by crater populations is shown in Figure 4.24. A curve lying above or to the right of another describes an older surface. Changes in the slope of a line are produced as a steady-state condition occurs in the smaller size classes, but "new" larger craters continue to be recorded. Crater counts on the rims of Orientale, Imbrium, and Humorum basins show that although Humorum is approaching steady-state values at small crater diameters, it has a greater total frequency of craters and is thus the oldest of the three basins. This can be easily seen by comparing the number of craters of a given size, for example 10 km, on each rim. Humorum has less than 100, Imbrium around 25, and Orientale only 10.

Radiometric Dates for Lunar Events

Some of the most critical information about the geology of the Moon was obtained from isotopic age determinations of lunar rocks. The passage of time is recorded in rocks by the accumulation of the products of radioactive decay. Most rocks contain several elements (usually potassium, uranium, or rubidium) that decay to other elements, called daughter products. If the rate of decay is known, careful measurements of the amount of daughter elements in a rock reveal its absolute (as opposed to relative) age—the time at which it started to accumulate decay products. Many expected that the lunar surface was old, but the fresh lava of the maria and the bright rayed craters appeared as if they formed as recently as the Ice Age on Earth, which ended only 20,000 years ago. The first lunar rocks to be dated were basaltic lavas from one of the maria; to the amazement of many, these gave an age of 3.65 billion years—older than almost all rocks found on Earth. Additional radiometric ages of mare lavas from other spacecraft landing sites indicate that the extrusions of lava that formed the maria began about 4.0 billion years ago and continued episodically for at least 800 million years. The lunar highlands, of course, are older, and samples collected by Apollo 17 astronauts show that some rocks crystallized over 4.5 billion years ago. This is nearly the age of the oldest meteorites. The lunar crust developed very shortly after the accretion of the Moon. The isotopic ages of most highlands rocks cluster around 3.9 to 4.0 billion years ago and probably reflect "resetting" of their radiometric clocks by intense bombardment that tapered off about this time.

Another important age was determined from ejecta thrown out of the Imbrium Basin. Fra Mauro breccias collected by the Apollo 14 astronauts were emplaced about 3.9 billion years ago. The age of Copernicus was determined from ray material collected from the Apollo 12 landing site. The age of this event, one of the most recent in lunar history, is 0.8 to 0.9 billion years. Other important ages determined from samples are shown in Figures 4.24 and 4.25.

By integrating these and other radiometric dates into the relative geologic time scale of the Moon determined from superposition and crater counts, it is possible to construct an absolute time scale for the Moon plus a graph showing the rate of

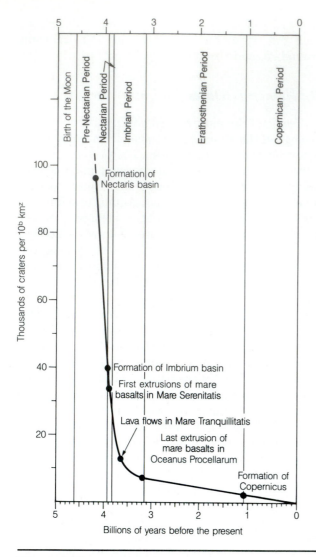

Figure 4.25

The variation in the number of craters formed on the Moon's surface during different periods of time shows that the rate of cratering on the Moon has not been constant but has decreased dramatically from pre-Imbrian times to the present. The most dramatic decline occurred 4 billion years ago.

cratering (Figure 4.25). This curve shows that early in lunar history the rate of cratering was hundreds or even thousands of times greater than today. Moreover, the rate of decline in impact was very rapid until about 3 billion years ago when it reached a low level. Since then, the rate of cratering has been relatively constant.

It is believed that other planets and satellites experienced broadly similar variations in cratering and that crater frequency may be a crude way to correlate planetary events. However, this notion is complicated by the large uncertainties in estimates of the rates of impact on individual bodies in the solar system. Impact rates may have varied dra-

matically from planet to planet or satellite system to satellite system.

Volcanic Features

Volcanic activity is especially important in studies of planetary evolution because volcanic products are a type of window into the planet's interior that provides valuable insight into how the planet operates. Many geologists long suspected that the lunar maria were composed of volcanic rocks, but it was not until orbiting satellites photographed individual flow fronts and the Apollo astronauts brought back rock samples that we knew for certain that the maria were formed by vast floods of basaltic lava and that the Moon had a spectacular volcanic history. The highlands, too, have some volcanic rocks, but volcanic landforms are not as obvious there.

The general nature of the lunar maria can be seen in Figures 4.1 and 4.26. The floods of lava fill the large multiring basins on the near side of the Moon and commonly overflow the rims, spilling out over the surrounding areas. From distant views it may appear that the maria represent one huge flood with the basins all filled to the same level, like the oceans on Earth. However, upon closer exami-

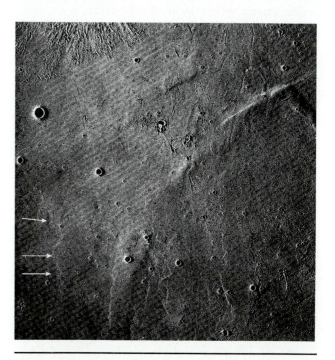

Figure 4.26

The lobate fronts of lava flows in southern Mare Imbrium are similar to the flow margins of flood basalts on Earth. These vast sheets were probably erupted from long fissures.

nation we find that the mare flows are not at the same level and were extruded over a relatively long period of time. The radiometric ages of lunar samples show that the lavas filled the large basins several hundred million years after the basins were formed. This is of considerable significance in considering the origin of the lava. Obviously the lavas cannot be melts generated by impact. Instead, they were produced by melting deep within the Moon. They were formed, however, during a specific interval of time of lunar history extending from about 4.0 to at least 3.2 billion years ago and represent a thermal "event" that lasted roughly a billion years. Although the youngest samples returned from the Moon yield radiometric ages of 3.2 billion years, crater frequencies on some unsampled lava flows in northern Oceanus Procellarum indicate that some flows may be much younger, around 2.5 billion years old. If this age is correct, the length of the episode of volcanism may be much longer than indicated by the samples we now have.

When we consider the relatively small volume of lava and the time span during which eruptions occurred, plus the fact that a given eruption occurred at a very high rate, the total period of time of lunar volcanic activity must have been interrupted by long dormant periods.

Although the maria appear to have flat, featureless surfaces, photos taken at low sun angle (Figures 4.27 and 4.28) show that there are numerous tiny craters dotting their surfaces. In places the mare surface appears to undulate in broad swells and depressions only a few meters high. The basalt in the interior of the mare basins completely covers the old cratered surfaces and therefore must be in excess of 1500 to 2000 meters thick. It is interesting to note, however, that some rims of old craters project above the mare material near the margins of the basins and show that here the lava is relatively thin. These "ghost" craters show that a considerable part of the margins of the basins are covered with only 200 to 400 m of basalt (Figure 4.20).

With the extensive lava fields of the maria, one might expect to find spectacular stratovolcanoes like Earth's Vesuvius, Shasta, and Fuji, together with numerous cinder cones and other volcanic landforms. If such features were present they would be easily discernible from the excellent photography we have because we can clearly see details of tiny impact craters across the surface of the maria; however, we see only a few volcanic shields and little evidence of the zones of fissures from which some of the lava was erupted. One of the reasons for this is that lavas in the maria were far more fluid than any lavas found on Earth.

Figure 4.27

The lunar maria are covered with small craters, as is shown by the long shadows cast by the setting Sun. Small craters are abundant and lava flow fronts appear as ridges.

Instead of flowing a short distance and developing well-defined margins, they flowed great distances and ponded in depressions, almost like water accumulating in lakes. There are few visible margins to the individual flows or other characteristic flow features seen on cooled lavas on Earth. The high fluidity of lunar lavas was confirmed when samples of the maria were melted in the laboratory. Their viscosity was similar to that of engine oil.

In the southern part of Mare Imbrium, however, lavas were viscous enough to develop well-defined flow margins. Excellent details of many individual flow units are apparent on photographs from the Apollo missions, which show that the flows are similar in many respects to the fluid basaltic flows on Earth. The flows shown in Figure 4.28 are typical. You will note that many flow units were extruded from a single source region and flowed down the regional slope. Each flow is characterized by lobate margins similar to basaltic flows on Earth. The Columbia River Plateau in the northwestern United States and the Deccan Plateau of India consist, in part, of a similar series of lava floods that were extruded quietly from a network of fissures. The individual lava flows on the Moon are much longer than those found on

Figure 4.28

The spectacular lengths of some lava flows in southern Mare Imbrium demonstrate that the lavas were highly fluid and were erupted at high rates. The flows, which terminate in the upper right corner of this photograph, are over 400 km long. The arcuate mare ridges that cross this region must not have existed when the lavas were flowing; otherwise the lavas would have ponded behind the ridges. A portion of a lava channel is visible near the lower right side of the photo. The "islands" formed by peaks on the floor of Imbrium Basin attest to the relative thinness of the lava accumulations in this region of the Moon. The sprays of irregular craters were formed by ejecta from the crater Copernicus, which lies 460 km to the south.

Figure 4.29

Sinuous rilles east of the Aristarchus plateau in Oceanus Procellarum are thought to be enlarged lava flow channels. Several of the channels head in low volcanic cones or shields which surround volcanic craters. In some cases it appears that the hot lava flowing through these channels has eroded the underlying surface, allowing the channel to incise itself into the surrounding terrain. These central vents contrast with the fissures, which are thought to have fed many of the mare lavas. The branching system of channels near the center of the photo are distributaries, which mark the end of a lava channel. Lunar rilles are distinct from terrestrial river channels but no close volcanic analog has been found on Earth. The rugged mountain chains in the upper part of the photo are old highland massifs, which protrude through the lava cover.

Earth. Some traveled as much as 600 km over slopes with gradients of probably less than 1 in 100. The distance lava flows depends on slope, viscosity and, to a considerable degree, on the rate of eruption. We can conclude, therefore, that the eruption rate for the mare lavas on the Moon was very high—much higher than any known on Earth. The flows were apparently fed from huge fissure systems. As the lavas were very fluid, they may have ponded over their vents and drowned the fissures from which they issued.

Other features associated with the maria and believed to be of volcanic origin are the meandering channels known as **sinuous rilles** (Figure 4.29). Most occur along the shallow edges of maria and on the flat floors of larger mare-filled craters. Many sinuous rilles begin at a volcanic crater and, when traced downslope, become progressively smaller until they disappear. Some have V-shaped profiles, but flat floors complicated by inner channels, craters, and irregular hummocks are more typical. Apollo 15 astronauts visited Hadley Rille, a large sinuous rille on the edge of the Imbrium Basin. Layered basalts crop out along its rim, and its V-shaped profile is produced as loose material falls from its walls. It is typical of many rilles in size: 135 km long, 1.2 km wide, and 370 m deep. At first glance, some sinuous rilles appear to be very similar to terrestrial stream valleys, but they lack many features characteristic of stream-cut channels, such as tributary systems, increase in channel size downslope, and such associated depositional features as deltas, flood plains, and alluvial fans. In all probability, the rilles originated in a variety of ways. Some are similar to collapsed **lava tubes**. Lava tubes are formed because a thin crust typically develops over the liquid interior of a lava flow. Pipelike zones of movement develop within the interior of the flow and may drain to form a long, hollow, cylindrical tube in the middle of the flow. The roofs of terrestrial lava tubes sometimes collapse, forming steep-sided troughs or chains of circular pits. Meandering **lava channels** with steep, high walls also form when the outer edges of a lava flow cool and solidify while the interior of the flow moves on downslope. This produces a lava channel or lava "gutter" (Figure 4.30). Sinuous lunar rilles are probably produced in similar fashion. Some rilles may be fault-troughs modified by flowing lavas. Many lunar rilles are, therefore, similar to common volcanic features on Earth.

Some smooth low domes occur in the lunar maria and are considered by most geologists to be the lunar equivalent of low-shield volcanoes (Figure 4.31). The small volcanic features are at first difficult to distinguish from impact craters because they sometimes have nearly circular craters similar in size to many impact craters. However, close examination indicates that they can be distinguished from impact craters: (1) They are positive geomorphic features rising above the surrounding surface. Impact craters are depressions. (2) Their slopes are commonly breached, presumably by lava flows. (3) As can be seen in Figure 4.31, some occur in groups to form a broad mound of domes and flows, and others occur as single isolated features. Foremost among the volcanic complexes are those in Oceanus Procellarum. The Rumker Hills (Figure 4.32) are an array of domes and cones that are similar in many respects to volcanic features on Earth. In addition, rilles ema-

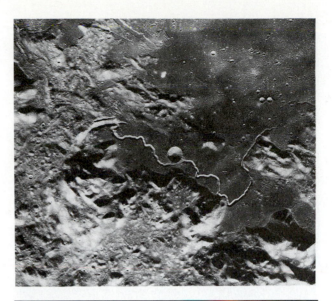

Figure 4.30

A sinuous rille, called Hadley Rille, found at the base of the Apennine Mountains in the eastern part of Mare Imbrium, is interpreted to be a lava channel. The channel was visited by the Apollo 15 astronauts. The rille starts from a small volcanic crater and follows a sinuous course downhill. The margins of the lava flow(s) from this crater are not apparent in this photo.

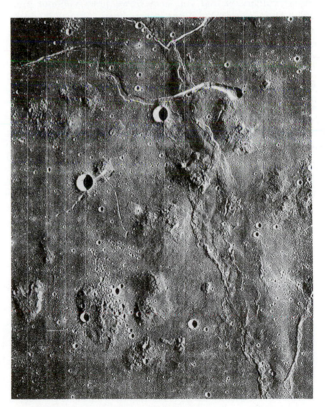

Figure 4.31

Low domes in this oblique view of Oceanus Procellarum are thought to be low-shield volcanoes. The Marius Hills are one of several such volcanic complexes in Oceanus Procellarum.

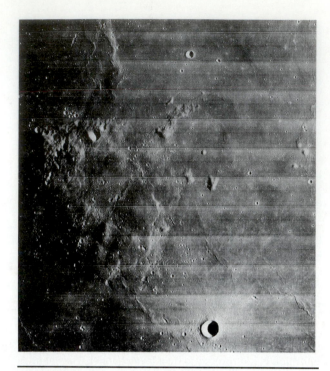

Figure 4.32

The Rumker Hills are believed to be another volcanic complex in Oceanus Procellarum. The low domes are low-shield volcanoes that fed lava flows both younger and older than the surrounding mare lavas. The volcanoes appear to have formed on a high patch of pre-mare ejecta from the Imbrium Basin.

central vents and low shields, has also been identified on Earth but is not generally well known to the layman. For example, in the Snake River Plain of southern Idaho, the apparently flat surface of these **basaltic plains** commonly consists of a series of coalescing lava cones or low-shield volcanoes (Figure 4.33). This type of volcanism occurs when magma supplies are small and are extruded from central vents, in contrast to the large volumes extruded from long fractures typical of flood basalt eruptions. Both types of volcanism (flood and plains) are found in the northwestern United States and produce edifices similar to those seen on the Moon.

The Moon has fewer types of volcanic landforms than does Earth. One reason may be that the physical environment on the Moon inhibits the development of many volcanic features found on Earth. If a typical cinder-cone eruption occurred on the Moon, the lack of atmospheric drag on the ejected particles and the lower gravitational attraction on the Moon would allow the ejected particles to fly far from the center of eruption. As a result, the ejected material would tend to form a thin, low, widespread blanket of ash rather than a steep cinder cone. Some distinctive patches of dark material surrounding craters with low rims (Figure 4.34) are possibly the lunar equivalents of cinder cones on Earth. Another reason for the lack of a diversity of lunar volcanic features lies in the small variability of volcanic rock compositions found there. No rhyolites or even andesites, which usually form stratocones on Earth, have yet been identified. It may be that water, extremely rare inside the Moon, plays an important role in the

nate from some of these low domes, further suggesting their volcanic origin. These domes may be a type of shield volcano and may be the source of the mare fill in some areas, particularly Oceanus Procellarum. This type of volcanism, associated with

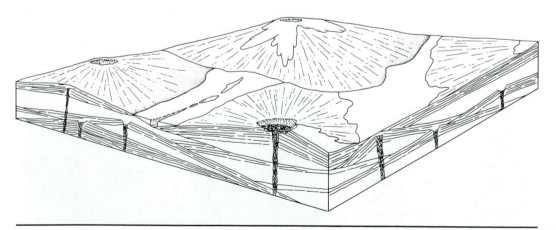

Figure 4.33

Accumulations of lavas erupted from low-shield volcanoes and short fissures typify basaltic plains like those of the Snake River Plain in southern Idaho. Each shield is less than 10 to 30 km across—much smaller than the large shields of the Hawaiian Islands. Some of the volcanic complexes of Oceanus Procellarum are probably composed of plains-style volcanism.

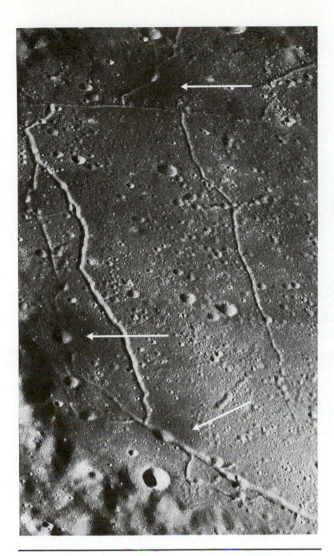

Figure 4.34

Dark-halo craters in this photograph are considered by some scientists as evidence of landforms produced by the accumulation of basaltic pyroclastic material—the lunar equivalents of cinder cones. These craters are centered on a fracture system that cuts the floor of the crater Alphonsus. The craters cap broad cones, and one has an irregular outline—features atypical of impact craters.

generation of silicic magmas that produce most of the high volcanoes on Earth.

In summary, the basalts of the lunar maria represent a major thermal period in lunar history that extended over a period of about one billion years, beginning about 4.0 billion years ago. During this period, heat sources from within the Moon partially melted parts of the interior, generating basaltic magma that migrated upward to the surface and was extruded through the fractured crust. There are a number of distinctly volcanic features associated with the mare flood basalts, but there are no spectacular stratocones or huge shields topped with calderas such as we find on Earth and

Mars. The extrusions occurred through low-shield volcanoes or as quiet fissure eruptions and appear to have terminated several billion years ago. As we will see, this type of thermal event occurred on other planetary bodies as well.

Tectonic Features

Tectonic landforms are created by the deformation of a planet's outer layers and surface. They arise from forces inside the planet. Young tectonic features show that a planet still possesses enough energy to be dynamic. If no young tectonic features are found on a planet, it is a sure indication that its internal heat energy has been dissipated. In general, there are two important types of tectonic forces—extension and compression. **Extension** causes stretching and fracturing of the shallow lithosphere. The lithosphere may fracture to form high-angle faults along which movement of blocks occurs. These extensional faults are usually long and straight (Figure 4.35). Rifts and grabens (fault-bounded valleys) are common expressions of extension. **Compression** is caused by shortening. If it is under compression, the lithosphere of a planet may buckle and fold, or it may break into thin sheets bounded by low-angle faults. These low-angle faults are descriptively named thrust faults; they usually form sinuous, overlapping ridges that are quite different from grabens (Figure 4.36).

There has been almost no significant tectonic activity on the Moon during the last 2.5 billion years, or since the extrusion of the flood basalts. We know this because the millions of craters that cover the lunar surface provide an excellent reference system for even the most subtle structural deformation of the crust. Just like lines on graph paper, the circular craters that cover the lunar surface would be deformed by any significant crustal movement, compression, or extension and would record even the slightest disturbances. The network of young craters, however, is essentially undeformed and the lunar crust thus appears to have been nearly fixed throughout time. There is no evidence of intense folding or thrust faulting and no indication of major rifts. The major features that may be attributed to crustal deformation are **linear rilles** and **wrinkle ridges**.

Linear Rilles

Many linear rilles are sharp, linear depressions, which generally take the form of flat-floored, steep-walled troughs ranging up to several kilometers in

Figure 4.35

Linear scarps seen in this picture are interpreted as expressions of faulting of the Moon's surface layers. These fault-bounded valleys, called linear rilles or grabens, are generally produced by tectonic extension, which may accompany the subsidence of a basin floor or global expansion.

width and hundreds of kilometers in length (Figures 4.35). The valley floors lie a few hundred meters below the surrounding region. The valley walls are straight or arcuate and stand at the same elevation as though they had been pulled apart while the floor subsided as a graben. Linear rilles are parallel or arranged in an echelon pattern. Some intersect and others form zigzag patterns.

There is little doubt that linear rilles are grabens that resulted from normal faulting. However, the origins are undoubtedly quite different from those that produce the great rift systems on Earth, which are related to the pulling apart of relatively thin lithospheric plates by movement over a plastic asthenosphere. Some rilles on the Moon may be the result of an early expansion of the entire sphere as it slowly heated. Others were probably caused by relatively local stress systems set up around lava-filled impact basins that subsided due to the added weight of the basalt. Still other linear rilles formed as the result of the massive impacts that created the large lunar basins. From careful studies of the terrain crossed by these grabens, it appears that few formed after about 3.5 billion years ago.

Wrinkle Ridges

Some of the most conspicuous structural features on the mare surface are long narrow ridges, sometimes called "wrinkle" or mare ridges (Figure 4.36). Typically, they have sinuous outlines and extend discontinuously for great distances across the maria. In some places, they transect highland surfaces as well. Individual segments may be several kilometers wide, a few hundred meters high, and hundreds of kilometers long. Systems of mare ridges are commonly parallel to the margins of the major mare basins, although some are parallel to structural trends in the highlands. In contrast to lunar grabens, these ridges disrupt even the youngest mare surfaces and must have continued to form until after the maria were completely formed less than 3.0 billion years ago.

Several explanations have been proposed for the origin of the ridges, and, as is the case for rilles, wrinkle ridges may originate in several different ways. Originally, some geologists thought these ridges marked the location of the fissures through which the mare lavas were erupted or that they were produced by intrusions of magma beneath a solidified crust, buckling and folding the overlying flows. These explanations are presently discounted for several reasons. Some circular ridges apparently formed as lavas settled or compacted over pre-existing crater rims. Other wrinkle ridges cross geologic units of different age and origin, for example, the highlands-maria boundaries. Where they do cross into adjacent highlands terrain, they

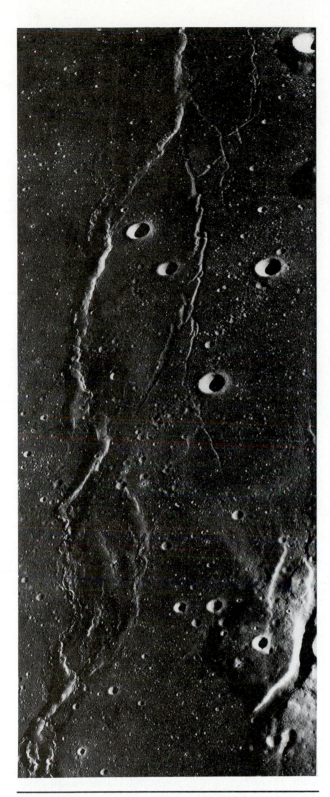

appear as simple scarps and are not associated with lava flows. Others cut craters like simple faults.

The evidence suggests that most mare ridges were produced by compression related to simple vertical adjustments of the lunar crust. It appears that these movements occurred along faults around and in the mare basins as the crust gradually adjusted to the load of the accumulating lava. A compressive stress system may have been established in the central part of the mare as the downwarping occurred, buckling the crust and producing the ridge systems. An instructive example, Mare Serenitatis, experienced earlier rille or graben formation followed by more recent mare ridge development (Figure 4.37). Some lunar scientists

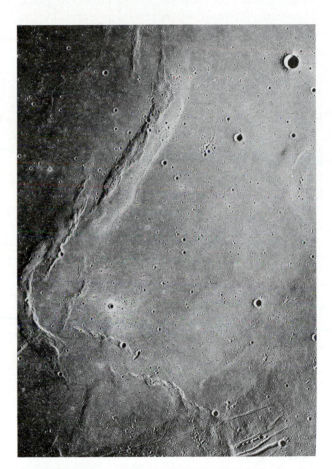

Figure 4.36

Wrinkle or mare ridges are visible in this view of one of the lunar maria. These ridges are thought to be the result of thrust faulting produced by compressional forces exerted at the Moon's surface. The compression may be the result of global cooling and contraction superimposed on local adjustments to the load imposed by the mare basalts. Fault-bounded linear rilles contrast sharply with these ridges.

Figure 4.37

Two distinctive types of mare surfaces are shown in this view of southeastern Mare Serenitatis. Older maria occur in the lower right part of the photo and are elevated relative to the younger maria within the basin interior. The older maria are cut by numerous linear rilles that are parallel to the basin margin and are also deformed by a number of broad mare ridges. The younger maria are not cut by linear rilles but mare ridges are prominent. These observations, also made elsewhere on the Moon, suggest that lunar tectonism changed from broadly extensional to compressional about 3.5 billion years ago—the age of the break between the two types of mare surfaces shown here.

think that the change in the style of tectonic deformation from extensional (graben formation) to compressive (ridge formation) stresses may mark a shift in global stress patterns, in turn accentuating the production of one or the other of these features. A change from extension to compression could occur when the Moon started to cool and contract after an earlier slight expansion caused by internal heating. This shift may have occurred when the last linear rilles formed, about 3.5 billion years ago.

In short, it appears that the preserved record of lunar tectonism is the result of small-scale vertical adjustments of the lithosphere to produce mare ridges and linear rilles. Many of these features can be related to the gradual subsidence of lava-filled impact basins superimposed on the effects of global expansion, followed by contraction early in the Moon's history. The Moon does not have an active tectonic system like that of Earth and represents a primitive body that has experienced relatively little tectonic activity.

Lunar Rocks

To some people, the rocks returned from the Moon were a disappointment. Few exotic minerals were found. Instead, Moon rocks are like the common rocks found on Earth. Yet these lunar rocks hold the key to understanding the thermal and chemical evolution of the Moon. Moreover, these rocks chronicle events in the early days of planetary evolution—lunar rocks are all much older than most rocks found on Earth.

The samples brought back from the Moon were obtained from a variety of geologic settings (Figure 4.38) in an effort to provide maximum information. They have been studied by hundreds of scientists from many countries and are still subjected to thorough and sophisticated analyses of their physical and chemical properties. This lunar material consists of several types of igneous rocks as well as rocks created by meteoritic bombardment.

Anorthosite

Anorthosite is a coarse-grained igneous rock composed almost entirely of the mineral plagioclase (Figure 4.39). This rock type was collected from the lunar highlands and is an important constituent of the soils and breccia described below. Along with other plagioclase-rich rocks, it forms a group of rocks that are the most abundant and oldest (greater than 4.4 billion years old) on the lunar surface. Part of the significance of anorthosite is that it records a thermal event very early in the Moon's history, long before the development of

mare basalts and even before the period of intense bombardment that formed the craters of the lunar highland.

Studies of the lunar samples indicate that shortly after accretion of the Moon, its outer layers melted to form a global "ocean" of molten rock, perhaps as deep as 100 to 1000 km. Crystallization within this "magma ocean" produced plagioclase feldspar crystals, which floated to the surface because they were lighter than the melt, much as ice cubes rise in a glass of water. Accumulation of these crystals produced masses of floating anorthosite "rockbergs," which coalesced to form the lunar crust. Most of these rocks have been significantly altered by subsequent metamorphism and melting caused by impact, and it is doubtful that much original, primordial crustal material remains; that which does remain appears only as small grains in other rocks.

One of the most exciting finds of recent years was the discovery that small pieces of lunar anorthosites occur in meteoritic material found on Earth, specifically in Antarctica. The distinctive characteristics of lunar anorthosite make their identification certain. Apparently, this material was delivered to Earth by an energetic impact event. Careful searches of meteorite collections may yield samples of lunar rocks that come from areas not visited by any spacecraft.

Lunar Basalts

Most of the igneous rocks collected from the Moon's maria are very similar to terrestrial basalt, the most common rock in Earth's crust. Like anorthosite, basalts were once totally molten, as is indicated by their vesicles (gas bubbles), interlocking crystalline textures, and compositions (Figure 4.40). Basalts differ from anorthosites in their mineral constituents. The principal minerals found in lunar basalt are plagioclase, pyroxene, ilmenite, and olivine, all of which are found in terrestrial basalt. Only minor amounts of a few minerals previously unknown on Earth were found. The most significant difference between lunar and terrestrial basalts is that the lunar basalts contain greater concentrations of refractory elements (titanium, zirconium, and chromium). Lunar basalts are also devoid of water and have much lower amounts of relatively volatile elements such as sodium and potassium than do terrestrial basalts. These chemical characteristics are significant in that they suggest that the material forming the Moon is poor in volatile elements—perhaps because it condensed at higher temperatures than did the material that formed Earth. This could explain both the concen-

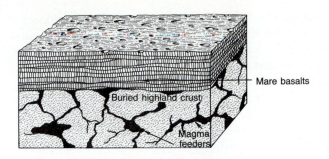

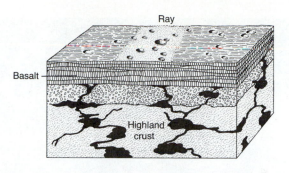

(A) **Apollo 11 landed on the flood basalt of Mare Tranquillitatis.** Here, the maria consist of numerous thin layers of basalt flows with an aggregate thickness of several kilometers that were extruded about 3.7 billion years ago. These mare basalts rest on older rocks of the cratered highlands.

(B) **Apollo 12 landed on a ray of Copernicus,** approximately 400 km south of the crater. The ray material rests on a sequence of basalt, which forms Oceanus Procellarum. Below the basalts is a layer of ejecta that was formed during the early stage of abundant meteorite impact and development of the lunar highlands.

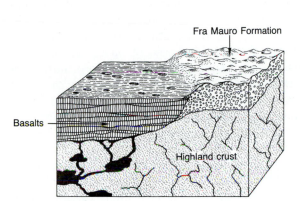

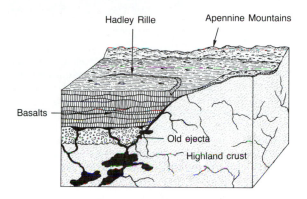

(C) **Apollo 14 landed on the Fra Mauro Formation,** which is composed of material ejected by the Imbrium impact event. Basalts from Oceanus Procellarum lap up against the ejecta, proving that the mare lavas are younger than the highlands. The rocks of the Fra Mauro Formation were thrown out of Imbrium Basin about 3.9 billion years ago.

(D) **Apollo 15 landed near Hadley Rille** at the base of the Montes Apenninus, which form the rim of Imbrium Basin. Hadley is a sinuous rille of volcanic origin. The mare basalts lap up against the ancient rocks of the lunar highlands, which have been dated as more than 4 billion years old.

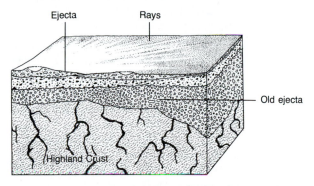

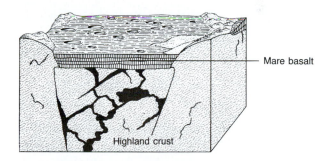

(E) **Apollo 16 landed in the highlands of the Descartes region.** The surface material is composed of debris churned up by North Ray and South Ray craters and overlies layers of breccia, which were formed by more ancient meteorite impacts.

(F) **Apollo 17 landed in the Taurus-Littrow Valley,** which cuts the rim of Serenitatis Basin. The high massifs are composed of ejecta thrown from the basin 3.9 to 4 billion years ago. The valley is floored by younger mare lava flows which were erupted 3.5 to 3.7 billion years ago. A prominent wrinkle ridge transects both the mare and the highland materials.

Figure 4.38

The structure of the Apollo landing sites are shown in these schematic block diagrams.

Figure 4.39

Lunar anorthosite, seen through a microscope, consists of a meshwork of plagioclase crystals. These are characteristically lath-shaped, with some pyroxene occupying the spaces between the plagioclase crystals. Olivine occurs in amounts as great as one percent, with small traces of opaque minerals and glass. This is an older lunar rock type which is abundant in soil and breccia from the highlands. Anorthosites are important because they record a major thermal event which formed its crust early in the Moon's history. The width of the field of view is approximately 4 mm.

Figure 4.40

Lunar basalt, seen through a microscope, is similar to terrestrial basalt but contains greater amounts of refractory elements (titanium, zirconium, and chromium). Lunar basalts are generally younger than the anorthosites. The width of the field of view is approximately 4 mm.

tration of refractory elements and the low proportion of volatile elements. The texture and composition of the basalts, along with melting experiments, suggest that most of the basalts crystallized at 1500 K to 1900 K and indicate that they were more fluid than their earthly counterparts. The high fluidity (low viscosity) of lunar basalts is also reflected in the broad sheets of basalt that fill the maria basins.

There are two basic types of lunar basalt. The older, less extensive basalts have been called **KREEP basalts** because of their relatively high concentrations of potassium (K), rare earth elements (REE), and phosphorous (P). Radiometric studies indicate that they were extruded at the surface around 4 billion years ago, prior to the formation of the Imbrium Basin, and are found mainly in highly fragmented rocks from the highlands. Indeed, it is difficult to determine conclusively that they represent lavas at all. These basalts may have evolved from the primitive molten layer that surrounded the Moon. As the magma ocean cooled, crystallization and removal of various minerals gradually changed the composition of the remaining melt. A crust of floating feldspar was created over the molten zone and an iron and magnesium-rich mantle accumulated beneath it. Residual liquids trapped between these thickening layers eventually became similar to the KREEP composition and were erupted on the surface through the fractured crust. Another theory about the origin of KREEP basalt holds that the entire magma ocean solidified but remelted around 4.0 billion years ago as the Moon warmed up by radiogenic heat. Partial melting of rocks believed to be part of the lower crust yields liquids very similar to the samples of KREEP basalt returned from the Moon.

The other basalts that are more common on the lunar surface are the **mare basalts**. Although there is evidence that mare basalts and KREEP basalts may have been erupted during overlapping epochs, mare basalts are generally less than 3.9 billion years old and are chemically distinct from earlier magmas. The simplest explanation for the origin of mare basalts employs remelting of portions of the mantle (formed earlier by crystal accumulation at the base of the magma ocean). Melting may have occurred in zones that deepened with time, but most mare basalts appear to have been produced at a depth of around 400 km.

Breccia

Breccia is a fragmental rock in which the individual particles are angular rather than rounded like particles of sand and gravel. Lunar breccias consist of fragments of rock and glass from a variety of sources (Figure 4.41). They result from meteorite impact and fragmentation. Some are consolidated regolith that have high proportions of glassy fragments. The mechanism by which regolith is consolidated into a coherent mass of breccia presents a problem. Two possibilities are immediately apparent: (1) shock lithification (compression of the grains together as a strong shock wave passes) and (2) welding of the deposits as they accumulate in a hot state after being ejected from impact craters.

Regolith

The Moon's surface is covered in most places by a thin layer of relatively loose, unconsolidated fragments of rock, crystals, and glass; particles vary in size from large boulders to fine powder. This layer is called the lunar soil or regolith. The presence of regolith was demonstrated on all lunar landings as dust thrown up by rockets of the landers, and the ever-present dust clinging to the astronauts and the lunar rover. The average thickness of the regolith depends upon the age of the surface on which it has been formed. The regolith on ejecta from very young craters such as Tycho

Figure 4.41

Lunar breccia, seen through a microscope, consists of angular fragments of broken rocks from a variety of sources. Typically, the fragments are angular and show essentially no evidence of modification by abrasion. Some lunar breccias contain large amounts of glass, with particles that are remarkably spherical. The dark gray materials are glass particles. Breccia results from fragmentation and subsequent compression due to repeated impact. Breccias are the main component of the lunar regolith.

may be only about 10 cm thick. On the maria, the average thickness is 5 m. The regolith in the older highlands is possibly more than 10 m thick. As a general rule, then, the older the surface, the thicker the regolith.

Regolith on the Moon consists of debris thrown out of craters. At any given place, most of the debris was derived from the local underlying layers of rock. As a result, the composition and texture vary considerably from place to place and reflect the history and the processes of the area where it is formed. For example, samples of regolith from the maria contain considerable amounts of basalt fragments and generally less than 50 percent glass, whereas regolith on the rays of the crater Copernicus contains 70 to 90 percent glass.

We can estimate the rate at which regolith forms on the Moon by measuring the thickness of the regolith and determining the age of the bedrock beneath. At Tranquillitatis, the lunar basalt is about 4 billion years old and the regolith is about 4 m thick. This gives an average rate of regolith formation of 1 mm per million years. This accumulation rate is somewhat misleading because regolith production must have proceeded at much higher rates during the early bombardment of the Moon and has since slowed to very low rates.

The surface of the Moon is modified by the churning action of impact that is the major factor in fragmenting solid rock and developing a regolith. There is, in addition, a type of microscopic weathering that occurs on the Moon. Since the surface of the Moon is unprotected by an atmosphere or a global magnetic field, the regolith is continually bombarded by micrometeorites, solar winds, and galactic cosmic rays. The net effect of this bombardment is to change the regolith slowly with time. One important change is the welding of particles together by glass generated by impact; glass-bonded aggregates are formed.

Thus, the outer few centimeters of the lunar surface is where interactions between space processes and the Moon take place. The core-tube samples of the lunar regolith brought by the returning astronauts are of particular interest in that the complex layers of brecciated rock record such interactions in the distant past. The Moon may therefore hold an important record of the Sun's activity billions of years ago.

Glass

Glass particles are abundant in nearly all samples of lunar regolith and are one of the features that distinguish lunar from terrestrial soils. The glass occurs as beautifully formed spheres and tear-drops and as spatter on other fragments (Figure 4.42). The lunar glass is distributed with ejecta and is believed to have formed by shock melting of rock debris during the process of impact. Samples from the rays of Copernicus, for example, consist of 70 to 90 percent glass. Other glass beads, orange and green in color, were probably produced by lava fountaining near the fissures from which the mare basalts were extruded. Under high magnification, the glass "beads" show pits, grooves, and spatter that resulted from micrometeorites striking the glass surfaces after the particles had cooled and accumulated in the regolith (Figure 4.43). Thus, practically every rock fragment in the regolith appears to have been involved in one or more impact events.

In summary, a variety of rock types has been found on the Moon, but the diversity is not as great as that found on Earth. Only trivial quantities of granite, peridotite, and so on have been identified. Apparently, the rock samples returned from the Moon tell exactly what one would expect from studies of the surface features. They record the details of impact and basaltic volcanism, the two major geologic processes that had operated on the Moon. Perhaps their greatest value is the information they provide concerning absolute dates of major lunar events. In so doing, they establish a radiometric time scale for another planet and inform us about the pace of its chemical and thermal evolution.

Figure 4.43
Micrometeorite craters on the surface of crystal fragments in lunar breccia show that the products of impact on the Moon range from the large multiring basins to microscopic pits. Practically every rock fragment on the Moon appears to have been involved in one or more impact events.

The Internal Structure of the Moon

Many of the major geologic processes that shape the surface of a planet are driven by forces from within the planet itself. Volcanic activity, tectonic movements, faulting and folding, and the generation of a magnetic field are examples of these internally derived processes. A thorough understanding of the interior structure of the Moon reveals much about its present state and past geologic history.

Our present understanding of the internal structure of the Moon is based on a variety of physical observations such as density, magnetism, and seismicity, as well as on the rock types and landforms present at the surface. Much remains uncertain, but several facts place significant constraint on what the internal structure may or may not be. First, the bulk density of the Moon is 3.34 g/cm³, whereas the mean density of lunar surface rock is about 3.3 g/cm³. Thus, the density of the surface material is only slightly less than that of the Moon as a whole, and there is little possibility for a significant increase in density with depth. (Earth, in contrast, has a mean density of 5.5 g/cm³ with the density of surface rocks being only 2.7 g/cm³, clearly indicating that the interior of Earth is much denser than the crustal material.) Nevertheless, from studies of lunar rocks, surface features, and lunar seismicity, it is clear that the Moon

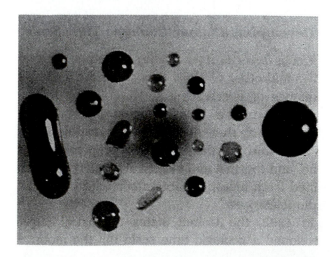

Figure 4.42
Glass particles are common in the lunar regolith. They are formed by the shock melting of rock fragments during impact.

is layered and that the composition of the interior differs from that of the rocks exposed at the surface. In addition, seismic energy on the Moon is 1 million times less than that on Earth, implying that internal convection is not occurring at shallow depths in the mantle.

The favored model of the Moon's interior is shown in Figure 4.44. The major units are (1) a crust 60 to 100 km thick, (2) a rigid mantle extending down to a depth of 1000 km, and (3) an asthenosphere that may surround a metallic core.

The Crust

Using instruments called seismometers, Apollo experiments recorded the strength of "moonquakes" at various points on the Moon, providing a way of "x-raying" the Moon and seeing its internal structure. Variations in the velocities of seismic waves with depth show that the crust is layered (Figure 4.45). In the maria, the surface layer is composed of a thin, 2- to 5-km layer of basaltic lavas.

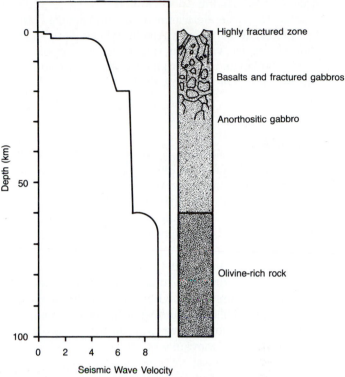

Figure 4.45

The structure of the lunar crust in the Oceanus Procellarum region has been interpreted from variations in the velocity of seismic (earthquake) waves. Discontinuities in seismic velocities indicate that the crust is layered. Seismic velocities increase very rapidly with depth to about 1 or 2 km beneath the surface. A very sharp increase occurs at a depth of about 25 km. Between 25 and 60 km below the surface, the velocities are nearly constant. A significant increase in velocity marks the base of the crust. From comparisons of these velocities with the velocities of seismic waves in major rock types, it appears that, from the near surface to a depth of about 2 or 3 km, the crust is composed of impact breccia and basaltic lava flows. Below this, to a depth of about 25 km, is a layer composed largely of brecciated and fractured anorthositic and gabbroic rocks like those exposed in the lunar highlands. Below a depth of about 25 km, rock fractures are not abundant and seismic velocities in the anorthositic crust increase markedly. Fractures produced by impact may have "healed" at this great depth or, alternatively, may never have formed. Below 60 km is the mantle of the Moon, which is believed to be rich in olivine and pyroxene.

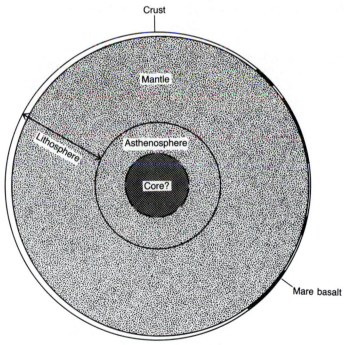

Figure 4.44

The internal structure of the Moon is interpreted from measurements of density, gravity, magnetism, and seismic properties. The thickness of the crust ranges from 60 to 100 km. The rigid mantle extends to a depth of about 1000 km and together with the crust makes up the lunar lithosphere. Most moonquakes originate in the region near the base of the lithosphere. The center of the Moon appears to consist of asthenospheric mantle, possibly surrounding a small iron-sulfide core.

Below this, the seismic velocities are low but gradually increase to depths of almost 25 km. This zone of shattered rock is the result of extensive fragmentation and fracturing of bedrock, caused by repeated impact of meteorites. The rapid increase in the velocity from the surface to 25 km is believed to be the result of fewer fractures in the anorthositic crust. From 25 km to 60 km, the velocities are similar to those determined for unfractured anorthosites. Results from Apollo experiments show that the crust varies in thickness from approximately 50 km on the near side to perhaps 100 km on the far side. The average thickness may be about 70 km.

The Mantle

At a depth of 60 to 100 km, a sharp increase in velocity occurs, marking the contact between the lunar crust and the mantle (Figure 4.45). The mantle rocks show a higher seismic velocity than crustal rocks and are believed to be rich in olivine, pyroxene, iron, and magnesium. The thick mantle is rigid and, together with the crust, makes up the Moon's lithosphere. The lunar lithosphere is much thicker (1000 km versus 100 km) and more rigid than Earth's. This thick lithosphere makes it nearly impossible for molten lava to reach the surface and prohibits lateral movements like those that have produced continental drift on Earth. In essence, the Moon is a "one-plate" planet. Most moonquakes occur at the bottom of this layer (800 to 1000 km deep) and are apparently triggered by Earth tides. Just as the Moon's gravitational attraction produces tides in Earth's oceans, Earth exerts a similar but stronger pull on the Moon's surface, slightly deforming the lithosphere.

The Core

As shown in Figure 4.44, an inner zone may exist beneath the moonquake zone. S-waves traversing the deep interior (below about 1000 km) are much weaker, suggesting that the rocks are partly molten, perhaps somewhat like Earth's asthenosphere. This lunar asthenosphere may extend to the center of the Moon or it may form a thin shell surrounding a solid metallic core, which may have a radius of about 400 to 700 km. As yet, evidence does not demand an iron core, but the detection of **remanent magnetism** in lunar rocks suggests that there may be a core composed of iron or iron sulfide. Remanent magnetism is acquired when hot magmas or rocks cool in the presence of a magnetic field. Geochemical studies of lunar samples also suggest that the Moon has a small metallic core. Even though the present-day Moon has no internally produced magnetic field, it appears that at one time thermal convection within the core produced a magnetic dynamo which set up a field that magnetized ancient igneous rocks as they cooled. Detailed studies show that strong remanent mag-

netism occurs only in rocks between 3.9 and 3.1 billion years old. This suggests that a magnetic field may have been produced at about the same time that the mare basalts were erupted. The lunar magnetic field may have been caused by heat released from the core once it was cool enough to begin to crystallize. Movement of material slowed subsequently, as the entire core cooled. This decreased the strength of the magnetic field.

The Geologic History of the Moon

Each planetary body has a time-varying set of dynamic geologic systems that modifies and shapes its surface and deep interior. On Earth, these include the hydrologic system, in which water deposits new rocks and erodes away older ones, and the tectonic system, which produces shifting lithospheric plates, crustal deformation, mountain building, volcanic activity, and the growth of continents. An entirely different set of geologic processes shaped the Moon's history and its surface. We have seen that billions of years ago huge meteorites impacted the surface; at other times great lava flows spread across its surface; and intermittently the surface was warped, cracked, or faulted. At present, only occasional small meteorites strike the surface, changing only very small areas, so the Moon is essentially geologically inactive—a striking contrast to dynamic Earth.

A sequence of events outlining the major events in the geologic evolution of the Moon can be constructed, utilizing the relative ages of lunar landforms and rocks, the absolute ages of lunar rocks, and the composition of lunar rocks and soils.

Stage I: Formation of the Moon (4.6 to 4.5 Billion Years Ago)

The Moon, like other planets in the solar system, is believed to have been formed by accretion of many smaller objects in a short period of time about 4.55 billion years ago. One scenario for the formation of the Moon calls for the accretion of material in orbit around a larger, but still accreting, Earth. This theory, however, does not satisfactorily explain why Earth is volatile-rich and the Moon is volatile-poor. Alternatively, the Moon may have formed in another part of the solar system, only to be captured by Earth's gravity. But this theory has severe dynamical problems regarding the Moon's present orbit. Another, presently popular, theory suggests that the Moon accreted near the ancient

Earth from material ejected from Earth's already differentiated interior by the impact of a Mars-sized object (see Figure 8.57). It is speculated that the impactor and part of Earth's mantle vaporized at temperatures between 1500 and 2000 K. This material cooled, condensed to solids, and reaccreted in Earth orbit. The giant impact hypothesis is attractive because it explains why the Moon has a composition similar to Earth's mantle for most elements but is poor in volatile materials. The volatile elements simply remained vapors after the impact, leaving the solids depleted in such materials. This hypothesis also explains why the Moon's metallic core is so small. The material that was ejected came principally from Earth's mantle and the mantle of the impactor. Apparently, core formation had already occurred inside both bodies.

In any case, the Moon probably heated up tremendously during the process of accretion, as the kinetic energy of impacting bodies was converted to heat. Additional heat may have come from the decay of short-lived radioactive elements. This may have resulted in the complete melting of the outer several hundred kilometers of the Moon, and the creation of an ocean of molten rock (Figure 4.46). As this global magma ocean cooled, minerals began to crystallize and separate according to their densities. Heavy mafic minerals that were rich in iron and magnesium—like the mineral olivine—sank, and lighter minerals, mainly plagioclase, migrated upward and were concentrated in a surface layer. This layer of anorthositic rock formed the lunar crust and is presently exposed in the highlands. It is easy to imagine a very active, mobile crust sliding across the convecting magma ocean during the period of crustal accumulation. The oldest anorthosites have ages of about 4.5 billion years. Some of the earliest volcanic rocks, the KREEP basalts, were erupted several hundred million years after the formation of the crust (Figure 4.46). They may have formed when residual pockets of magma trapped beneath the thickening crust broke through to the surface. These early developments led to a crust composed of anorthosite, gabbro, and KREEP basalt. From a geologic point of view, the development of a global crust, which is now exposed in the lunar highlands, was the first important event in the evolution of the Moon.

Gradually the crust cooled. A rigid lithosphere formed, and thickened with time. As the lithosphere thickened, lateral movements were inhibited and eventually halted completely. Tectonic disturbances must have been obvious until this time, but because of the intense meteoritic bombardment that the Moon experienced, no evidence of them has been preserved. The development of a lunar plate tectonic system was hindered by the rapid development of this thick globe-encircling shell of material. The uniformly low density of anorthosite precluded Earth-style subduction of lithosphere back into the mantle; lithospheric subduction appears to be driven by the gravitational sinking of dense lithospheric plates.

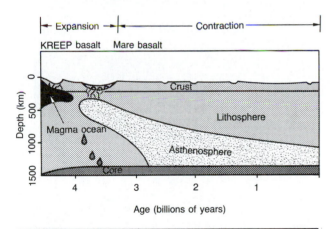

Figure 4.46
The thermal history of the Moon is summarized in this schematic time-versus-depth diagram. Accretion of the Moon led to widespread shallow melting and the creation of a magma ocean hundreds of kilometers deep from which the anorthositic crust crystallized. Subsequent heating by radioactive decay led to partial melting of the mantle (producing an asthenosphere) at increasingly greater depths. The differentiation of the interior is shown as including the formation of a small lunar core overlain by the asthenosphere. The thickness of the lithosphere steadily increased with time, achieving a present-day thickness of 1000 km. The early history of the Moon appears to have been marked by slight expansion, followed by cooling and later contraction.

Stage II: Pre-Nectarian and Nectarian Periods (4.6 to 3.9 Billion Years Ago)

The next major event (perhaps a continuation of accretion) was a period of intense bombardment of large and small bodies. The impact of these bodies formed a densely cratered terrain over the entire surface of the Moon, a terrain that is now preserved in the lunar highlands (Figure 4.47a). Meteorite impact may have churned the ancient magma ocean or later fragmented the crust to facilitate the eruption or exposure of KREEP basalts. Events in Stage II include the impact of large asteroid-size bodies that produced most of the multiring basins. These large objects may have been small proto-moons formed at the same time as the Moon, their infall resulting from the larger gravitational pull of

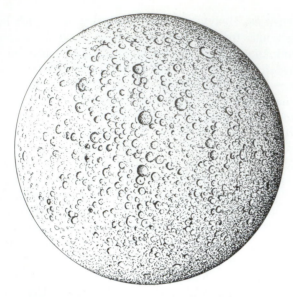

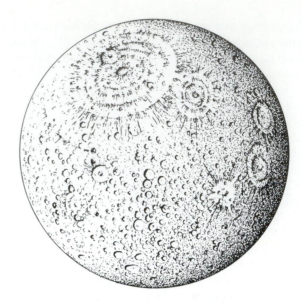

(A) Stage I. Formation of Moon by accretion (about 4.55 billion years ago) created densely cratered terrain over the entire surface of the Moon. The outer layers of the Moon may have been completely molten before this surface was shaped.

(B) Stage II. The formation of multiring basins (Imbrium Basin formed 3.9 billion years ago) is attributed to the impact of asteroid-sized bodies. The infall of these meteoritic bodies may represent the final stages of accretion. Remnants of this cratered surface are preserved in the lunar highlands.

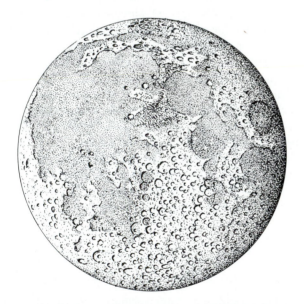

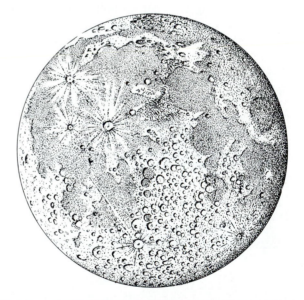

(C) Stage III. Extrusion of the mare basalts (from about 4 billion to perhaps 2.5 billion years ago) was a manifestation of a major thermal event in lunar history, which occurred when the lithosphere was still relatively thin. Lava flows filled many of the multiring basins on the Moon's near side and in some areas they covered parts of the highlands that lack obvious multiring structures.

(D) Stage IV. Relatively light meteorite bombardment (from 3.2 billion years ago to the present) formed some craters with bright rays, but the rate of cratering has been greatly reduced. The lunar landscape has changed little during the last 3 billion years.

Figure 4.47

The major events in lunar history include: intense meteorite bombardment during an early period, formation of multiring basins, extrusion of mare basalts, and, subsequently, light meteorite bombardment.

the Moon. Alternatively, they may have been asteroids nudged from beyond Mars by the gravitational effects of Jupiter. In any event, their collisions significantly modified the surface of the Moon (Figure 4.47b). At least 40 basins with diameters ranging from 300 km to over 1000 km are older than Imbrium Basin. During and after the formation of the multiring basins, impacts of smaller meteorites formed craters, such as Plato, Archimedes, and Sinus Iridum, on the large basins. These craters show that there was a significant time lapse between the formation of the multiring basins and their filling with lava.

During this stage of development, some planets outgassed envelopes of fluids at their surfaces. Because of its bulk composition and relatively small mass, the Moon developed neither an atmosphere nor a hydrosphere. No water has been found in any of the lunar rocks, and other volatile elements are found only in very low abundances. By contrast, terrestrial rocks found at the surface commonly contain several percent water by weight. Possibly the Moon accreted from materials that were depleted in volatiles as a result of vaporization during the hypothetical impact mentioned above. The low lunar gravity, determined by its mass, is also important for the present lack of an atmosphere. Even if tremendous amounts of volatiles had been released from the interior during an early melting episode, they would have been able to escape quickly into space. Larger planets with larger gravitational attractions have retained these gases to form atmospheres and hydrospheres that continually change the surface landforms and create new rock bodies.

Stage III: Imbrian Period (3.9 to 3.2 Billion Years Ago)

The next major events in lunar history were the extrusions of the mare lavas that cover large areas of the Moon (Figure 4.47c). The lavas erupted episodically during an interval of at least 800 million years and probably over a much longer time. There is even some evidence for pre-Imbrian basaltic volcanism. The basalts were generated at depths of about 400 km in the lunar mantle. Heat was most likely provided by radioactive decay. The zone in which basaltic magma was generated migrated inward as the Moon slowly cooled and its lithosphere thickened. Similar plains are found on Mercury and Mars. It is therefore possible that an early thermal event and the eruption of basaltic

lavas represent basic elements in the evolution of all terrestrial planets.

A fundamental problem in lunar geology is the distribution of the mare lavas in near-side basins underlain by thin anorthositic crust (less than 60 km thick). By contrast, the far side of the Moon has fewer maria and typically thicker crust, from 60 to 100 km thick. The near-side crust may have been preferentially thinned by the excavation of several basins formed coincidentally in the same region. As the Moon heated by radioactive decay and mare basalts were produced, the magmas would have been able to escape to the surface much more easily through these areas of already thin crust. By the time the basalts formed, the lithosphere was too thick to be much deformed but could still be punctured by these hot liquids. Although the lithosphere bent slightly beneath the loads exerted by accumulations of these lavas (producing ridges and rilles), it was strong and thick enough to prohibit their complete isostatic compensation. Calculations indicate that a switch from early graben and rille formation to compressive faulting and mare ridge formation occurred about 3.5 billion years ago. This change is attributed to the Moon's cooling and contraction.

Stage IV: Eratosthenian and Copernican Periods (3.2 Billion Years Ago to Present)

The internal differentiation of the Moon was essentially complete by the end of Stage III, and most of the Moon's present features were developed. The most significant events to occur since that time were the impacts of meteorites to form post-maria craters (Figure 4.47d). The influx of meteorites was greatly reduced, and most authorities consider the post-maria craters to have been formed by bodies from the asteroid belt or from comet nuclei.

Minor local volcanic activity, such as is found in the domes in the Marius Hills, has probably occurred since the maria formed. As the lithosphere thickened, the depth of basalt sources also migrated downward; by about 2.5 to 2.0 billion years ago, this rigid layer was too thick to allow the extrusion of lavas. It appears that the tectonic evolution of the Moon was then complete. Volcanism ceased and the Moon entered a terminal quiet state. Today the lithosphere is probably 1000 km thick, so the Moon is geologically active only at great depths. Aside from the occasional impact of a small meteorite, landscape evolution is largely complete.

Conclusions

The vast amount of new knowledge obtained about the Moon during the period of lunar exploration permitted geologists to study, for the first time, the details of another planet. In a sense, the Moon is a controlled experiment that shows a planet evolving without a hydrologic system and with a single global tectonic plate. Cratering, which is responsible for the formation and modification of most of the lunar landscape, was the most important surface process. Major thermal events did occur on the Moon, however, during differentiation and when basaltic lavas were extruded in a series of eruptions to form the lunar maria. Only minor tectonic features have been found that appear to result from relatively small down-warping of the lithosphere, combined with modest expansion, and followed by contraction of the Moon.

A geologic time scale has been developed for the Moon, using the principles of superposition originated by geologists studying Earth. Radiometric dates of lunar rock samples provide a base of absolute time for events in lunar history. Perhaps the most important aspect of the Moon's geologic evolution is that most of it occurred during the early history of the solar system, before even the oldest rocks on Earth were formed. The Moon thus provides important insight into planetary evolution that is unobtainable from studies of Earth.

Soon after the planet was formed by accretion, the crust of anorthositic rock and minor basalt formed; it is preserved in the highlands. A mantle and perhaps a small iron core formed at the same time. The crust was then subjected to a period of intense bombardment. The lunar maria formed from vast floods of lavas subsequent to the heavy cratering. Since then (about 3.0 billion years ago) the Moon has been inactive, and the surface has been only slightly modified by relatively few impact craters.

Today, the Moon does not have a significant source of internal energy nor a tectonic system like Earth. It has no continents nor ocean basins and no deformed rocks resulting from mountain building. Moreover, because of its volatile-poor character, it has no atmosphere nor surface fluids, so it lacks a hydrologic system to modify its surface. The Moon apparently lacked sufficient mass to experience an extended differentiation history. Small bodies radiate away their internal heat at a much higher rate than larger bodies because they have larger surface-area mass ratios. This is because all heat must eventually escape through the surface. Small planetary bodies like the Moon cooled much faster than larger ones. Their thermal and tectonic evolution proceeded at an accelerated pace and terminated when the lithosphere became so thick that it could no longer be deformed.

Review Questions

1. Draw a diagram of the internal structure of the Moon, and briefly describe the core, mantle, asthenosphere, lithosphere, and crust.
2. Outline the stages in the production of a crater by the impact of a meteorite. What geologic features are produced by impact?
3. Contrast the origin and features of an impact crater with one produced by volcanic activity.
4. Explain two ideas for the origin of a multiring basin.
5. How are lunar craters modified as time passes?
6. Why are there fewer craters on the lunar maria than on the highlands?
7. What are the major differences between the highlands and maria on the Moon?
8. In what way is the Moon asymmetric?
9. Why are there so few impact craters on Earth as compared to the Moon?
10. What do linear rilles and wrinkle ridges tell us about the tectonic history of the Moon?
11. Discuss some of the possibilities to explain why cinder cones are so rare on the Moon.
12. If the Moon's crust formed from a magma ocean about 4.5 billion years ago, why are rocks that have radiometric dates this old so rare on the Moon?
13. The Moon's highland crust consists largely of plagioclase feldspar. Why does this imply that a lunar magma ocean once existed? How was plagioclase separated from the pyroxene and olivine crystals forming in the magma at the same time?
14. Could Earth and the Moon have once been part of a larger body?
15. Outline the major events in the history of the Moon.
16. Explain how a geologic time scale was developed for events in the Moon's history.

Key Terms

Anorthosite

Ballistic

Basalt

Basaltic Plains

Breccia

Compression Stage

Copernican Period

Crater Degradation

Ejecta

Eratosthenian Period

Excavation Stage

Imbrian Period

Impact Melt

Isostatic Adjustment

KREEP Basalt

Lava Channel

Lava Tube

Linear Rille

Mare Basalt

Maria

Megaterrace

Modification Stage

Multiring Basin

Nectarian Period

Nested Crater

Peak-Ring Basin

Pre-Nectarian Period

Rarefaction

Rays

Regolith

Remanent Magnetism

Sinuous Rille

Terrae

Wrinkle Ridge

Additional Reading

Bowker, D. C., and J. K. Hughes. 1971. *Lunar Orbiter Photographic Atlas of the Moon.* NASA SP-206. Washington, DC: National Aeronautics and Space Administration.

Hartmann, W. K. 1977. "Cratering in the Solar System." *Scientific American.* Vol. 235, No. 1, pp. 84–99.

Masursky, H., G. W. Colton, and F. El-Baz. 1978. *Apollo over the Moon: A View from Orbit.* Washington, DC: National Aeronautics and Space Administration.

Mutch, T. A. 1972. *Geology of the Moon.* Princeton, NJ: Princeton University Press.

Schultz, P. H. 1976. *Moon Morphology.* Austin: University of Texas Press.

Taylor, S. R. 1975. *Lunar Science: A Post-Apollo View.* New York: Pergamon Press, Inc.

Taylor, S. R. 1982. *Planetary Science: A Lunar Perspective.* Houston: Lunar and Planetary Institute.

Wilhelms, D. E. 1987. *The Geologic History of the Moon.* U.S. Geological Survey Professional Paper 1348. Washington DC: U.S. Geological Survey.

Wood, J. A. 1975. "The Moon." *The Solar System.* New York: W. H. Freeman and Co., pp. 69–80.

TABLE 5.1

Physical and Orbital Characteristics of Mercury

Mean Distance From Sun (Earth = 1)	0.387
Period of Revolution	88 d
Period of Rotation	59 d
Inclination of Axis	28°
Equatorial Diameter	4,880 km
Mass (Earth = 1)	0.055
Volume (Earth = 1)	0.06
Density	5.44 g/cm^3
Atmosphere (main components)	O, Na, K (thin)
Surface Temperature	100 to 700 K
Surface Gravity (Earth = 1)	0.37
Magnetic Field (Earth = 1)	0.01
Surface Area/Mass	23 × 10^{-11} m^2/kg
Known Satellites	0

CHAPTER 5

Mercury

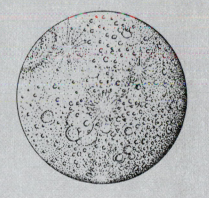

Mercury

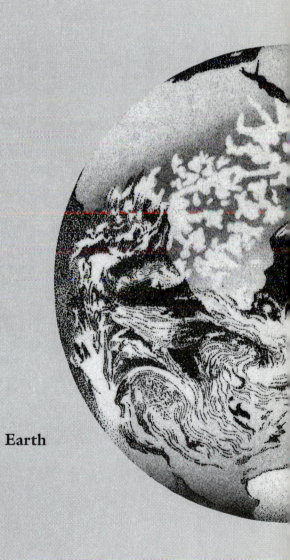

Earth

Introduction

The primary goal of the Mariner 10 mission was to obtain data about Mercury, a planet that had never before been visited by a spacecraft. Twin television cameras and six other instruments constituted the vehicle's 675-kg scientific payload. They provided new insight into the nature of the planet closest to the Sun and how it fits into the overall picture of the solar system. On March 23, 1974, Mariner 10 began photographing Mercury, and by April 3 it had collected an unprecedented store of scientific data, including more than 2000 high-resolution television pictures. The spacecraft passed within about 725 km of Mercury's surface at the point of closest approach. By a lucky coincidence, Mariner 10 was placed in an orbit around the Sun that returned the spacecraft to Mercury twice more at six-month intervals. Nearly complete photographic coverage of the illuminated half of the planet was obtained, with some photographs showing features as small as 150 m in diameter. Although years of study of these photographs and more thorough investigation by future space probes will be required before a detailed picture of the geology of Mercury can be devised, the images available reveal much about the nature of the small planet closest to the Sun.

Major Concepts

1. The processes that shaped the surface features of Mercury were remarkably similar to those that shaped the Moon. The major landforms are as follows: (a) impact craters and cratered terrains, (b) intercrater plains, (c) multiring basins, and (d) sparsely cratered smooth plains presumably flooded with lavas.

2. Impact craters range in age from old, highly eroded features to young, rayed craters surrounded with halos of bright ejecta and prominent systems of secondary craters.

3. There are at least two generations of plains on Mercury, both of which are probably lava flows.

4. Prominent fault scarps extending across the surface of Mercury are believed to be the result of global contraction that occurred as the planet cooled. Grabens are rare, and large rift valleys have not been observed.

5. Mercury has a large metallic core, compared to its size, that may be partly molten, so its internal structure differs significantly from that of the Moon.

6. Mercury's geologic systems were driven dominantly by thermal energy from within the planet and the infall of meteoritic debris. Apparently the lithosphere is now thick and immobile. Lacking surface fluids, the surface of Mercury has changed little in the last billion years.

7. A preliminary interpretation of the major events in the history of Mercury includes (a) accretion and initial differentiation, (b) a period of intense bombardment, (c) the start of crustal shortening, (d) impact of large meteorites to form multiring basins, (e) formation of plains material—presumably by floods of basalt, and (f) subsequent meteorite impact at a much lower frequency. Because we have no rock samples from Mercury that can be radiometrically dated, there is no absolute time scale for Mercury.

The Planet Mercury

Several of Mercury's physical features distinguish it from the other planets, as is summarized in Table 5.1. Icy Pluto, in the outer solar system, is the only principal planet smaller than Mercury, but the solar system contains many objects that are much smaller (the asteroids and planetary satellites). Mercury is the planet closest to the Sun, and stored in the compositions of its rocks is much important information about the chemical composition and early differentiation of the inner solar system. This information remains largely untapped, but some models of the formation of the solar system suggest that Mercury should be rich in refractory materials. Mercury is also much denser than would be expected by strict analogy with the Moon; it is probably the most iron-rich planetary body. Mercury represents an extreme in another respect as well. Its surface environment is very harsh; with essentially no atmosphere to moderate its surface, temperatures may rise to almost 700 K during the day and at night drop to less than 100 K. Some areas near the north and south poles may get little or no sunlight and are permanently frigid. The rotation period is 59 terrestrial days long and a year is 88 days long. This represents a 2:3 ratio, where there are 2 mercurian years to exactly 3 mercurian days. Such coincidence is called **spin-orbit coupling** and it probably evolved during the early history of the planet as a result of the constant tidal tug of the Sun on the planet.

Because of Mercury's small size and proximity to the Sun, its surface features were almost totally unknown before the Mariner 10 voyage. Now, with images of essentially half its surface, we are able to interpret the major events in the planet's geologic history. We now have important information about the general mode of planetary development because Mercury represents a unique "end-member" with a small size and a presumed refractory element-rich composition. The concepts of planetary evolution that were developed earlier for the Moon will be tested here; new principles will be developed that can then be applied to larger, more complex planets such as Mars.

Major Geologic Provinces

The pictures of Mercury beamed back to Earth by the Mariner 10 spacecraft show large tracts of heavily cratered terrain and broad areas covered by lightly cratered smooth plains like the lunar maria. These and other similarities with the Moon are immediately obvious from the photo mosaics of Mercury shown in Figure 5.1. Indeed, it is difficult for many nonspecialists to tell the surface features of the Moon apart from Mercury. The mosaics were made from a series of computer-enhanced pictures taken at a distance of approximately 230,000 km and are similar in resolution to telescopic pictures of the Moon.

Several terrain units with distinctive histories are very extensive (Figure 5.2). They include cratered terrains, intercrater plains, the Caloris Basin, and smooth plains.

Figure 5.1

The cratered surface of Mercury is similar in many respects to that of the Moon. This photomosaic, taken by Mariner 10, the only spacecraft to visit the innermost planet, shows densely cratered terrains, a large multiring impact basin (Caloris Basin), younger smooth plains, and rayed craters. In comparison to other planetary bodies, Mercury has not had an exciting history, but it occupies a special place at one end of the spectrum of planetary types.

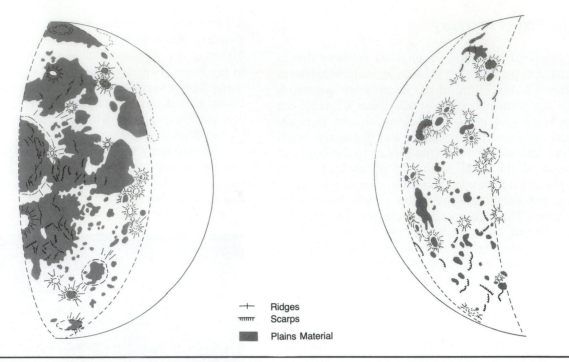

Ridges
Scarps
Plains Material

Figure 5.2

The major geologic provinces of Mercury are shown on this map prepared from Mariner 10 photographs. The rims of impact basins more than 200 km in diameter are shown with dashed lines. Plains are shaded gray. The ejecta and secondary craters surrounding the major craters are shown by radial lines. Scarps appear as hachured lines.

Cratered Terrains and Intercrater Plains

Craters on Mercury appear similar to their lunar counterparts, but the heavily cratered regions of Mercury have broad areas of gently rolling plains, impact craters, and basins; the lunar highlands are more evenly cratered. This mercurian terrain is called the *intercrater plains* and is the most widespread type of terrain on Mercury. Clusters of impact craters are very common in these areas. Secondary craters are distinct in shape, are relatively shallow, and are aligned in long chains. Their rims are commonly ill defined and form a linear or grooved fabric. Primary craters of the same size are circular, bowl shaped, and deeper with well-defined sharp rims (compare Figure 5.3 with Figure 5.7). The topographic complexity of these areas is particularly well developed around some of the ancient mercurian basins. In some places, these heavily cratered units are transected by high scarps, somewhat like the lunar wrinkle ridges. Also in places, huge, bright streaks, apparently unrelated to craters, extend for thousands of kilometers (Figure 5.1). As on the Moon, this heavily cratered terrain must represent one of the oldest surfaces on the planet in spite of its simpler appearance (caused by the large areas of intercrater plains). Although no absolute ages can be determined for Mercury's features, the heavily cratered

regions probably formed at the same time as the lunar highlands and record the same period of intense bombardment. The color of the light reflected from the surface also suggests a similar composition to the feldspar-rich highland crust of the Moon.

The relative ages of the intercrater plains and the oldest, most degraded craters and basins are not firmly established. In some places, the plains clearly overlie ancient craters; in other areas, the craters and their ejecta are obviously younger than the plains. It appears that although most craters are younger than the intercrater plains, a few are older. These relationships could be explained by a major thermal event in Mercury's early history in which widespread volcanism occurred at the same time as intense bombardment, or possibly even by a thermal event involving planetwide surface "softening." Viscous flow of the surface might rapidly obliterate craters as they formed. One fact is clear: The intercrater plains represent a significant period of time in the early history of Mercury, a time during which many of the earlier impact structures were erased and the planet was resurfaced.

Caloris Basin

The huge multiring Caloris Basin, 1300 km in diameter, dominates much of the photographed

Figure 5.3
Intercrater plains are the most widespread terrains on Mercury. They consist of smooth to gently rolling plains with a high population of craters less than 15 km in diameter. Many form chains or clusters suggestive of secondary origin. Plains occur between and around areas with larger impact structures that form the densely cratered terrain. It is believed that the intercrater plains are of volcanic origin, although there is insufficient evidence to make the conclusion certain.

area of Mercury (Figure 5.4). Like the lunar basins, it was created by the impact of an asteroid-sized object early in Mercury's development. Caloris is half again larger than Imbrium Basin on the Moon but is very similar to it in general form. The perimeter of Caloris Basin is defined by a rugged ridge rising 2 km above the floor (Figure 5.5). A subdued outer scarp has a diameter of about 1450 km (Figure 5.6). Although it is discontinuous, this outer ring appears to separate an inner zone of hilly or blocky ejecta deposits from an outer, distinctly lineated zone. These lineated areas consist of a well-developed system of valleys and ridges and are similar in many ways to the ejecta that surround the Imbrium basin. Both the lunar and mercurian terrains were probably formed in the same way, by erosion and deposition of ejecta thrown from the crater. The lineated terrain extends beyond the rim to a distance roughly equal to the diameter of the basin and is there buried or embayed by extensive smooth plains that completely surround the eastern half of the basin. The lineated terrain is best expressed to the northeast of Caloris Basin and is possibly the most rugged topography on Mercury. An extensive field of secondary crater chains, clusters, and gouges has been mapped beyond the lineated ejecta.

Younger smooth plains material covers the basin floor inside the main scarp that forms the rim (Figure 5.6). If Caloris has inner rings like lunar basins, they are buried beneath the fill. The Caloris plains are extensively ridged and fractured and are unique among the planets—similar features have not been found in other basins of Mercury, the Moon, Mars, or the satellites of the outer planets. Morphologically, the ridges resemble wrinkle ridges on the lunar maria but are much higher and have a polygonal pattern from the intersection of crudely radial and concentric networks. The fractures range in width up to 9 km, appear to be flat floored and grabenlike, and cut the older ridges. The width of the fractures increases toward the center of the basin. There are few arcuate trends that could be interpreted as flooded craters on the floor of the basin (ghost craters) and that are common in some of the lunar basins. This relationship suggests that the basin was covered with plains material soon after it was formed and, unlike the Imbrium Basin on the Moon, was not modified by impact before filling.

The impact of the body that formed Caloris was so great that the effects were apparently felt on the opposite side of the planet as well. A peculiar terrain of hills and linear valleys (Figure 5.7) occupies a region more than 500 km across, centered on the exact opposite side of the planet from Caloris. Perhaps the intercrater plains and pre-existing crater rims were broken up by focused seismic waves originating from the impact site. At a point opposite the basin, vertical movements of several kilometers may have been caused by this event. Smooth plains have partially buried this terrain and are therefore younger.

Smooth Plains

Another major geologic unit of Mercury consists of scattered areas of smooth plains that resemble the

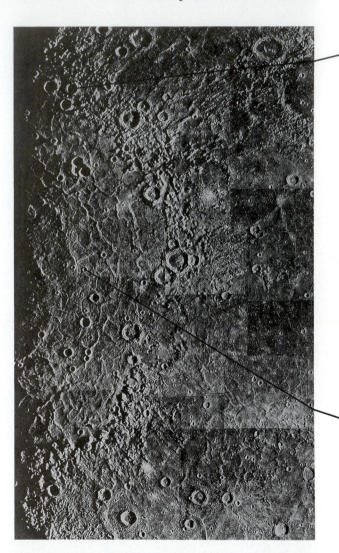

Figure 5.4

The Caloris Basin, with its concentric rings and radial ejecta, is the largest impact structure viewed by Mariner 10. It is 1300 km in diameter and in many ways is comparable with the Orientale Basin on the Moon. Unlike other multiring basins the floor is filled with smooth plains and is highly ridged and fractured. Both ridges and fractures display a radial and concentric pattern. The impact that created Caloris Basin was a key event in Mercury's history because it modified the landscape over an enormous area and probably led to the formation of an unusual hilly terrain on the opposite side of Mercury, as a result of focusing of seismic waves in that area.

Figure 5.5

The rim of Caloris Basin is marked by an annulus of hilly or knobby topography and linear ridges formed by ejecta. Two rings are visible in this view. Caloris Basin appears to have formed near the end of the period of intense bombardment on Mercury.

Figure 5.6

The floor of Caloris Basin is covered with smoother plains material, probably of volcanic origin. It is unique in that it is deformed by a complex system of ridges and fractures that form a polygonal pattern. The fractures transect the ridges, indicating that they are younger. The largest crater is about 10 km in diameter.

(A) This region is broken into valleys and hills up to 2 km high that are interspersed with smooth plains. Similar terrains have been found on the Moon's antipode to Imbrium and Oriental Basins. The large crater to the left is Petrarch.

(B) It is believed that this terrain is the result of focused seismic waves caused by the impact that formed Caloris Basin.

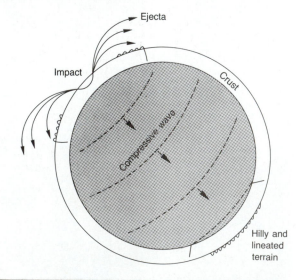

Figure 5.7

The hilly and lineated terrain located on the antipode of Caloris Basin is one of the most peculiar areas viewed on Mercury.

lunar maria. This type of terrain covers about 15 percent of the photographed portion of the planet and is distinct in that it is very smooth and only sparsely cratered. Impact structures larger than 10 km in diameter are rarely developed on this geologic unit. The smooth plains are quite level and often fill major depressions. Marelike wrinkle ridges are common. The largest area of smooth plains lies in and around Caloris Basin (Figure 5.5). Numerous other small patches in and around other large craters are scattered across the planet (Figures 5.8 and 5.9). A concentration of smooth plains lies in the northern hemisphere, possibly reflecting a global asymmetry similar to the distribution of maria on the Moon. Although the smooth plains resemble the lunar maria, they lack a strong contrast in brightness with the surrounding terrains, so their boundaries are not always clear and distinct. Stratigraphic relations between the smooth plains and other terrain types are clear, however, and indicate that the smooth plains are the youngest major terrain on the surface of Mercury. This conclusion is, of course, supported by its sparse crater population. In addition, the frequency of superposed small craters is approximately the same wherever the smooth plains occur, indicating that most of the terrain is about the same age. However, this is also true for the lunar maria, which were erupted as lava flows over a 1-billion-year time span. Throughout large areas, the plains material is probably relatively thin, much thinner than the mare basalts that fill the lunar basins. This conclusion is based on the fact that parts of the rims of numerous craters protrude through the cover of plains material so that on a regional scale the cover appears to be incomplete and discontinuous (Figure 5.9).

Suggestions for the origin of the mercurian smooth plains include their formation by ballistic erosion and deposition of ejecta associated with the formation of major impact basins, notably Caloris. As impact-energized debris surged away from Caloris, it may have ponded in depressions, creating smooth plains in the same way that the lunar highland plains were formed (like those that fill Ptolemaus, Figure 4.18). In addition, the plains are light colored like the plains in the lunar highlands and unlike the basaltic lavas seen on other planets. However, most of the plains are younger than Caloris, the youngest known impact basin large enough to create plains of ejecta. Small patches of smooth plains within craters may also arise by mass wasting from the walls.

On the other hand, many geologists believe that the smooth plains were formed by the extrusive outpouring of lava much like that which formed

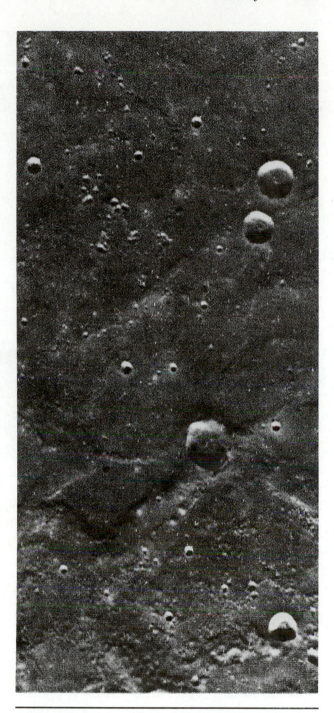

Figure 5.8

Smooth plains on Mercury closely resemble the lunar maria but lack a strong color contrast with their older surroundings. Their surfaces are only sparsely cratered and are commonly deformed by ridges. Most geologists believe that the smooth plains were formed by the extrusion of basaltic lava.

the lunar maria, although evidence for this is not conclusive. Evidence favoring a volcanic rather than impact origin for the smooth plains includes (1) the large volume of material that accumulated to form smooth surfaces, (2) differences in the

Figure 5.9
Crater rims protruding above the smooth plains indicate that on a regional basis the smooth plains material is relatively thin and discontinuous.

volume of plains material in craters and basins of the same size, (3) the striking similarity in morphology and distribution of the smooth plains and the lunar maria, (4) the age differences between the smooth plains and the basins they occupy, and (5) the spectrum of light reflected from the surface suggests that basalt is present. The major obstacle to accepting a volcanic origin is that there are no obvious associated volcanic features (vents, flow fronts, or sinuous rilles). This is possibly due to the fact that even on the best Mariner 10 photographs, details of lava domes or thin flow fronts cannot be resolved. However, some small hills that dot the surface of the plains may be low, shield-type volcanoes. If volcanic eruptions formed the smooth plains, the process of extrusion must have been similar to that which produced the lunar maria—quiet fissure eruptions of fluid basaltic magma that ponded in depressions, covering most of the vents through which lava rose to the surface.

The ultimate origin of the mercurian smooth plains will not be known until we have samples of rocks to examine. In the meantime, it is probably safe to conclude that the plains of Mercury were produced by several of the processes described above.

Impact Craters and Basins

The dominant landforms on Mercury are craters of all sizes and states of degradation (Figure 5.10). These range in size from small pits at the very limit of resolution (about 1 km in diameter) to large multiring basins like Caloris, 1300 km across. In many ways, these impact features are similar to those found on the Moon. The craters represent a wide range in age—from older, highly degraded depressions to young, fresh craters surrounded by halos of bright ejecta and extensive ray systems. However, close examination reveals that the mercurian craters differ from lunar craters in several important aspects.

Figure 5.10

Craters on Mercury are similar to those on the Moon. They range in size from less than 100 meters (highest resolution obtained by Mariner 10) to large basins over 1000 km in diameter. As shown in this image, small craters are simple and bowl shaped, but with increasing size, craters develop central peaks, terraced inner walls, peak rings, ejecta deposits with radial structures, and swarms of secondary craters.

Impact Cratering and Gravity

Perhaps the major physical differences between the Moon and Mercury that would influence crater morphology are their diameters and masses. Mercury is much larger and more dense and therefore has a surface gravity about twice lunar gravity. After the initial studies of the Mariner 10 photos, it was thought that Mercury's fresh craters were substantially shallower than lunar craters with the same diameters. The differences were attributed to Mercury's greater gravitational pull. It was also thought that the progression of changes in crater morphology (for example, the transition from simple to complex craters) occurred at smaller diameters on Mercury. Although controversy continues, neither of these early conclusions has been completely borne out by later studies, and significant differences (within about 10 percent) in depth-to-diameter ratios or morphologic relations do not appear to exist. Apparently, the effect of the planet's gravitational field is not as important as many other factors. Small mercurian craters are simple and bowl shaped. With increasing size, terracing of the crater walls becomes apparent and central peaks develop; then irregular clusters of peaks appear. The largest impact features are basins with inner rings (Figure 5.10), just as on the Moon.

However, other gravity-induced differences for impact craters appear to be real. The extent of the ejecta blanket and secondary craters around a primary crater is systematically smaller for a given crater size on Mercury than on the Moon (Figures 5.11 and 5.12). The fields of secondary craters appear to be better preserved on Mercury as well. The distance traveled by material ejected from a crater is due in part to the higher gravitational attraction of Mercury as compared to the Moon, which pulls these ejected objects down to the surface faster and produces shorter travel distances, thus explaining the more pronounced clustering of secondary craters. The ejecta blanket surrounding a mercurian crater must also be thicker, as it is spread over a smaller area and will have an increased ability to degrade or bury nearby craters. The zone of secondary craters is often marked by long linear grooves, which usually radiate from the center of the crater (Figure 5.12). These grooves are produced by the impact of closely spaced ejecta fragments, occur much closer to the crater, and are more pronounced than their lunar counterparts. Mercury's higher gravitational pull may likewise give ejected blocks higher velocities and produce larger, more prominent secondary craters when the blocks hit the surface. Although the resultant feature is slightly different, the cratering process is fundamentally the same on both planets.

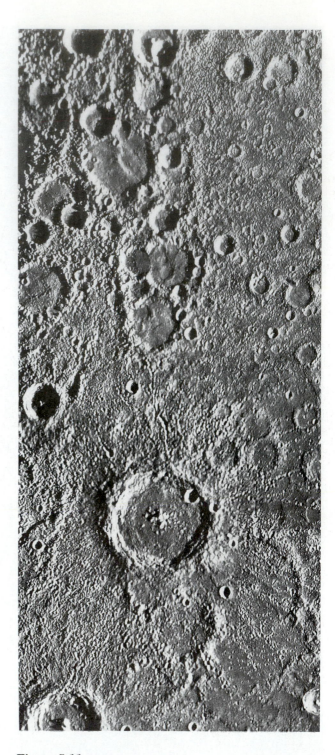

Figure 5.11

Secondary craters on Mercury are relatively small but are well preserved. As shown in this image, secondary craters appear in linear chains radiating from the point of major impact. Note also the terraced inner rim, central peak, and ejecta blanket of the large crater, which is similar to the morphology of equivalent-size craters on the Moon.

Figure 5.12

Secondary impact craters are commonly elliptical and in some areas form elongate grooves and ridges that impart a "wormy" texture to the surface. They appear to be preserved better than their lunar counterparts, perhaps because they were formed by ejecta with higher velocities and are therefore deeper. Because Mercury's gravity is stronger than the Moon's, impact ejecta travels only half as far on Mercury for an impact of similar size.

Multiring Basins

Large multiring basins, similar to those on the Moon, are also found on Mercury. The most common are relatively small, ranging from 200 to 600 km in diameter (Figure 5.13). These craters usually have two well-preserved rings and an ejecta blanket with numerous secondary craters. The largest impact structure photographed by Mariner 10 is the Caloris Basin discussed earlier.

Two important observations have been made regarding the impact basins on Mercury. First, the intercrater plains are not saturated with small craters; second, there is a similar lack of large impact basins on Mercury. Even accounting for the

Figure 5.13
Small ringed basins on Mercury represent a transition from large craters with clusters of central peaks to large multiring basins over 1000 km in diameter.

part of Mercury that has not been photographed, almost twice as many basins over 400 km in diameter have been found on the much smaller Moon. These observations can be explained by several competing hypotheses. For example, there are several indications that the numbers of meteorites passing through all areas of the solar system were not the same. Some scientists have concluded that the lack of large basins on Mercury indicates that fewer meteorites were available to impact Mercury than the Moon. This may explain why many old surfaces are not saturated with craters and why secondary craters and other small features have not been eroded by subsequent impact. Another way to explain this apparent lack of large basins centers on an observation regarding the state of isostatic adjustment of old basins. Many basins, like

the one in Figure 5.14, are very shallow and show clear evidence of advanced isostatic compensation. This has led some to believe that Mercury's crust cooled more slowly than the Moon's, remained plastic longer, and was able to adjust rapidly to erase all signs of impact, just as thick mud oozes to remove signs of disturbance. A third alternative to explain the small number of mercurian basins was alluded to in the description of the intercrater plains. These plains appear to have formed during the early bombardment, and their emplacement may have destroyed many older basins.

Crater Degradation

As on the Moon, the dominant erosional processes on Mercury are caused by cratering. Degradational

Figure 5.14
Shallow craters on Mercury show evidence of advanced isostatic adjustment, possibly because Mercury's lithosphere was warm and plastic during the period of bombardment. The secondary crater field around the double-ring crater Ma Chih-Yuan (170 km across) is nonetheless well preserved. Arrows show crater rim.

sequences have been established for mercurian craters that show the morphologic changes with increasing age. The freshest craters have well-defined rims, hummocky ejecta blankets, and systems of bright rays comparable to Copernicus and Tycho on the Moon. Numerous rayed craters from 1 to 50 km in diameter dot the surface (Figure 5.1). Subsequent bombardment breaks down the crater rim and churns up the ejecta blanket or completely buries it beneath other ejecta deposits. Ultimately, the crater is transformed into a low-rimmed depression with large numbers of younger, superposed impact features. Many of the original crater features become completely obliterated or barely recognizable (Figure 5.14). Degradation of mercurian craters by impact from secondary fragments does not extend as far from the primary crater as on the Moon because of shorter ballistic ranges. In spite of the shorter range, cratering processes active over billions of years have produced a thick layer of soil or regolith on the surface. Optical and radar measurements made from Earth indicate that it is similar in composition and physical properties to the lunar regolith.

In summary, mercurian impact features differ from lunar craters and basins in three important ways. First, the ejecta thrown out of mercurian craters does not appear to have traveled as far as on the Moon. Considering the larger strength of the mercurian gravitational field (almost twice the Moon's) this appears to be logical. Second, even the densely cratered terrain of the surface is not saturated with craters. The population of projectiles may have been smaller at Mercury's orbit or basins formed early may have been removed by some process. Third, many of the ancient mercurian basins are very shallow and ill defined.

Tectonic Features

Even though some of its surface features are quite similar to those on the Moon, Mercury appears to have been subjected to a style of tectonic deformation not found on the Moon or other terrestrial planets. Evidence of extensional stress is found only in small areas near Caloris Basin. On Mercury, the dominant tectonic features are a series of lobate escarpments or **scarps**—steep, clifflike slopes. These are often more than 1 km high and hundreds of kilometers long. The general nature of these scarps is shown in Figure 5.15. They may be irregular, arcuate, or lobate in outline and generally have rounded crests that greatly differ from the sharp crests and straight ridges formed by vertical faults and graben margins on the Moon. Mercurian scarps are similar to but larger than the wrinkle

ridges found on the Moon. These characteristics seem to indicate that the faulting or flexing occurred as the result of compression in the mercurian crust. Thrust faults in which one block of rock is pushed or thrust over another seem to have been produced. Thrust faults usually have shallow dips as compared to normal faults. The scarps transect the older intercrater plains and craters, whereas younger craters and portions of the smooth plains cross the scarps. These cross-cutting relations indicate that the scarps began forming sometime near the final phase of the heavy bombardment and continued developing after some smooth plains were formed.

An important characteristic of the scarps is their global distribution. Maps prepared from Mariner 10 photos show that the scarps extend from pole to pole over most of the visible surface (Figure 5.2) and trend in a more or less north–south direction. The relatively uniform global distribution of the scarps suggests that the entire planet was subjected to compressive forces that resulted in crustal shortening after the early period of differentiation and intense bombardment.

There are several ways in which this global deformation may have occurred. As mentioned earlier, Mercury's rotational period has probably changed substantially over the course of geologic time as it evolved toward the 3 days per 2 years stable relationship. This **despinning** or slowdown may have induced substantial changes in the shape of the planet, creating compressive forces in the surface layers near the equator and causing the flexures and faulting observed in the form of the scarps. An important difficulty with this model is that it predicts extension in the polar regions, but no evidence of extension in the form of grabens has yet been found. Another method for planetwide compression is suggested by calculations of Mercury's probable thermal history, which predict that substantial contraction occurred as it cooled after differentiation. Cooling of a large metallic core or cooling of silicates in the lithosphere and consequent contraction is adequate to explain many of these features. A change in the radius by only 2 km (0.1 percent) is sufficient to cause crustal compression of approximately the same magnitude as that observed on the surface.

The smooth plains inside Caloris Basin display features obviously produced by structural deformation including the ridges and grabenlike fractures. Both the ridges, with a compressive origin, and the fractures, formed by tensional stresses, have similar radial and concentric patterns (Figure 5.6) and were most likely caused by minor vertical movements of the interior of the circular basin. Initial subsidence of the basin as it filled with smooth

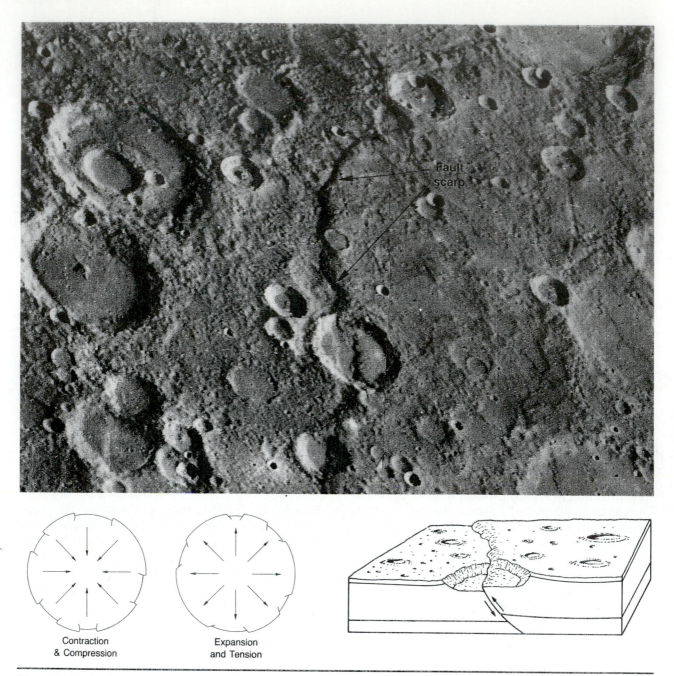

Fault
scarp

Contraction
& Compression

Expansion
and Tension

Figure 5.15

Fault scarps on Mercury transect and offset craters, clearly indicating crustal deformation. The large scarps are probably thrust faults, some of which are 2 km high and 500 km long. They were probably caused by crustal shortening associated with cooling and contracting of the planet after the period of intense bombardment. Heating and planetary expansion cause extension and grabens to occur at the surface.

plains material probably formed the ridges. Subsequently, it appears that the floor was uplifted and fractured. In most cases, the fractures cut the ridges and are therefore younger, which is consistent with this model. An explanation for the uplift is not in hand. However they were formed, these structural features were most likely produced by relatively local stress concentration similar to the type of tectonics operative on the Moon that pro-

duced the linear rilles and wrinkle ridges and probably are not closely related to global processes like those that produced the global scarp system.

The Internal Structure of Mercury

From the results of the Mariner 10 mission and other Earth-based studies, the radius (2439 km) and the density (5.44 g/cm^3) of Mercury are well known. Data returned from Mariner 10 also demonstrate the presence of a significant magnetic field, with a strength of less than 1 percent of Earth's at the surface. These few facts and a knowledge of the surface history help scientists to determine the type of internal structure Mercury possesses. The high density implies that the planet has a very dense interior, most likely as a result of high concentrations of iron. Using mathematical models to determine how the interior of Mercury may have evolved, scientists think that the core may be around 3500 km in diameter and that it developed in the first billion years of Mercury's history. This large core occupies about 75 percent of the radius (Figure 5.16). If these calculations are correct, Mercury's iron core occupies the largest fraction of any planetary volume.

The discovery of Mercury's magnetic field came as a surprise and its origin is still unknown, but the most likely explanation is that the outer portion of the core is still molten. Convective movements within this electrically conductive liquid zone may create a magnetic dynamo that produces the magnetic field. Heat released from a crystallizing inner core may drive convection. A pure iron core should have solidified long ago. Perhaps the core in Mercury is iron sulfide, which has a lower melting point than pure iron. A less likely possibility is that the core is at this time entirely solid, but a remanent or inherited field is present. An earlier molten core

may have established a magnetic field that permanently magnetized a relatively thin surface layer several hundred kilometers thick. Orbiting spacecraft may be able to determine which of these two possibilities is the case.

The rigid outer shell of Mercury, consisting of a mantle and crust, may be called the lithosphere. It is probably around 500 to 600 km thick and consists of iron and magnesium silicates (Figure 5.16). Spectral data obtained from Earth indicate that the crust may be quite similar to the lunar highlands, with an impact regolith of anorthositic composition. This is also consistent with the bright nature of the mercurian crust and may indicate that the early crust developed by crystallization of a magma ocean like that of the Moon.

Mercury possesses a very tenuous atmosphere consisting of oxygen, potassium, and sodium vapor. The pressure of the atmosphere is not sufficient to support wind-related processes. This discovery suggests that Mercury is not totally devoid of moderately volatile elements like sodium, but then neither is the volatile-poor Moon. Potassium and sodium may come from feldspars in the crust. The only other body in the solar system with a similar sodium atmosphere is Io, which develops its atmosphere by active volcanism. Another puzzling observation regarding volatiles on Mercury comes from recent radar observations of Mercury. With the use of radio telescope dishes, radio beams can be bounced off Mercury; maps of the brightness of the reflected energy show a very bright area at the north pole. Although other materials could be responsible, a smooth polar cap of water ice would have similar characteristics. Is it possible that Mercury, the closest planet to the Sun, has polar caps of water ice? Some areas near the poles may be permanently in shadow and never warm up, providing a cold sink for the accumulation of volatiles. But where did the water come from? The other information we have about Mercury tells us that it is depleted in volatile materials like water.

In summary, the presence of a magnetic field and the high density of the planet indicate that Mercury is differentiated, with a large iron core that is probably still molten, unlike the Moon's, and a silicate crust and mantle like the Moon's. Much more remains to be discovered about the structure and composition of Mercury. Such data should be eagerly sought because of Mercury's unique position in the solar system.

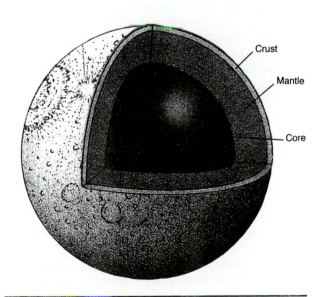

Figure 5.16

The internal structure of Mercury is postulated on the basis of its size, density, composition, and surface features. The best models predict an iron core 3600 km in diameter, containing 80 percent of the planet's mass. The overlying silicate rock layers are probably differentiated into a mantle and a crust.

Geologic Evolution of Mercury

The geologic history of a planet depends on many factors, including its size (mass and radius) and its chemical composition (determined by its position in the solar nebula). As it ages, each planet passes through three general stages: (1) a highly active period of crustal formation and mobility; (2) a volcanic stage accompanying a thickening sub-crustal lithosphere; and (3) a terminal quiescent state when the lithosphere is too thick to allow magma to puncture it or to move laterally. The rate at which a planet evolves through these steps depends on how quickly it cools, which in turn depends on the planet's size and composition. In this sense, Mercury provides geologists with an important reference marker: (1) It is the planet closest to the Sun and thus it may have an "extreme" chemical composition dominated by high temperature, refractory elements, and a lack of more volatile materials like water. (2) Mercury is larger than the Moon and should have evolved at a slightly different tempo. (3) Although smaller in radius, Mercury has nearly the same mass and surface gravity as Mars and almost the same bulk density as Earth, thereby showing how differences in these qualities affect a planet's subsequent development.

Only part of the surface of Mercury has been photographed, but geologists, utilizing the same methods and techniques as those used to study the Moon, have been able to establish a preliminary geologic time scale and develop a working hypothesis for the geologic evolution of Mercury. The large impact basins, such as Caloris Basin, like the Imbrium basin on the Moon, provide a useful reference for the major geologic events. It is clear from superposition and from crater frequencies that a period of intense bombardment occurred before the formation of the Caloris Basin, and the volcanic (smooth plains) material was emplaced afterward, followed by minor cratering. These periods of time have been given formal names taken from prominent craters of various ages. From oldest to youngest they are pre-Tolstojan, Tolstojan, Calorian, Mansurian, and Kuiperian. Comparisons of the crater densities on these terrains with those on the Moon can be used to estimate the absolute ages of these stages: pre-Tolstojan (4.6 to 4.0 billion years ago), Tolstojan (4.0 to 3.9 billion years), Calorian (3.9 to 3.5 or 3.0 billion years), Mansurian (3.5 or 3.0 to 1.0 billion years). Careful study of the figures in this chapter will reveal these fundamental relative age relationships. Figures 5.17, 5.18, and 5.19 show schematically how the interior and surface of Mercury may have evolved.

Stage I: Accretion and Differentiation.
Mercury probably accreted from materials that condensed at high temperature from the nebula

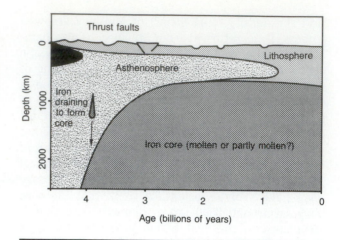

Figure 5.17

A graphic representation of Mercury's thermal history shows that a massive core must have formed early during the period of accretionary heating. At the same time much of the mantle probably melted. Rapid cooling later produced a thick rigid lithosphere and resulted in contraction of the planet.

that gave birth to the Sun and the rest of the planets. Where Mercury formed, only refractory elements were condensed as minerals and much of the iron was metallic—not in less-dense silicates minerals. Thus, the high density of Mercury could be explained by a large proportion of metallic iron. Moreover, Mercury is apparently water-poor. Silicate condensates that contain water, a volatile substance, formed farther from the Sun, apparently in the vicinity of the asteroid belt. Ices condensed only in the outer solar system.

Heat deposited in Mercury by accretionary impacts, and radioactive decay drove the internal differentiation of the planet. By analogy with the Moon, much of the outer part of Mercury probably began to melt soon after its formation about 4.5 billion years ago. Light silicate minerals eventually crystallized and formed the crust. Denser silicates accumulated as the mantle. For a planet with the size and density of Mercury, this silicate shell (the crust and mantle) could only be 600 to 700 km thick.

Mobile crustal plate interactions may have been limited to this early period of crustal formation. A rigid lithosphere must have developed well before the end of heavy bombardment because craters that formed during this episode are preserved. As heat was radiated from the planet into space, the depth of the molten zone increased and the rigid lithosphere formed above it.

During this epoch the core was formed, as a "rain" of metallic droplets concentrated in the center of the planet. This redistribution of mass from its initially more homogeneous state provided more

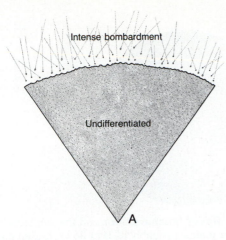

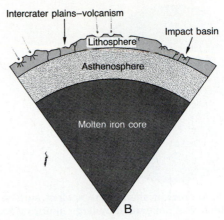

Stage I. Accretion and differentiation resulted in the formation of a planet with a large iron core. The molten core and silicate mantle caused global expansion and tensional fracturing in a thin, solid lithosphere.

Stage II. Period of intense bombardment and formation of intercrater plains.

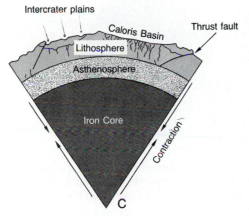

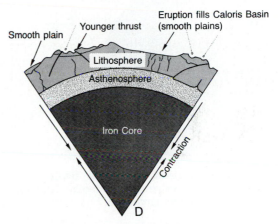

Stage III. Excavation of Caloris Basin and formation of the associated hilly and lineated terrain on the opposite side of the planet. Convection in the mantle had already allowed the planet to cool sufficiently to cause global contraction, resulting in compressive stress and thrust faulting at the surface.

Stage IV. Formation of the smooth plains, probably from volcanic extrusions.

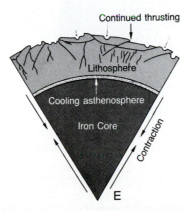

Stage V. Cooling and contraction are completed and the planet became tectonically inactive as the lithosphere thickened. The only process to modify the surface significantly is the occasional impact of meteorites, which create rayed craters.

Figure 5.18
The geologic history of Mercury.

Stage I. Accretion shaped the ancient surface of Mercury which was dominated by large multiring basins and large tracts of heavily cratered terrain.

Stage II. Heavy bombardment and the emplacement of intercrater plains, probably as lava flows, buried many of the earlier features.

Stage III. Formation of Caloris Basin, late in the period of heavy bombardment, modified the surface of the planet over a very large area.

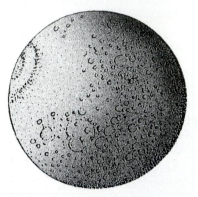

Stage IV. Emplacement of smooth plains continued after formation of Caloris Basin during a period of declining impact rates that extended to the present.

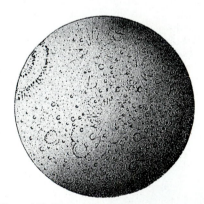

Stage V. The present surface of Mercury is modified only by occasional impact craters. Cooling and contraction formed scarps.

Figure 5.19

The surface of Mercury has changed dramatically over the course of its history as illustrated in this sequence of diagrams.

heat that may have aided in forming magmas that extruded on the surface as the intercrater plains during the early heavy bombardment and enhanced isostatic adjustment of craters. This difference may be one of the most important Mercury–Moon contrasts, explaining why intercrater plains are more expansive on Mercury. Even if core formation occurred on the Moon, the amount of energy released would have been small by comparison with Mercury.

Theoretical models of condensation and geologic evidence suggest that Mercury never contained abundant volatiles and probably never outgassed an atmosphere or hydrosphere as the interior differentiated. Mercury has sufficient gravity to retain an atmosphere at least as thick as that of Mars, but its envelope of sodium vapor pales by comparison. Nor is evidence of a past eolian regime visible in Mariner 10 photos. Subsequent research may determine if an atmosphere ever formed by release from the interior and, if so, may explain what happened to it.

Another hypothesis that might explain Mercury's high density and lack of a thick atmosphere appeals to the possibility of a large impact during this stage of its early evolution. A large, perhaps Moon-sized body may have collided with Mercury after core formation. If the impactor was large enough (20 percent of Mercury's mass), it may have stripped away the outer shell of less-dense silicates, leaving Mercury smaller and richer in dense iron. The mass of precatastrophe Mercury could have been twice as large before impact. Therefore, the high density of Mercury could reflect its accretion history and not necessarily a high condensation temperature. Such a large collision may have purged the volatiles from Mercury's outer portions, making the formation of an atmosphere less likely. If the giant collision scenario holds true for Mercury, as it appears to for the Moon, condensation of solids in a thermal gradient around the ancient Sun may not be required to deplete volatile elements.

Stage II: Heavy Bombardment and Formation of Intercrater Plains (Pre-Tolstojan and Tolstojan).

A period of intense bombardment is recorded by the clusters of densely packed large craters and basins. The oldest surfaces on Mercury do not have as many large craters as those on the Moon, and it appears that periods of heavy bombardment on Mercury occurred during the emplacement of the intercrater plains (Figure 5.19). The material that forms the intercrater plains could be volcanic, or it may be basin ejecta. Similar deposits occur in the lunar highlands; based on crater statistics and stratigraphic relations on both planets, they appear to be volcanic rocks emplaced during the late stages of the heavy bombardment. The mercurian intercrater plains may be more voluminous than their lunar counterparts because core formation in Mercury produced much more melting during this early period than was possible for the Moon with its small core. There is no evidence preserved of planetary expansion, which presumably would have accompanied this thermal event. Subsequently, the lithosphere cooled, thickened downward, and in time contracted. Calculations of this development show that Mercury's radius may have decreased by 2 km. This contraction may have decreased the surface area and caused global thrust faulting, producing the scarps and ridges so characteristic of the mercurian surface. Many scarps appear to have formed before the Caloris Basin formed.

Stage III: Formation of Caloris Basin (Early Calorian Period).

The formation of the large multiring Caloris Basin was a major event in the geologic development of Mercury. Ejecta from this basin extends more than 1000 km away from the rim. The excavation of the basin modified the landscape over much of the photographed surface, forming large ejecta deposits and radial ridges and valleys far beyond the outer ring of mountains. Hilly and lineated terrain on the opposite side of the planet probably formed as a result of this tremendous impact. There may be other large multiring basins not seen on Mariner 10 photographs. If a lunar analogy can be drawn, Caloris probably formed about 4 billion years ago.

Stage IV: Formation of Smooth Plains (Middle to Late Calorian Period).

The smooth plains material that fills Caloris Basin and parts of the surrounding areas represents flooding of the earlier basins at a time when the heavy bombardment had greatly decreased. This material is probably of volcanic origin and was extruded over a period of time, but small variations in crater densities between different areas imply that the period during which flooding occurred was relatively short. The smooth plains were emplaced as the final product of the volcanic stage of Mercury's evolution, probably by 2 or 3 billion years ago, shortly after the decline in the cratering rate. Even though the lithosphere was thickening, magmas were apparently still able to reach the surface. Mercury's lack of water (which would have lowered the melting points of many of its component materials, allowing them to stay liquid over a longer period of time) may have increased the rate of lithospheric thickening relative to Earth; and the timing of the volcanic events may approximately coincide with similar events on the Moon.

Structural modification of the smooth plains in the Caloris Basin produced large ridges and open fractures that may be related to isostatic adjustment of the basin's interior. Smooth plains elsewhere are wrinkled by lobate scarps formed as Mercury continued to contract or are undeformed.

Stage V: Light Cratering (Mansurian and Kuiperian Periods).

After the period of smooth plains formation, the surface of Mercury was subjected to light cratering, which formed the bright-rayed craters. The density, distribution, and morphology of these craters resembles the postmare cratering on the Moon with slightly degraded but still relatively fresh craters formed during the Mansurian Period and rayed craters formed during the Kuiperian Period.

The absence of subsequent modification of the surface of Mercury by tectonism, volcanic activity, or atmospheric processes is significant because it indicates that after the period of basin flooding, the geochemical and tectonic evolution of Mercury was essentially completed. The extrusion of the plains material was apparently the end of Mercury's dynamic history. Mercury's lithosphere may be rigid all the way to its core, with no intervening asthenosphere. Nonetheless, Mercury appears to have remained warm enough to maintain a convecting iron core. There is no observable deformation of the outer silicate shell of Mercury such as would arise from recent movement caused by the postulated fluid core. Although the unexplored 50 percent of Mercury's surface could reveal evidence of recent internal activity, it is very likely that the only processes available to modify Mercury after the end of its final period of volcanic activity are degradation of slopes by gravity-driven mass movement and the occasional impact of objects ranging from small meteorites through micrometeorites and cosmic particles.

Conclusions

The cratered surface of Mercury is strikingly similar to that of the Moon and attests to the importance of meteorite impact as a general process in the solar system. The largest impact structure photographed on Mercury is the multiring Caloris Basin, similar in form, and probably in age, to the Moon's Imbrium Basin. This large basin is younger than a heavily cratered terrain (similar in many ways to the lunar highlands) that contains interspersed smoother plains. Still younger plains fill the Caloris cavity and are found scattered across the rest of the photographed part of the planet. Both generations of plains were probably produced by lava flows—an indication of the importance of volcanism in the development of the planets. These terrains are transected by distinctly mercurian scarps that appear to be thrust faults created as the planet cooled and contracted.

The most significant differences between the Moon and Mercury are the result of Mercury's larger size and enrichment in iron. Impact crater ejecta are distributed closer to the craters than on the Moon. Perhaps more important, Mercury appears to have cooled more slowly so that plains-producing volcanic activity during the period of intense bombardment was more long-lived and perhaps more vigorous than on the Moon. Moreover, the interior must be relatively hot to this day because Mercury has a magnetic field that is thought to be generated by convection within a molten metallic core. If the Moon has a metallic core, it solidified completely billions of years ago. Mercury's iron-rich composition and large core may be the consequence of condensation and accretion of its constituents near the forming Sun. The absence of a significant atmosphere or any surface fluids on Mercury was predetermined by its conception in this part of the solar system.

The geology of Mercury reinforces the notion that the tectonic and volcanic activity on a planet depend on the thermal state of the interior (the temperature distribution at depth). Since most planets were initially quite hot as a result of their accretion, much of their thermal history is dominated by cooling. Small planets, like Mercury, with large surface-area mass ratios, cool rapidly and have short thermal histories. Mercury, with a ratio higher than Mars and lower than the Moon, may have had a thermal history intermediate to these planets.

In short, the history of Mercury produced a Moonlike planet whose development was modified in pace and tenor by the distinctive properties of this, the innermost of the planets.

Review Questions

1. Compare and contrast the surface of Mercury with the surface of the Moon.
2. In what ways do the impact craters on Mercury differ from those found on the Moon? Why do they differ?
3. Is it possible to determine the absolute ages of surfaces and features on Mercury?
4. Why do surface temperatures on Mercury range from very hot to very cold?
5. Why is spin-orbit coupling a common phenomena in the solar system?
6. What composition of volcanic rocks would you expect to find at the surface of Mercury? Why?
7. How do the plains on Mercury differ from the lunar maria? What was their probable mode of origin and age?
8. What is the principal evidence that Mercury experienced global contraction during its history? When did this happen—early or late? Why did Mercury contract rather than expand?
9. How does the interior of Mercury differ from the interior of the Moon? Is there any evidence that the interior is still molten?
10. Mercury formed very near the early Sun. What is the evidence that Mercury is rich in refractory elements as a result? Are there other processes that could explain its iron-rich composition?
11. Outline the major events in the history of Mercury and compare them to the major events in the Moon's history.
12. Your job is to make recommendations for a manned mission to Mercury. Where should the space ship land to obtain the most information about Mercury? What tasks should the astronauts perform? What instruments should they take with them? What should they bring back with them? Assume the astronauts have a small "rover" and will be on the planet for two weeks.

Key Terms

Antipode

Despinning

Scarps

Spin-Orbit Coupling

Additional Reading

Davies, M. E. et al. 1978. *Atlas of Mercury*, NASA SP-423.

Journal of Geophysical Research. 1975. Vol. 80, No. 17. (This entire issue is devoted to analysis of data returned from Mercury by Mariner 10.)

Murray, B. C. 1975. "Mercury." *The Solar System*. New York: W. H. Freeman and Co., pp. 37–48.

Strom, R. G. 1984. "Mercury." *The Geology of the Terrestrial Planets*, NASA SP-469, pp. 13–55.

Strom, R. G. 1987. *Mercury: The Elusive Planet*. Washington, DC: Smithsonian Institution Press.

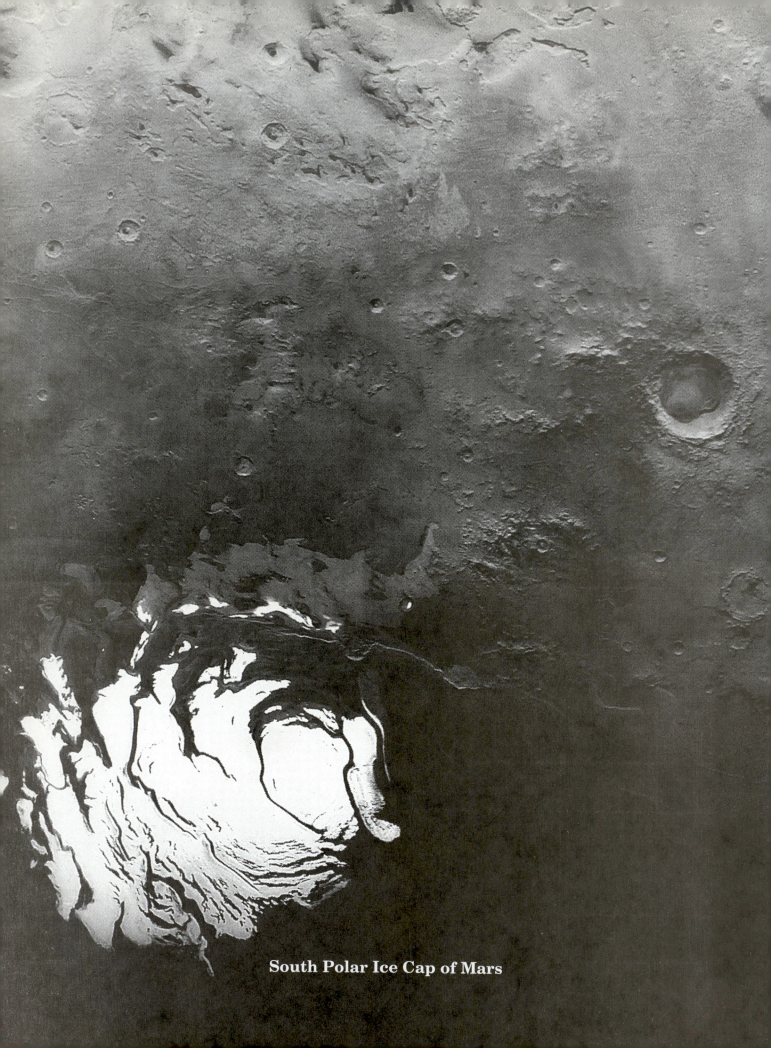

South Polar Ice Cap of Mars

TABLE 6.1

Physical and Orbital Characteristics of Mars

Mean Distance From Sun (Earth = 1)	1.52
Period of Revolution	687 d (1.881 y)
Period of Rotation	24 h 37 m
Inclination of Axis	24
Equatorial Diameter	6,787 km
Mass (Earth = 1)	0.108
Volume (Earth = 1)	0.15
Density	3.93 g/cm^3
Atmosphere (main components)	carbon dioxide
Surface Temperature	190 to 240 K
Surface Pressure	6 to 10 mb
Surface Gravity (Earth = 1)	0.38
Magnetic Field (Earth = 1)	< 0.00003
Known Satellites	2
Surface Area/Mass	22 × 10^{-11} m^2/kg

CHAPTER 6

Mars

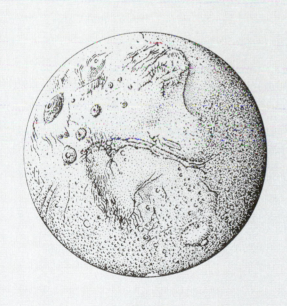

Mars

Earth

Introduction

The past 30 years, American and Soviet spacecraft have produced a wealth of information about Mars. Photomosaics of the entire planet have been made, topographic and surface relief maps have been published, and geologic maps compiled in a variety of scales. Moreover, we may also have pieces of Mars here on Earth, delivered free of charge in small meteorites. Consequently, our understanding of the planet generations of humans thought to be most like Earth has increased dramatically.

Compared to the landforms on the Moon or Mercury, the surface features of Mars have origins more diverse and complex. Almost everything on Mars is not only big but gigantic. Impact shaped about half the surface; the largest crater is about twice the size of the largest crater on the Moon. Other areas have been formed by volcanic activity, with some volcanic shields reaching more than 20 km above the surrounding surface. The largest volcano on Mars has ten times more volume than the largest volcano on Earth. Mars even has a tenuous atmosphere and bright polar caps. Photographs show the details of advancing and retreating frost shrouds. Wind action is a major geologic process, and moving sand and dust altered many features on the planet. To the surprise of all, ancient channels are found on the surface; these appear to have been formed by running water. Tectonic movements fractured parts of the planet and produced a great canyon system that has been enlarged by erosion. It is apparent that the geologic agents operating on Mars not only differed from place to place, but also varied throughout the planet's long history.

Because it is larger than the asteroids, the Moon, and Mercury, Mars provides another reference point in the progression of planet size for understanding the fundamental principles of how and why the planets evolve.

Major Concepts

1. The surface of Mars can be divided into two major regions: (a) the densely cratered, more ancient highlands in the southern hemisphere and (b) the younger, lower plains in the north.

2. Mars probably possesses an internally differentiated structure with a metallic core, a thick mantle composed of iron-rich silicate minerals, and a thin crust.

3. Cratering has been a major geologic process on Mars, and a record of an early period of intense meteorite bombardment has been preserved.

4. Volatiles outgassed from the interior formed an atmosphere and a simple hydrologic system. Later, as temperatures lowered, liquid water became locked up in the polar caps and in the pore spaces of rocks and soil, as groundwater or ice, and was only occasionally released in large floods.

5. Eolian processes have been observed in action on Mars, and many surface features have been modified by wind erosion or deposition.

6. Volcanism is revealed by three types of features: (a) huge shield volcanoes, (b) volcanic patera possibly unique to Mars, and (c) volcanic plains.

7. Large crustal domes and graben are the major tectonic features and may be produced by thermal convection in the mantle.

8. Mars experienced a long and complex geologic history; it is not a primitive sphere dominated by impact scars, as are the Moon and Mercury. The larger size of Mars and its compositional differences may be responsible for its extended thermal evolution.

The Planet Mars

The planet Mars has fascinated observers for many centuries, partly because of its distinctive red glow but also because it is so near Earth. Indeed, many of its physical characteristics suggested that Mars might even harbor life (Table 6.1). Fortuitously, the rotational axis of Mars currently tilts almost exactly the same amount as Earth's, causing cyclical season changes (which last about twice as long on Mars because of its longer period of revolution about the Sun). Polar caps of carbon dioxide and water ice advance and retreat in response to the temperature changes. (Some imaginative scientists early in the twentieth century suggested that the color changes accompanying the season changes showed that plant life existed on the surface, advancing like a front toward the poles as each spring brought warmer temperatures.) Mars also possesses a thin atmosphere composed mainly of carbon dioxide. Nitrogen, oxygen, and a seasonally variable amount of water are also present in the atmosphere. The water-vapor content is much less than that in the air over terrestrial deserts. Tenuous clouds and migrating storms indicate that Earthlike processes occur within the atmosphere. Although temperatures are generally below freezing and the atmospheric pressure is a hundred times lower than that on Earth, the fourth planet from the Sun seems hospitable compared to the airless Moon or the searing heat on Mercury.

Mars is a fairly small planet, larger in diameter than Mercury or the Moon, but smaller and less massive than Earth (one-tenth mass of the Earth). The gravitational force exerted on objects near the surface is slightly more than one-third of Earth's constant tug and about the same as Mercury's. At 24 hours and 37 minutes, the martian day is only a bit longer than a day on Earth. Mars has two natural satellites. These tiny moons, called Phobos (about 20 km across) and Deimos (only 10 km across), are the size of small asteroids and were described in Chapter 3.

Geologically, Mars is vastly different from Earth in spite of some outward similarities mentioned above. It is an "intermediate" planet, with an array of Moonlike, Earthlike, and uniquely martian features, which we will examine closer.

Major Geologic Provinces

The surface of Mars was completely photographed during the epic flight of Mariner 9 in 1971 and 1972. Subsequently, remarkably detailed photographs were obtained by the orbiting satellites of the Viking mission, which operated from 1976 to 1980. From these data, topographic and geologic maps of the whole planet have been made and an intriguing, increasingly complex interpretation of the martian surface is developing.

The generalized geologic map of Mars (Figure 6.1) shows the location of the major provinces and the distribution of the important terrain types. The surface of Mars may be divided into two major geologic provinces: (1) the old, densely cratered terrain in the south and (2) the low, young, sparsely cratered plains in the north. The boundary separating the two almost coincides with a great circle inclined at 35° to the equator. The landforms in these major provinces were modified by a surprising variety of surface and internal processes to produce terrain types not seen on the Moon or Mercury.

Densely Cratered Southern Highlands

The densely cratered terrain in the southern hemisphere of Mars is a highland somewhat like the ancient cratered terrain of Mercury or the highlands of the Moon, where craters of all sizes and shapes dominate the landscape. Several huge multiring basins, larger than Imbrium or Caloris, are the major features in parts of the highlands. Elevations are generally in excess of 1 km above the mean martian radius except in the smooth interiors of the Hellas and Argyre basins. Much of the martian cratered terrain contains large areas of relatively smooth intercrater plains and a few old shield volcanoes.

Although the nature of the intercrater plains on Mercury is still in question, some martian plains are definitely volcanic. Some large channels that appear to have been carved by **rivers** or huge floods cross this ancient battered surface as well, giving it a decidedly nonlunar appearance.

The Northern Plains

The low plains in the northern hemisphere are separated from the southern highlands by a prominent cliff that is 2 to 3 km high in places. If Mars had an ocean 2.5 km deep, the ocean would cover the northern plains and leave most of the heavily cratered terrain high and dry as one huge landmass. The brightness of the plains is highly variable, reflecting the diverse origins of the different smooth surfaces. Photographs of this area show prominent mare-type lava flows, small lava or cinder cones, and ridges, all indicative of a volcanic

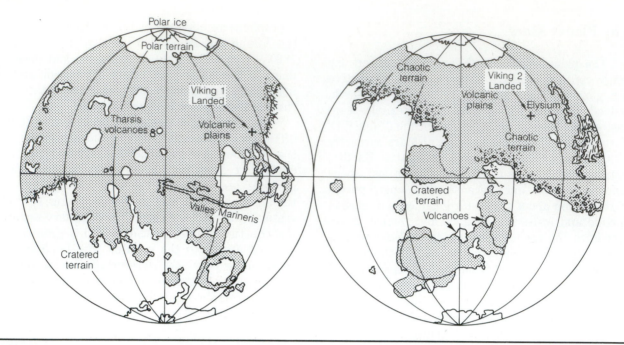

Figure 6.1

A geologic map of Mars shows the major regions: (1) The southern hemisphere is an old, densely cratered highland with a crater distribution more like Mercury's intercrater plains than like the lunar highlands. This hemisphere is furrowed by small, dendritic valley systems. (2) The northern hemisphere is dominated by younger, relatively smooth plains, apparently composed of sedimentary deposits and vast lava flows. A well-defined escarpment separates the two provinces, except where it is buried by younger lavas of the Tharsis region. Several distinctive terrain types have been formed by erosion and slope retreat along the escarpment. The major flood channels cross the escarpment and empty into the northern lowlands.

origin. Dunes and wind streaks that formed behind topographic obstacles attest to eolian (wind) modification and blanketing of large tracts. Since many of the major channel systems empty into basins in the north, stream sediments may be interlayered with the other deposits. It was in this area of complicated geology that the two Viking landers touched down.

Crustal Upwarps

Two continent-sized domical highs or upwarps capped with huge shield volcanoes are located in the northern hemisphere near the escarpment. The largest is in the Tharsis region of the western hemisphere, where the volcanic field appears to sit astride the global escarpment. The smaller field, Elysium, is located well within the northern plains of the eastern hemisphere.

The large structural dome in the Tharsis region has created a system of radial fractures that extends nearly halfway around the globe. Faulting and erosion along an east-west segment of the fracture system has developed Valles Marineris, the "Grand Canyon of Mars," which extends from

the fractured bulge near Tharsis across the cratered plains and breaks into several north trending channels near the escarpment.

Global Escarpment

The ultimate origin of the scarp separating the two physiographic provinces is still unknown. It is well defined in the eastern hemisphere but is masked in the west by lava flows from the Tharsis volcanoes. The escarpment is an erosional contact between the younger northern plains and the ancient cratered highlands in the south. It has been dissected by various types of stream erosion and mass movement that are unique to Mars. This erosion gradually forced the margin of the highlands to retreat southward. In places, the cratered highlands south of the escarpment are fractured and modified by slumping. This landscape is accentuated as the scarp is approached so that only chaotic masses of angular blocks, or **mesas**, are left standing above the plain—creating a fantastic landscape called the fretted terrain. Farther into the plains, only small, rounded knobs are left; these eventually disappear to the north. Some major channels arise in this complexly eroded transition zone and modify the global escarpment.

Polar Regions

The bright polar caps of Mars were first identified from Earth with the use of telescopes. Large permanent caps of water ice are located at each pole. The caps rest on a sequence of layered sedimentary deposits. Carbon dioxide ice is present at least temporarily at both poles. Revealing the power of the atmospheric circulation, sections of the surrounding terrain have been pitted and etched by wind-related erosional processes, and a discontinuous ring of dunes surrounds much of the northern cap, which was photographed during the Viking mission.

On the Surface of Mars

Certainly one of the most important accomplishments during the last two decades of space exploration was the soft landing of a pair of spindly Viking spacecraft on Mars in 1976. The landers performed amazingly well and returned panoramic vistas of boulder-strewn plains from two widely separated spots on the planet, as well as chemical analyses of the regolith. Although both sites appear superficially the same, there are some significant differences. The sites were chosen because of their smoothness as seen from orbit and because of the strong indications that water may be present in the soil. They thus do not show us a view of the more spectacular landscapes of Mars found elsewhere.

Chryse Planitia

Viking 1 landed in a low, smooth basin in the northern hemisphere several hundred kilometers from the global escarpment. Several large channels that drained water from the highlands to the west and south flow into Chryse Planitia (Gold Plain), and it was thought that water from them might have helped to support some sort of biological activity. The cameras on Viking 1 showed a dusty, orange-red surface strewn with boulders of all sizes and shapes (Figure 6.2). Drifts of windblown sand have collected behind and on top of many rocks; the wind has also fluted some pebbles and scoured away the sand to expose the light-colored bedrock beneath the soil.

One striking feature photographed by the lander is the bright sky (Figure 6.2). The Moon, because it has no atmosphere, has no sky and is surrounded only by dark space. The martian sky has a reddish tint imparted by dust particles suspended in the thin, carbon dioxide atmosphere. The dust may be red because of the oxidation of iron-rich minerals in the soil—a process similar to rusting.

There are no indications from the photographs in Figure 6.2 that Chryse Basin was formed by impact. Instead, it may be a shallow sediment- and lava-filled trough. Although no impact craters occur near the spacecraft, several appear on the horizon, and views from orbit show that large channels extend to within about 60 km of the landing site. The history of the area probably began with the formation of a densely cratered surface that was later covered with lava flows. Channel formation along the borders of the basin may have flooded the area and deposited sediments eroded from the highlands to produce the rubbly landscape. Subsequent light bombardment and continual eolian modification have rendered further changes.

Utopia Planitia

Far to the north, on the other side of Mars, Viking 2 set down in Utopia Planitia about a month after Viking 1. In spite of the low temperatures, abundant water vapor had been detected from orbit. During the summer days temperatures hovered at about -10° C (260 K) and the low at night reached a frigid -50°C (220 K). The photos returned from Utopia show an orange-red plain littered with porous, spongelike rocks, similar to some terrestrial volcanic rocks (Figure 6.3). Drifts of sand and wind-shaped cobbles and boulders are not as common here as on the Chryse plain. A polygonal system of troughs that is not seen at the other landing site developed in the soil. The troughs may have formed as ice wedges in the soil that cracked it and formed shallow linear depressions on the surface.

The view from orbit shows that Utopia Planitia is a relatively smooth plain, locally dotted with low volcanic domes and small craters. The crater frequency is lower here than at Chryse and must represent a younger surface. The simplest history of the area maintains that Viking 2 landed on the far edge of an ejecta blanket formed when the crater called Mie was excavated. Other theories hold that the rubbly surface was shaped by massive floods or as debris was dropped by a formerly more extensive ice cap.

Using various instruments, the Viking landers alternately poked, prodded, scooped, hammered, cooked, and analyzed the martian soil, determining

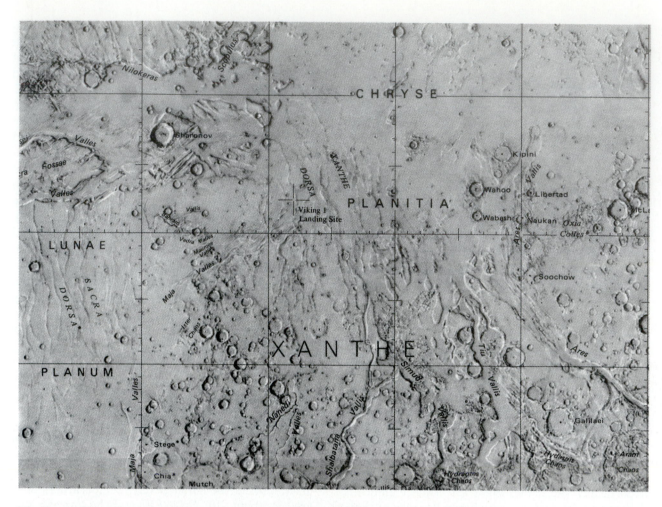

(A) **Chryse Planitia,** where Viking 1 landed, consists of deposits shaped by vast floods, which flowed north across the highly cratered southern hemisphere and east across Lunae Planum.

(B) **The panoramic view of the Viking 1** site shows the meter-high dunes and abundant angular rocks. The large boulder in the foreground is about 2 m across. Many of the features here indicate the importance of wind activity in shaping the details of the martian landscape. Some of the sand dunes have been eroded, so their internal cross-stratification is exposed. The gravel surface was formed as a lag deposit resulting from deflation.

Figure 6.2
The Viking 1 landing site on Mars is located near the mouth of several large channels which converge on Chryse Planitia in the northern hemisphere of Mars.

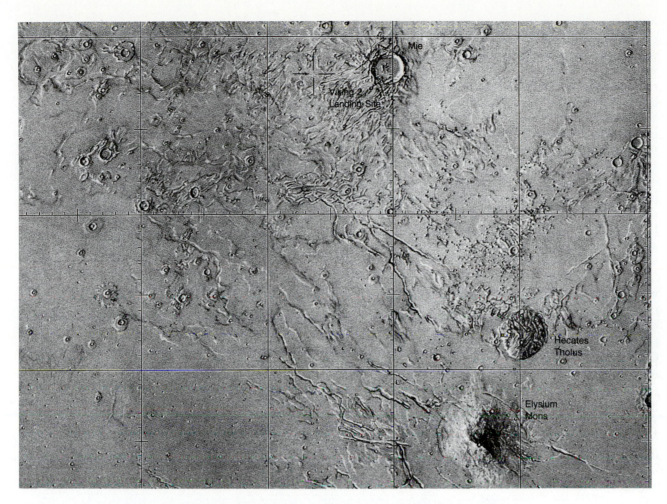

(A) **Viking 2 touched down in Utopia Planitia** about 200 km west of Mie, a large and relatively young impact crater. The volcano Elysium Mons lies far to the south. High concentrations of atmospheric water vapor had been measured from orbit over this area.

(B) **The view from Viking 2** reveals a vast plain littered with angular blocks and low drifts of sand. The blocks may be ejecta from the impact crater, or they may be the weathered remnants of a lava or debris flow. Like those at the Viking 1 site, many of these rocks are pitted. The scene is tilted because the lander is resting on a tilted surface.

Figure 6.3

The Viking 2 landing site is located on the opposite side of Mars but is similar to the Viking 1 site.

its composition and physical characteristics and trying to find out if it contained any evidence of life. Although no unambiguous signs of martian biota were found, we now know much more about the chemistry of martian soil. Comparison with terrestrial materials indicates that the martian soil may have originated from hydration or alteration of iron-rich basaltic rocks, consistent with the hints of volcanism in both areas.

The recent conclusion that a few meteorites are samples of Mars, blasted away by a low angle impact, has opened a new door for our exploration of Mars. We discussed these meteorites briefly in Chapter 3. These **SNC meteorites** (short for *Shergotty, Nakhla, and Chassigny)* are igneous rocks with basaltic compositions very similar to that found by the Viking landers. The gases trapped within them have the same composition as that of the martian atmosphere as analyzed by several spacecraft. Their ages (1.3 billion years) are extremely young for any asteroid and suggest that they must have formed on a large planet that cooled slowly and had a long volcanic history. Their compositions tell us that there is significant water in the martian interior and at the surface; the abundances of moderately volatile elements like potassium suggest that Mars has a larger endowment of volatiles than Earth itself. Their ages give us a few punctuation points for the long martian history.

The Internal Structure of Mars

We have seen in previous discussions that both the Moon and Mercury have differentiated interiors made up of shells with different chemical compositions. Unfortunately, the Viking landers did not detect "marsquakes" strong enough to define the thickness of the various shells precisely, but much has been inferred about their nature from variations in the gravity field, comparison with better-understood planets, and calculations to predict the thermal evolution of the martian interior (Figure 6.4).

Mars is larger in diameter than Mercury and about twice as massive, but it has a lower bulk density (3.93 g/cm^3). Apparently, Mars is composed of lighter materials than Mercury (5.44 g/cm^3), which is consistent with nebular condensation models that predict significant incorporation of light volatile materials (water for example) in Mars that did not condense where Mercury formed. This compositional implication and its larger size, and consequently a greater ability to retain heat, lead one to suspect that the internal structure of Mars

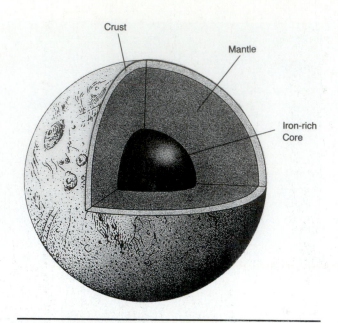

Figure 6.4

The interior of Mars is probably dominated by a thick mantle composed of iron and magnesium silicates. The core, probably composed of iron and sulfur, does not occupy as large a portion of Mars as metallic cores occupy in Earth and Mercury. The crust of Mars and most other planets is very thin and probably resulted from the original differentiation of the planet. The crust may be 25 to 70 km thick. An asthenosphere of partially molten mantle material probably exists at a depth of about 250 km, marking the base of the lithosphere.

may be more like that of Earth than that of the Moon or Mercury.

Theoretical models of the interior structure predict just such a situation (Figure 6.4). Mariner 9 returned data demonstrating that there is a significant density contrast between the shallow layers of Mars and its deep interior, including evidence for a core. The radius of the core is probably about 1500 to 2000 km. It is not composed just of iron but probably contains some sulfur or oxygen, which lowers its density. The SNC meteorites are also depleted in elements that have an affinity for sulfur; we speculate that these elements were removed into the sulfur-rich core by differentiation. Several Mars probes have shown that the martian magnetic field is very small. This is somewhat surprising in that Mars is larger than Mercury and presumably warmer, with a convecting core. It also spins much faster on its axis. Both features should create a stronger magnetic dynamo for Mars, and yet Mercury has a magnetic field and Mars does not. Perhaps the martian core is completely solidified now, after eons of cooling. The larger size of the mercurian core, a unique composition, or tidal energy inputs may have kept it molten and convecting for much longer.

The bulk of the mass of Mars probably lies in a thick silicate mantle about 1300 to 1800 km thick. It is very likely that the interior is still actively convecting, which may have created forces that tectonically deformed the crust. Moreover, some models of the thermal evolution of Mars predict that a thin layer within the mantle could still be partially molten and might serve as a source for active volcanism at the surface. This ductile asthenosphere probably occurs at a depth of about 250 km and marks the base of the overlying rigid lithosphere. The lithospheres of the Moon and Mercury are probably much thicker (1000 km and 600 km respectively); partial melt zones, if they occur at all, are very near the compositional boundary between mantle and core in these other planets.

The martian lithosphere is probably composed of two units: the upper mantle just described and the crust formed during the primordial differentiation of the planet. From comparisons with the Moon, the crust is thought to be rich in aluminosilicate minerals and is very thin compared to the mantle. The thickness of the Moon's crust varies from 60 to 100 km; Earth's varies from 5 to 80 km, with an average thickness in the range 30 to 35 km. Analysis of the gravity field of Mars indicates that its crust averages 25 to 40 km thick and that it is probably thinner in the northern hemisphere. The crust may reach a maximum of around 70 km under the high Tharsis plateau; however, some models suggest a very thin crust or lithosphere beneath Tharsis. Resolution of this problem awaits further investigation.

Internal differentiation of a planet is thought to occur when the planet partially melts from either accretionary heat or radioactive decay early in its history. As rocks melt, gases and vapors are usually released first and, because they are much lighter, rise rapidly to the surface. The relatively large gravity field of Mars has enabled the planet to retain a tenuous outer shell of such gases. A planet's atmosphere may be considered to be its outer layer, an important part of its structure. The martian atmosphere is thin, exerting a pressure of only about 0.01 that of Earth's atmosphere, and is composed mainly of carbon dioxide, but it has played a very significant part in the geologic evolution of Mars.

Impact Craters and Basins

Although photographs from recent space missions have revealed a far greater variety of landforms on the surface of Mars than were seen on the Moon or Mercury, cratering still appears to have been the dominant process in shaping the surface of Mars, especially during the early history of the planet. The craters on Mars have many distinctive features, however, that reflect the gravitational attraction, surface processes, and erosional history of the planet. Many craters are shallow, flat-floored depressions that show evidence of much more erosion and modification by sedimentation than do those on Mercury or the Moon. Rayed craters are rare, but fresh craters do occur and are unique in that some are surrounded by lobate scarps, indicating that the ejecta may have moved like a mudflow after it fell to the surface.

The densely cratered highland surface of Mars was apparently created by the impact of many meteorites during the episodes of cratering that shaped the surfaces of the Moon and Mercury. In terms of size and shape, the craters themselves are nearly the same on all three bodies. Craters smaller than 10 to 15 km in diameter are simple, bowl-shaped depressions with raised rims and smooth walls and floors. Craters larger than this are complex and usually have central peaks and terraces on the walls. Concentric rings, instead of peaks, are characteristic of most craters larger than 100 km in diameter; the same transition occurs in lunar craters when they exceed 200 km in diameter, possibly due to the effect of the lower surface gravity on the Moon. Hummocky ejecta blankets and fields of secondary craters are less prominent around martian craters, and classic rayed craters like Tycho are rare (Figure 6.5).

The morphology of crater ejecta on Mars is unique. Ejecta blankets of lunar craters are usually blocky near the rim and grade outward to finer

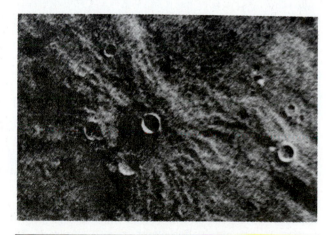

Figure 6.5

Rayed impact craters like those found on the Moon and Mercury are rare on Mars. One in Acidalia Planitia is shown here, as photographed by Mariner 9. The crater is about 3 km across. Bright halos, not rays, surround other smaller craters.

particle sizes until the blanket merges imperceptibly with fields of secondary craters and rays. All of these features are consistent with the ballistic emplacement of the ejecta. Many martian craters, in contrast, have ejecta deposits that appear to have flowed over the surface like mudflows and lack the delicate rays seen around craters on other planetary objects. The continuous ejecta deposit commonly extends from 1.5 to 2 times the radius of the crater—on the Moon this value is only about 0.7 and on Mercury (which has a gravity similar to Mars) only about 0.5. Apparently, the ejecta from those martian craters became fluidized in some fashion, allowing it to flow long distances after excavation. These craters are called **rampart craters**, fluidized craters, or splash craters.

A spectacular example of a rampart crater is the 18-km-diameter crater named Yuty shown in Figure 6.6. The ejecta consists of several relatively thin layers or sheets of material with tongue-shaped projections. It appears that the debris flowed outward like huge splashes of mud that surged outward when they hit the ground. A prominent ridge was produced at the front of each ejecta lobe. To the right, the ejecta flow overlaps an older degraded remnant of a larger crater and actually moved up and over the wall formed on the eroded ejecta blanket. To the south, a smaller and older crater separates two large lobes on the surface of the ejecta but was also filled with liquified debris. Liquid water may have been incorporated into the material as it was thrown out of the crater, creating a rapidly moving mudflow. The collision of the meteorite with the surface may have melted ice in the regolith and mixed the water with the ejecta.

Craters with diameters less than 15 km usually have a single ejecta sheet (Figure 6.7) that only extends out to about one crater radius. The surface of the sheet is typically marked with many concentric ridges and ends abruptly at a ridge or escarp-

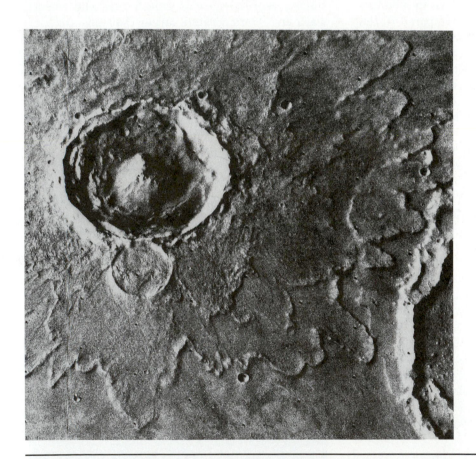

Figure 6.6

The mudlike ejecta of the crater Yuty is typical of the nature of the ejecta of many martian impact craters. The ejecta consists of a series of overlapping lobes. The smooth, rounded fronts of the lobes and their diversion around the small crater rim suggest that the debris moved close to the ground as a surge of mud. The ejecta deposits have nonetheless overtopped the eroded remnants of an older crater's ejecta (on the right side of the photo). The ejecta was fluidized at the time of impact, probably by the melting of near-surface ground ice. Yuty lies in the flooded portion of Chryse Basin.

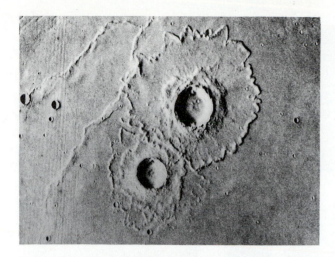 

(A) **The complex lobate and radial patterns** of the inner ejecta deposits are strikingly displayed. A relatively thick platform of ejecta surrounds the rim. A steep outward-facing scarp is variably developed around the innermost ejecta.

(B) **The lobes of lineated debris** that extend beyond the scarp were probably emplaced as mudflows across the fractured northern plains of Mars. Piles of debris at the base of the central peak formed as material slumped off the steep walls of the crater.

Figure 6.7

Craters with fluidized ejecta are common at the surface of Mars and present a variety of forms.

ment. Locally, rays or fields of secondaries or hummocks related to the crater extend beyond the edge of the ejecta lobe. The ejecta sheet appears to be thicker and more viscous than those associated with Yuty-type craters. Terraces are present on the walls of many of these craters and are significantly displaced from the crest of the rim.

Another type of rampart crater, which has elements of both types described above, is shown in Figure 6.8. Like the two craters just described, it is surrounded by an annulus of thick ejecta with a patterned surface. However, unlike the previous craters, radial ridges are an important component of the pattern. Although an escarpment bounds this lobe, a distinct ridge is usually absent. Beneath and extending to greater distances are sets of thin lobate ejecta sheets similar to those around Yuty. Occasionally this outer set of lobes is missing, but this may be a result of modification of the primary landform.

Rampart craters commonly have large central peaks or pits, which may result from the explosive expansion of ice as it vaporizes during impact. Terraces are absent on the walls of some rampart craters. For example, in Arandas (Figure 6.8) only a small amount of material lies in a heap around a large central peak.

If rampart craters really are the result of the incorporation of water into ejecta, detailed studies of their distribution on the planet may eventually show regional variations in **ground ice** (or **ground water**) and the smallest craters that have fluidized ejecta patterns may reveal the depth to which water lies beneath the surface.

Over twenty large, circular basins that are similar to the multiring lunar basins such as Orientale and Imbrium have been discovered on Mars. Some of them are readily apparent on the geologic map (Figure 6.1) and, although they have been significantly modified by both erosion and deposition from processes unique to Mars, they are in many ways similar to their lunar counterparts (Figure 6.9). The largest martian basin is Hellas, which is almost 2000 km across, much larger than either Imbrium or Caloris. Several vaguely defined rings have been identified, but they are highly degraded and not continuous. Much of the basin is covered by plains, so parts of the inner rings may be buried by basalt or sediment. The northern and eastern rims of Hellas are composed of belts of rugged, peaked mountains that resemble the Apennine Mountains that form the rim of the Imbrium basin on the Moon. The high rim has probably been eroded and much of the material has been deposited in the center of the depression. Other large basins, like Isidis and Argyre, are modified by erosion and are partly covered with a sedimentary blanket, but similarities between them and the lunar basins are great enough to support the conclusion that they, too, were formed by the collision of asteroid-size bodies with Mars.

Figure 6.8

The crater Arandas has features, including its distinctive ejecta pattern and its central pit, that suggest the involvement of ground ice in its development.

The early period of intense bombardment and basin excavation is recorded primarily in the rugged southern highlands. The only way to estimate absolute ages on Mars is by comparison with the cratering record of the Moon. If this comparison is valid, the most intense cratering occurred from 4.5 to 3.9 billion years ago. The frequency of large impact basins on Mars (about 1 per 10 million km^2) is much less than on the Moon (8 per 10 million km^2) or Mercury (14 per 10 million km^2). Even assuming that the processes that produced the northern plains destroyed half of the martian basins, Mars would still have a much lower frequency than the Moon. The progression in basin density from the Sun outward suggests that perhaps the numbers of large impactors decreased outward from the Sun during the early history of the solar system. However, even for smaller craters this heavily cratered region is not as densely cratered as the Moon, a sure indication that burial (by lavas or sediments) and substantial erosional modification obliterated some of the earliest traces of the planet's bombardment.

Water on Mars

In contrast to the Moon and Mercury, Mars is not a dry planet. Volatile materials rich in water were probably incorporated in the planet during accretion. During differentiation, these volatiles were partially outgassed and accumulated at the surface. Several arguments can be made for outgassing of the equivalent of a global **ocean** several hundred meters deep. Evidence for these ideas are the many channels, which closely resemble dry river beds on Earth. Many enormous channels originate in the southern highlands near the erosional escarpment and flow northward, emptying into the low plains. Other smaller channels have furrowed much of the old highlands. If any of these channels had been seen on Earth, no one would hesitate to call them dry rivers, but their presence on Mars presents some most perplexing questions about the planet. With the present temperature and atmospheric pressure on Mars, water cannot exist for long in a liquid state at the surface; it either evaporates or is frozen. But the presence of **river** or **fluvial channels** shows that running water and even huge floods occurred in the past. When did liquid water exist, where did it come from, and where has it gone? These problems are not easy to solve and will continue to be discussed for some time.

Instruments on the Viking spacecraft showed that there are large quantities of water ice and vapor on Mars. Parts of the ice caps are water ice,

Figure 6.9

Argyre Basin is a highly degraded multiring basin in the southern hemisphere of Mars. This photomosaic of Argyre shows the southern and eastern rim of the basin. Argyre is about 1000 km across—about the same size as Imbrium Basin on the Moon. The interior of the basin is filled by a much younger, relatively smooth deposit of probable sedimentary origin. The deposit buries craters and has been deformed into wrinkle ridges on the right. High-standing massifs mark one ring of the basin; an ill-defined outer ring is visible to the right. Galle is a large (200 km in diameter) peak-ring basin shown in the upper part of the photo.

and at night the martian atmosphere is saturated with water. Water is apparently also present in the pore spaces of rocks and soil. Most water in the martian hydrologic system is locked up as ice in the polar regions, as ground ice in frozen soils, or as ground water (liquid) at a depth of several hundred meters where the temperature is higher. A well-integrated hydrologic system composed of oceans and rivers (or their frozen equivalents—**ice caps** and **glaciers**) that transfers water back and forth between its components through evaporation, rainfall, and surface flow does not exist at present on Mars. However, the martian climate must have been very different at some time in its geologic past and a better-developed hydrologic system must have existed.

Dendritic Valleys

Perhaps the best evidence for the former existence of a hydrologic system on Mars comes from the networks of filamentous channels in the southern highlands. These channels or valleys show many characteristics of **dendritic river systems** on Earth (Figure 6.10). Many erode the rims of ancient craters or the slopes of volcanoes. Individual segments are commonly less than 50 km long and less than 1 km wide, but whole systems, consisting of many branching tributaries, may be up to 1000 km long. In some cases, they resemble stream networks that result from the collection of rainfall as it flowed down slopes. Deposits concentric to the rims of some large impact basins may be

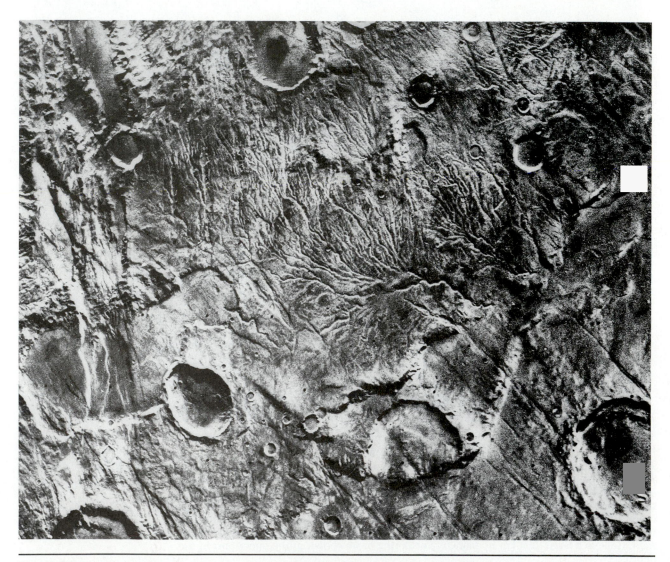

Figure 6.10

Dendritic valley networks are common in the southern highlands of Mars. This area is about 250 km across. Large areas between the channels are not dissected, suggesting to many scientists that the valleys were not formed by the accumulation of precipitation. Instead, some valleys appear to have been produced by sapping and the release of groundwater to the surface.

the remnants of deltas or alluvial fans where these rivers dropped the sand and gravel they were carrying into lakes or seas, but some valley systems end abruptly, with little evidence of a sedimentary deposit.

Crater frequencies on these valleys show that most are very old and probably formed 3.5 to 4.5 billion years ago during the heavy bombardment of Mars. During this early part of martian history, the atmosphere may have been denser. Occasional warm periods, resulting from the greenhouse effect or orbitally induced climatic change, may have allowed rain to fall, collect in small streams, and dissect portions of the ancient terrain. Seas or oceans may have developed in the low northern plains or in deep impact basins. The formation of dendritic valleys seems to have ceased by the end of the heavy bombardment, suggesting that a persistent ocean, if it existed, disappeared by this time. Although this may be the simplest explanation for the formation of valley networks, it is not consistent with some features of the channels. For example, some channels show strong evidence of structural control, have large areas of undissected plains between tributaries, and occasionally cut across craters (Figure 6.11). Groundwater seepage may explain these features. Channel networks could form by growth upslope from the point where water emanates from the ground (a spring) and would be strongly controlled by local faults and fractures. Whatever the source of the water, the early history of Mars seems to include substantial fluvial erosion and the development of a hydrologic cycle. Water was cycled from the atmosphere, to the surface, into the ground, back to the surface, and by evaporation back into the atmosphere. By the close of this era, water most have become locked into the upper martian crust as ice and deeper as ground water.

Outflow Channels

Other apparently fluvial features also exist on Mars. The most spectacular of these are the large channel systems, called **outflow channels**, which cross the eroded escarpment that separates the cratered highlands from the low northern plains. Figure 6.12 is a sketch map of the escarpment, showing the size and locations of major channels. Many arise at the eastern end of Valles Marineris and converge on Chryse Planitia, where Viking 1 touched down.

The system of channels in Tiu Vallis is a good example of these broad outflow channels. Tiu Vallis lies at the eastern end of Valles Marineris; it is well over 600 km long and, in some places, over 25 km wide (Figure 6.13). The floor of the valley is covered by a network of interlacing channels and streamlined erosional forms that may have been islands. Similar patterns are found in braided streams on Earth (Figure 6.14), which usually form in rivers heavily laden with erosional debris. It would be easy to believe that these pictures of martian channels were of parts of the desert areas of the southwestern United States, were it not for the craters and the huge size of the channel. Many layers of rock, which apparently were eroded by running water, are shown in the walls of the valley. Downstream, the channel broadens, the valley floor becomes flat, and the channel appears to empty into Chryse Planitia. In the area west of the Viking 1 landing site, the floodwaters apparently were dammed behind several wrinkle ridges until they overflowed and cut narrow gaps in the ridges (Figure 6.15). The ejecta blanket of the crater in Figure 6.15, shown on page 158, has been completely stripped away in many places. The entire region has been sculpted by fluid flow, as channels twist through the area and cut across one another.

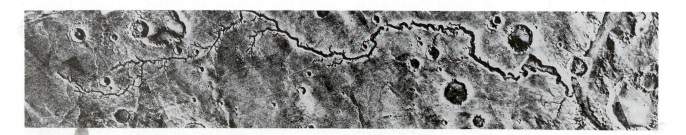

Figure 6.11

Nirgal Vallis shows many of the features associated with sapping. The channel is 800 km long but does not have an extensive drainage basin. Its numerous tributaries are short and confined to a narrow band adjacent to the main channel; many are stranded high on the valley's walls. By terrestrial standards, the valley network has large undissected areas between the channels. All of these features suggest an origin by sapping and indicate that runoff from precipitation was not important.

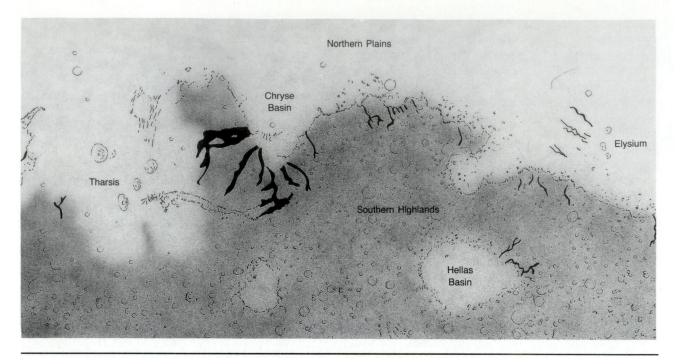

Figure 6.12

The major outflow channels of Mars are related to the global escarpment that separates the northern lowlands from the southern highlands.

Teardrop-shaped mesas that were once islands with craters atop them have been streamlined by fluid flow and occur over broad portions of Chryse Planitia, showing how extensive these floods were (Figure 6.16). The craters formed effective barriers to the flow and as the water streamed around them, teardrop-shaped mesas were eroded into the former plain. Terracing, produced as sediment built up around the islands or as rock layers were stripped away, is visible on the sides of many islands.

The major channels that flow from the southern highlands across the escarpment to the northern plains are large features: some are over 1000 km long, 100 km wide, and 4 km deep. Such tremendous amounts of erosion must have involved huge quantities of water, but where did the water come from? These outflow channels do not have large, integrated drainage systems with many branches resulting from continuous rain-fed flow. The few tributaries that are present are short and stubby and lack the intricate networks of progressively smaller tributaries that are characteristic of terrestrial streams. Instead, many martian channels have their headwaters in highly fractured regions or in jumbled masses of large blocks called **chaotic terrain**. As shown in Figure 6.17, the terrain consists of a complex mosaic of broken slabs and angular blocks with intervening valleys. Chaotic terrain at the eastern end of Valles Marineris is shown in Figure 6.17; the sources of many of the outflow

channels, including Tiu Valles, are visible as elongate depressions over 100 km across. Stream channels extend away from the chaotic terrain toward Chryse Planitia.

A possible mechanism for the formation of the chaotic terrain and the large channels involves melting of ice and escape of water from within the layers of the martian crust near the surface. As the ice melted, it may have drained rapidly when the meltwater reached a cliff face. The removal of water in the pore spaces could cause collapse of the overlying rock, and the flowing water could cut the channels. Heat to melt subsurface ice reservoirs may have come from intrusions of molten rock; this relationship between volcanoes and channels is especially clear on the western slopes of Elysium Mons (Figure 6.18). Other, larger floods may have resulted from the breakout of confined groundwater. For example, high pore pressure could have been achieved by gravity-driven flow of groundwater down the eastern slope of the Tharsis bulge. This latter suggestion for the source of the liquid water is consistent with the apparent absence of volcanic features near the sources of the Chryse (and most other) outflow channels.

The best terrestrial analog of the outflow channels occurs in Washington's Channeled Scabland. The Scabland consists of a complex network of deep channels cut into basaltic lava flows (Figure 6.19). The braided channels are 15 to 30 m deep,

Figure 6.13

Details of the 600-km-long Tiu Valles. Near the bottom of the map, chaotic terrains form the sources of Tiu Valles and other outflow channels. Streamlines mark the flow direction of what must have been a series of vast floods derived from the outbreak of water from subsurface reservoirs. Farther downstream a deep gorge with erosional terraces on its walls developed from the Tiu floods. Evidence of early flow is seen near the top of the map. See Figure 6.17 as well.

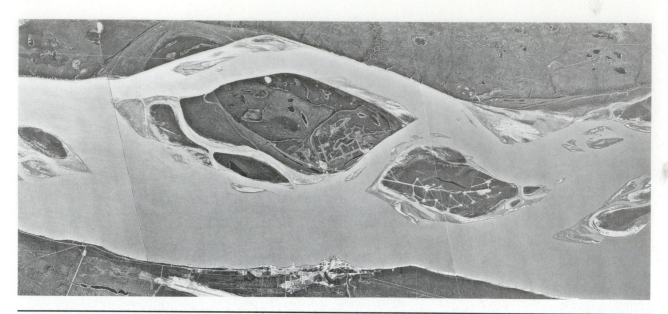

Figure 6.14

Braided streams on Earth develop a network of interlacing channels that are similar to those seen in Tiu Vallis.

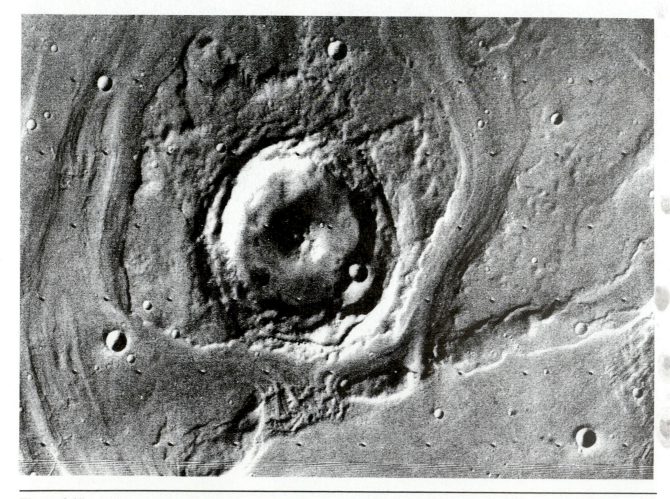

Figure 6.15

Wrinkle ridges and craters dammed and diverted the flow of the floods that spread across Chryse Planitia from Tiu Vallis and other outflow channels. The ridge was eventually overtopped and narrow gaps were eroded. The erosional streamlines demonstrate the diversion of floodwaters. The crater shown is about 10 km across, and its ejecta deposits have been extensively eroded.

Figure 6.16

Teardrop-shaped islands were formed where the floods were diverted around high-standing impact craters. Erosional terraces on the flanks of these streamlined forms demonstrate the progressive stripping of the surrounding surface. The large crater is about 10 km across.

with steep walls and abandoned waterfalls, and cover an area of 40,000 km². Giant ripple marks and huge bars of sand and gravel were created by catastrophic erosion and deposition. This spectacular topography was caused by the failure of a glacial ice dam, which released millions of cubic meters of water in tremendous floods over the plain. The area was flooded several times as glaciers advanced, formed dams, and then failed.

On Mars, the large unconfined outflow channels apparently did not originate from regional precipitation and collection of rain but represent the products of several huge floods released at high velocities. During the waning stages of floods, streams were unable to transport all of their sediment and deposited it on the floors of channels to form the braided segments, bars, islands, and terraces. The outflow channels postdate the heavily cratered highlands. In fact, the frequency of fresh craters on the floors of the outflow channels indicates that they are young compared to the valley networks, as well as compared to features on the Moon or Mercury. Nonetheless, in an absolute sense they are still very old; water has not flowed through them for millions of years. Some have suggested that the channels around the Chryse Basin are 1 to 2.5 billion years old.

These tremendous floods of water from the outflow channels periodically filled and refilled a martian ocean, Oceanus Borealis, in the low northern plains. The rise of the Tharsis and Elysium bulges and the development of volcanoes were triggers for the draining of water from the crust. The oceans may have been as deep as 2000 m. Each ocean gradually dried up by evaporation, by sublimation of sea ice, and by water seepage into the floor of the basin to form reservoirs of groundwater. When the ocean was filled, water vapor was readily lost to the atmosphere by evaporation. This water vapor would effectively trap heat from the Sun in what is called the greenhouse effect and would temporarily raise the surface temperature.

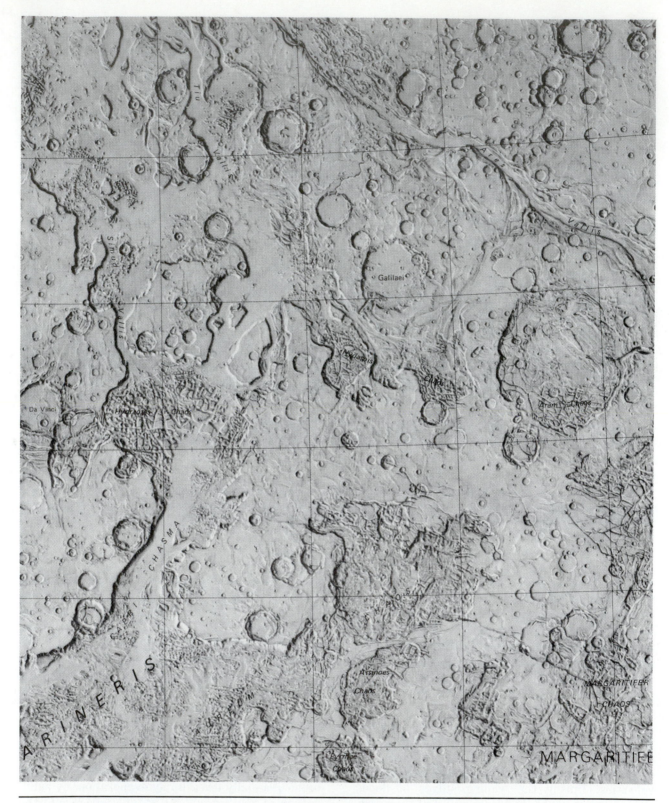

Figure 6.17
Chaotic terrain is well developed in the source regions for Tiu Valles and other outflow channels southeast of Chryse Planitia. The channels with lineated floors extend from regions with large polygonal blocks and masses of smaller rounded knobs. Some areas of chaos are not connected to channels. This fantastic landscape was probably developed as groundwater catastrophically breached to the surface. Collapse occurred over some areas. These blocks of chaos were then extensively modified to form knobs by the north-flowing floods that carved these channels. Figure 6.13 shows enlargements of the channels at the center of this map.

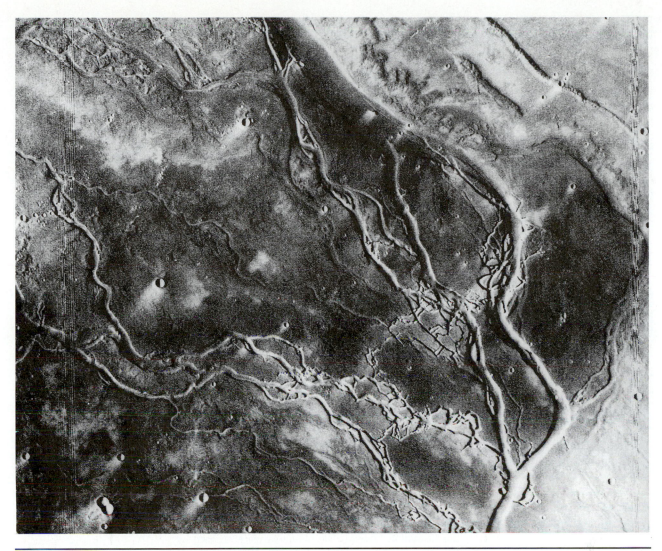

Figure 6.18

Sinuous channels issue from fractures on the flanks of the volcano Elysium Mons. The channels form a series of distributaries at the base of the steep northwestern flank of the Elysium dome. The water that carved these channels was probably released from reservoirs of ground ice beneath the volcano. Heat from the volcano provided the energy for melting. The scene is about 700 km across.

Evaporation from the seas and the later precipitation of rain or snow may have fed a few small, young valley networks. Evaporation from the ocean may have stabilized and fed the growth of a huge glacial ice cap at the south pole. This south polar cap was at one time much larger than it is today. Moreover, carbonate minerals form with ease in standing bodies of shallow water. Eventually, much of the atmospheric carbon dioxide was precipitated, leaving the planet cold and dry with but a thin atmosphere. It seems that Mars has seen long periods of little change punctuated by brief periods of exceptional erosion, climate change, and carbonate deposition.

Ground Ice

There are numerous indications that significant amounts of water as ice or liquid may occur beneath the martian surface. As we saw in a previous section, the ejecta deposits around many fresh martian craters appear to have been shaped by surface flow (Figure 6.6). Apparently the ejecta was wet because impact occurred in ground that contained ice or water and started to flow when it hit the surface. In other areas on Mars, low plateaus may be formed and progressively destroyed as ground ice melts or evaporates and the surface collapses (Figures 6.20 and 6.21). Small depressions

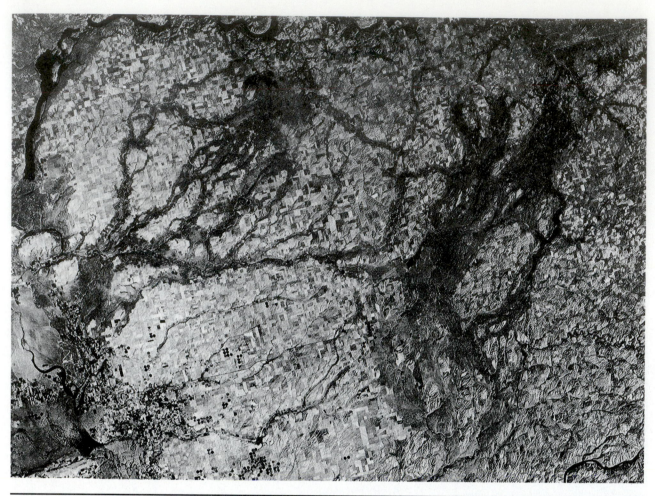

Figure 6.19

The Channeled Scabland of Washington consists of large anastomosing channels that cut through the loess and basalt of the Columbia Plateau. The channels were created by catastrophic floods that streamed across the plateau after the failure of a glacial dam that had blocked a river to the east of this satellite image.

eventually grow and merge, forming the scalloped edges of the shrinking plateaus. Moreover, some channels appear to have been fed by springs at the head of box canyons. These springs may have worn away the soft layers, removed support for the upper mass, and allowed the canyon to extend up the regional slope (Figure 6.22). This process is known as *sapping* and may be particularly active along the edges of channels that were cut previously. Extensive areas of fractured plains occur in the northern hemisphere (Figure 6.23) and may be similar to the **patterned ground** that occurs in Earth's polar regions as water in the soil freezes and thaws, forming the characteristic polygonal cracks at the surface. Although there are alternative explanations for some of these features, taken together they indicate that ice is present in the near-surface layers of much of the planet and that

it has played a significant role in modifying and shaping martian landforms.

Polar Regions: Ice and Wind

The unique geologic character of Mars is accentuated by its polar **ice caps**, which are large enough to be visible from Earth. They are of special interest because they are centers of present-day geologic activity and are not relics produced billions of years ago. The polar caps are also a major reservoir of the water on Mars. The structure, rock types, and terrain of the north and south poles are very similar and in some respects, resemble Earth's ice caps.

Cratered plains surround both polar areas and apparently extend beneath the ice. In the southern hemisphere, the plains are pitted or etched by

Figure 6.20

Erosional processes on Mars may include the sublimation of ground ice to create lowlands, which grow at the expense of plateaus with irregular margins. Sublimation or melting occurs at cliff faces. This process is similar to **karst** processes, and in Earth's polar regions this terrain has been called *thermokarst*. Landscapes that evolve in this fashion must have volatile-rich surface layers.

Figure 6.21

The northern flanks of Elysium Mons, near Hecates Tholus, are being modified by the collapse of lava-capped polygons. This type of chaos occurs in alcoves on cliff margins, presumably as ice sublimates or melts, allowing the collapse of overlying material. The knobby plains at the top of the photo are littered with remnants of a formerly more extensive upland, which retreated southward by this process. The largest crater shown in this mosaic is about 15 km across.

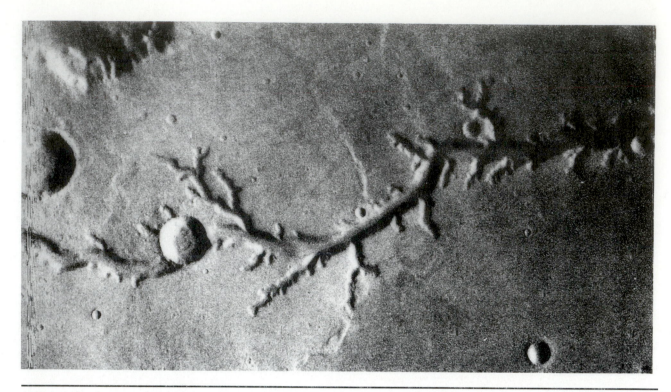

Figure 6.22
Spring sapping at the base of cliffs in the box canyon terminations of Nirgal Vallis is probably responsible for its headward growth. Large areas between the tributaries are not dissected by valleys, and the tributaries terminate in amphitheaters or alcoves. Old wrinkle ridges cross the plain. This scene is 80 km across.

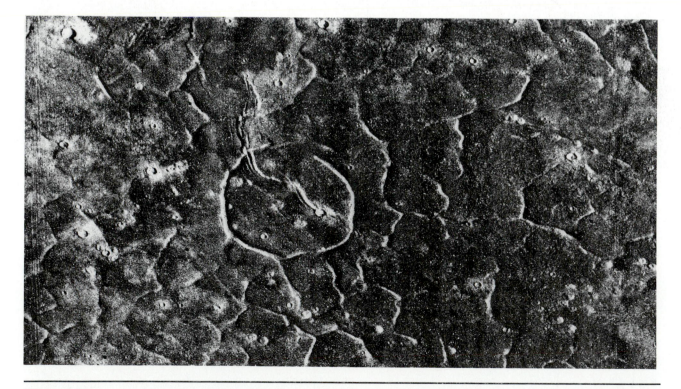

Figure 6.23
Polygonally fractured plains cover vast areas of the northern lowlands of Mars. The fractures may be produced by freeze-thaw processes, or they may have formed as desiccation or shrinkage cracks analogous to, but much larger than, those that develop when thin deposits of wet sediment dry. This scene is about 50 km across; individual troughs are about 1 km across.

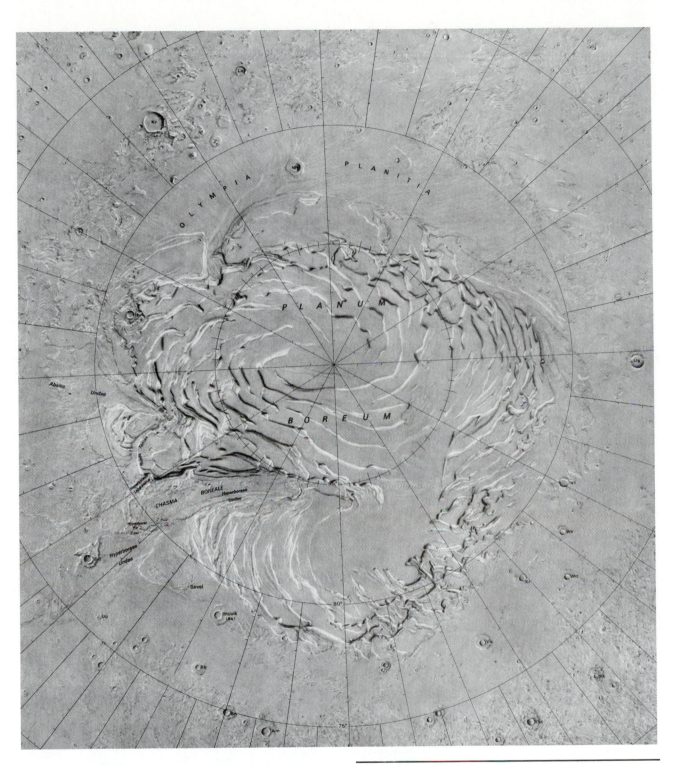

Figure 6.24

The vast dune field that surrounds the polar ice cap of Mars is larger than any found on Earth. (A) **The perennial ice cap** of the north pole is virtually surrounded by a sea of sand as shown on this map. Transverse dunes and barchan dunes are the most common types. The spiral pattern in the ice cap is defined by exposures of layered deposits beneath the perennial ice cap.

irregular depressions. The pits were probably hollowed out by the wind. Extensive wind activity in the northern hemisphere has created a vast circumpolar dune field (Figure 6.24). Dunes are present but are less common at the south pole. No young craters have been found in the area covered by dunes because the actively shifting sands have buried or obscured them.

(**continued**)

Figure 6.24   (continued)
(B) **A sand sea** consists of a variety of dune types. Transverse dunes, shown here, are parallel to one another, with only minor branching or merging. The small, white patches are accumulations of frost. The average distance between dune crests is about 500 m. The outlines of several craters buried by the sand are still visible; no craters younger than the dunes are visible.

Closer to the poles, a thick series of layered deposits buries the cratered plains. These deposits are a type of sedimentary blanket, cut in many places by deep, spiraling channels that reveal its internal stratification (Figure 6.25). Elsewhere, it is smooth and nearly featureless except where dunes may lap over its margins. The surface is very young and lacks fresh impact craters. The layers are remarkable for their uniformity and continuity and may contain a record of physical processes on Mars in much the same way as sedimentary rocks preserve the geologic history of Earth. As many as 50 beds have been observed in a single slope; their total thickness may be more than a kilometer.

These sedimentary deposits consist of a series of alternating light and dark layers that are essentially horizontal and are thought to have originated from the combined result of ice and wind activity. Possibly the brighter layers are water frozen out of the atmosphere and the darker stripes are interlayered dust and sand blown in by the wind. If these laminated deposits have such a composition, they may contain tremendous amounts of water, which at one time must have resided in the atmosphere or elsewhere on the surface.

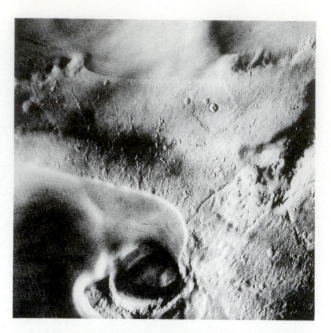

(A) **Layered deposits in the south polar region** form a thick, craterless blanket burying an older cratered surface. The margin of the deposits is obvious in this photo, which is 200 km across. The older surface displays the pitted or etched appearance of much of the surrounding terrain.

(B) **The delicate layering of the north polar deposits** is exposed in cliffs cut through the terrain. Patches of wind-sculpted frost on top of the layered deposits make it difficult to see the topography. This scene is 90 km across.

Figure 6.25
Layered deposits of the polar regions consist of alternating light and dark beds of sedimentary rock deposited by ice and wind.

The rotational axis of Mars is presently inclined almost exactly the same as Earth's. As a consequence, Mars experiences seasonal changes. The white polar caps grow in the winter as frost and ice collect on the surface and shrink as the frosty hood sublimates during the summer (Figure 6.26), leaving only a smaller residual ice cap. This temporary frost layer is probably made of dry ice (frozen carbon dioxide) and may be as much as 50 cm thick during the winter. An important observation from the Viking orbiters showed that the residual or permanent cap in the north consists of dirty water ice and is not composed solely of frozen carbon dioxide as once thought. Carbon dioxide ice may dominate the south polar cap, but at the low temperatures encountered there it is difficult to detect water ice, even if it is present. However, the residual ice caps are thin—possibly only a few meters thick and probably less than 1 km. Movement of glacial ice cannot occur until the weight of the ice exceeds its strength; to initiate movement in the low gravitational field of Mars requires greater thicknesses of ices than would be needed on Earth. Thus, at present, the ice caps of Mars may lack the great erosive power of Earth's continental glaciers. Permanent ice covers most of the layered deposits and also occurs as isolated patches in craters or surrounded by dunes farther from the poles. One of the most interesting discoveries is that both polar caps have spiral patterns of dark, frost-free valleys that extend through the residual ice cap and layered terrains. They could be erosional channels formed as the wind sweeps across this bleak, craterless expanse (Figure 6.27). Alternatively, the pinwheel pattern may reflect selective deposition and sublimation of ice, determined by the atmosphere's circulation.

Although the present ice sheets appear to be too thin to flow as glaciers, there is some evidence that thick glaciers formed repeatedly. Glacial ice transports sediment by dragging it along the base of the ice. The sediment may be piled up into long ridges, called **moraines**, at the front of the glacier. When the ice sheet retreats, these ridges are left as evidence of the former extent of the ice sheet. Sinuous ridges on Mars have been interpreted to be glacial moraines that formed when an ancient glacier scoured the landscape. Although their very existence is still extremely controversial, such glaciers may have formed at the same time that the seas were created by the outflow channels.

Whatever the exact nature and timing of water or ice activity on Mars, it is clear from the presence of stream channels, ground ice, and polar ice caps that water is and has been available to shape the surface. Water resides on Mars at present as vapor in the atmosphere, as ice at the poles, and probably as groundwater or ice in rocks near the surface.

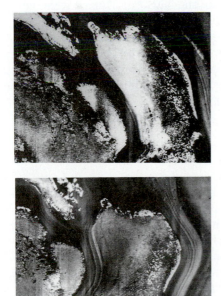

Figure 6.26
Channels in the polar ice caps change with the seasons as the ice caps expand and contract over the polar regions.

Figure 6.27
A spiral pattern of valleys cuts the residual ice cap and the layered deposits of the North Pole. A similar spiral pattern is also found at the South Pole.

The Martian Hydrologic System

The circulation of water on Earth's surface takes place within a huge hydrologic system that operates continually. The oceans form a vast reservoir of water, which evaporates and moves with the atmosphere. It is then precipitated as rain and snow and constantly bathes the surface of the land. Water may move as surface runoff in streams, as groundwater through pore spaces in rocks, in glacial ice, and eventually as water vapor in the atmosphere before it returns to the ocean and is recirculated. Heat from the Sun drives this system and determines the physical state of the water.

Apparently Mars developed portions of a similar hydrologic system during part of its history. We have seen that Mars appears to have a differentiated interior that resulted from melting and redistribution of the elements within it. Interior melting has other important consequences. When rocks with appropriate compositions are melted, some of the elements incorporated in the rock—such as oxygen, carbon, hydrogen, argon, nitrogen, and neon (all volatile elements)—are released and form other compounds that are stable under the new conditions. Some of these compounds ultimately form light gases (such as water and carbon dioxide) that easily rise with the molten rock and escape to the surface to form a secondary atmosphere. If the gravity of the planet is sufficient to retain the gas, and if temperatures are high enough, water will remain in a gaseous state; at lower temperatures it may become a liquid and collect in rivers, lakes, and oceans, forming a hydrosphere and initiating circulation in a hydrologic system that shapes surface features, deposits new rock bodies, and has important implications for the evolution of life. If liquid water were stable at the surface of Mars, it would have collected in craters and impact basins. However, the composition of the early martian atmosphere probably did not remain in its primordial state. Chemical reactions of the atmospheric gases with the surface materials modified the composition of the atmosphere and extracted carbon dioxide; additions to the bulk of the atmosphere occurred from continued melting and differentiation in the interior. Some light gases slowly but inexorably escaped into space, while others, such as water vapor and carbon dioxide, froze out of the atmosphere when the temperatures became low enough and formed polar ice caps and ground ice.

Estimates of the total amount of water that may have been released from the martian interior range from a globe-encircling layer 10 m thick to one a kilometer or more in thickness; a reasonable estimate is probably 50 to 200 m. (Earth outgassed enough water to cover its surface, if it were a perfect sphere, to a depth of about 3 km.) In any case, a large quantity of water has been released from the interiors of both Earth and Mars. From the features we have discussed in this section, it is at least conceivable that during the very early history of Mars, a relatively well-integrated hydrologic system existed. The small channels in the ancient highlands may record an epoch of rainfall and collection in rivers and possibly small seas or lakes. If so, Mars must have been warmer then and must have had a thicker atmosphere that made it possible for liquid water to flow across the surface. Since that time, much of the globe has been resurfaced, erasing most of the evidence of this erosional epoch and any evidence for an ancient ocean. About 4 to 3.5 billion years ago, the atmosphere of Mars must have cooled and ice began to form—first in the polar areas, forming ice caps, and later at lower latitudes. Water-saturated ground froze in some places and created ground ice deposits. Later, liquid water, from reservoirs beneath the surface, may have been released when the huge outflow channels and chaotic terrain were formed on the flanks of the Tharsis and Elysium domes. Exposures of ice in cliffs or scarps may have melted or sublimated, forming the box-canyon streams and scalloped plateaus. Meteorite impact, though greatly reduced in this later period, catastrophically melted pockets of ice and incorporated the water into ejecta that flowed like mud as it returned to the surface. During this later, colder episode, the hydrologic system, with its host of components and processes, lost much of its recycling capability (Figure 6.28). Presently, a rudimentary circulation system transfers small quantities of water and carbon dioxide from pole to pole with seasonal changes.

Eolian Features

The effects of flowing fluids (gases and liquids) at the surfaces of planets are dramatic. Compare for example, the surfaces of the airless, waterless Moon with Mars. Having considered the role of water on Mars, we now turn our attention to the role of the fluids in the atmosphere.

When Mariner 9 entered orbit around Mars in the fall of 1971, a planetwide dust storm was raging, completely obscuring the surface of the entire globe. The great storm had begun some two months earlier and was observed with Earth-based telescopes as a yellowish cloud that rapidly expanded over the entire planet. The magnitude of

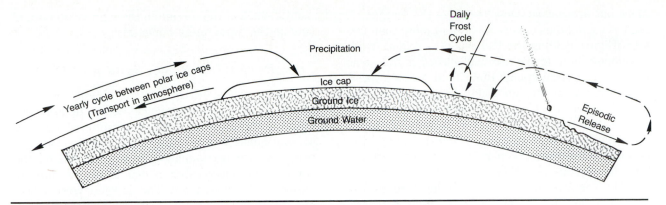

Figure 6.28

The martian hydrologic system involves the polar ice caps and water stored as a liquid or as ice beneath the surface. Occasionally, water released to the surface carved huge outflow channels. The present hydrologic system consists of seasonal cycling of carbon dioxide and water frosts between the polar regions and the atmosphere. Ground ice is only rarely mobilized by meteorite impact.

such a storm is hard to imagine. Winds comparable to those in a strong hurricane on Earth raged continuously for several months, and dust was blown many kilometers into the air, covering the entire globe. Similar dust storms developed during the Viking missions.

The pictures taken by Mariner 9 and the Viking spacecraft revealed that eolian activity has played an important and active role in shaping the surface features of Mars. A variety of wind-shaped landforms has been observed, including (1) groups of long, parallel streaks, (2) dune fields, (3) large stretches of dust- and sand-mantled terrain, and (4) linear grooves and ridges. The features collectively indicate that wind activity on Mars is quite likely the dominant surface process presently in action and that wind is constantly moving and redepositing loose surface material.

Streaks

The most obvious eolian features on Mars are the systems of parallel plumes or streaks that originate at craters, ridges, or cliffs and that extend hundreds of kilometers across the surface of the planet (Figure 6.29). These features are the result of wind erosion and deposition. Erosion may occur behind some obstacles that create a turbulent flow of the wind. In this case, fine particles may be swept away. In other places, a crater rim may be associated with a wind shadow, a pocket of quiet air downwind from the crater, where sand or dust can accumulate. There are light and dark streaks. Pictures taken of the same area at different times show that the shapes of some markings are always changing.

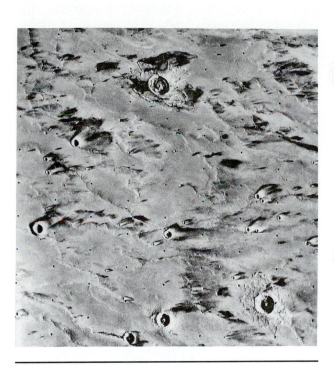

Figure 6.29

Bright wind streaks with splotchy, irregular dark halos form behind obstructions to the wind and reveal the principal wind directions, in this case from left to right. Large and small craters on this lava plain southwest of Tharsis were responsible for these streaks. This frame is 230 km across.

Dunes

Dunes are great moving piles of sand. Individual grains are moved by the drag of the wind blowing across them. Dunes will not form from dust-sized particles but require larger grains typical of what we call sand. Dune fields are present on

Mars but are not as conspicuous as the streaks. A field of such dunes covers an area of more than 200 km² (Figure 6.30) in the Hellespontus region. Well-developed dune forms are widespread in the polar regions; the circumpolar dune field forms a **sand sea** larger than any on Earth (Figure 6.24). The identification of dunes is important because it proves that martian winds are strong enough to lift loose particles and transport them in spite of the tenuous atmosphere. When these airborne grains hit other surfaces, erosion can occur.

The surface pictures from Viking landers further show the importance of wind activity. Small drifts, with features remarkably similar to many seen in terrestrial deserts, are seen in the spectacular picture of the martian landscape in Figure 6.2. The blocky, angular rocks that mantle the area resemble "desert pavements" produced on Earth where the wind has selectively transported fine sediments, leaving behind a coarse rubble. Drifts of sand have collected behind some rocks where the wind was diverted and the particles it was carrying dropped.

Other evidence for rather large amounts of eolian deposition are the layered deposits of the polar regions, which cover thousands of square kilometers. A thin layer of wind-blown debris may mantle much of the terrain in the high latitudes and subdue the topography beneath it.

Deposits of Wind-Blown Sand and Dust

Large tracts of the martian highlands show strong evidence that a thick eolian deposit has mantled and subdued the topography of an older surface beneath it (Figure 6.31). The deposits appear to have accumulated in craters and on the surrounding intercrater plains after the intense meteoritic bombardment. The deposits were subsequently stripped away in part, leaving mesas with steep cliffs to mark their former extent. The mantlelike nature of the deposits and their easily eroded character may indicate that they were deposited as extensive sheets of **loess** during the middle history of Mars. Loess forms when dust-sized particles suspended by the wind settle uniformly over a landscape. Similar layered deposits exist in the polar regions (Figure 6.25), where they cover thousands of square kilometers. The polar deposits appear to be younger and may be interlayered with the deposits of ice or frost.

Eolian Erosional Features

The effects of eolian processes are seen on many of the martian craters. Although they resemble lunar craters in general appearance, they have been distinctly modified by processes other than impact and in this sense they are degraded. Careful study of Figure 6.30 shows that some crater rims appear to be abraded or worn down. There is no sharp upturned lip at the crater margin, as is typical of fresh craters on the Moon, and the rays and ejecta blankets have been eroded or buried. Some investigators have suggested that the wind, in concert with other processes, has subdued the expressions of these craters. Elsewhere, wind erosion is strongly suggested by the pronounced alignment of long, narrow ridges or spines that project from low plateaus (Figure 6.32). The ridge crests are sharp and keellike, and the ends taper sharply. Similar features, called yardangs, occur in desert regions on Earth; their shape and alignments show that they were formed by wind erosion. It appears that large regions near the equator of Mars were stripped of soft surface layers by this process. Etched terrain, which is particularly common near the south pole, is likewise the product of a different, but less intense, type of wind erosion (Figure 6.33). The basins appear to be the result of deflation. Some believe that the pinwheel shape of the ice caps is due to erosion by outward-spiraling polar winds.

Figure 6.30

Small dune fields have formed on the floors of several craters in this region of the southern highlands, west of Hellas Planitia. The largest dune field covers an area of about 60 by 30 km.

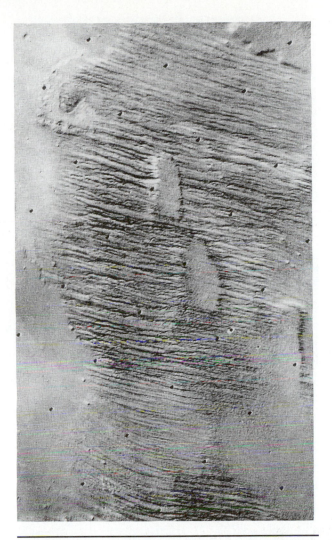

Figure 6.32
Yardangs are linear ridges formed by eolian erosion of the intervening valleys. Yardangs emanate from an incompletely stripped mesa consisting of easily eroded material. The yardangs formed in a deposit that is younger than lava flows that surround Olympus Mons.

Figure 6.31
A partially eroded blanket of loess (wind-deposited dust) may explain the appearance of this part of the heavily cratered terrain. An easily eroded deposit fills craters and adjacent intercrater plains. Deposition and erosion of this sedimentary cloak were probably caused by eolian processes.

The Martian Eolian Regime

The surface features described above, together with studies of the martian atmosphere, have permitted scientists to draw some conclusions about many parts of the eolian system on Mars. Comparisons of the system can be made with Earth and other planets. Even though a variety of cloud formations has been observed on Mars, the martian atmosphere is still very thin (about 1/100 the pressure of Earth's) and cold. The pressure exerted by the atmosphere at the surface of Mars corresponds to the pressure found at heights of 30 to 40 km above sea level on Earth. Several important consequences result from these differences. For example, the velocity of the wind necessary to start the movement of grains is estimated to be 10 times greater on Mars than on Earth, but once the wind lofts these grains they could have a hundred times more kinetic energy (kinetic energy = $1/2 \times$ mass $\times$ velocity^2). Mariner 9 data indicating that dust storms in the atmosphere move at velocities in excess of 200 km/hr and gusts of 500 to 600 km/hr

Figure 6.33

The pitted or etched appearance of this south polar terrain is probably the result of eolian deflation. Vast regions of similar landforms surround the south pole. This scene is about 200 km across.

(half the speed of sound in Earth's atmosphere) do not seem unreasonable, far exceeding wind velocities on Earth. The Viking landers measured wind velocities up to 30 m/sec. Moreover, the thin atmosphere on Mars produces practically no cushioning effect as grains collide. These two factors permit very small particles to act as very effective instruments of erosion. These considerations suggest high erosion rates. In addition, the relatively weak gravitational field plus the thin atmosphere on Mars combine to permit dust and sand to reach heights three to four times greater than on Earth, increasing the opportunity to form huge loess mantles.

Fine, unconsolidated material similar to the lunar regolith, with glass beads and rock fragments, is undoubtedly produced from the impact of meteorites, so loose material is readily available to be picked up and transported by the strong winds. In addition, sand and dust may be produced by volcanic eruptions and mass wasting and in the stream channels, as well as by continued wind erosion. Ice particles may also be moved by the wind.

Wind action may have continued on Mars without significant interruptions since the origin of its atmosphere. Only rarely have volcanic activity, tectonics, or running water interrupted the processes of the ceaseless winds. On Earth, plate tectonics continually create and destroy the crust, and flowing water has been the dominant surface process from the very beginning. The products of wind action on Earth have been largely restricted to the desert regions and have often been masked by or intimately mixed with the effects of the more universal processes of running water. This has not been the case on Mars; the action of running water has been very limited in scope and duration. All of these factors collectively suggest that eolian processes may have been more intense and that they proceeded faster on Mars than on Earth. Indeed, wind action is considered by many geologists to be the principal cause of surface changes, shaping martian landforms as universally as running water affects the surface of Earth. It is perhaps amazing that any evidence of the very early history of Mars is preserved at all.

Thus, a major feature that distinguishes Mars from Mercury or the Moon is the existence of a martian atmosphere. This tenuous shell of carbon dioxide and nitrogen, which may at first seem insignificant from a geologic point of view, has markedly changed the surface features of Mars. Even so, the present atmosphere is probably only a fraction of the total amount of volatile gases that were outgassed from the planet. Several estimates based on the present composition of the atmosphere suggest that a layer of water 50 to 200 m thick spread evenly over the planet may have been outgassed; an amount of carbon dioxide equivalent to 1 to 3 bars (1 bar equals the pressure exerted by Earth's atmosphere at sea level) may also have been expelled during the differentiation of the interior.

Much of the water released from the interior is now locked up in the polar caps and within the regolith. Even though the formation of carbonates (an important sink for atmospheric volatiles) at the surface is much less efficient than on Earth, the process may have removed some of the carbon dioxide from the martian atmosphere. Carbon dioxide is also locked in the regolith and absorbed onto mineral surfaces. Some nitrogen may have escaped from the top of the atmosphere into space. These processes and the lack of a biologic cycle, which both form free oxygen (O^2) and help tie up carbon dioxide in solids, have created a martian atmosphere that is similar in composition to that of Venus but much thinner—95 percent carbon dioxide, 3 percent nitrogen, and 1.6 percent argon with only traces of water vapor, oxygen, and other gases.

Mass Movement

The Viking orbiter photography, with its great detail, provides dramatic evidence that **mass movement**, the gravity-driven downhill movement of unconsolidated material, is an important process in the evolution of the landscape of Mars and is especially important in the enlargement of the canyons and development of chaotic terrain. The variety of mass-movement features on Mars includes those produced by the rapid and devastating effects of tremendous landslides as well as features produced by the slow, downslope creep of loose material.

Figure 6.34 shows a section of Valles Marineris and illustrates the role that mass movement has played in slope retreat and enlargement of the canyon. Here the canyon is about 5 km deep, and several massive landslides are visible. Resistant rock layers form a steep cliff at the top of the plateau and appear to have broken into a series of relatively coherent slump blocks, whereas the lower slopes appear to have flowed away in a surge of moving debris. Some of these tongue-shaped deposits extend across the canyon floor, at least 50 km from their point of origin.

Figure 6.35 provides a broader view of Valles Marineris and the gigantic landslides along the walls. The canyon was originally formed by subsidence along a series of parallel faults extending from the crest of the Tharsis upwarp. Many small grabens can be seen on the flat plateau surface; earthquakes along these faults may have triggered the great slumps that widen the walls of the major canyons.

Mass movement is, of course, of prime importance in the development of the chaotic terrain discussed in a previous section. As ground ice melts or as aquifers break through to the surface, the support for the overlying rock is removed. It may then collapse and form huge, muddy debris flows and slump blocks on the steep walls (Figure 6.17).

Similar mass-movement processes probably sculpted the fretted and knobby terrain skirting the global escarpment (Figure 6.36). The fretted terrain consists of a maze of flat-topped buttes and linear valleys, which are probably controlled by fractures in the crust. This photograph shows how the escarpment retreats—material is shed off the cliffs by mass movement and is eventually removed by fluvial or eolian processes or buried by later lava flows or other debris flows. Melting or sublimation of ice in these materials may substantially contribute to this mode of erosional evolution. Eventually only small knobs or hills are left to mark the former extent of the plateau.

The detailed photograph shown in Figure 6.37 reveals another type of slow mass movement in the fretted terrain. On the floors of these trenchlike valleys, surface materials appear to have moved or flowed slowly downhill, possibly aided by freezing and thawing of ice in the spaces between fragments. In this area, mass movement causes slope retreat and also transports the material downslope, like a valley glacier on Earth.

Although landforms that were produced by the direct, gravity-driven movement of slope materials have been observed on the steep walls of lunar craters, the process has not proceeded as dramatically as on Mars. The role of melting ground ice on Mars is probably the most important reason for the great difference.

Martian Volcanic Features

The discovery of huge volcanoes in the northern hemisphere of Mars was one of the most significant results of the flight of Mariner 9 in 1971. Previous Mariner missions (4, 6, and 7) sent back pictures of the southern hemisphere, which, for the most part, is densely cratered and superficially resembles the lunar highlands. Therefore, the existence of enormous volcanoes on Mars was entirely unexpected. Yet when the rest of the planet was photographed by Mariner 9, it became clear

Figure 6.34

Landslides in Valles Marineris are shown in this photo. The landslide on the far wall has two components—an upper blocky portion, which is probably disrupted cap rock, and a finely striated lobate extension, which is probably debris derived from the old cratered terrain exposed in the lower canyon walls. Similar lineations are found on terrestrial landslides and show the direction of movement. This part of Valles Marineris is about 5 km deep.

that prominent volcanic features occur over a significant part of the planet, and their lack of superposed craters suggests that Mars continued to experience volcanism relatively late in its history (during the last billion years). Later, detailed studies revealed that there are many older volcanoes as well, although they are highly eroded and difficult to recognize. It appears that Mars has had a long and interesting volcanic history.

Three major types of volcanic features, including several types not found on Mercury or the Moon, are found on Mars. The most striking are the giant **shields** concentrated primarily in the north-

ern hemisphere. There are also large volcanic structures with very low profiles and large central craters called **patera** (saucers). The third type of volcanic feature, less spectacular but nonetheless very significant, includes the **volcanic plains**, which form much of the sparsely cratered regions in the northern hemisphere and are similar to the plains of Mercury and to the lunar maria.

Volcanic Shields

The most spectacular volcanic features on Mars are the enormous shield volcanoes, which have no

Figure 6.35
Fault-bounded troughs of Valles Marineris are extensively modified by huge landslides. Most of these landslides lack the lineated debris lobes shown in Figure 6.34 because the canyon is much narrower here.

analog on the Moon or Mercury. Most of the shield volcanoes occur in the Tharsis region, where twelve large and several smaller volcanoes developed (Figure 6.38). The Elysium region is smaller and has only three large volcanoes (Figure 6.39). The largest volcano is at least twice the size of the largest volcano on Earth. The most dramatic is Olympus Mons, which lies west of the Tharsis ridge. It is about 550 km in diameter (five times larger than the largest on Earth) and rises 25 km (82,000 ft) above the surrounding plain. This is half again the distance from the depths of the Mariana Trench to the top of Mt. Everest, the deepest and highest points on Earth. The mosaic in Figure 6.40 shows that Olympus Mons has a complex crater at its summit called a **caldera**. These large calderas are not vents but depressions, produced when the peak of the cone collapsed and subsided along circular faults as magma was withdrawn from shallow reservoirs in the volcano. High-resolution photography of the flanks of Olympus Mons shows many of the individual flows that make up the cone

(Figure 6.41). There are many long ridges radial to the central caldera, narrow channels that resemble collapsed lava tubes or lava channels, and fingerlike flows. All of these flow types are found on terrestrial shield volcanoes, suggesting that the physical nature of the lava that made them is similar to terrestrial basalt.

Portions of the base of Olympus Mons form a steep cliff several kilometers high, which is receding by slumping and gully erosion. Originally, the volcanic cone probably graded smoothly into the surrounding plain. In places, the scarp has been flooded by younger lavas, re-establishing a smooth profile (Figure 6.40).

The other large shield volcanoes in the Tharsis and Elysium areas resemble Olympus Mons but differ in size and detail. All are relatively young features, as is indicated by the remarkably fresh surface features of the lava flows, sharp rims of their summit caldera, and the lack of impact craters. Olympus Mons is probably the youngest of the large shield volcanoes; according to some crater

Figure 6.36
Fretted terrain, made up of isolated high-standing mesas separated by intervening troughs and plains, occurs along the escarpment separating the two distinctive terrains on Mars. Piles of debris slumped off the mesas and accumulated at the cliff bases. Eolian processes may eventually remove some of this material, but sublimation of ice may also contribute to the retreat of these scarps.

Figure 6.37

Mass movement off the walls of this portion of the fretted terrain has created a deposit with lineated surfaces. There may be some slow, glacierlike, down-valley movement that is parallel to these lineations, but much of the movement is in a direction perpendicular to the lineations.

counts, it may be as young as 200 million years old and would have been built while reptiles dominated life on Earth.

The large shield volcanoes of the Tharsis and Elysium regions appear to be associated with large domes in the lithosphere, both of which may be the result of thermal processes within the deep interior, which bulged and fractured the lithosphere and produced the magmas.

In addition to the large shield volcanoes, a number of smaller volcanic shields are scattered across the planet. They are typically slightly steeper than the large shields but commonly have summit calderas and radiating channels (Figure 6.42). They may simply be small shield volcanoes produced over a limited magma source, but some geologists believe that because they have such large calderas they are the summits of nearly buried shields.

Patera

Possibly the largest volcanic structure on Mars is not a shield volcano at all, but a very low profile feature with a large caldera or vent area called Alba Patera (Figure 6.43). It may be over 1500 km in diameter. The central portion of the structure is surrounded by a circular set of fractures along which the volcano may have subsided somewhat. Extremely long lava flows that appear to be as young as those in Tharsis emanate from its center and extend for more than 1000 km away from their vents. Small hills around the margins of this huge structure may be cinder cones or lava domes. In short, it appears that Alba Patera was formed by a complex series of volcanic and tectonic events. No similar volcanic structures have been found on any of the other planets.

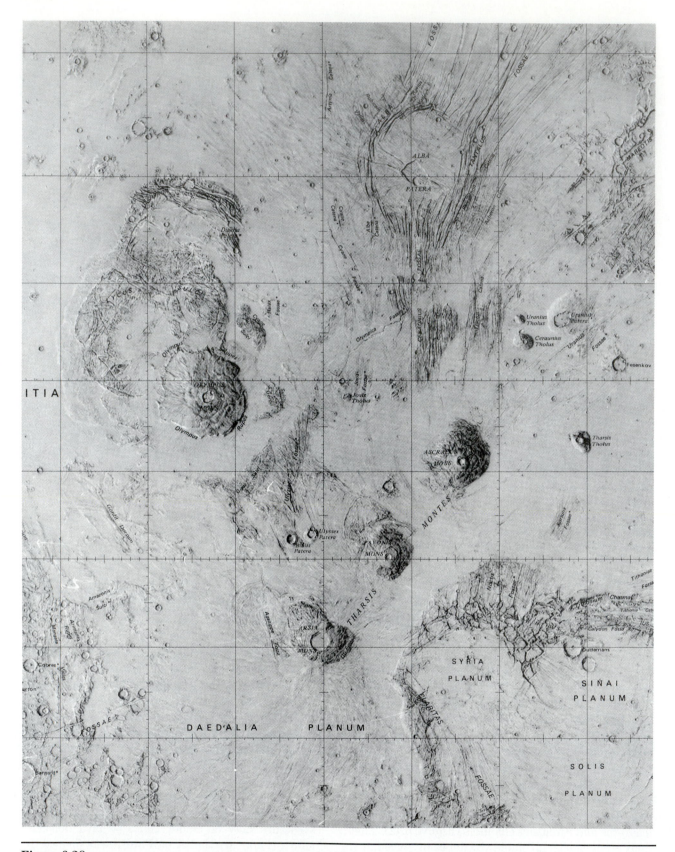

Figure 6.38

The Tharsis region of Mars is dominated by a dozen relatively young volcanoes and vast lava-covered plains. These features developed around a huge crustal bulge. Long fractures and graben systems radiate away from the uplift and cut older lava plains as well as the heavily cratered southern highlands.

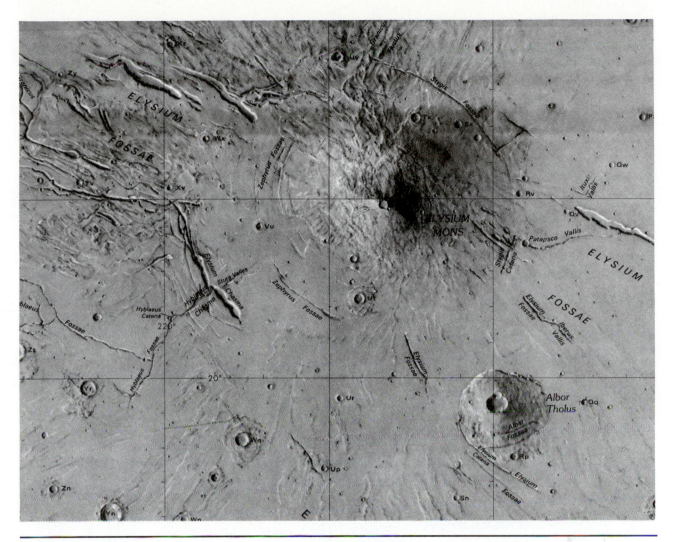

Figure 6.39

The Elysium region contains only three large volcanoes, but like Tharsis it sits atop a crustal swell and is associated with a long fracture system. Much of the surrounding plain is covered by lava.

Other patera volcanoes occur in the southern hemisphere and appear to be very old (Figure 6.44). Tyrrhena Patera (Figure 6.45), northeast of Hellas Basin, is extremely degraded and appears to be surrounded by younger plains. By crater counts, Tyrrhena Patera is estimated to be over 3 billion years old. Thus, eruptions from central volcanoes, shields, and pateras have apparently extended over a large span of martian history. Some pateras may have experienced an early phase of explosive volcanism that emplaced thick sheets of volcanic ash. The explosions may have been generated as hot magma rose through the water-saturated regolith of the ancient highlands. The tremendous change in the volume of water as it passes from liquid to gaseous states may trigger such eruptions. On Tyrrhena, these initial explosions were followed by

the quieter eruptions of lavas, which appear to have formed the deep channels on its flanks. Alternatively, the channels may have been cut by water released from the regolith by volcanic activity.

Volcanic Plains

Although the great shield volcanoes present the most spectacular evidence of volcanic activity on Mars, the lava flows of the plains regions may represent much greater volumes of volcanic material extruded from the interior and are certainly important because of their similarity to the flood basalts on other planets. Indeed, over 60 percent of the planet is covered by plains, a substantial portion of which may be of volcanic origin. Geologic evidence shows that volcanic plains were formed in

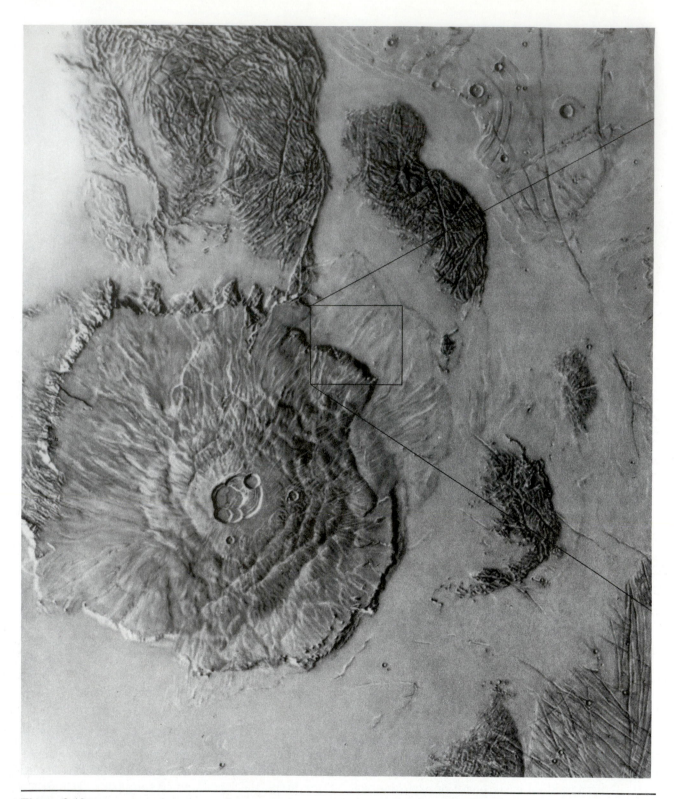

Figure 6.40

Olympus Mons is the largest of the Tharsis volcanoes; it rises 25 km above the surrounding plains. A complex collapse caldera marks its summit. This map of the 550-km-diameter volcano shows the radial texture of the flanks created by lava flows. The low semiconcentric ridges may mark different stages in the growth of the volcano. The erosional scarp (up to 10 km high) is buried on the northeast by younger lavas. A large aureole of strongly ridged lobes surrounds the base of the volcano. The origin of the lobes is controversial—suggestions range from accumulations of pyroclastic flows to debris sloughed off the volcano to form the scarp.

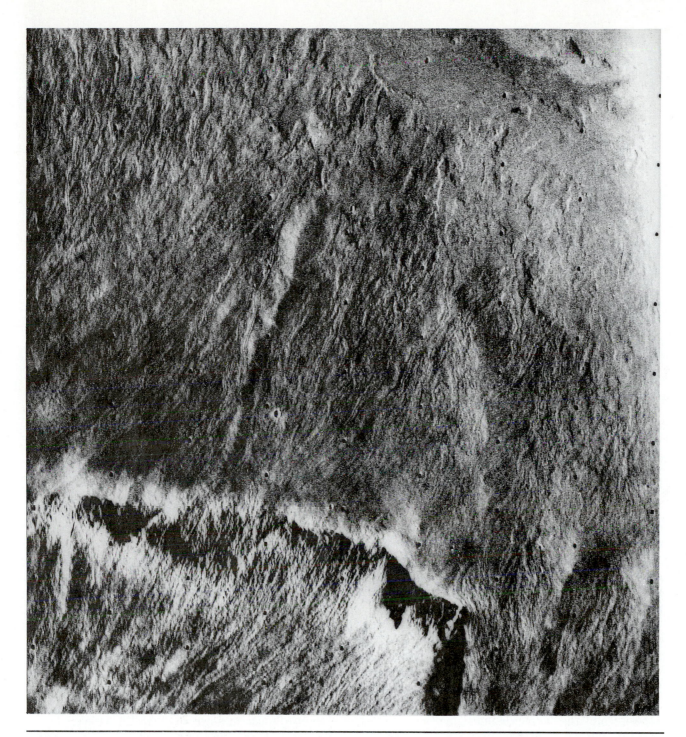

Figure 6.41

Countless lava flows form the flanks of Olympus Mons. Many have narrow channels. The lavas have cascaded down the steep, faulted margin of the volcano to create lava deltas at the base of the cliffs.

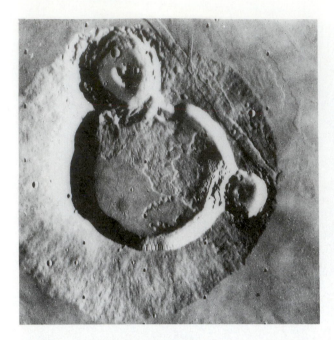

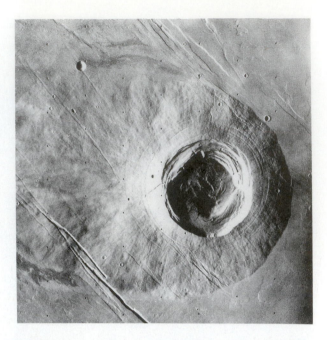

(A) **Ulysses Patera** has a caldera that is 55 km across. The volcano has two sizable impact craters on its flanks and is cut by graben radial to the Tharsis uplift.

(B) **Biblis Patera** has a caldera 50 km in diameter from which several large lava channels issue. These channels appear to have developed by thermal erosion, caused by hot lavas flowing down the flanks of the volcano. Smaller channels are also visible. Young grabens cut the lava covered plains and the shield.

Figure 6.42

Small Tharsis volcanoes have steep sides and relatively large calderas. They have been at least partially buried by younger flows.

the ancient highlands (similar to Mercury's intercrater plains) and in the vast northern lowlands. Their emplacement appears to have spanned the entire length of martian history. Knowledge of their absolute ages will be important for determining the thermal evolution of Mars.

Various features are used to identify volcanic plains, the most obvious being the presence of flow fronts (Figure 6.46A). Less definitive are the wrinkle ridges, which are ubiquitous on the highland plains, which are thought to be volcanic (Figure 6.46B). The sheetlike nature of most of these deposits and the large volumes involved suggest that the eruptions occurred from large fracturelike vents with high flow rates typical for flood lavas. Some plains (Figure 6.46C) are dotted by low conical mounds several kilometers across, with depressions at their summits. These low shields and the surrounding plain are the result of small eruptions from numerous pipelike vents that may be localized along large fracture systems and are representative of basaltic plains.

The detailed nature of the northern plains remains unresolved, but much of the area probably consists of lava flows. Information collected by the

Viking landers suggests that the bedrock may consist of iron-rich, basaltic lavas. Moreover, the rocks seen on the surface by the landers are frothy, like those produced in gas-rich lava flows on Earth. Flow margins are also evident in some areas and resemble those in the lunar maria. Some of the youngest plains on the planet, lying near the equator, have been interpreted as large ash-flow or ash-fall deposits (Figure 6.46D), but no vents have yet been identified. In short, the geology of the martian plains is much more complex than that of the lunar maria, as the plains were probably built up by fluvial and eolian deposits as well as by volcanic flows. Subsequent modifications by ice-related phenomena have further complicated our attempts to understand the origin of the northern plains.

The large range of ages for the various volcanic features clearly shows that Mars has been thermally active during most, if not all, of its history. In addition, we have seen evidence for several types of volcanic processes that did not develop on the Moon or the portion of Mercury photographed by Mariner 10. The unique style of volcanic processes indicates significant planetary differences, which

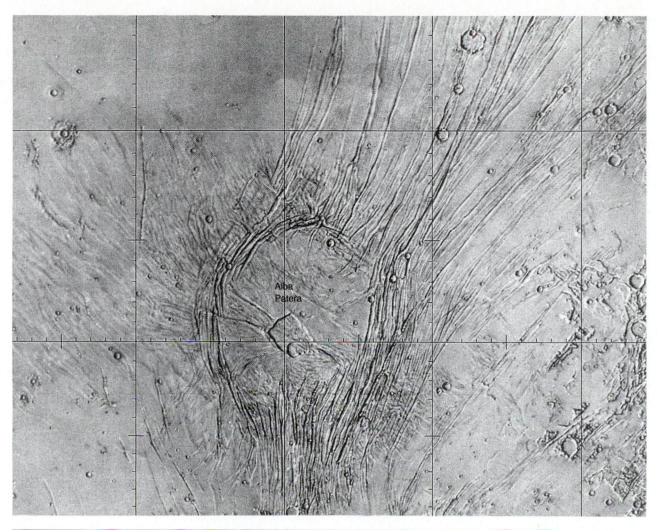

Figure 6.43

The volcano Alba Patera is surrounded by a large set of arcuate grabens that partially encircle the summit region of this low volcanic edifice and cut the youngest lavas. The caldera is located at the center of the map; lavas that erupted from this vent have partially filled a larger, older caldera just to the west (left). See Figure 6.38 for a regional perspective.

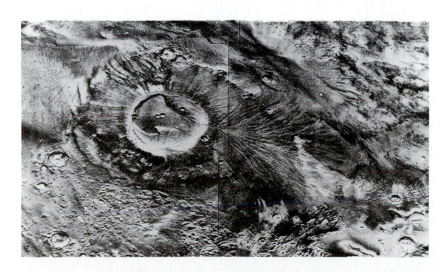

Figure 6.44

Apollinaris Patera, an ancient volcano in the southern highlands, may have erupted ash, as is suggested by the easily eroded deposits on its flanks.

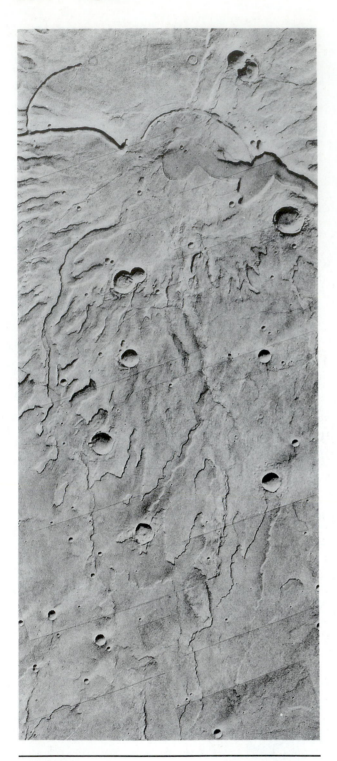

Figure 6.45
Tyrrhena Patera, located 1500 km northeast of the rim of Hellas impact basin, may have been the site of explosive volcanic activity during the early history of Mars. The deep channels may have been cut by younger lavas. This frame covers an area 280 km across.

must be, at least in part, the result of the size and composition of Mars.

Martian Tectonic Features

The presence of undeformed craters across the surface of Mars clearly indicates that the crust has not been subjected to extensive horizontal compression since the period of intense bombardment. There are no mountain ranges made of folded layers of rock and no active system of moving plates. Yet there are some very important tectonic features on Mars, produced by extension of and compression of the rigid lithosphere. These are convincing evidence that significant tectonic deformation has occurred throughout martian history.

The unique tectonic character of Mars is marked by large domal upwarps in the crust and extensive fracture systems associated with the production of these bulges. Other small wrinkle ridges and graben are also present but were probably the result of local vertical adjustments.

Crustal Warping

Several large domal upwarps occur on Mars; these are features not found on the Moon or Mercury. The two largest are in the Tharsis and Elysium regions, near the global escarpment. Both are capped with large volcanic cones (Figure 6.1). The Tharsis dome is a broad bulge 4000 km in diameter and 6 to 7 km high. A row of volcanoes, some with an extra 15 km of relief, cross the crest of the dome. This upwarp is much steeper on the northwestern side, where it rises from the low northern plains and therefore has an asymmetric profile. The Elysium dome is smaller, only 1500 to 2000 km across and 2 to 3 km high. The volcanoes there are less numerous and smaller.

Fracture Systems

The domes in the lithosphere on Mars are associated with large fracture systems that make spectacular patterns on the surface. The most extensive fracture system is northeast of Tharsis, where a fanlike array of grabens and fractures converge toward the row of shield volcanoes. This system of faults and fractures extends across a third of the planet. The fractures form magnificent sets of grabens that are typically 1 to 5 km wide and that may be several thousand kilometers long (Figure 6.47). Wherever older terrain is exposed, the rocks are intensely fractured by the intersection of several

(A) **Flow fronts for the vast sheets of lava** that have flooded parts of the cratered terrain are well developed in this region south of Tharsis. The large crater with the remnant central peak is 100 km in diameter.

(B) **Wrinkle ridges** form when compression buckles a mechanically strong surface layer. Even though they are not formed by volcanic processes, some suggest they are indicative of lava-covered plains.

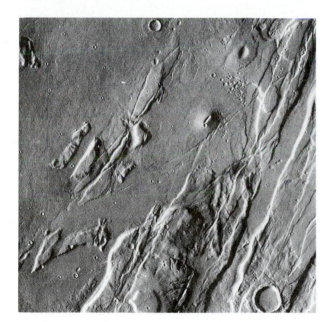

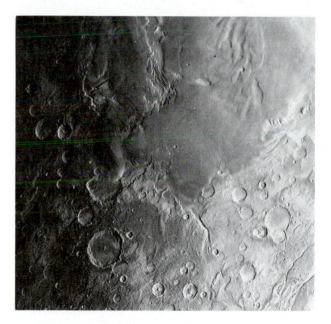

(C) **Plains-style volcanism** develops small, low-shield volcanoes like the three distributed across the middle part of this photo. The vents are one-half to 1 km across and cap very low shields with diameters of only 5 km or so. Smooth lava plains surround the shield and bury an older complexly faulted terrain.

(D) **Young, smooth plains,** like those in the upper left part of this photo, have been interpreted to be deposits of volcanic ash. The deposits are easily eroded and mantle underlying terrains, but no vents have yet been identified. The deposits occur near the equator of Mars.

Figure 6.46

Lava plains are not the most obvious volcanic features on Mars, but they represent the largest volume of volcanic material extruded on the surface.

Figure 6.47

Fractures, grabens, and wrinkle ridges that surround Tharsis cover almost one-third of Mars. The nearly radial orientation of the fractures is apparent and is probably the result of lithospheric fracturing during the rise of the Tharsis bulge.

sets of grabens with different orientations. It is almost certain that these fault systems extend beneath the young volcanic terrain. They represent a complex history of structural deformation by extension of the lithosphere.

The fractures appear to have been produced as the brittle lithosphere bulged upward and cracked. The location of the volcanoes on Tharsis ridge was probably controlled or facilitated by these fractures. The association in space and time of large domes in the martian lithosphere, graben and rift formation, and volcanism suggests that they may all have a common cause that is rooted in the flow of material in the mantle. Convective movements within the mantle may have pushed up and arched the overlying lithosphere to produce the domes. If solids in the mantle rise very far, they may become partially molten in the low pressure environment. This type of melting will occur even in the absence of any temperature rise. Once a molten magma is formed, it may rise even farther because of its low density and eventually feed a volcano. It seems that mantle convection may occur by the buoyant rise of less-dense (warm) mantle material in cylindrical pipes of upwelling mantle. Such feature have been called **mantle plumes**. The return flow of cold dense mantle probably occurs in cylindrical plumes of downwelling mantle.

From crater densities on the faulted surfaces, it seems quite likely that the major tectonic events that produced the Tharsis rise ended about 1 billion years ago and were followed by a series of volcanic eruptions that formed the shield volcanoes and young plains. Apparently, Mars remained tectonically active long after the smaller Moon had cooled and "died."

Canyons

A better insight into the type of tectonism that operated on Mars can be derived from a close inspection of the western part of the Tharsis fracture system. Earlier, we mentioned the vast system of interconnecting canyons called Valles Marineris; their location and size is controlled at least in part by fractures developed around the Tharsis dome. The huge dimensions of this feature are vividly shown in Figure 6.48, in which an outline map of the United States is superimposed for scale. This great canyon system is more than 4000 km long, 700 km wide, and as much as 7 km (20,000 feet) deep, dwarfing terrestrial river canyons. By comparison, the Grand Canyon of the Colorado River would be a minor tributary. The canyons of Valles Marineris consist of a series of parallel depressions with steep walls that drop abruptly from an upland plain. The troughs are highly irregular in detail, with sharp indentations caused by landslides and scallops and side canyons that may have been shaped by running water derived from ground ice. Much of the relief, however, is clearly the result of faulting along the fault system radial to the Tharsis bulge. Considerable erosion has widened the canyon walls, but the trace of the faults, along which vertical movement has occurred, is clear at the base of the cliffs.

Valles Marineris can be divided into three major divisions from west to east. Nearest the dome in the Tharsis region is an intricate labyrinth of intersecting canyons called Noctis Labyrinthus (Figure 6.49). The canyons that form this section are controlled by a set of intersecting fracture systems and are characteristically short and narrow. They thus divide the cratered plains of the plateau into a mosaic of blocks forming a huge maze. The labyrinth may have developed directly over the dome that produced the fractures and stands at an elevation of about 10 km. This set of canyons appears to have formed by localized collapse as the floors subsided rather than by erosion and transportation of material out of the canyons.

The central portion of the canyons consists of a system of long parallel troughs that extend a distance of 2400 km. They are the deepest depressions, with some walls rising 7 km above the valley floor. Throughout this section, Valles Marineris consists of chains of pits, all oriented in an east-southeast direction. In many places, the floor is covered with landslide debris or cracked by movements along faults, suggesting that the topography was produced by subsidence of the floor along faults at the foot of the canyon walls. However, the floor is usually smooth and featureless. In some of the broader canyons, plateaus of layered rocks have been found on the floor. Some scientists suggest that these are the eroded remnants of sediments deposited in ancient lakes that formed within the canyons; others suggest that they are merely the eroded remnants of the ancient martian crust isolated within the canyons.

The far eastern part of Valles Marineris is a series of irregular depressions that merge into chaotic terrain even farther east (Figure 6.50). Large areas of collapsed terrain have probably resulted from the removal of ground ice. Faults are not as obvious here, and the canyons are much more shallow and the floors are hummocky; apparently, these canyons were formed more by flowing water than by subsidence along faults. Valles Marineris is continuous with several chaotic areas, the largest being Aureum Chaos and Hydroates Chaos, which connect with the large channels that drain into the northern lowlands.

The enormous canyons of Valles Marineris are complex, involving tectonic activity and a variety of erosional processes. Faulting related to the creation of the Tharsis bulge appears to have played a key role in its formation by providing differential relief and zones of weakness, both necessary conditions for significant erosion to occur on Mars.

The features on Earth most comparable in size and nature to Valles Marineris are the Red Sea and the east African rift valleys. A **rift valley** forms where the crust is pulled apart, creating a system of parallel faults that allow an inner zone to subside, creating a large valley. Valles Marineris may have originated in a similar manner, as fractures developed along the crest of a large flexure in the crust and allowed the interior block to subside. Subsequently, the rims of the valleys were sculpted by erosional processes, producing the present topography. A much smaller set of fractures occurs around the Elysium dome and may reflect similar events that occurred there.

Other areas of Mars are cut by smaller grabens and faults (Figure 6.47). In fact, much of the global escarpment is transected by linear patterns,

Figure 6.48

Valles Marineris is related to the development of the Tharsis uplift. Downfaulting or rifting along faults radial to Tharsis created a system of troughs nearly 4,000 km long. Modification by fluvial and eolian processes and by mass movement has shaped this area's present appearance. The central part of Valles Marineris consists of Ius (south) and Tithonium (north) Chasmas. The linear bounding faults are clearly visible at the base of faceted cliffs in some places, but in many places they are obscured by ridged landslide debris. Narrow graben systems cut the upland plain and are parallel to the main canyons. Branching valleys, probably developed by seepage erosion, are common on the south walls of the canyon; landslides are more abundant on the north.

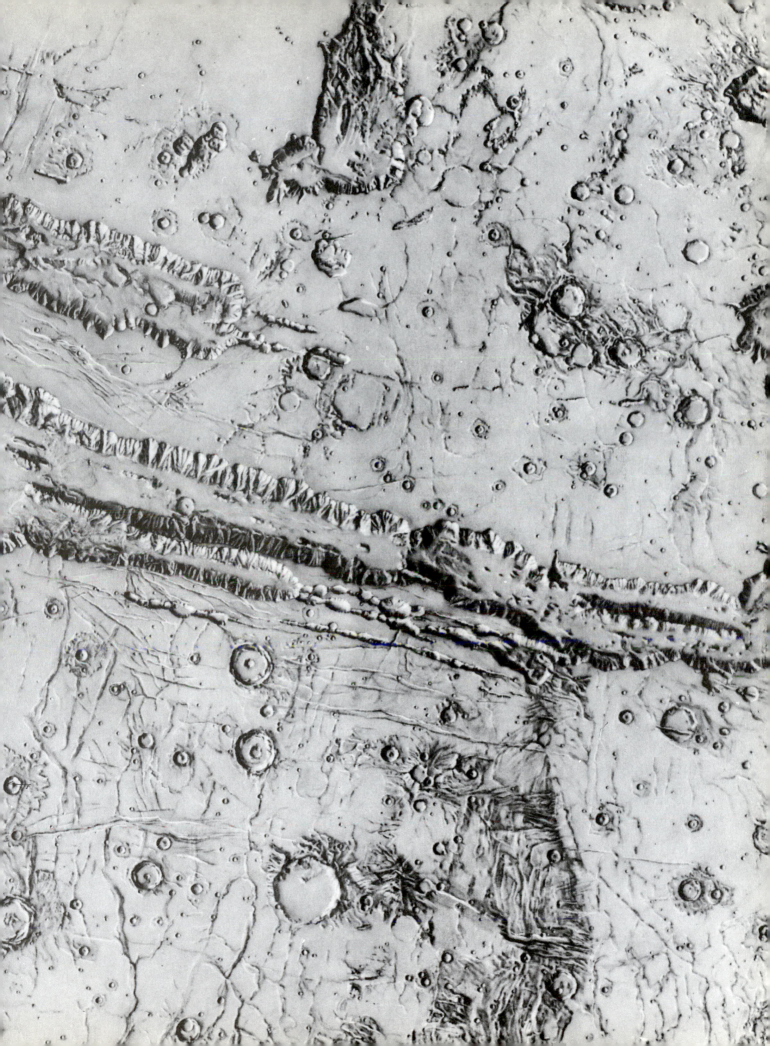

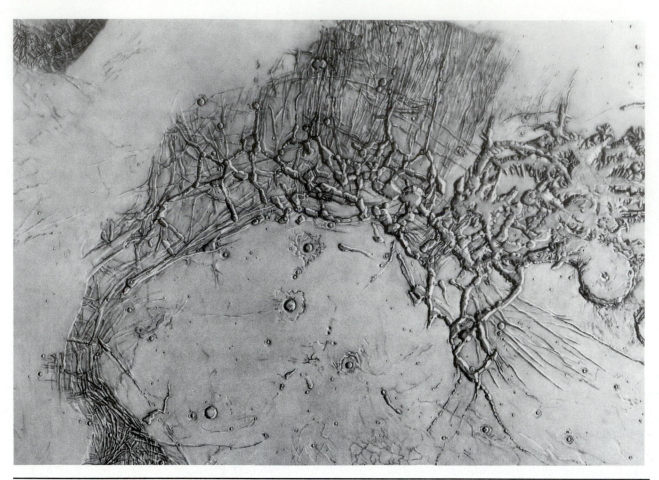

Figure 6.49
Noctis Labyrinthus is a network of interconnecting grabens centered on the highest
nonvolcanic part of the Tharsis uplift. The smaller, north-trending fracture system to the
southwest is much older than Noctis Labyrinthus.

suggesting that its evolution has been controlled in
part by a set of fractures and joints that have accelerated its recession. The thinner crust in the
northern plains may have been more extensively
fractured early in the history of Mars, possibly even
as a result of thermal expansion of the entire planet.

Wrinkle Ridges

In the plains, wrinkle or mare-type ridges are
very well developed and extend for many kilometers across the surface (Figure 6.51). Generally, the
ridges occur in sets that parallel regional depressions, as in the Chryse area, which may indicate
that they were produced by buckling of the near-
surface deposits (perhaps lava flows) just as has
been postulated for the Moon.

The development of domal upwarps and associated fracture systems clearly represents major
tectonic events in the geologic history of Mars,
events not shared by the Moon or Mercury. It

appears that this tectonism had some relationship
with the focus of the youngest volcanic activity on
the planet. On the Moon, we saw that tectonic
activity and the deformation of the outer layers of
the planet ceased between 2 and 3 billion years ago.
Even then, most of the changes were only minor,
related to more or less passive changes, such as the
vertical adjustments of mare basins. Mercury experienced global contraction as it cooled. Mars is a
more dynamic planet that experienced some tectonism in the form of crustal doming and rift
formation at least until a billion or so years ago.
Apparently because of the larger size of Mars, it
progressed more slowly through its thermal evolution. The duration of its tectonic systems was
longer, producing young features. Mars has a mixture of Moonlike features modified by some Earthlike processes. Even though no evidence of past
plate motions has been preserved on the surface,
mantle convection has produced some very obvious
changes to the cratered landscape.

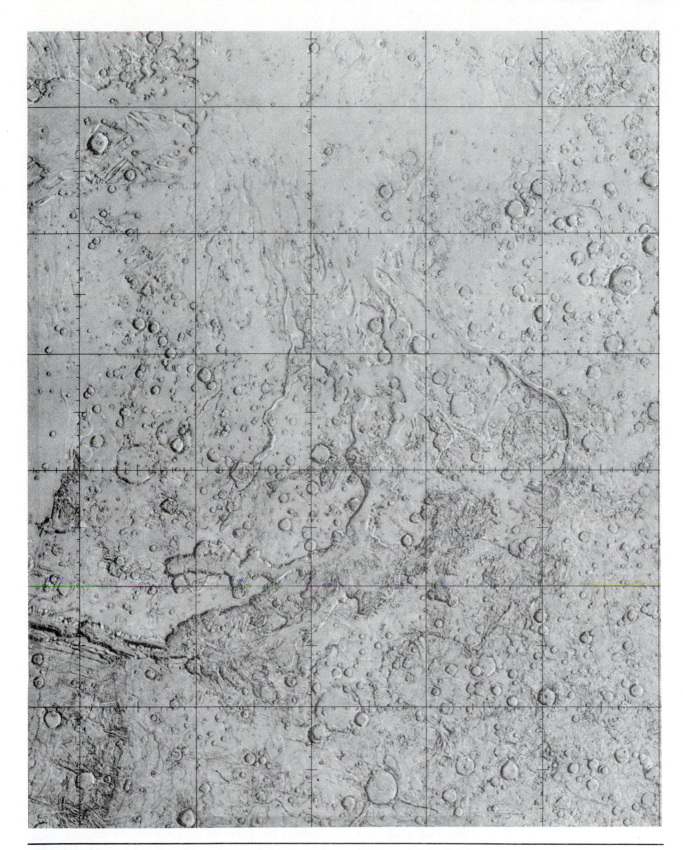

Figure 6.50

The eastern canyons of Valles Marineris broaden dramatically at the eastern end of Coprates Chasma. The location of the walls no longer appears to be controlled by faulting, and the canyon is not as deep as its western sections. The canyon is continuous, with large areas of chaotic terrain, which appear to be the sources of several of the outflow channels that emptied into Chryse Planitia.

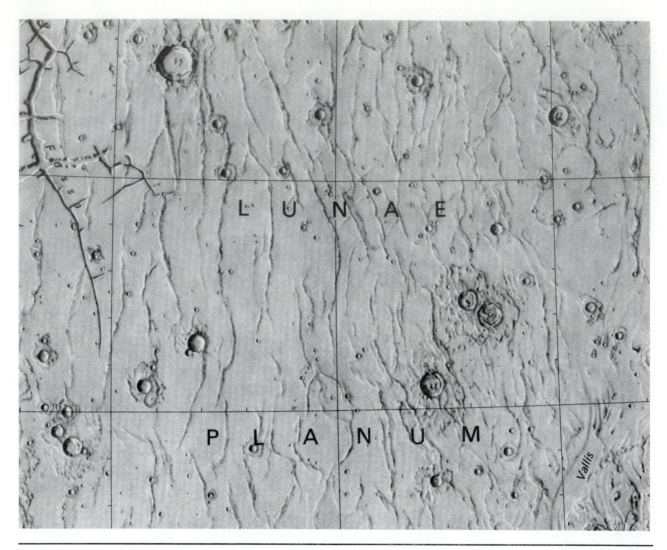

Figure 6.51
Wrinkle ridges cross this region on the western flanks of Chryse Basin. East-flowing catastrophic floods have extensively modified this area.

The Geologic History of Mars

Careful mapping of the surface of Mars has led to the identification of distinctive terrain units on the basis of their topography, color, and brightness. Using the principles of superposition and crater frequency, these geologic units and the processes which formed them can be placed in their proper time sequence. The formal names applied to these periods of time are, from oldest to youngest, Noachian, Hesperian, and Amazonian. The names are taken from prominent geographic regions on Mars. The absolute ages of these geologic periods are subject to great uncertainty because no rocks have been collected from the surface of Mars and because the rate of meteorite impact on Mars is not known and cannot be simply tied to the dated cratering record on the Moon. Even though the actual ages of these developmental stages are not known precisely, they provide an important framework for developing the history of Mars.

Six major stages in the evolution of Mars are recognized. Each stage involves distinct events or processes, but they may overlap somewhat in time. The effects of the major events in the geologic evolution of the interior and surface of Mars are shown in the series of diagrams in Figures 6.52 and 6.53.

Stage 1. Accretion and Differentiation.
Mars, like all the other planets, is believed to have formed by accretion of a myriad of smaller bodies over a fairly short period of time about 4.6 billion years ago. As the planet grew and developed a larger gravity field, the infalling debris was collected at higher speeds. Consequently, the impacts released large amounts of heat. Probably much of the planet was melted by this accretionary heat

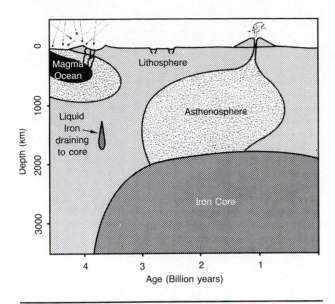

Figure 6.52

The thermal history and internal structure of Mars are shown on this schematic diagram. Accretion of Mars may have led to widespread shallow melting and the creation of a magma ocean hundreds of kilometers deep. The primitive crust may have crystallized from this magma ocean. Core formation was probably nearly simultaneous with the epoch of crust formation, as iron and iron sulfide became molten and, because of their great density, sank to form a metallic core. Radioactive decay added heat to the interior of the planet and sustained an asthenosphere, which fed volcanism at the surface. The Tharsis and Elysium domes also formed at this time. The lithosphere thickened with time and the zone of partial melting in the mantle may be entirely absent at present.

that was augmented by radiogenic heat produced internally. As a result, Mars may have differentiated internally, forming an atmosphere, crust, mantle, and core. If the initial heating of Mars was great enough, a magma ocean may have formed from which the primordial crust crystallized. This crust was probably basaltic and not anorthositic, as on the Moon. Great variations in crustal thickness may have been inherited from this time. On Mars, the crust in the northern hemisphere appears to be thinner than that in the south.

The gases and water that are presently at or near the surface were probably released from the interior of Mars during this early stage to form a primordial atmosphere and hydrosphere (Figure 6.53A). The presence of these fluids near the surface of Mars resulted from its more volatile-rich composition as compared to Mercury or the Moon.

Stage 2. Noachian Period.

During and following its formation as a planet, Mars was subjected to a period of intense meteoritic bombardment. Crust formation predated the end of the catastrophic bombardment, and a densely cratered surface such as that shown in Figure 6.53B, including several multiring basins, was formed. In fact, a few geologists have speculated that the ultimate cause of the north–south dichotomy was the impact of a huge body that thinned the crust of the northern hemisphere. The asteroids, the Moon, Mercury, and presumably even Earth experienced similar impact-dominated early histories.

An early period of widespread lava flooding is represented by extensive tracts of plains between and within the ancient craters. The plains are similar in appearance to the mercurian intercrater plains and in some cases show distinctive features of volcanic activity. Thus, the oldest rocks on Mars must consist of interlayered and overlapping ejecta blankets and lava flows formed shortly after the planet's accretion. By about 3.5 billion years ago, the rate of meteorite impact had probably declined to very near the present low rate.

The early martian atmosphere was quite likely more dense, and temperatures were probably higher, perhaps sufficient to allow liquid water to flow across the surface. If so, rainfall and flooding would have caused significant erosion of the highlands, and the runoff may have collected in local basins in the southern highlands and in the northern lowlands as ephemeral lakes or seas. The small filamentous channels of the highlands may be remnants of these ancient river systems. However, evidence for a thicker atmosphere early in martian history is at best equivocal. An alternative explanation for the valley networks holds that they were formed when groundwater, stored in the crust at some earlier time, seeped to the surface and created valleys by sapping at springs.

Stage 3. Early Hesperian Period.

Vast sheets of flood lava formed some of the ridged highland plains. Lavas are probably interlayered with eolian sediments in the huge areas of highlands that were resurfaced at this time. Several large volcanic fields, such as Tyrrhena Patera, formed in the southern hemisphere around volcanic patera.

Regional uplift in the Tharsis area began and was accompanied by radial faulting and volcanism, presumably because of plume convection in the mantle. Valles Marineris began its development along a structural trough in the fracture system. Much of the crust in the northern hemisphere may have been extensively fractured at this time, possibly because it was thinner there than in the southern hemisphere. In addition, slow expansion of the entire planet due to radioactive warming of its deep interior may have contributed to the tectonic breakup of the northern hemisphere (Figure 6.53C).

(A) **Stage 1. Accretion and planetary differentiation** form a layered internal structure as well as an atmosphere and a primitive hydrosphere.

(B) **Stage 2. Intense meteorite bombardment** and the formation of multiring basins.

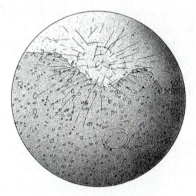

(C) **Stage 3. Uplift of the proto-Tharsis region** with radial extension and fracturing. This or other events disrupted the cratered terrain of the northern hemisphere and initiated formation of the bounding escarpment.

(D) **Stage 4. Widespread volcanic and sedimentary blankets** were emplaced in the northern hemisphere.

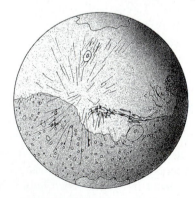

(E) **Stage 5. Renewed uplift and radial extension** in the Tharsis region. Valles Marineris continues to develop; Chryse Basin flooded as large outflow channels form in chaotic terrain of the eastern canyons.

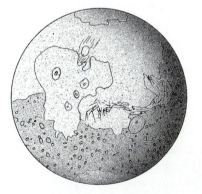

(F) **Stage 6. Emplacement of relatively recent plains** and formation of the large Tharsis volcanoes, including Olympus Mons. Eolian activity has been continuous throughout martian history, but the eolian features we see today were generally shaped during this epoch.

Figure 6.53

The geologic evolution of Mars can be summarized in six stages. Schematic maps of the western hemisphere are shown.

Probably during this or an earlier stage, the temperature at the surface began to drop as less heat was being released from the interior. Water became locked up beneath the surface as ground ice or in the polar caps. Carbon dioxide from the atmosphere may have been trapped in carbonate minerals formed at the surface. Atmospheric pressure probably dropped, partly as a result of the removal of water vapor and carbon dioxide from the air, but also because of the slow, inexorable loss of gas into space. Both of these processes made it very unlikely for water to exist as a liquid on the surface. Much of the erosion along the global escarpment occurred just before or during Stage 4.

Stage 4. Late Hesperian Period.

Widespread volcanic activity and deposition of eolian and perhaps fluvial sedimentary rocks occurred in the northern hemisphere to form great, low, sparsely cratered plains (Figure 6.53D). Eruptions of flood lavas formed the lunar maria and plains of Mercury and then stopped, probably several billion years ago, but on Mars volcanic activity continued intermittently over a much longer period of time, possibly to the present. The amount of volcanism appears to have declined with time. The large volcanoes in Elysium, sitting atop a large crustal swell, began to erupt during this epoch. Highland volcanic centers also developed.

Stage 5. Late Hesperian and Early Amazonian Periods.

Recurrent uplift in the Tharsis region formed additional radial faults and was accompanied by episodic volcanic outpourings to create the Tharsis volcanoes. Valles Marineris was progressively modified by filling with sediments and then enlarged by continued faulting and slumping. Catastrophic outbreaks of groundwater, possibly as a result of the formation of the Tharsis rise, released floods of water to form the chaotic terrain and large outflow channels of the eastern canyon system. These catastrophic floods carried sediment downslope to Chryse Basin in the northern lowlands and filled temporary seas. The chaotic terrain enlarged and the global escarpment continued to retreat southward. Major volcanism and related outburst of groundwater occurred in the Elysium province as well. Eolian and volcanic processes also resurfaced portions of the northern plains (Figure 6.53E).

Stage 6. Middle and Late Amazonian Period.

Volcanic extrusions partly covered the Tharsis area with fresh lava, and the most recent lava flows on Olympus Mons were erupted at this time (Figure 6.53F). However, the volume of magma erupted was significantly lower than in previous epochs, a clear indication that Mars was cooling to the critical level beyond which no volcanism can occur. Channels, probably related to the melting of ground ice by volcanic activity, formed west of Olympus Mons. Eolian activity persisted, modifying the entire surface while the polar layered terrains and surrounding dune fields continued to develop.

Mars has certainly passed its peak in geologic activity; it is likely that even the volcano Olympus Mons has been inactive for tens or hundreds of millions of years. As Mars continues to cool, radiating its energy away to space, active geologic processes such as faulting and volcanism will continue to wind down. At present, eolian processes are the dominant active geologic agents shaping martian surface features, although sublimation of ice and mass movement may continue to cause slope retreat along the global escarpment and along steep crater or canyon walls.

Although critical data are missing, particularly radiometric dates of the various rock units, a complex and fascinating history of Mars is arising from the study of relative ages. Tantalizing hints of further complexities and geologic paradoxes remain for generations of scientists to decipher and integrate into this framework for the geologic history of Mars.

Conclusions

The geological exploration of Mars has been a highlight of the space program and has revolutionized our knowledge of the red planet. Eight American spacecraft have successfully returned thousands of photographs and other data pertinent to deciphering its geologic history. Recent space missions have revealed that Mars is an enormously exciting place of vast geologic interest. We have seen Mars close up, landed on its surface, mapped its terrains, and found out what Mars is really like. Some of the more important discoveries are summarized below.

Much of the martian surface is cratered, some of it so intensely that it may be inherited from the close of the heavy bombardment (on the Moon these regions are about 4 billion years old). The craters on Mars are similar in size to those on the Moon but show evidence of greater modification by erosion and burial. Many martian craters have a unique appearance, caused by the motions of fluid ejecta that flowed across the surface.

The great shield volcanoes in the Tharsis and Elysium regions and the vast volcanic plains clearly indicate that Mars has experienced more volcanic activity later in its history than the Moon. Moreover, distinctive central volcanoes are important on Mars but are not seen on Mercury or the Moon.

Two large domal upwarps in the Tharsis and Elysium regions stand out as the major tectonic features of Mars. A pattern of radial fractures extends out from the Tharsis region over much of the western hemisphere. The great canyons of Noctis Labyrinthus and Valles Marineris formed as part of this system of extensional faults. Compressive forces have also deformed large regions of Mars, buckling the crust to form wrinkle ridges. Even so, no folded mountain belts, like those on Earth, formed and no system of mobile lithospheric plates was sustained. Tectonic (domes and graben) and volcanic features (localized on and near the domes) on Mars suggest that mantle plumes are an important kind of mantle convection.

Systems of stream channels exist on Mars. The largest originate in the southern highlands, cross the global escarpment, and terminate in the low plains to the north where shallow temporary seas waxed and waned. Their headwaters originate in chaotic terrain marked by collapse structures. Smaller, more ancient channels occur on old volcanic mountains and crater rims and as small branching systems. They may represent a major erosional interval early in martian history when liquid water was available at the surface, when temperatures were higher, and when the atmosphere was thicker. Much of this water now resides in polar ice caps or as ice within the surface layers.

Huge landslides and debris flows and other types of gravity-driven mass movement have been important processes in the development of the martian landscape; these enlarged or modified the canyons of Valles Marineris. The chaotic terrain is another expression of large-scale mass movement, involving collapse as subsurface ice melted and flowed away.

Winds gusting up to 200 km/hr sweep much of the surface during the global dust storms that periodically rage across the planet. Numerous streaks, dunes, and grooves show that the wind may be the dominant geologic process still active. Surface photographs from the Viking landers show many small eolian features such as sand ridges, dunes, and a type of desert pavement.

Superposition and crosscutting relationships show that Mars has had an eventful geologic history involving, in sequence, (1) accretion and internal differentiation, including outgassing of an atmosphere; (2) formation of a densely cratered surface, perhaps during a period when the atmosphere was thicker and a hydrologic system operated; (3) uplift

and radial fracturing of the crust in the Tharsis region; (4) widespread obliteration of the cratered terrain in the northern hemisphere, perhaps accompanied by extrusion of lava and deposition of sediment to form the northern plains; (5) renewed uplift and volcanism in the Tharsis region with the development of Valles Marineris and episodic release of groundwater in vast floods to create the outflow channels; and (6) modification of the surface, mainly by eolian deposition and erosion in a cold, dry environment.

Although Mars is still a relatively primitive planet in the sense that it has large tracts of heavily cratered terrain, it is far more Earthlike than the other inner planets we have examined. Mars has an atmosphere and a partial hydrologic system that worked together to provide a diverse arrangement of landforms. Mars is intermediate in size between the small terrestrial planets (Moon and Mercury) and the larger ones (Earth and Venus), and its thermal evolution also appears to have been intermediate, resulting in the production of mantle plumes, lithospheric domes, and an extensive and varied volcanic history.

Important factors that affect the thermal evolution of a planet and in turn help determine the nature of its geologic processes include its mass, diameter, and composition. Mars has almost twice the mass of Mercury but a larger diameter and lower bulk density. Thus, it has a slightly smaller surface-area/mass ratio. Calculations suggest that the martian cooling rate was slower, allowing for high internal temperatures for billions of years. Perhaps as a result of this slower cooling rate, Mars experienced a longer period of volcanic activity than is apparent on Mercury or the Moon (Figure 6.52).

Some scientists think that both the low density and long thermal evolution of Mars may have been caused, in part, by significant differences in the composition of Mars (and not just its size), as compared to either Mercury or the Moon. The presence of an atmosphere and water-related features support this assumption, as both are absent on Mercury or the Moon. Mars accreted at a distance farther from the early Sun than the rest of the inner planets and possibly at lower temperatures or pressures. These conditions probably allowed volatile-rich materials to condense from the nebula and become incorporated in the planet. If so, internal melting and convection in Mars may have occurred at lower temperatures than in Mercury, for example, and could have persisted to more recent periods of time. Other compositional differences, especially regarding the proportions of radioactive elements, may have helped slow the cooling history of Mars as compared to the Moon and Mercury.

Mars is more complex and geologically diverse than the Moon and Mercury in other ways as well. Tectonic disturbances, such as the Tharsis uplift, have altered the face of the planet dramatically. No contraction on a global scale is indicated. Internal circulation, or convection, in the hot, plastic mantle is thought to be the driving force behind much of this tectonic diversity. These internal motions provided forces that arched and cracked the thickening lithosphere, producing the vast fracture systems and localizing volcanic activity. However, the lithosphere of Mars apparently lacked lateral density contrasts and thickened more rapidly than Earth's. Thus, the lithosphere was not broken into a system of moving plates. Instead, persistent melting beneath the lithosphere produced huge volcanic piles, not the volcanic chains formed on a moving lithosphere like Earth's.

Review Questions

1. How may the interior of Mars differ from that of Mercury and the Moon?
2. Why are we unlikely to find liquid water on the surface of Mars? Does liquid water occur anywhere on or in the planet?
3. What does the layered nature of the martian polar deposits imply about the history of the martian climate?
4. Why is Mars red?
5. Describe the craters on Mars. How and why do they differ from those formed on the Moon?
6. Describe the volcanoes on Mars. Why are they so much larger than those found on Earth?
7. What are the principal differences between the northern and southern hemispheres of Mars?
8. Describe the origin and evolution of the atmosphere of Mars.
9. What is the evidence that Mars once had a warmer climate and denser atmosphere?
10. What tectonic features are found on Mars?
11. Describe Valles Marineris. How did it form? Does it have a close analog on the Moon or Mercury?
12. Describe the fluvial features on the surface of Mars. Are they caused by episodic or long-lived processes?
13. What is the evidence for "catastrophic" flooding? Describe the birth, growth, and death of a northern sea.
14. What landforms are produced by wind erosion and deflation?
15. Outline the geologic history of Mars. In what ways is it similar to the history of the Moon and Mercury?
16. Explain the cause of the differences between the surface features of the Moon and Mars.
17. If you were planning the next mission to Mars, what would be important objectives? What two landing sites would be best? Why?

Key Terms

Caldera	Ground Ice	Outflow Channel
Chaotic Terrain	Groundwater	Patera
Dendritic Pattern	Ice Cap	Rampart Craters
Dunes	Loess	Rift Valley
Eolian Processes	Mantle Plume	River
Fluvial Processes	Mass Movement	Sand Sea
Fretted Terrain	Mesa	Stream
Glacier	Moraine	Wind Streak

Additional Reading

Arvidson, R. E., A. B. Binder, and K. L. Jones. 1978. The Surface of Mars. *Scientific American*, Vol. 238, No. 3, pp. 76–89.

Baker, V. R. 1982. *The Channels of Mars.* Austin: University of Texas Press.

Batson, R. M., P. M. Bridges, and J. L. Inge. 1979. *Atlas of Mars.* NASA SP-438.

Carr, M. H. 1981. *The Surface of Mars.* New Haven, CT: Yale University Press.

Journal of Geophysical Research. 1977. Vol. 82, No. 28.

Science. Vol. 193, No. 4255, pp. 759–815.

Science. Vol. 194, No. 4271, pp. 1274–1353.

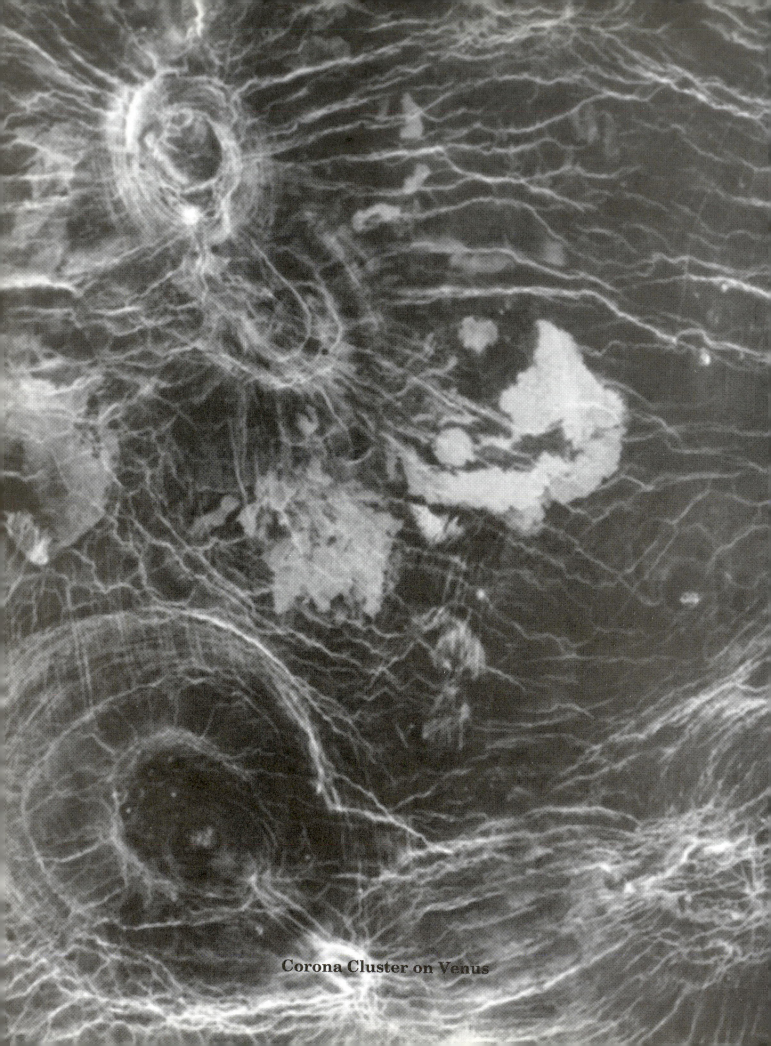

Corona Cluster on Venus

TABLE 7.1

Physical and Orbital Characteristics of Venus

Mean Distance From Sun (Earth = 1)	0.723
Period of Revolution	224.7 d
Period of Rotation	243.0 d
Inclination of Axis	3°
Equatorial Diameter	12,103 km
Mass (Earth = 1)	0.815
Volume (Earth = 1)	0.88
Density	5.2 g/cm^3
Atmosphere (main components)	carbon dioxide
Surface Temperature	760 to 650 K
Surface Pressure	109 to 46 bars
Surface Gravity (Earth = 1)	0.88
Magnetic Field	small
Surface Area Mass	9.45×10^{-11} m^2/kg
Known Satellites	0

CHAPTER 7

The Venus System

Earth

Venus

Introduction

Long considered to be Earth's twin because of its similar diameter, density, and distance from the Sun, the planet Venus is turning out to be quite different from the image evoked by this analogy. An impenetrable shroud of swirling clouds encircles the planet, while temperatures on the dimly lit surface exceed 750 K. Rains of sulfuric acid fall through a dense atmosphere of carbon dioxide, and lightning discharges continuously. Today, Venus is a waterless, sterile inferno with no hydrologic system. The evolution of Venus since initial accretion has diverged from the path taken by Earth—a temperate sanctuary for life.

Although over twenty spacecraft have examined the planet and its environs, Venus has yielded its secrets slowly. Orbital photographs of the surface of Venus are unobtainable because of its cloud cover (Figure 7.1). Our understanding of Venus is based on radar maps obtained by Pioneer Venus (1978 and 1979), Venera 15 and 16 (1984), and Magellan (1990–1995). We now know that Venus is an amazing place with abundant volcanoes, complex tectonic features, and relatively young terrains with few impact craters. Ultimately, as the clouds are penetrated and the interior probed we will understand how and why Earth and Venus, sisters of the inner solar system, have such different personalities.

Major Concepts

1. Venus is a rocky planet about the size of Earth. It probably differentiated into a crust and mantle, composed of silicate minerals, and a metallic core. Nonetheless, Venus has no detectable magnetic field and may lack an asthenosphere.

2. The secondary atmosphere of Venus is a dense envelope of carbon dioxide with traces of strong acids and very little water. The greenhouse effect creates surface temperatures (about 750 K) too high for liquid water to exist. Weathering and eolian transportation are, however, important geologic processes.

3. Radar imagery of Venus shows large areas of lowlands, upland plateaus, and volcano-capped swells. Impact craters are not abundant, suggesting the surface is on average 0.5 billion years old. Tectonic features include lithospheric domes cut by rifts, belts of compressional folds, ridges, mountain ranges, distinctive volcano-tectonic rings called coronae, and intensely disrupted highland plateaus. Evidence of volcanism is abundant and includes flood lavas, lava channels, shield volcanoes, and lava domes. Venus has been resurfaced many times by tectonic and volcanic processes.

4. Judging from the surface features, it is unlikely that Venus has a plate tectonic system like Earth. Like Mars, Venus may be a one-plate planet expelling heat through rising mantle plumes. Downwelling plumes may produce lowlands that evolve into deformed highland plateaus.

5. Accretion, internal differentiation, and outgassing of the atmosphere mark important events in the early geologic history of Venus. At some point, high surface temperatures were produced by greenhouse heating. Water may have been lost as a result, forcing dramatic differences in the tectonic evolution of the planet compared to Earth.

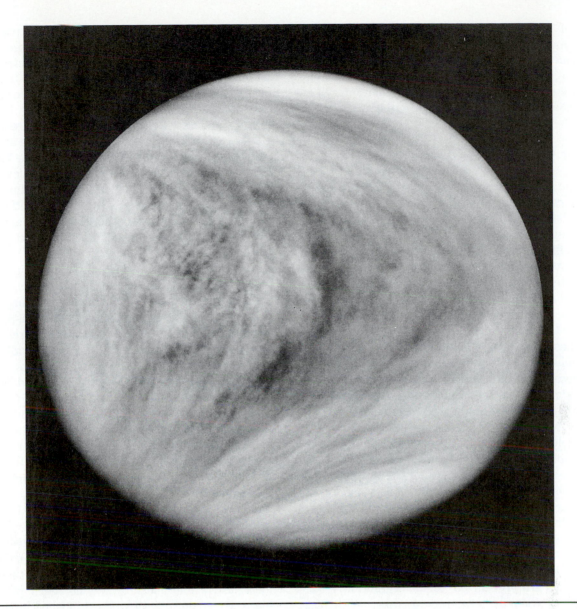

Figure 7.1

The clouds of Venus perpetually mask its surface; they consist of droplets of sulfuric acid, instead of water. Indeed, the atmosphere and surface of Venus are extremely dry. Absent are the great cyclonic storms typical of Earth's atmospheric circulation. The atmosphere of Venus is mostly carbon dioxide and is extremely dense; the surface pressure is about 90 times that at the surface of Earth. This dense atmosphere maintains a near inferno on the surface where temperatures reach 750 K.

The Planet Venus

A glance at the table of physical and orbital characteristics of Venus (Table 7.1) shows some of the most obvious reasons for making direct Earth-Venus comparisons. Both are about the same size and density and are approximately the same distance from the Sun. Presumably, they accreted from similar materials in the early solar nebula and should be similar in chemical composition. Therefore, the rates of internal heat generation and the

energy to drive volcanism and tectonism should be similar. Further examination of the physical data, however, reveals striking differences between Earth and Venus. For example, the average temperature at the surface of Venus is almost 750 K (hot enough to melt lead) and the atmospheric pressure is 90 times that to which we are accustomed. Venus is almost devoid of water. Thus, the processes of weathering and erosion on Venus are strikingly different than on Mars and Earth. Venus rotates very slowly in the direction opposite that of most of the planets. Moreover, Venus lacks a moon and a measurable magnetic field. Thus, Venus provides an interesting planetary contrast with Earth because, although it is like Earth in size and density, its surface environment and evolutionary development were dramatically different.

Major Geologic Provinces

The cloudy atmosphere has kept the surface of Venus completely hidden from view (Figure 7.1), but radar observations from orbiting American and Soviet spacecraft and from Earth-based radio telescopes finally parted the veil of clouds and gave us our first comprehensive glimpse of the surface of our sister planet. The major landforms have been mapped over 98 percent of the planet's surface. The topographic map derived from radar altimeter data reveals this global view (Figure 7.2).

These maps revealed that most of the surface of Venus consists of relatively smooth lowlands with local relief of less than 1000 m. Two continent-sized highlands and several smaller ones rise above the uplands. These plateaulike highlands are complexly deformed by tectonic processes and have few volcanoes. Large lithospheric domes with abundant volcanic features make up the uplands with elevations between those of the lowlands and the highlands. Therefore, Venus looks somewhat like Earth drained of its seas but with larger lowlands and smaller highlands. Moreover, the Venera spacecraft and Earth-based radar studies showed that volcanoes, linear mountain belts, rifts, complexly deformed plateaus, and enigmatic ring structures are present on Venus. Significantly, Venus lacks heavily cratered terrain. With the completion of the Magellan mission, the resolution and coverage of the radar images exceeds that available for much of Earth's solid surface; 70 percent of Earth is covered by oceans. Objects as small as a football stadium are visible in the new radar images of Venus. With these magnificent images, scientists pieced together the first global view of Venus, but many questions about the potential for active volcanism and tectonism on Venus remain.

Lowlands

As can be seen on the radar map, rolling plains of the **lowlands** are found across the entire planet and are relatively flat. They cover most of the surface of Venus. The mean radius of the planet (6051 km) is used as the global point of reference for altitudes, just as sea level is used on Earth.

Elevations below this are given as negative numbers and those higher are positive numbers. Lowlands are those areas less than 0 km elevation. Basins as deep as -2 km are present, but throughout most of the plains, the surface varies locally by only about 500 m. The largest of these lowland plains, Atalanta Planitia, is about the size of the Gulf of Mexico (Figure 7.3). Other large lowland regions include Lavinia Planitia and Sedna Planitia (Figure 7.2). Volcanoes are sparse in the lowlands, but lava flow fronts and long sinuous rilles cross them. The plains appear to consist of basaltic lavas, as indicated by their smoothness and vast expanses and by the compositions of rocks analyzed by several Venera and Vega landers.

The topographic basins show no evidence of being formed by impact and are probably structural basins produced by downward bending of the lithosphere. Some of the most revealing aspects of the lowland plains are their tectonic features. They are marked by a multitude of strongly deformed zones that are similar in many respects to the mountain belts of Earth (Figure 7.3). Evidence for strong compressional deformation such as found on these lowland plains is unique to Earth and Venus, the two largest planets of the inner solar system.

The lowlands of Venus constitute over 60 percent of the planet and could be called its great plains. These provinces of Venus resemble the ocean basins of Earth, the mare basins on the Moon, or the northern plains of Mars in that they are low, smooth, and covered in large part by volcanic deposits. Notable for their near absence, however, are impact craters. The plains have impact crater densities significantly less than the lunar maria and most of the northern plains of Mars. By analogy with well-established lunar cratering rates, the terrain must be quite young (less than a billion years old and perhaps as young as a few hundred million years old). This is evidence that the lowlands (as well as the rest of Venus) were resurfaced one or more times since the period of heavy bombardment. They are much younger than the Moon's maria and most of the martian plains. The relatively young, tectonically modified venusian plains may be more similar to Earth's sea floor than to these other surfaces.

Uplands

Isolated domes and broad swells rise as **uplands** above the venusian lowlands (Figure 7.4). The uplands are transitional between the highlands and lowlands, lying between 0 and 2 km in elevation. In contrast to the lowlands, tectonic features

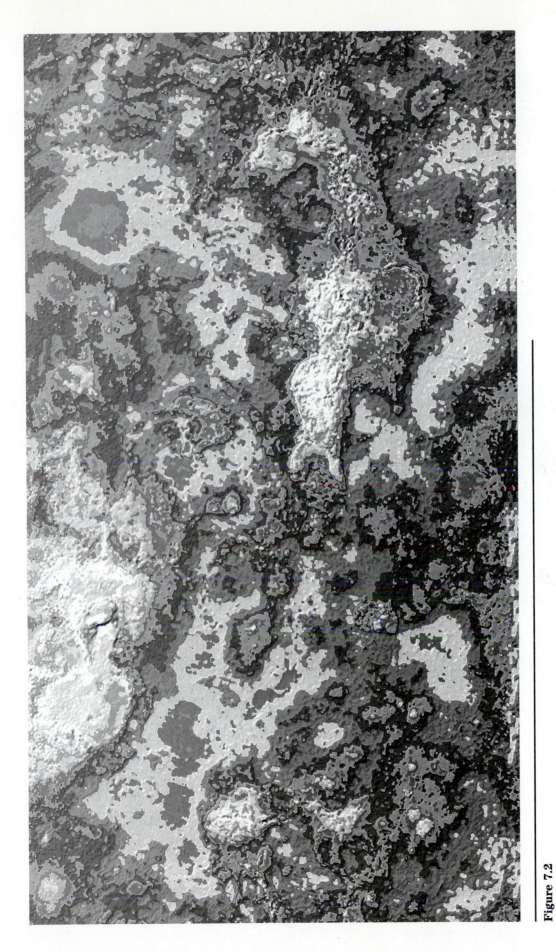

Figure 7.2

A shaded relief map of Venus was constructed from radar altimeter data collected by the Magellan and Pioneer Venus orbiters. A mercator projection shows two large highlands named Ishtar and Aphrodite rising above vast expanses of rolling lowlands. The locations of the landing sites for Veneras 8, 9, 10, 11, 12, 13, and 14 are shown.

Figure 7.3

Atalanta Planitia is typical of the low plains on Venus, as shown in this radar image obtained by the Soviet Venera spacecraft. Long belts of mountains cross the plains and stand several hundred meters above them. Each belt is separated by several hundred kilometers. These distinctive belts of slightly sinuous ridges are most similar to folded mountain belts on Earth, indicating compression. The nearly circular rings of ridges, called coronae, are distinctive volcano-tectonic features.

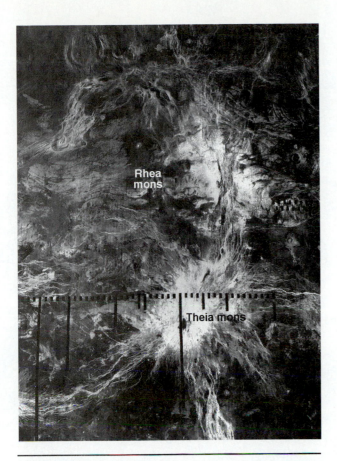

Figure 7.4

A Venusian upland is visible in this radar image of Beta Regio, a domed area with a large rift and volcano. The northern bright splotch is Rhea Mons; the southern one is Theia Mons. Theia Mons had radiating flowlike features on its flanks and a summit caldera. The volcanoes are closely associated with Devana Chasma. This fault-bounded trough cuts Rhea and the margins of Theia and extends southward for another 2500 km. The width of this image is about 1750 km.

are dominantly extensional in character and include long fracture belts, troughs, and rifts. Some rises are broad domes capped by fault-bounded rifts and shield volcanoes. Many volcanoes have summit craters and bright flow features around them. A wide variety of other volcanic features are found on these swells, or volcanic rises. Low-relief calderas, fissure-fed lavas, and central volcanoes are common in the uplands. The ringlike **coronae**, apparently formed by a combination of volcanic and tectonic processes, are distinctive features common in the uplands (Figure 7.3). Each corona consists of a nearly circular wreath of grabens and horsts. Some coronae are surrounded by flowlike features. Others have collapse calderas, lava channels, clusters of small shield volcanoes, and patches of smooth

plains showing the importance of volcanic processes in creating these unique landforms.

The most prominent of the upland domes is Beta Regio (Figure 7.4). Beta Regio is several thousand kilometers across and contains a large shield volcano and several coronae. The rift at Beta Regio, called Devana Chasma, is several hundred kilometers long and is flanked by the large shield volcano.

This volcanic rise, and other similar uplands and their rifts, may be the result of tectonic doming and stretching of the lithosphere above upwelling mantle plumes. In many ways, they resemble the lithospheric domes on Mars, Tharsis and Elysium, with their fault-bounded rifts and associated volcanoes, but the venusian features also have some similarity to the rift valleys on Earth, such as the East African rift.

Highlands

Perhaps the most intriguing features of Venus are the continentlike **highlands**, which typically stand at elevations of 3 to 5 km (Figure 7.2). The highlands are like plateaus in that they are high but relatively flat and not dome shaped, like the volcanic rises. Mountainous regions on the plateaus have peaks as high as 11 km; by comparison, the peaks of the Himalayas are about 12 km above Earth's average radius. One highland is surrounded by steep escarpments that resemble the continental slopes on Earth. These highlands cover less than 15 percent of the venusian surface; in contrast, Earth's continents occupy over 30 percent of its surface. No samples of rocks from the principal highlands were analyzed by Soviet spacecraft. Direct comparison between the composition and origin of the highlands of Venus and the continents of Earth is still premature. However, from radar studies, we understand something about the sizes, shapes, and surface features of the highlands. We also learn how they are related to surrounding terrains.

The highest and most spectacular highland region, Ishtar Terra, is in the northern hemisphere (Figure 7.5). It forms a partial arc around the north pole. Ishtar Terra is about the size of Australia and is wrinkled by large chains of mountainous ridges that include Maxwell Montes, the highest mountains on Venus (Figure 7.5). These rugged mountain belts dominate central Ishtar Terra, rising to an elevation of 11 km. They resemble the mountain belts found on the venusian plains in that they consist of slightly sinuous valleys and ridges,

Figure 7.5

A shaded relief map of Ishtar Terra, constructed from radar images of part of the north polar region of Venus, dramatically portrays the nature of the planet's highlands. The high plateau of Lakshmi Planum is fringed by mountain belts including Maxwell Montes, which rises 10 km above the plains. Ishtar Terra is thought to be an area of major compression with resulting mountain belts and volcanic activity. The high elevation and steep scarps are reminiscent of continental blocks on Earth. Volcanic calderas, Colette and Sacajewea, 250 km across, are as large as any on Earth. The surrounding lowlands are cut by complicated tectonic features that represent large scale disruption of older terrains by faulting that produces the tessera terrains. Cleopatra is a large impact crater. The number of craters suggests that the average of the surface is about 0.5 billion years old. Compared with Earth, one of the most significant differences is that Venus lacks valleys produced by stream erosion. Surface temperatures are far too high for water to exist as a liquid; moreover, the present atmosphere of Venus contains very little water vapor. Thus most of the surface features on Venus are produced by tectonism and volcanism but are unmodified by fluvial erosion.

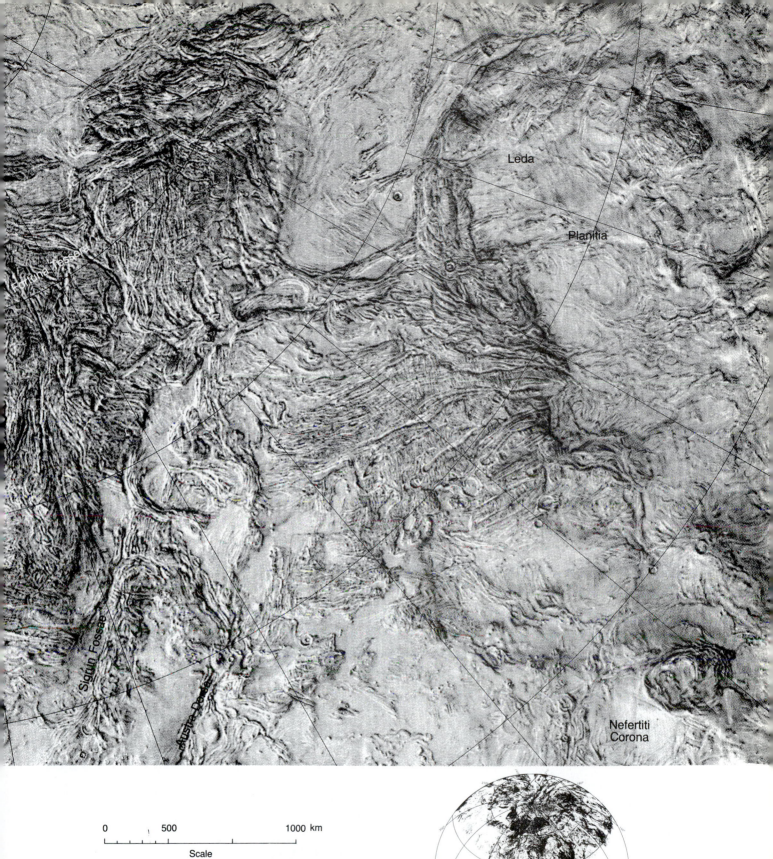

Leda

Planitia

Fortuna Tessera

Sigrun Fossae

Austra Dorsa

Nefertiti
Corona

0   500   1000 km

Scale

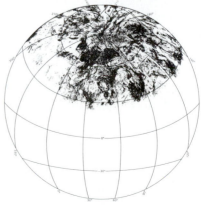

but they are much higher. These valleys and ridges are interpreted by most geologists to be compressional folds, ridges, and troughs, indicating lateral movement in the lithosphere of Venus. The mountainous terrain of Maxwell Montes is marred by a large 100-km-wide impact crater, Cleopatra Patera, whose ejecta deposits are not obvious. The valley and ridge mountain belts grade out into lower regions of extremely complex disruption (Figure 7.5) called **tesserae** (Greek meaning "tiles" because of the similarity of the texture of these surfaces to a tiled floor). To the west of Maxwell Montes lies Lakshmi Planum, a huge, relatively smooth plateau (Figure 7.5). Rimmed on all sides by rugged linear mountain belts, Lakshmi Planum rises abruptly 3 to 4 km above the plains province. Lakshmi is remarkably smooth and flat but is broken by two volcanic calderas. These calderas may be the source of the volcanic materials that form the plateau's surface. Small volcanoes are much rarer on Ishtar Terra than on the surrounding uplands.

The largest highland on Venus is Aphrodite Terra, named after the Greek goddess of love and beauty (Figure 7.6). It is centered just south of the equator and extends in an east-west direction more than 9500 km. Aphrodite is approximately the size of Africa and is dominated by three plateaus of tessera terrain, separated by two saddlelike depressions. One saddle is crossed by deep canyons or troughs (Diana and Dali Chasma). It is also cut by a remarkable arcuate trough (Artemis Chasma) that defines an almost complete circle on Aphrodite's southern borderlands (Figure 7.6). These troughs are probably of tectonic origin. The mountainous plateaus are quite rough with complex ridges and intersecting fractures like other tessera; isolated peaks rise as high as 8000 m above the surrounding low plains. A few large volcanoes dot the eastern plateau of Aphrodite Terra (Atla Regio) including Maat Mons which reaches an elevation of 9 km, but overall Aphrodite has fewer volcanoes than the adjacent regions. Expansive uplands with many more volcanic features extend east of Aphrodite to the dome at Beta Regio, a volcanic region discussed below.

Like the lowlands, tectonic features in the highlands are dominantly compressional. This observation suggests that the highland plateaus are related to extensive crustal thickening over mantle downwelling. It may seem strange to think that downwelling can produce a highland, but downwelling in the mantle could pull crustal material together, piling it up to great thicknesses over a region of downflow. The low density crust thus rises to higher altitudes because it is thicker.

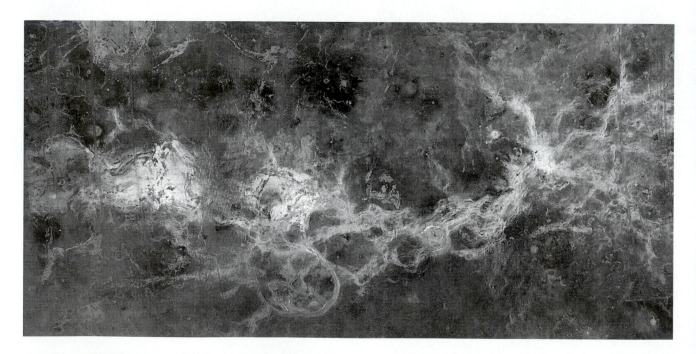

Figure 7.6
Aphrodite Terra contains some of the most rugged topography on Venus. It is marked by interconnected troughs and ridges called **tesserae** and isolated patches of highlands. This terrain may result from extension of the lithosphere. The most remarkable feature, however, is the nearly circular trough, Artemis Chasma.

On the Surface

Although no actual photographs of the surface can be obtained from orbit, television pictures taken by Soviet landers provide an entirely different perspective of the surface of Venus. In spite of the dense cloudy atmosphere, there is enough light at the surface to take photographs. In fact, it may be as well lighted as a cloudy day on Earth. Soviet spacecraft landed at four different locations: Venera 9 and 10 soft-landed in October 1975, and Venera 13 and 14 touched down in March 1982. The landers are all in the uplands region south and east of Beta Regio. Three other Soviet landers also analyzed rocks and assessed the surface environment but had no cameras.

The *Venera 9 landing site* (Figure 7.7) consists of a field of boulders, some with angular and others with rounded edges. They range from 20 to 70 cm across. Small, darker particles fill the depressions between the boulders. The slabs of rock are layered, and most lie unburied on the surface. The ground beneath the lander slopes away at an angle of about 20°, and the boulders may comprise some sort of talus slope derived by mass-wasting from farther up the slope. The pictures are not clear enough to decide if the slabby layered rocks are impact breccias, lavas, or some type of consolidated sediment; but clearly some dynamic process has fractured, rounded, and possibly moved the rocks. Small particles may be weathering products of boulders emplaced earlier (as impact ejecta or a lava field). Such fine particles could easily be moved by the wind.

The photograph returned by Venera 10 (15.4°N 291.5°E) shows an extensive plain composed of patches of varying brightness (Figure 7.8). The lighter material appears to be smooth outcrops and the dark areas may consist of unconsolidated fine particles. A few protrusions are apparent; most of the edges are rounded, possibly the result of a more extensive period of weathering than at the Venera 9 site.

The Venera 13 and 14 landing sites (Figures 7.9 and 7.10), at 7.55°S 303.7°N and 13°S 310°E) are more like the Venera 10 site than that of Venera 9, consisting of smooth surfaces with few pebbles or cobbles. Fine-grained, moderately dark regolith surrounds slabs of lighter-colored rocks at the Venera 13 site. The rocks are fractured and show signs of stratification on a scale of several centimeters. The Venera 14 site appears to have no fine particles. Instead, the surface consists almost entirely of indurated slabs of stratified rocks. Russian scientists interpret this as a volcanic plain where thin sheets of lava created the layered rocks.

Precise interpretation of this information is not possible. The areas resemble, in some respects, those photographed on Mars, especially the Venera 9 site. They contrast strikingly with the lunar surfaces because of the apparent thinness of the regolith. The surfaces could be composed of lithified eolian sediments or they may be the surfaces of lava flows. However, the pictures clearly show that weathering and erosion have modified the surface materials of Venus. The rocks show evidence of layers, fracturing, and chemical or physical weathering. Transportation of the finer particles by the wind may be the process that exposes the slabby outcrops to the view of the cameras. Thus, some of the same geologic processes we see on Mars and Earth also take place on Venus—with the important exception of hydrologic processes.

Elemental analyses of venusian rocks, performed by instruments on Soviet spacecraft at seven different localities (not all of the landers had cameras), show that they may be similar to terrestrial basalts. Most are similar in composition to basalts erupted from ocean islands or hot spots on Earth. Some analyses show that the rocks are rich in potassium but still are similar to other types of terrestrial basalts from the continents and ocean islands. Granitic rocks, like those that form Earth's continents, were not found by these landers, but as discussed below, some volcanoes look as if they were made by eruption of viscous magma. The most common kinds of viscous magmas on Earth are granitic or rhyolitic magmas with high SiO_2 contents.

Internal Structure

Earth and Venus formed in the same general area of the primordial solar nebula. Therefore, if the composition of nebular condensates depended on their distance from the Sun, it is quite likely that Earth and Venus have similar concentrations of the major elements—iron, magnesium, calcium, silicon, and aluminum. This notion is supported by density measurements determined from the motions of passing spacecraft. The bulk density of Venus (5.24 g/cm^3) is only slightly less than that of Earth (5.52 g/cm^3).

We have seen that most of the planets and the larger asteroids underwent extensive internal differentiation into layers of different chemical composition and density. In all probability, Venus experienced a similar differentiation process, which produced a metallic core composed predominantly of iron, a mantle of dense iron and magnesium silicates, and a crust of lighter silicate minerals

Figure 7.7

The Venera 9 landing site, on the eastern flanks of Beta Regio, consists of a jumble of closely spaced slabby or layered rocks. The horizon is visible in the upper corners of this and succeeding Venera photos. The arc at the bottom of the photo is part of the spacecraft.

Figure 7.8

The Venera 10 landing site, in the lowlands south of Beta Regio, lacks the boulders of the Venera 9 site and appears to be an extensive plain covered in a few places by fine-grained regolith.

possibly enriched in aluminum, alkalis, and the radioactive elements uranium and thorium (Figure 7.11). As this occurred, volcanic activity released gases (mostly CO_2 and H_2O), which formed the thick venusian atmosphere. There is, however, little direct evidence for these differentiation events and the nature of the interior of Venus.

For example, the magnetic field of a planet is probably due to the effect of the planet's rotation on a liquid, metallic core; therefore, the size and strength of this field measured from space probes should tell something about internal structure and temperature. Unfortunately, Venus has no measurable magnetic field. Perhaps this is because Venus rotates so slowly that it does not produce a significant magnetic field. Thus, the absence of a magnetic field does not preclude the possibility of a liquid metallic core at the center of Venus today.

Beside comparisons with other solar system bodies, there is indirect evidence that Venus is a differentiated planet. Volcanoes and vast volcanic plains prove that internal heating and even melting occurred. The young age of the surface shows that high temperatures within the planet persisted long after the time of heavy bombardment and may extend to the present. Measurements of the chemical composition of surface materials performed by

Venera 8 showed that rocks with high potassium and uranium concentrations may occur on the surface. Although these are only partial analyses, terrestrial rocks with similar high concentrations of these elements (alkali basalts or granites) have experienced extensive chemical differentiation. The more complete analyses returned by Vegas 1 and 2 and Veneras 9, 13, and 14 are similar to basalt found on other planets. These, too, show differentiation of the planet because basalts are different from primordial meteorite compositions like those that accreted to form the planets. The simplest way to produce basalt is by partial melting of olivine and pyroxene in a mantle, one of the products of differentiation. Linear chains of mountains and deep canyons, discussed shortly, are probably products of lithospheric mobility, driven by substantial convection of the deep interior. Therefore, an internal heat engine must be fueled by radioactive decay. If our interpretations of these surface structures are correct, the interior of Venus has undergone significant internal differentiation.

The atmosphere of Venus shows another strong evidence of internal differentiation. It does not have the same composition as the solar nebula gases (H_2 and He) and cannot be an inherited atmosphere. Instead, its carbon dioxide-rich char-

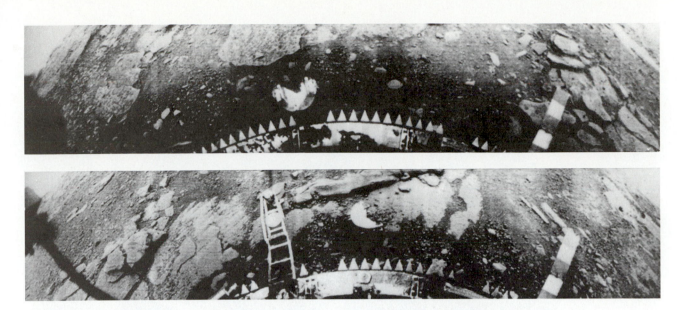

Figure 7.9

The Venera 13 landing site, on an elevated peninsula east of Phoebe Regio, consists of bare rock exposures of slabby fractured rocks covered in most places by unconsolidated rock fragments. Rocks analyzed from this location are similar in composition to potassium-rich terrestrial basalts.

Figure 7.10

The Venera 14 site, 1000 km southeast of the Venera 13 lander, also displays bare fractured surfaces of slabby rocks, but the regolith here is poorly developed. Venera 14 rocks are also similar in composition to terrestrial basalts, but they are not as rich in potassium as those from the 13 site.

acter is exactly what is predicted for an atmosphere made by secondary processes of differentiation and outgassing.

Thus, it appears from several independent lines of evidence that Venus must be differentiated. The thicknesses of the differentiated layers within Ve-

nus are not known. Venus has a slightly smaller radius than Earth, as well as a lower density. Perhaps the dense core of Venus is not as large relative to the rest of the planet as Earth's (or Mercury's, for that matter). Some estimate that the core would contain 25 percent of the mass of Venus

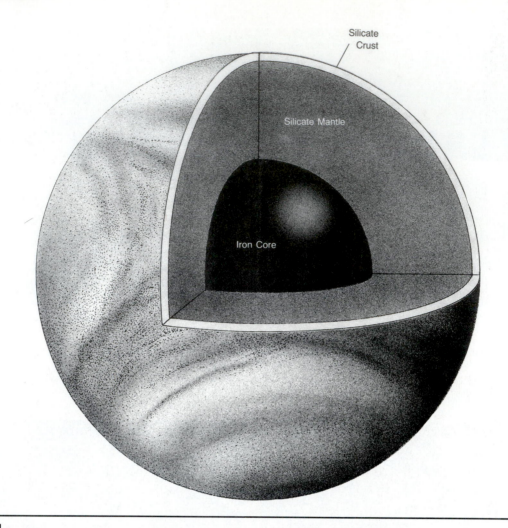

Silicate
Crust

Silicate Mantle

Iron Core

Figure 7.11

The interior of Venus is probably similar to that of Earth, since their sizes and densities
are nearly identical. In addition, the planets probably have similar bulk compositions since
they formed in the same part of the solar nebula. The core of Venus may be liquid, is certain
to be iron-rich, and probably contains sulfur and oxygen as well. The planet's mantle is
probably similar to Earth's—rich in silicates of Fe and Mg—with a crust composed of
feldspar-rich rocks. In spite of the inferred presence of a liquid metallic core, Venus has no
magnetic field. The slow rotation of Venus (nearly 225 days) may explain this fact. The
present thickness of the lithosphere is a matter of speculation.

(33 percent of Earth's mass lies in its core). One
theoretical model for the internal structure of Venus is shown in Figure 7.11.

One surprising result of study of the interior
Venus is that it may be substantially different from
Earth. Examination of the variations in the planet's
gravity field suggest, but do not demand, that
Venus lacks a shallow asthenosphere like Earth's.
Earth's asthenosphere permits the lithosphere to
move separately from the underlying mantle. As
noted in the last chapter, deep mantle circulation on
Earth may be dominated by the movement of
mantle plumes, whereas convection in the shallow
mantle is the result of the independent movement
of slabs of lithosphere. If Venus lacks an asthenos-

phere, then flow in the deep mantle may be directly
reflected in the surface features of the planet.

The thickness of the mechanically strong lithosphere of Venus is also unknown, but it is important to realize that even the surface of the lithosphere can deform and flow, by solid-state creep, at
the temperatures measured on Venus. The temperature is a significant fraction of the melting
temperature of many silicate minerals. At such
high temperatures, the mechanical strength of
rocks is dramatically reduced and they may experience viscous flow (a process like the flow of glacial
ice or viscous mud) to reduce the elevation of high
scarps or mountains (Figure 7.12). One set of calculations suggests that the topography of large fea-

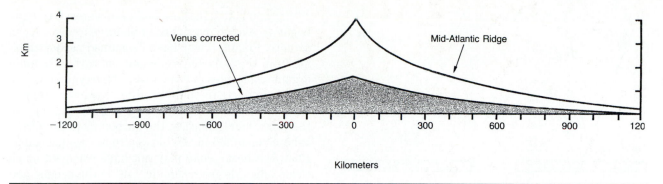

Figure 7.12

Idealized midocean ridge profiles for Venus and Earth. The actual smoothed topography of the mid-Atlantic spreading center is shown by the upper curve; the lower curve shows what the same structure would look like under the high-temperature conditions on Venus. The mechanical strength of rocks declines with increasing temperature, so they may flow in response to topographic loads.

tures such as volcanoes or impact basins would be completely erased after a billion years or so. We can conclude from the presence of high elevations on Venus (for example, Ishtar Terra—mountains up to 12 km high—or Beta Regio—4 km high) that either these calculations are incorrect or that many of the landforms of Venus were produced relatively recently.

The Atmosphere

Venus is totally enshrouded by clouds that extend vertically for more than 50 km (Figure 7.1); as a result, the atmosphere superficially resembles Earth's. Yet the atmospheric temperature, pressure, and composition of the "twin" planets are strikingly different. Because of the continuous global cloud cover and runaway greenhouse effect, surface temperatures on Venus approach 750 K. As a result of these high temperatures, which would vaporize water, and the inherent lack of water even as a vapor in the atmosphere today, there is no hydrologic system on Venus. The mass of water vapor in the atmosphere of Venus is nearly 100,000 times less than in Earth's fluid envelope. Atmospheric pressure at the surface of Venus is 90 times that found at the surface of Earth. At high altitudes, like the top of Maxwell Montes, the pressure drops to less than 50 bars and the temperature to about 650 K. Studies of the spectral characteristics of light reflected from Venus to Earth tell us much about the composition of the atmosphere; more sensitive analyses of the atmosphere also were performed by spacecraft passing through the atmosphere. Most of the atmosphere consists of carbon dioxide, and nitrogen comprises most of the rest.

So far, these studies have identified carbon dioxide (CO_2, 96.4 percent), carbon monoxide (CO, 30 ppm), nitrogen (N_2, 3.5 percent), water vapor (H_2O, 100 ppm), hydrogen sulfide (H_2S, 2 ppm), sulfur dioxide (SO_2, 185 ppm), oxygen (O_2) hydrogen chloride (HCl), and hydrogen fluoride (HF). The contrast with the much cooler, thinner, nitrogen-, oxygen-, and water-rich atmosphere of Earth is striking.

The Greenhouse Effect

The explanation for the extremely high surface temperatures on Venus is attributed to the **greenhouse effect**. The nearly transparent atmospheres of Earth and Mars allow much of the heat received from the Sun to escape by being reflected or reradiated into space. But on Venus, carbon dioxide and sulfuric acid in the atmosphere absorb much of this reflected energy. The energy is eventually re-emitted but not until the surface is much warmer than it would be in the absence of CO_2. In essence, Venus is like a well-insulated house in the winter. This planetary greenhouse entraps heat received from the Sun and distributes it planetwide so there are no cold polar zones as on Earth and Mars (Figure 7.13). The temperature anomaly is particularly striking because, due to Venus's reflective cloud cover, the surface of Venus actually receives less solar energy directly from the Sun than Earth does.

The clouds of Venus are not composed of water vapor or ice crystals, as on Earth, but are probably made of micron-sized droplets of sulfuric acid (H_2SO_4). The main deck of venusian clouds lies at altitudes between 45 and 75 km because this is the level at which sulfuric acid can condense because of low temperatures. This potent acid drizzle does not

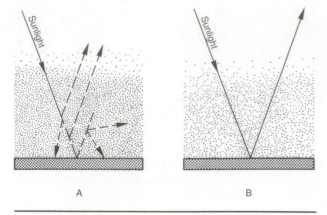

Figure 7.13

The high surface temperature of Venus may be explained by the presence of a greenhouse effect produced by the selective absorption of heat by the atmosphere.
(A) Much of the solar energy that falls on Venus is reflected back by the bright clouds (only about 20 percent is absorbed), so the surface of Venus receives less solar energy directly from the Sun than Earth does. The heat absorbed by the surface is reradiated at a different, longer wavelength (thermal infrared). Radiation of these wave lengths is readily absorbed by the gases (carbon dioxide) and clouds (sulfuric acid) of the atmosphere and cannot escape directly into space. As a consequence, the temperature at the surface rises. Since the clouds are not perfectly opaque to the reemitted thermal radiation, a little heat leaks out to space. Eventually, the temperature at the surface stops rising and stabilizes at some high temperature where the energy input and output are balanced.
(B) A hypothetical transparent atmosphere allows the heat to escape readily from the planet back into space.

result in the weathering of rocks because the sulfuric acid rain evaporates at an altitude of about 30 km above the surface. Although the clouds obscure the venusian surface, they are quite tenuous, and visibility within them is several kilometers. It is the thickness, not the density, of the cloud layer that hides the surface. Thin hazes occur both above and below the main cloud deck. Lightning discharges crackle through the venusian atmosphere, but they appear to originate below the main cloud deck, unlike terrestrial lightning.

Differences in the compositions of the atmospheres of Earth and Venus, especially regarding water and carbon dioxide content, are probably due to several factors, foremost among which may be the greater distance of Earth from the Sun. The factor may bear on both the water-poor and carbon dioxide-rich character of the venusian atmosphere. Some nebular condensation models suggest that Venus and Earth had similar amounts of water and carbon dioxide. The differences could be explained as follows: Earth may have been just enough cooler than Venus to allow water vapor to condense to a liquid, rain out of the atmosphere, and collect into

oceans and lakes. In these bodies of water, Earth life began to evolve over three billion years ago. Even primitive plants must have consumed carbon dioxide from the atmosphere, using the carbon to build organic molecules and releasing the oxygen gas as a waste product. In this way, the oxygen content of the atmosphere on Earth was gradually built up. On the other hand, much of the carbon from the atmosphere remained in the biosphere as organic solids. Photosynthesis alone may not have removed much carbon dioxide from the air; but as the oceans and life continued to evolve, carbonate minerals—calcite ($CaCO_3$) or dolomite ($MgCa(CO_3)_2$)— began to crystallize from Earth's oceans to form layers of limestone rock on the sea floors. Some types of ancient plants help to extract calcite from sea water. In addition, organisms with carbonate shells eventually appeared. When the organisms died, their shells accumulated on Earth's sea floor and, along with carbonate minerals precipitated directly from the oceans, formed limestone. As a result of the removal of carbon dioxide from the atmosphere and the ocean, Earth's crust contains great deposits of carbonate rock—mountains of rock, which could be considered Earth's "fossilized" atmosphere. If all the carbon dioxide could be released from terrestrial carbonate rocks, the atmosphere of Earth would be similar in composition to that of Venus, although it would be about half as dense. Even if carbonate minerals formed on Venus in the distant past, temperatures now are high enough to break down these minerals and release the carbon dioxide to the atmosphere again.

Water

The apparent lack of water on Venus is more puzzling than its thick atmosphere of carbon dioxide. Obviously missing are features produced by the circulation of water through the atmosphere and on the surface of Venus. River valleys, oceans, groundwater, glaciers, and ice caps—all important elements of change on Earth and Mars—are not apparent in radar images of Venus. At least three explanations for the absence of water-related features need to be considered:

1. The bulk of Venus is inherently water-poor but not carbon-poor.
2. Venus contained water that outgassed. However, the water never condensed to form a liquid because of the high atmospheric temperatures.
3. Water vapor outgassed, condensed to a liquid, and flowed across and shaped the landscape, but then disappeared and all ancient landscapes were subsequently destroyed.

The first explanation assumes that Venus did not accrete materials with the same amount of water-forming elements. Some models of the temperature gradient in the primeval solar nebula predict that minerals containing water did not form until well past the orbit of Venus (or Earth for that matter). If hydrated phases did not condense near Venus, neither could they be accreted into planetesimals that formed the planet nor later be outgassed.

The second explanation assumes the loss of water after outgassing. In this model of planetary evolution, both Earth and Venus started with similar contents of volatile elements. This could be true if the volatiles were added to the planets by late impacts of volatile-rich materials that condensed farther from the Sun than either Venus or Earth did. Some speculated that the inner planets received a substantial part of their volatile endowment when they were peppered by comets late in their accretion histories (see Chapter 2). But because it is closer to the Sun, Venus may have been warm enough to keep all of its water in a gaseous state. Moreover, under proper conditions of pressure and temperature, ultraviolet radiation from the Sun may break the hydrogen-oxygen bonds in molecules of water vapor. The light hydrogen may then leave the atmosphere and escape into space. Thus, the atmosphere of Venus may have been desiccated. If this happened, the oxygen either escaped to space as well or combined chemically with surface rocks; if not, the oxygen content of the atmosphere of Venus would be much higher than observed. In fact, the reddish color of the surface of Venus as seen through the eyes of the landers suggests that it is oxidized. Measurements of the isotopic compositions of hydrogen in the venusian atmosphere are consistent with this notion and suggest that the equivalent of a shallow venusian ocean may have been lost over the course of geologic time.

The third alternative for the history of water on Venus is much like the second but allows for the possibility that liquid water existed on the surface of a pre-greenhouse Venus. If temperatures were low enough, outgassed water vapor may have condensed as a liquid and shaped surface features for a time during the early history of Venus. Some time later, greenhouse heating vaporized the liquid water, driving it into the atmosphere, where water molecules could be disrupted and the hydrogen lost into space. Perhaps ancient water-related features were destroyed by the active volcanic and tectonic processes on Venus because no evidence has yet been found for an early hydrologic system on the planet. Future studies of Venus may be able to tell us which of these alternatives is correct.

The lack of water on the surface has important implications for the evolution of Venus. The important role of water and ice in shaping the surface of Earth cannot be understated. Movement of water occurs continually through an extensive, well-integrated system of oceans, rivers, lakes, and glaciers to produce an endless variety of landforms and rock types. Many common rocks on Earth—shales, sandstones, and limestones—owe their existence to some portion of the hydrologic system. Water and ice were also important in shaping the geologic evolution of Mars, as evidenced by the rampart craters, ancient streams, and huge channels produced by catastrophic floods. However, Venus lacks these surface features and rocks produced by water. No evidence of ancient water-related landforms has yet been found on Venus within the limits of resolution on the available radar images. At present, Venus is far too hot for water to exist either as a liquid or a solid, so water cannot now be a major geologic agent there.

The lack of water may be critically important to other aspects of planetary development as well. A small fraction of a percent of water in a planet's mantle can lower melting temperatures by 200 K, make the mantle fluid at lower temperatures, and consequently aid development of an asthenosphere at shallow depth. A weak, lubricating asthenosphere is vital for the lateral movement of large plates of lithosphere. A brief statement of this idea is: no water, no asthenosphere; no asthenosphere, no plate tectonics. The generation of light granitic magma (rich in silicon and poor in iron, magnesium, and calcium), by extracting a partial melt from rocks initially more rich in iron, magnesium, and calcium, also seems to require that melting take place in the presence of some water. Dry melting of silicates produces less silica-rich magmas. From these facts, some have inferred that the production of light granitic magma is prohibited inside dry Venus, thus prohibiting the development of Earth-like continents: no water, no granite; no granite, no continents.

Some of the most important future observations may clarify the role water has played in the history of Venus. If liquid water (which can exist to a temperature of almost 570 K at the present atmospheric pressure) at one time existed before greenhouse heating, why aren't water-related features preserved? Has Venus always been devoid of liquid water as a consequence of condensation near the Sun? If not, how did Venus lose its water? Is there granite on Venus? Does the lack of water make a shallow asthenosphere impossible? The answers to these questions will be important in determining the nature of the materials in Venus.

The Eolian Regime

An atmosphere is also a surface fluid. Its movement as wind plays an important role in the geologic evolution of surface features on several planets and moons. The nature of a planet's atmosphere helps determine, among other things, the surface temperature, the type of weathering or sediment cementation, the kind of hydrologic system, and the character of the eolian regime. Certainly the atmosphere of Venus, with its host of strong acids, high temperatures, and high pressures, has played an important role in the evolution of surface features and the types of rock exposed and produced at the surface. But is the wind also an important agent of change?

The Soviet landers photographed loose soil that could be a source of particles small enough to be moved by the wind. They also revealed bare rock surfaces swept clean of fragmental debris, perhaps by the wind. More conclusively, sequential television images taken by the Venera landers show that particles move on the surface. Other images show accumulations of particles in wind shadows behind rocks. Other affects of the dense, soupy atmosphere on eolian processes are likely to be significant. The wind-related landforms produced under high atmospheric pressure (90,000 mb; an equivalent pressure occurs in terrestrial oceans at a depth of about 1 km) could be quite different from those produced from Earth's moderate atmospheric pressure (1000 mb) or from Mars's light pressure (10 mb).

Direct measurements at the surface by Venera spacecraft show that, although generally low, wind velocities (1 to 36 km/h) are high enough to move small particles like sand and dust. Because the atmosphere is so dense, it can transfer kinetic energy from flowing gas to solid particles more easily than can a thin atmosphere like the one on Earth. Only one-tenth of the wind velocity necessary to start sand grains bouncing and moving on Earth is required to start eolian transportation on Venus. Impact processes may also produce massive but temporary wind storms that transport particles on Venus.

Thus, it was with great anticipation that the first high-resolution images returned by the Magellan spacecraft were searched for any evidence of wind activity on Venus. A host of landforms were found (Figure 7.14). Although no global duststorms were discovered on Venus, many of its eolian landforms are like those on Mars. The wind has created streaks, dunes, and the erosional features called yardangs. Wind streaks are the most common and are found at all latitudes and eleva-

(A) The most common are wind streaks. Area shown is about 100 km across.

(B) Yardangs, erosional ridges, have also been identified. Area shown is about 200 km across.

Figure 7.14
Wind-related features on Venus are abundant.

tions, demonstrating that eolian processes operate widely on Venus.

Streaks. The most obvious eolian features on Venus are groups of parallel plumes or streaks (Figure 7.14). The streaks can be as much as 100 km long and they can be bright or dark. Streaks commonly originate at topographic obstacles that acted as wind shadows or to enhance wind turbulence and hence scouring of the area immediately behind the obstacle. On Venus, ridges, tectonic hills, volcanic cones or shields, and impact craters all act as obstacles to the flow of the wind and many have wind streaks behind them. In fact, several thousand streaks were identified on the surface.

Before we can understand of the origin of wind streaks on Venus, it is important to remember that these features were revealed by radar reflections and were not revealed by the reflection of visible light. Fortunately, the amount of energy in reflected radio waves is sensitive to the surface roughness and composition. For example, a rough surface with irregularities that are about 10 cm across shows up as a bright region in a radar image. Smooth plains that lack these small irregularities reflect less energy and look dark in the radar images. Therefore, a thin, smooth layer of dust accumulated in a wind streak will look quite different than rougher surroundings (Figure 7.14). Alternatively, surfaces of a slightly different mineral composition could have different physical properties, including those that control how radar is reflected (the dielectric constant of the material). Just as some materials reflect light much more readily than others, some materials also reflect much more radar energy than others. The Magellan scientists concluded that surface roughness is the most important of the two for wind streaks.

These characteristics leave us with several ambiguities regarding the origin of specific wind streaks, but they are the same ambiguities we faced in interpreting the origin of streaks on Mars. Namely, are the streaks erosional features or depositional features? In general, streaks can be caused by both processes. For example, erosion behind a turbulence-enhancing crater or volcanic cone could strip away rough materials to reveal an underlying rock layer with a smooth, radar-dark surface. Alternatively, erosion of a smooth blanket of weak, smooth sediment could reveal a rough (radar-bright) surface beneath it. In addition, deposition of a smooth patch of dust in a wind shadow would likewise create a radar-dark streak. Nonetheless, the streaks show the activity of the wind. Because the streaks are effectively local wind vanes, they reveal the dominant wind directions that can in turn be tied to the global circulation pattern of the atmosphere and the climate. They show that small particles are produced and transported. Thus streaks are important because they show that some process of physical or chemical weathering occurs on Venus to create the small particles. Most fine particles are probably created by impact processes. Some of the winds that created the streaks may also be in response to the explosions or impact of meteorites on to the surface. Atmospheric blasts set up by inclined impacts may be especially effective at scouring fine particles downwind.

Dunes. Dunes are important because they show that sand, and not just small dust-sized par-

ticles, are present. Two small dune fields have been identified thus far on Venus. One field covers an area of over 1000 km². As seen in Figure 7.15, the crests of the dunes are aligned transverse to the dominant wind direction. Impact or explosive volcanism may have created the sand-sized particles for this dune field. Some impacts may also create air blasts that make transverse dunes upwind from a crater. Perhaps the dark haloes that surround some impact craters are dune fields created by impact winds.

Yardangs. Long ridges carved by wind erosion have been tentatively identified on Venus (Figure 7.14). The impact of wind-born particles on soft sediments can cause erosion of valleys that are parallel to the wind direction. Yardangs are rare on Earth and Mars, but they reveal that wind can be powerful agent of erosion, where soft sediment is exposed to strong winds. A field of about 100 possible yardangs lies near the impact crater Meade. The yardangs here are slightly sinuous parallel ridges separated by narrow grooves. A typical ridge is 25 km long and only 0.5 km wide. The similarity of these ridges to tectonic features makes it difficult to be certain that they are caused by wind erosion.

Venusian Eolian System. The variety of wind-related features on the surface of Venus is similar to those found on other planets and satellites that have atmospheres. The ubiquitous flow of the atmosphere in response to pressure and temperature differences is an important part of the present geologic modification of the surface. This

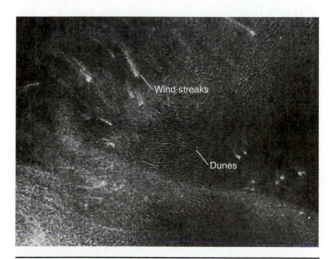

Figure 7.15

Dunes have formed as Venusian winds sweep across the surface transporting sand. Area shown is about 60 km across.

occurs even though the atmosphere of Venus is much denser and very different in composition, compared to Earth's. The streaks and dunes also attest to the role of physical weathering processes in creating small particles. This inference is strengthened by observations made by the landers that photographed the surface of the planet. However, the rate of erosion is quite low on Venus, especially so because it lacks a hydrologic system. As persistent as the wind may be, it is unable to produce the dramatic effects that running water and moving ice wrought on the surfaces of other planets.

Radar images from orbit attest to the effectiveness of some erosive process, probably the wind, in removing the rough, radar-bright ejecta patterns found around some impact craters. Only the youngest of the Venusian craters possesses these bright haloes of ejecta. However, the surface of Venus is not completely covered by eolian sediments. Analysis of the radar data shows that less than one quarter of the surface is covered by unconsolidated soillike materials.

Weathering

Weathering processes, wherein atmospheric gases react with rocks to form new minerals stable at the surface, must occur on Venus. We expect that weathering on Venus is very different than weathering on Mars or Earth. The main differences result from the absence of water, the extremely high temperature and pressure at the surface, and the composition of the atmosphere. Each of these factors exerts some control on chemical reactions between gas and rock. Minerals that crystallized at high temperatures in lava flows and that are now exposed to CO_2 and SO_2 gas in the venusian atmosphere are inherently unstable and should decompose to form new minerals. The weathered zone probably consists of a mixture of incompletely reacted minerals and newly formed weathering products. The growth of new minerals may destroy the fabric of the original rock and pry grains apart that can then be moved by the wind or gravity. For example, it is predicted that as weathering decomposes basaltic lavas, iron oxides and sulfur-rich minerals form as iron silicates are destroyed. However, carbonate minerals, which could remove significant amounts of carbon dioxide from the atmosphere, are not stable on the hot, dry surface of Venus.

Weathering rates on Venus are probably quite low because of the lack of liquid water, which on Earth mobilizes many elements. The presence of joints, such as those produced in cooling lava flows, may also aid weathering by allowing atmospheric gases to penetrate a meter or so below the surface.

Some of the reactions that could occur on Venus may be capable of weathering rock to a depth of 1 meter over the course of several hundreds of millions of years. This is a slow rate, but weathering of rocks on Venus probably is more important than the action of the wind in modifying the surfaces of the lava flows, impact craters, and volcanoes.

Direct evidence for the action of such weathering processes is meager, but promising indications were found in some radar images. Radar-bright areas are common at high elevations on volcanoes and other types of rises. The transition between radar-bright and radar-dark terrains lies at an elevation of about 3 to 4 km. In some regions, the boundary between the bright and the dark zones is amazingly consistent and cuts across various types of terrains. In places, the boundary varies by as little as 100 m over hundreds of kilometers. Some scientists have inferred that weathering at high altitude must produce minerals that strongly reflect radar waves. Weathering reactions at lower altitude create different, nonreflective minerals. Why would the weathering process produce different minerals at different altitudes? The explanation lies in the dramatic temperature and pressure variations in the atmosphere. At high altitudes (11 km) on Venus the temperature is as low as 650 K; at low altitudes (–2 km) it may be as high as 750 K. Atmospheric pressure changes as well from as low as 49 bars at high altitudes to as much as 109 bars. The stable mineral assemblage thus changes with altitude because the temperature and pressure change, even if we regard the atmosphere and the unweathered rock types to be uniform (Figure 7.16). One highly reflective mineral is pyrite, an

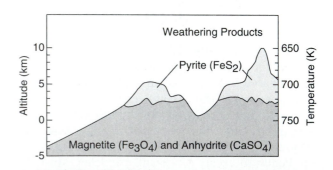

Figure 7.16

The style of weathering varies with altitude on Venus. At high altitudes and low temperatures, sulfur in the atmosphere reacts with lava flows to form the mineral pyrite (FeS_2). Pyrite is highly reflective of radar waves and may explain why mountain peaks and high plateaus of Venus are radar bright. At lower altitudes and higher temperatures, magnetite (Fe_3O_4) and anhydrite ($CaSO_4$) may be the stable minerals produced by weathering. Magnetite is not as reflective as pyrite.

iron sulfide mineral common in ore deposits on Earth and called fool's gold because of its bright yellow color and metallic sheen. Is it possible that weathering has created a thin coating of pyrite crystals across the surfaces of lava flows at high altitudes on Venus? This would be a spectacular site if seen from the ground. To test this hypothesis, thermodynamic calculations can be used to predict what minerals are stable and produced by weathering. They show that the stable minerals at high altitude include pyrite. At low temperatures, pyrite is created when iron-rich lavas, like basalt, react with the sulfur in the atmosphere. At higher temperatures like those found in the lowlands, magnetite, an iron oxide, is the stable iron mineral in the weathering zone. At low altitudes, sulfur is probably held mainly in a calcium sulfate mineral called anhydrite.

What other processes provide the small particles moved by the wind? Some particles are probably produced by eolian abrasion itself; as particles collide with each other and with the surface, new fragments may be broken from exposed rocks. Locally, tectonic processes such as faulting could also produce small particles by mechanically breaking down larger particles. Other processes that could create fine dust and sand-sized particles include explosive volcanism and impact processes. The association of wind-related features to impact craters may indicate that this is the dominant way in which small particles are created on Venus.

Impact Craters

Impact craters are ubiquitous on all of the planetary bodies we have examined thus far, from the tiniest asteroid to planets the size of Mars. In addition to being spectacular landforms in their own right, impact craters reveal much about the unique history of each planet. Such is the case on Venus as well.

Age of the Surface

Close examination of the high-resolution radar images of Venus shows that meteorite impact craters are present (Figure 7.17). Impact craters appear as radar-dark circular features with bright central spots and bright rims (bright areas on radar images correspond to rough surfaces). The planet lacks large tracts of heavily cratered terrain like those found on the Moon, Mercury, and Mars. The entire surface of Venus must have been resurfaced by volcanism and tectonic deformation many times

during its long history. In this important way, Venus is more like the Earth than any of the other inner planets.

The topographic maps produced from Pioneer Venus radar data were not of high enough resolution to detect impact craters with certainty. Venera spacecraft later explored the northern quarter of the planet and revealed that impact features were not abundant. These data were supplemented by Earth-based radar images that confirmed the paucity of impact craters on Venus but could only resolve craters larger than about 8 km across. Global coverage at sufficient resolution to see essentially all of the planet's impact craters was finally acquired in the opening years of the 1990s by the Magellan spacecraft. Spectacular images show about 1000 impact craters with diameters that range from 1.5 to 280 km across.

In contrast to the Moon and some other inner planets, no striking differences in crater frequencies between different regions were found on Venus. Instead, the craters are fairly evenly distributed across the plains, uplands, and highlands. Moreover, there are no heavily cratered terrains that date back to the time of heavy bombardment in the inner solar system (Figure 7.18). Because the distribution of impact craters is so uniform, all of the craters can be used to calculate an average age for the entire surface of Venus. Using the number of comets and asteroids that presently cross the orbit of Venus, the radiometrically calibrated lunar cratering rate can be adjusted to a value appropriate for Venus. If such estimates of the Venus cratering rate are correct, then the venusian plains are only about 0.5 billion years old. Of course, there is a large uncertainty on this estimate; nonetheless, the real age appears to lie somewhere between 0.7 and 0.2 billions years. This dramatic conclusion shows that the venusian plains are much younger than the lunar maria, the plains of Mercury, or even the youngest volcanoes on Mars. In fact, the surface of Venus may be almost as young as the vast plains on Earth that we call the sea floor. As discussed in the next chapter, Earth's oceanic crust is less than 0.2 billion years old and, like the venusian plains, is covered by tectonically disrupted basaltic plains.

It is important to remember that this estimated age represents the average age of the entire surface and does not exclude the presence of very young surfaces as well. Based on the abundant evidence of volcanic features on Venus that formed the plains and modified the highlands, some Soviet and American geologists believe volcanism is probably still active. In fact, lava flows on one of the high volcanoes on Venus are not radar bright like

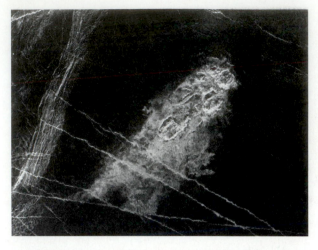

(A) Cluster of small craters is less than 13 km across.

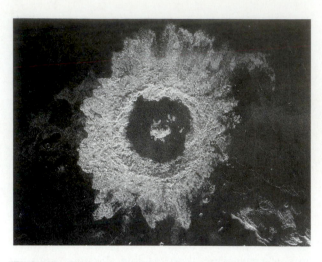

(B) Saskia (38 km across) has a central peak, a flat floor, and terraces.

(C) Wheatley has a peak ring (72 km across).

(D) Meitner (150 km across) is a multiring crater.

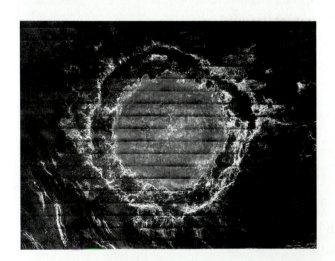

(E) Multiring basin Meade (280 km across) is the largest on Venus.

Figure 7.17

Impact craters on Venus show many of the same changes in appearance with size that are seen on other planets. Small craters with irregular shapes occur in clusters created by the breakup of a meteorite in the atmosphere. Progressively larger craters have terraces, central peaks, and peak rings, and the largest are multiring basins.

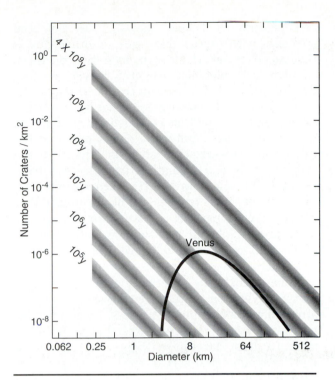

Figure 7.18

Crater frequencies on Venus are low compared with those on the surface of the Moon and Mars. This shows that the surface is young, perhaps no older than about 0.5 billion years old.

most high regions. This has been taken as evidence that the lavas are so young that they have not weathered to form the reflective minerals common in high regions elsewhere on Venus.

Crater Morphology. Venusian impact craters display many of the same changes in shape as lunar craters (Figure 7.17). From largest to smallest, Venusian craters show the following features: multiple outer rings, peak rings, and central peaks and terraces. There are only 6 multiring craters on Venus, and all are between 100 and 300 km in diameter. Although smaller, they are similar to multiring basins on Moon, Mercury, and Mars. Taking into account the young surface age, it is not surprising to find that Venus has so few large impact craters. There are no large multiring basins like Orientale (900 km in diameter) on the Moon, Caloris (1300 km) on Mercury, or Hellas (2000 km) on Mars, a sure sign that the surface of Venus was completely made over in the years after the heavy bombardment. Meade, the largest of the craters on Venus, is 280 km across (Figure 7.17). It has a well-defined inner terrace that was apparently formed by collapse of the steep wall of the crater immediately after excavation. The crater is less

than 2 km deep and has a broad featureless floor. The smooth floor is probably covered by lava that welled up into the center of the crater long after formation. Impact craters larger than about 40 km in diameter have peak rings that rise above relatively flat smooth floors. About 5 percent of the craters on Venus fall in this category. Craters with central peaks or mounds that rise above broad flat floors are the most common type of crater on Venus, accounting for about 40 percent of the craters. They are most common for craters between 15 and 35 km in diameter. Wide, terraced walls encircle these craters. Many of them have very smooth, radar-dark floors, implying that they were flooded by smooth lava flows. Whether these smooth floors result from the emplacement of impact-generated melt or from eruption of volcanic materials is still controversial. Unlike the Moon, the smallest craters (less than 40 km in diameter) are not simple bowl-shaped craters. These small craters typically are irregular in shape and occur in clusters.

Impact Craters and the Atmosphere

The effect of the dense, hot atmosphere even extends itself into the realm of impact crater morphology on Venus. As noted above, small craters on Venus typically are found in clusters of from 2 to 5 separate depressions. Each crater may have a diameter of 2 to as much as 40 km. The rims of the depressions in these clusters may overlap and the ejecta deposits are irregular and splotchy. These craters seem to be the result of the impact of a group of fragments that struck the surface nearly simultaneously. What is it about Venus that leads to this great difference?

The fragments may have been part of a single larger body that broke apart as it passed through the thick atmosphere of Venus. As a meteor passes through a planet's atmosphere, it will heat up as a result of friction with the air molecules. In the dense atmosphere of Venus, heating and expansion are probably dramatic and sufficient to create enough stress to rip the incoming meteor into several fragments that strike the surface simultaneously. Asteroids smaller than about 3 or 4 km across probably cannot survive the trip through the atmosphere of Venus in one piece.

The effects of the dense atmosphere on impact processes reveal themselves in a number of other ways. Perhaps the most striking is the complete absence of craters smaller than 1.5 km in diameter. The radar cameras on Magellan are good enough to see craters as small as a few hundred meters across. The atmosphere of Venus acts as an efficient "filter" and eliminates the smallest meteors before they can hit the ground to form a crater. For example, about 88,000 objects capable of forming a crater as big as Arizona's Meteor Crater (1.2 km across) should have entered the atmosphere of Venus in the last 0.5 billion years. Moreover, all craters smaller than about 35 km in diameter are less abundant than we would predict them to be. The cratering record preserved on the airless Moon shows that the number of smaller impacts greatly exceeds the number of large impacts. We also know that the small asteroids greatly outnumber those that are progressively larger. As a result, we expect the number of impact craters on a planet to increase progressively for smaller and smaller crater diameters (Figure 7.18). On Venus, just the opposite is true, at least for the smaller crater diameters.

Even though an enormous number of asteroids and comets that entered the atmosphere of Venus left no craters, many left some record of the final moments of their histories. Nearly 400 diffuse splotches with no central crater mar the surface of Venus (Figure 7.19). The splotches are from 10 to 70 km across. Haloes, similar to these splotches, surround many legitimate impact craters on Venus

and provide the clues necessary to link the splotches with impact processes. Another clue is that some crater clusters include two or three circular dark splotches. The craters were created by fragments that survived transit through the atmosphere and the nearby splotches by fragments of the same meteor that exploded before they hit the surface. These craterless haloes are probably related to pulverization and smoothing of the ground by blasts emanating from meteor fragments that explode in the air. Such an explosion could also generate tremendous blasts of wind that sweep or scour fine particles away from the center of devastation. An air blast with an initial velocity of 360 km/h could surge outward from a body only 300 m across.

The dramatically different ejecta deposits that surround Venusian craters are another effect of the atmosphere on the impact process. Long irregular tongues of bright materials extend away from the rims of at least 200 craters (Figure 7.20). These are not the symmetrical sort of eject deposits we see around lunar craters, nor are they like the thick lobate splashes of ejecta around many craters on Mars. Rather, in addition to a blanket of hummocky continuous ejecta, there are flows of what appears

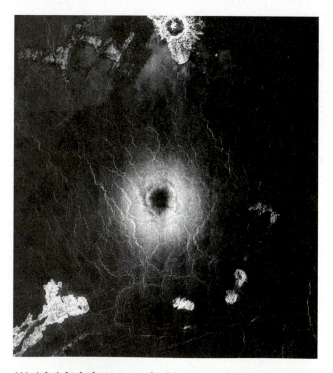

(A) A bright halo surrounds this 25-km diameter crater.

(B) A bright ring and an inner dark splotch surrounds this small impact crater.

Figure 7.19

Dark and bright splotches such as these are probably made when small asteroids and comets explode in the atmosphere before they hit the surface. Shock waves from the explosion sweep across the surface and create these splotches.

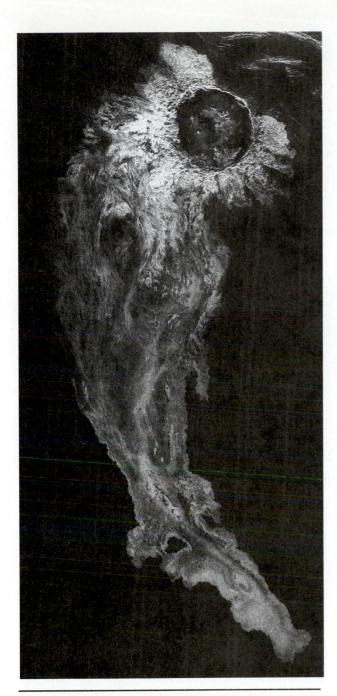

Figure 7.20

Long lobes of ejecta flowed away from many impact craters on Venus. The flows are more dramatic than those on Mars and may be related to melting of surface materials or to the fluidizing effect of the thick venusian atmosphere. Adams Crater is 90 km across and the bright flow extends 600 km away from its rim.

on these thin deposits. However, these features occur on many types of low viscosity flows, including some dry avalanches, mud flows, and pyroclastic flows. The direction of flow is consistently downhill and some flows are block by hills, fractures, or other high obstacles. Larger craters are more likely to have outflow deposits. Moreover, elongated craters made by low-angle impacts have more pronounced outflows. This is surprising in that oblique impacts should produce less impact melt. The outflow is concentrated in the direction parallel to the movement of the impactor. Two quite different hypotheses have been offered to explain these outflows. Perhaps these outflows are mobilized because they include a larger proportion of impact melt than ejecta on the cooler planets. The second explanations holds that the ejecta in the outflow is a turbulent mixture of solids and hot vapors created during impact and accompanied by entrainment of atmospheric gas. These flows would be much like the huge pyroclastic flows found around some calderas on Earth. Of course, we are not forced to choose between one of these mechanisms; both methods of outflow generation may be operating on Venus.

Other young craters are partially surrounded by huge, dark parabolic arcs that open to the west (Figure 7.21). These arcs may be caused when fine material, moved by high altitude winds, settles out after being ejected from the crater. Immediately after impact, all ejecta particles move ballistically through the atmosphere. However, in a planet with an atmosphere, winds force particles to move downwind. The wind affects the smallest particles most strongly and transports them downwind where they eventually fall out in a triangular area with the parent impact crater at one apex. This thin blanket of fine ejecta should be smoother than the surroundings and, therefore, should be dark in radar images. The wind-blown ejecta may pile up and become thicker on the margins of the parabola.

In summary, the atmosphere strongly influenced impact processes on Venus. Large impactors passing through the atmosphere were relatively unaffected and left sizable craters, with dramatic ejecta flow deposits. Intermediate-sized impactors fragmented and produced overlapping or multiple craters, and even smaller ones produced shock-induced splotches but no craters. The smallest projectiles were completely destroyed in the atmosphere and left no traces on the surface.

Crater Degradation

On most other planets, the age of a crater is directly reflected in its morphology. Crater degradation processes act progressively to modify the

to be lava away from the central impact site. One crater with a diameter of 90 km has a bright outflow that extends over 600 km from one side of the crater rim. The brightness patterns on the surface of the outflow look just like those on many lava flows. Channels, streamlined islands, and tributary and distributary patterns all are visible

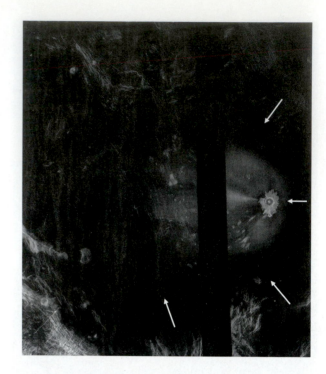

(A) Dark arc surrounding Stuart Crater (70 km in diameter).

(B) Adivar Crater (30 km across) is nested in a bright and a dark parabola. The jetlike streak west of the crater may mark the path of the incoming meteor.

Figure 7.21

Dark arcs like these encompass many young craters on Venus. They may be created by the deposition of fine particles swept out of the ejecta plume by the wind.

crater's ejecta, rim, and floor. Repeated meteorite impact, burial by volcanic flows, tectonic disruption, and erosion by wind and water are the principal mechanisms of degradation. The Moon, Mercury, and Mars show a wide range of craters, from those that are highly degraded and barely discernible to those that are extremely fresh. The range of degradation reflects the range of ages of the craters, with the most degraded craters generally being very old and fresh craters being young.

A striking phenomenon about craters on Venus is that most are very fresh. According to one interpretation, only a few highly degraded craters exist (Figure 7.22). What are the implications of this conclusion? Some geologists concluded that this means that Venus was completely resurfaced in a single event about 0.5 billion years ago and that little resurfacing has taken place since (Figure 7.23). In this scenario, the resurfacing event was short, perhaps even catastrophic. Other geologists looking at the same planet came to a very different conclusion (Figure 7.23). They point out that many of the "pristine" craters actually have smooth, dark floors, indicating that they were partially buried by later lava flows. Likewise, evidence for complex superposition relationships in volcanic terrains and the variation in the extent of weathering on various lava flows suggest that a range of surface ages does exist on the surface of Venus.

Tectonic Features

Two decades of exploration of the surface of Venus have shown that a wide variety of tectonic features are ubiquitous on the planet. Venus is not like the slightly deformed Moon. Its surface is strongly deformed at a variety of scales. Its lithosphere has been extended and compressed, domed and depressed. Apparently, deformation has been an important process on Venus for eons and may persist even today, driven by the churning of its mantle and the gravitational spreading of its highlands. There are several important questions we should ask about the tectonic system. Is the tectonic system on Venus similar to that of Mars, dominated by mantle plumes that create vertical movements of a laterally immobile lithosphere? Or is it Earthlike in character, with plate tectonics and lateral movement of thick plates of lithosphere that produce rifts and folded mountain belts? Or does Venus have a tectonic style of its own?

(A) Barrymore Crater (50 km in diameter) is cut by younger wrinkle ridges.

(B) This crater doublet (40 km across) is almost completely buried by smooth (dark) lava flows.

Figure 7.22

Crater degradation on Venus is the result of tectonic disruption and burial by lava flows.

Much of Venus is covered by tectonic landforms that indicate that the lithosphere may be mobile, at least to a limited extent. There are suggestions of tectonic features resulting from doming, extension, and fracturing of the lithosphere. In other areas, large mountain belts and wrinkle ridges suggest compression. These tectonic features occur in many areas, but so far there is no global tectonic pattern that can be discerned. There are no features that can definitely be called subduction trenches, volcanic arcs, or spreading ridges. If present, such features would indicate lithospheric recycling, with production and consumption related to a system of plate tectonics.

Domes and Rift Valleys

Broad crustal swells and rift valleys are common tectonic features on Venus. Beta Regio (Figure 7.24) is a striking example of this association. It is a large domical upland about 2500 km across that is crisscrossed by many faults. The gentle rise is about 4 km high and is crossed by a central trough. A multitude of nearly parallel linear scarps shows that the depression is a fault-bounded rift valley, formed as the dome was pulled apart by extension of the lithosphere. The fault scarps are spaced regularly 10 to 20 km apart. The valley has a variable width, like rift valleys on other planets.

Called Devana Chasma, this spectacular rift valley has three arms that extend for hundreds of kilometers across the plains to the south, then across the crest of the dome as a series of troughs. The rift is a deep as 6 km in places. A large impact crater, Somerville, is cut by the rift. The crater was originally about 40 km across but rifting has stretched part of it 10 km away (Figure 7.25).

Flanking the southern part of the rift and partly filling it is the large shield volcano Theia Mons (Figure 7.24), which rises 4000 m above the surrounding plains of the western hemisphere. The three arms of Devana Chasma converge on Theia Mons. Radar observations show that Theia Mons has a gently sloping conical shape with a radar-dark circular region near the summit, which is presumably a collapse caldera. The flanks of the volcano are composed of a complex network of overlapping lava flows. Theia Mons is about 350 km in diameter, not quite as large as Olympus Mons on Mars (550 km across). To the north, another large radar-bright region is also cut by the rift. Rhea Mons is not a volcano but a region of intense faulting called a tessera. Smooth plains bury some of the fractured region.

Beta Regio shows many similarities to the Tharsis or Elysium domes and volcanoes on Mars; the Valles Marineris rift cuts the Tharsis rise. Devana Chasma is similar in size and gross mor-

Catastrophic Resurfacing

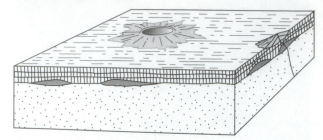

Vertical Equilibrium

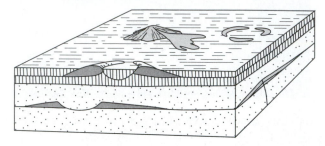

Regional Resurfacing

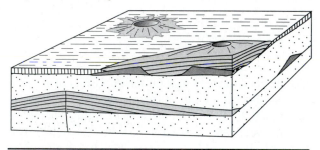

Figure 7.23

Volcanic or tectonic resurfacing of Venus may have occurred in a variety of ways. One suggestion is that catastrophic burial of the globe was associated with the geologically instantaneous overturn of the lithosphere about 0.5 billion years ago. This may have been the last in a series of older resurfacing episodes. Alternatively, dramatic changes in the surface of Venus may have been wrought by the progressive burial of small areas at a time as new volcanic centers developed. Ultimately, the whole planet would be resurfaced, but over a much longer time period.

phology to such features as Earth's Red Sea or the East African rift. On Earth, such rift valleys are formed by extensional forces, related to plate movements or to doming caused by upwelling hot mantle. Often, they are bounded by a series of parallel normal faults and are associated with basaltic volcanism. The association of plumes, rifting, and volcanism is common because the fractures and thin lithosphere found in rifts make it possible

Figure 7.24

Beta Regio is a large dome in the lithosphere of Venus. It rises 5 kilometers above the surrounding plains, but is transected by a deep rift known as Devana Chasma. A large shield volcano formed on one flank.

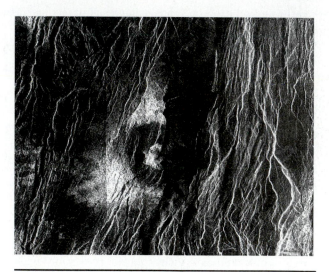

Figure 7.25

A large rift cuts an impact crater in Beta Regio. The crater, Sommerville, was about 40 km across before rifting, but the eastern rim has been moved about 10 km to the east by extension along Devana Chasma.

for basaltic magma to form in the plume and erupt. The interpretation that the Beta Regio rift is related to basaltic volcanism is supported by information obtained by Venera 9 and 10, which landed just east of Beta Regio. Data sent back from the landers indicate that the rocks in this area have concentrations of radioactive elements similar to those in basalts. Thus, a volcanic origin is suggested by both the nature of the landforms and information about the composition of surface material. Based on its features and comparisons with other planets, Beta Regio must be a region of uplift, rifting, and volcanism that probably overlies upwelling hot mantle material in a mantle plume.

Near Beta Regio lies another large upland dome called Atla Regio (Figure 7.2). It lies near the equator at the easternmost end of Aphrodite Terra. Like Beta, it rises about 3 km above the lowlands. A complex set of nearly radial fractures is centered on the dome. Three large volcanoes, Ozza Mons (4.5 km high), Sapas Mons, and Maat Mons, lie on the crest and flanks of the dome. Other volcanic centers, including several coronae and many small shields and volcanic domes, lie in the same rise. Cross-cutting relationships show that faulting and volcanism were contemporaneous.

Large domes, with associated rifts and volcanoes, occur elsewhere on Venus. Sif and Gula Mons are volcanoes situated on a rise in Western Eistla Regio (Figure 7.26). The large dome is about 3000 km across and reaches an elevation of 1.4 km. Sif is an extensively fractured shield volcano over 100 km in diameter that rises another 2 km above the top of the dome. The summit caldera is about 40 km across and includes several smaller calderas nested within it. Apparently Sif had a long and active history of volcanism and repeated fracturing related to the rise of less dense mantle beneath it. Likewise, Gula Mons has a complex volcanic and tectonic history with multiple volcanic vents. Gula Mons is much higher than Sif Mons. It is also the center of a radial set of fractures, suggesting uplift of the dome, but it also sits astride a rift valley. Just

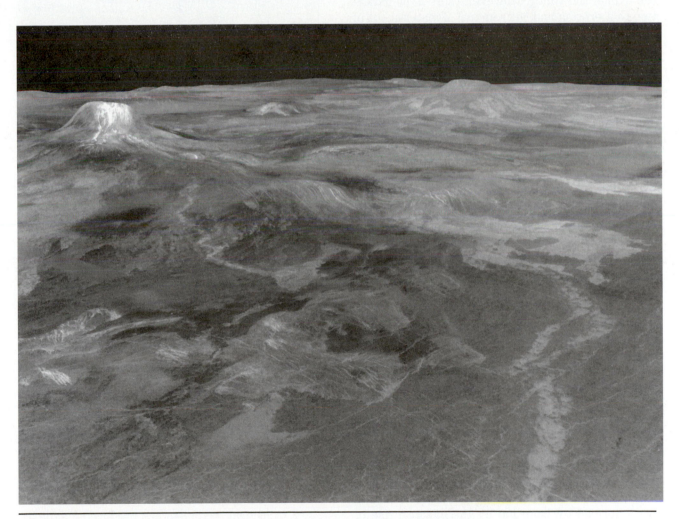

Figure 7.26

A large dome transected by a rift and capped by volcanoes comprises Western Eistla Regio. Gula Mons (left) and Sif Mons (right) are in the background and a large corona lies between them. Some long lava flows erupted from vents on the plains; others flowed down the flanks of the volcanoes. Like other large swells on Venus, the dome probably overlies a site where hot mantle wells up in a plume.

north of Gula is a large corona that rises only 600 m above the flanks of the dome. Like other coronae, it is surrounded by a wreath of grabens sitting within a semicircular trough. Large lava flows erupted from the troughs. Other large coronae are strung like beads along the eastern extension of the rift through Eistla Regio.

Other riftlike valleys developed in the eastern part of Aphrodite Terra (Figure 7.4). Here the terrain is particularly rough and complex, cut by a number of relatively straight valleys and adjacent high ridges with steep vertical walls nearly 4 km high. Individual valleys can be traced at least 2500 km across the surface. In places, the main valley is 150 km wide and 5 km deep. Like terrestrial rift systems, it branches into two similarly sized depressions. A ridge, very similar to portions of the larger martian canyon, Valles Marineris, runs up the axis of the valley. However, there is no indication that erosion and the removal of material out of the valley system is important on Venus, as is the case for portions of Valles Marineris.

The clear association of lithospheric doming, fracturing or rifting, and volcanism lead us to the conclusion that mantle plumes are common within the interior of Venus. They may be the primary means for Venus to convect and exhaust internal heat. Mantle upwelling on Venus operates on several scales, with coronae representing the smaller, short-lived upflows and major volcanic rises representing larger, long-lived convective upwellings. These plumes may be broad upwellings, with individual coronae, novas, arachnoids, and other types of volcano-tectonic centers representing shallow plumes associated with the main upwelling or smaller isolated plumes (Figure 7.27). Because the venusian lithosphere is not moving across the mantle plumes, the volcanic and tectonic effects of the plumes are concentrated in one area. The volcanoes grow to be larger; the uplifts are higher and broader.

The life history of a plume may proceed in the following manner. A plume initiates in the deep mantle of Venus and rises because of its low density. As the plume traverses the mantle, it should have a large head and long narrow tail. Partial melting of the plume occurs at shallow depth because of lower pressure. As the partially molten plume head impinges on the base of the lithosphere, it will flatten and create a broad dome in the crust, formed by heat and by the dynamic rise of the plume. The uplift may cause the rigid crust to crack and form radial and concentric fractures or large through-going rifts. As magma from the plume head reaches the surface, volcanoes grow at the surface. Eventually, the plume head loses its thermal energy as it cools and solidifies. At this

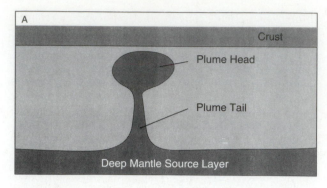

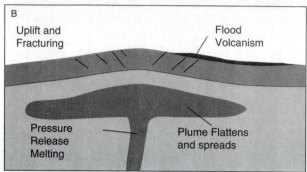

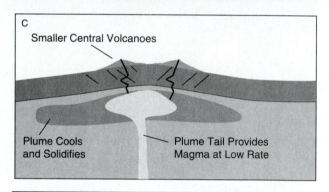

Figure 7.27
Mantle plumes embedded in large upwelling regions of the mantle of Venus appear to make the large domes, rifts, and volcanic fields that are scattered across the surface of the planet. Their progressive development is shown in these three schematic cross sections.

time, the dome may partially subside because of eruption of voluminous magma from the plume head or by contraction upon cooling. Subsequent volcanism will proceed at a slower rate, fed by magma from the still rising plume tail.

Fracture Belts

Extensional tectonism has produced long belts of deformation marked by abundant fractures and grabens (Figure 7.28). These belts persist over hundreds of kilometers. The grabens are usually narrow, typically less than a kilometer across. Some are so narrow that their floors cannot be seen even at the resolution of the Magellan radar cam-

Figure 7.28

Fractures like these are common on the plains of Venus and show that extension of the lithosphere is important.

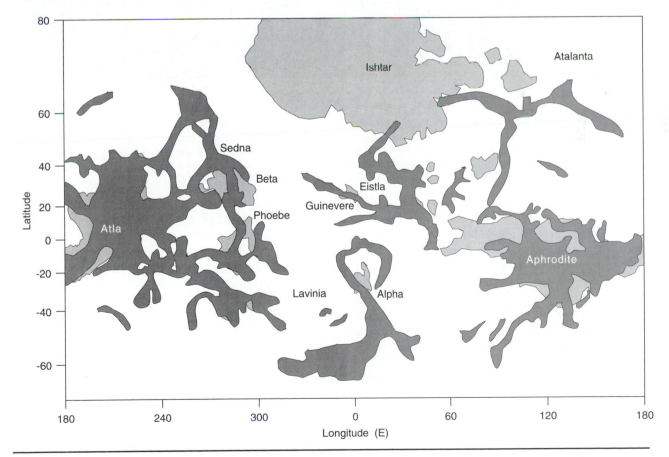

Figure 7.29

Belts of extensional fractures are common in the equatorial region of Venus and occur in areas with lithospheric domes and abundant volcanic edifices.

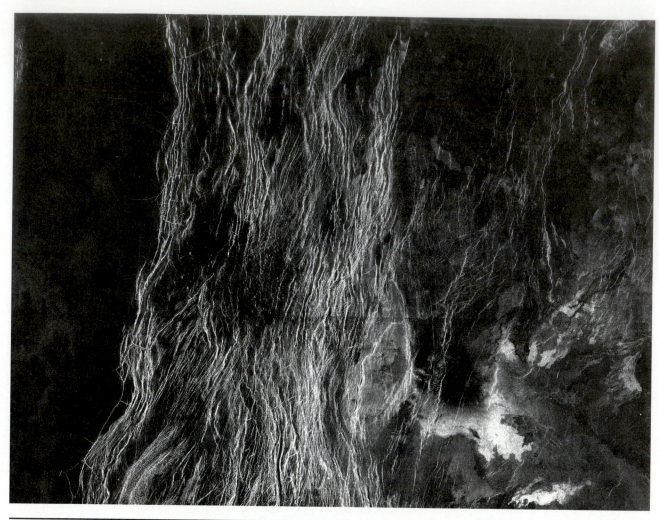

Figure 7.30

Belts of sinuous ridges are the most striking features in this Magellan image of Pandrosos Dorsa. The sinuous belts form nearly continuous chains up to 5000 km long, rivaling the lengths of the great folded mountain belts of Earth. Nonetheless, the Venusian ranges are not as high as those on Earth; most are only about 1 km above the plains. Presumably these mountain ranges represent fold and thrust belts formed by compression of the lithosphere.

eras. These are grouped with the fracture belts. The narrowness of the grabens implies a shallow depth of deformation. The amount of stretching indicated by such features is modest, only about 1 percent. These fracture belts are most common in the uplands and surrounding low-lying plains. The region around Beta Regio and Atla Regio is especially rich with fracture belts of diverse orientation (Figure 7.29). Because of the extensional nature and abundance in the uplands, it is inferred that they form in response to uplift of the lithosphere over mantle plumes.

Ridge Belts

The lowland plains of Venus are crossed by narrow belts of ridges that rise as much as 1 km above the surrounding plains (Figure 7.3). Where they are best displayed in Atalanta Planitia, ridges and valleys alternate with spacings of about 10 km (Figure 7.30). Individual ridges are gently sinuous and appear to braid together with others to form belts 100 to 200 km wide and thousands of kilometers long. Individual ridge belts are separated by expanses of less deformed plains 500 km or so across. These mountains are almost certainly tectonic rather than volcanic features. The ridges are somewhat like the wrinkle ridges of the Moon and Mars or the scarps of Mercury, but they are much more closely spaced and form longer belts. As a result of these differences, the mountain belts of Venus have been variously interpreted as folded mountains, like the Appalachians of North America, or as areas of extensional faulting form-

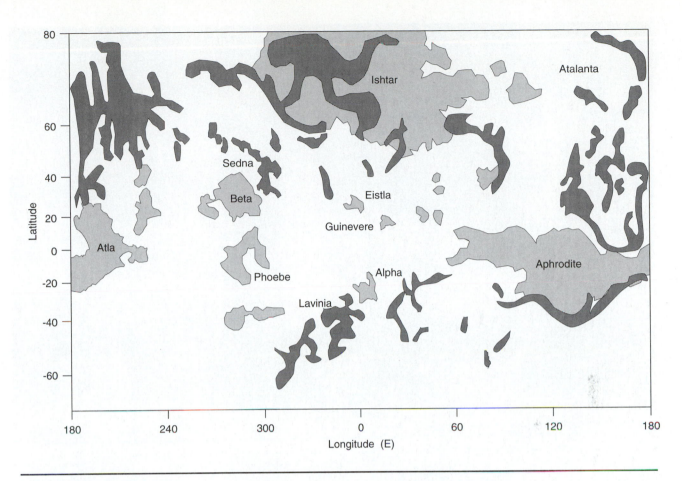

Figure 7.31

Belts of compressional ridges and mountains are common in the lowlands of Venus where young volcanoes and extensional fractures are not common.

ing regions like the Basin and Range province on Earth. Several detailed analyses have concluded that the ridge belts of the venusian plains are compressional in nature, partly because of the sinuous, overlapping, and branching character of the ridges. Extension-produced mountain ranges have straighter fronts and do not appear to overlap like these venusian features. Perhaps the ridge belts are folds or stacks of thin sheets bounded by low angle thrust faults (see Chapter 8). If so, the ridge belts must be the result of compression of the crust as a result of shortening.

Ridge belts are most common in or near the broad topographic basins in the lowlands (Figure 7.31). Lavinia and Atalanta Planitia are decorated with multiple ridge belts. They are rarely found in regions that have fracture belts. The distribution of the ridge belts and their presumed compressional origin give us a few clues to how they formed. They appear to be formed by large-scale downwelling in the mantle. Just as there are rising plumes of warm material inferred to be present beneath many topo-

graphic rises and volcanic provinces, there must be a pattern of return flow for cool, dense mantle. The large lowland plains of Venus may lie above such "cold spots," cylindrical regions of downward flowing mantle. Mantle downwelling causes the crust to bend downward in much the same way that mantle plumes dynamically uplift the crust. Such a flow field could pull on the overlying crust to crumple it into belts of small folds and thrust faults called ridge belts. Subsidence of a region on a sphere decreases its surface area; compressional tectonism is expected.

This hypothesis is attractive because: (1) the lowlands are concentrated in circular bowl-shape depressions, corresponding to what we expect to form above a cylindrical sinking flow of cold, dense mantle; (2) no hotspot volcanism, such as shield volcanoes or coronae, are seen in the lowlands; (3) no extensional rifts like those on hot spot swells have been identified; and (4) compressional features dominate the plains.

Figure 7.32

Tesserae terrains consist of complex systems of intersecting fractures and grooves. Some suggest that these deformed regions, common on the plains of Venus and along the margins of Ishtar Terra, are the result of gravitational flow and destruction of former domes. The area in this image is 200 km across.

Highland Tesserae

Other strongly deformed terrains consist of densely packed systems of ridges and grooves that crisscross one another and form a disorderly pattern of polygonal blocks (Figure 7.32). These are the tesserae that cap the highland plateaus and that rim the folded and thrust-faulted mountains of Ishtar Terra at lower elevations. The radar-bright feature called Alpha Regio (Figure 7.2) is an example of this type of highland. Others form the high plateaus of Aphrodite Terra. Magellan data show that these complexly ridged terrains consist of compressional ridges and troughs (folds) upon which are superposed numerous younger grabens.

Their appearance and association with mountain belts on Ishtar Terra suggests that these regions were shaped by tectonic forces.

Prior to acquisition of the high-quality images transmitted to Earth by Magellan, some scientists suggested that linear features in Aphrodite Terra were transform faults that cut across the east-west axis of the irregular highland. These same scientists also suggest that the patterns of mountains and valleys are symmetrical north and south of a central high. As a result, they concluded that Aphrodite Terra might be a zone of lithospheric spreading and production of new lithosphere like Earth's divergent midocean ridges. If new lithosphere is

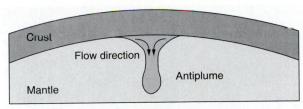

1. Downwelling plume develops in mantle and drags on crust

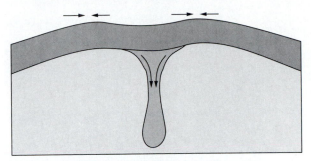

2. Crust buckles in response to compression

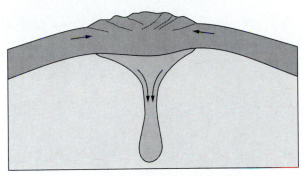

3. Crust thickens and a highland plateau develops

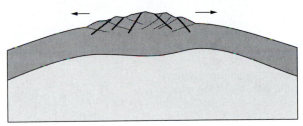

4. Downwelling ceases and highland spreads gravitationally

Figure 7.33

Downwelling plumes of cold mantle may have a significant effect on the surface of Venus, as shown in these cross sections.

produced in Aphrodite, then it must also be consumed somewhere else on the planet. This tantalizing hypothesis was rejected with the new data. No transform fault zones or symmetrical patterns were found in the new images. No subduction zone trenches or volcanic arcs appear on the new radar maps of Venus. Instead, Aphrodite is a string of highland tesserae.

The cause of the deformation within the highland plateaus of Aphrodite and elsewhere is controversial, but the sequence of compression followed by extension implied by their features eliminates some hypotheses and strengthens others. The folds indicate that crustal shortening must be involved; models that involve simple upward movement of the crust are thus unlikely because they predict extension to be the dominant process. One theory is that the tesserae are simply long-lived blocks of thick crust that record a longer history of deformation than the surrounding plains. An elevated terrain is less likely to be buried by volcanic flows and could preserve the changing pattern of regional stress over several billion years. Adjacent lava-covered plains would only record the effects of the last one or two deformation events. Another plausible alternative holds that tesserae are initiated by compression and crustal thickening caused by downwelling of a mantle cold spot (Figure 7.33). Mantle downwelling could cause the crust to be deflected downward, forming a more-or-less circular lowland with compressional ridges and ridge belts. If downwelling continued, the low density crust would buckle and thicken to form an elevated plateau replacing the lowland. During further downwelling, extension and graben formation begin in response to the gravitational failure of the highland. Once downwelling ceases, the high plateau should continue to spread under its own weight at the high temperatures found in the venusian crust.

Mountain Belts

The highlands and plateaus of Ishtar Terra are bounded by long belts of mountains that are similar to, but higher than, the ridge belts of the lowland plains (Figure 7.5 and Figure 7.31). These mountain belts represent more extensive deformation and crustal thickening than the ridge belts. The mountain belts of Venus, as on Earth, show widespread evidence for lateral extension both during and following active crustal compression. The deformed areas that border Lakshmi Planum have rugged ridges and valleys that parallel the margin of the plateau but stand 2 or 3 km higher (Figure 7.34). Maxwell Montes of central Ishtar Terra are the highest of these mountain chains, rising 6 kilometers above Lakshmi Planum and up to 12 km above the lowland plains (Figure 7.35). Farther from Lakshmi Planum, the pattern of parallel ridges breaks down, and the mountains consist of intersecting ridges. Some even have V-shaped bends, merging with the chaotic fault patterns of the tesserae described above.

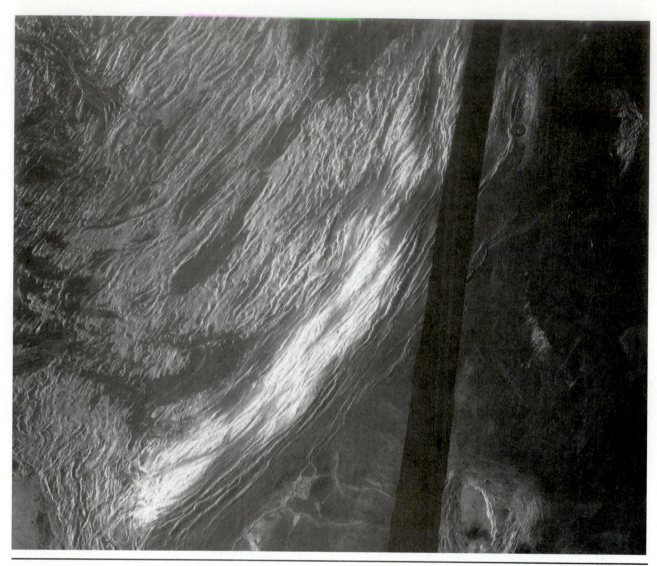

Figure 7.34

Mountain belts, apparently formed by compression and folding, surround Lakshmi Planum. Colette caldera and surrounding volcanic flows lie near the bottom of the photo. This image shows an area about 700 km across.

Figure 7.35

Maxwell Montes (right), which rise 6 km above Lakshmi Planum, are marked by a series of rugged northwest-trending ridges and troughs, as shown in this radar image obtained by the Magellan spacecraft. Most of the ridges are nearly parallel, but they pinch and swell and appear to overlap one another near the western side of the mountain belt. These relationships strongly suggest that Maxwell Montes are the result of compressional forces, not extension and rifting. The massif is capped by a circular impact crater, Cleopatra (upper right). A volcanic caldera is visible near the bottom of the image.

It is unlikely that the ridges and valleys that surround Ishtar were produced by some type of impact or volcanic process. The height and patterns of the ridges imply that they were created by the pileup and warping of the crust as a result of lateral compression. Could it be that they were made by compressional forces as mobile lithospheric plates pressed against one another and eventually buckled into parallel ridges? Probably not, for the parallel mountain belts occur along the entire margin of a highland, not between two colliding blocks of lithosphere. Moreover, no subduction trenches bound Ishtar Terra. Perhaps the ranges are the result of gravitational spreading or slumping of a large block or mountain, as suggested for other highland plateaus on Venus. The high temperatures at the surface of Venus may accentuate this type of deformation by enhancing lateral flow across the surface. The high plateau at Ishtar may have originally formed as a tectonic convergence zone marked by subsidence and crustal thickening over a center of downwelling in the mantle. Such a model may also explain the presence of the two calderas and smooth volcanic plains on Lakshmi Planum that make it different from other tessera plateaus. Crustal thickening may have occurred to the extent that the temperature in the crust became high enough to cause it to melt partially. This magma may have risen buoyantly and accumulated beneath the large calderas where it was periodically erupted to make the smooth plains. Such crustal partial melts may be broadly granitic in composition. Perhaps Ishtar Terra represents a feature similar to those volcanic welts speculated to have formed on ancient Earth, which by their subsidence and basal melting led to the development of granitic continents on Earth.

Venusian Tectonic System

Even after analysis of the thousands of spectacular images returned by the Magellan mission, it is still impossible to provide a conclusive integration of all of its tectonic features. But we can make some conclusions and some reasonable speculations about the global tectonics of Venus. We realize now that a variety of tectonic landforms exist, some like those on Earth and Mars, and some distinctive of only Venus. Its crust is more mobile and deformed than that of Mars, but no compelling evidence for a terrestrial-style plate tectonic system has yet been found on Venus. Crustal shortening or extension by several percent seems to be demanded by the mountain belts and rifts. Volcanism has been voluminous and diverse.

Most of these features can be explained by appealing to a fairly simple mode of mantle convection dominated by plumes and counterplumes. Upwellings underlie the rifted, volcano-capped domes and regions rich in volcanic features (Figure 7.27). Mantle downwellings underlie the lowland plains and the deformed highland plateaus, where crustal thickening has been caused by extreme downwelling (Figure 7.33).

Volcanic Features

The presence and age of volcanic landforms are key indicators of the evolution and internal differentiation of planetary bodies. The Magellan radar images show that there are thousands of large and small volcanoes on Venus and that as much as 80 percent of the surface is covered by volcanic deposits. Moreover, as noted above, this volcanic terrain was formed much later than the period of heavy bombardment that shaped most of the surfaces of the Moon, Mercury, and Mars.

The most common volcanic deposits on Venus are shield volcanoes, calderas, flows of flood lava, steep-sided lava domes, sinuous lava channels, and coronae (distinctive volcano tectonic features found only on Venus).

Small Shield Volcanoes. The most common volcanoes on Venus are small shield volcanoes less than 20 km across, formed by the eruption of fluid lava (Figure 7.36). There are probably tens of

Figure 7.36

Small shield volcanoes dot the plains of Venus. This field contains more than 50 small volcanoes. A graben cuts across the center of the volcanic field. This group of volcanoes is about 120 km across.

thousands of these features on the planet. Many occur in clusters several hundred kilometers across, and others occur on the flanks of larger shield volcanoes and are associated with coronae. These small volcanoes commonly have small collapse pits at their summits. Some of the fields of shields are centered on longer fractures that may form part of their magmatic plumbing system. A few fields form the focus of extensive lava flows. They are similar to the plains-style volcanism typified by small, low, shield volcanoes and intervening plains, found on the other inner planets.

Large Shield Volcanoes

About 500 more-or-less isolated volcanoes larger than 20 km in diameter have also been identified. Theia Mons on Beta Regio (Figure 7.4) is a good example. Individual volcanoes are as large as 1000 km across, much larger than any volcano on Earth and even larger than Olympus Mons, the largest volcano on Mars. They show a variety of features that demonstrate that they are volcanoes, including radial lava flows and summit calderas. Individual lava flows were fed from central vents or from vents on the flanks of the volcanoes. The sloping sides of some of these volcanoes have collapsed to form debris flows and scallops on the outline of the volcano.

The larger volcanoes in this category are dominated by complex multiple-vent volcanic centers that accumulate into large shieldlike edifices. However, a few are relatively simple shields, with calderas, radial and concentric fractures, and many of the characteristics of the large shields in the Tharsis region of Mars (Figure 7.37). There is a transition between some of the large shields and the coronae, described below.

Large volcanoes are almost totally absent from the tessera terrains and are sparsely scattered in the lowland plains. Plains with abundant ridge belts also lack large shield volcanoes, even partially buried shields. The large shields are concentrated in the fractured region centered on the broad topographic rises of Beta and Atla Regio. This is the same pattern displayed by the distribution of the small shield volcanoes.

Coronae

Dotting the uplands of Venus are several hundred fascinating ring features called **coronae** (Figure 7.38). A corona is a system of concentric fractures and ridges surrounding a central plain. In some ways, a corona resembles a mountain belt that wraps around to form a circle. The rugged mountainous ring is usually higher than the sur-

Figure 7.37

Large shield volcanoes are constructed of multiple overlapping lava flows that erupted from central vents that cap the summit of the volcano. Collapse calderas mark the sites of magma chambers that emptied. These volcanoes are about 500 km across, almost as large as Olympus Mons on Mars and larger than any shield volcano on Earth.

(A) Aramaiti (400 km across) is a circular depression with a central dome. Lavas have flooded the floor.

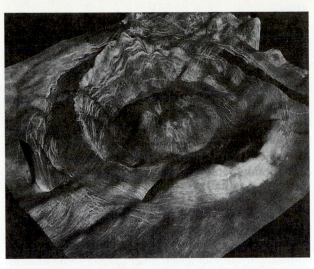

(B) Fractures mark the crest of the rim and the central dome, as seen in this perspective view of Aramaiti.

(C) Arachnoids have fractures that radiate away from a central corona. This feature lies in northern Atla Regio and is about 100 km across.

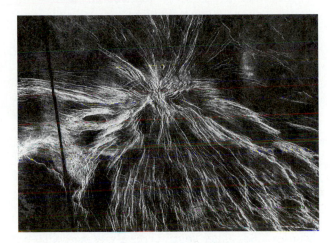

(D) This nova has a radial fracture pattern but lacks a central corona; it is about 250 km across.

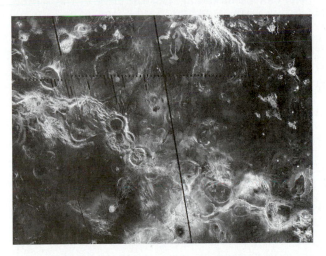

(E) A chain of about 15 coronae is aligned along a region set of troughs and fractures. The image is about 3600 km across.

Figure 7.38

Ringlike features called coronae are among the distinctive features on Venus. Found on the plains, they consist of nearly circular chains of rugged mountain belts created by faulting and fracturing that surround a central plain or dome. The largest discovered so far is 600 km across. A great variety of small volcanoes and radiating flows that emanate from the rings suggest they are volcano-tectonic features.

rounding plains. Fractures with other orientations are also common (Figure 7.38), but they are not as abundant as in the tesserae described below. In most cases, the wreath of deformed terrain has been pulled apart by extension to create grabens and horsts, but in some cases broadly sinuous ridges and warps cross cut earlier formed graben. These are probably wrinkle ridges and folds formed by compression. Volcanic features are commonly located in and near the coronae. Small volcanic shields, lava domes, massive flood lavas, lava channels, and small collapse calderas commonly mark the interior and exterior flanks of a corona. A typical corona has a diameter of about 250 km, but coronae range in size from less than 100 km to more than 1000 km across.

Coronae are widely distributed across the planet and are among the dominant volcanic landforms. About 400 have been identified; many more are partially buried by lava flows or are otherwise degraded. Most coronae lie in an area of lithospheric domes and rifting centered on the equator near the Atla-Beta region. There are markedly fewer coronae in the lowlands, such as Atalanta, Guinevere, and Lavinia planitia, than in the uplands or highlands. Many other coronae lie in chains along fracture belts.

Other volcano-tectonic features have many of the characteristics of coronae. **Arachnoids** have a radial set of fractures superimposed on the concentric set, giving them their spiderlike appearance (Figure 7.38). The "legs" typically merge with regional fracture belts. Their overall appearance is that of multilegged spiders sitting on webs of interconnected fractures. **Novae** have prominent radial fracture patterns leaving starburst patterns on the radar images (Figure 7.38). They lack the wreath of grabens that surround a typical corona. A small central caldera may cap the broad dome. Novae and arachnoids are typically smaller and less abundant than coronae.

The origins of coronae, arachnoids, and novae are probably similar, differing only in the style of fracturing, an indication of differences in their uplift histories. The evolution of coronae can be related to three general stages (Figure 7.39). The earliest phase includes uplift, radial fracturing, and volcanism. Radial fracturing accompanies domical uplift over buoyant plumes or hot spots in the mantle of Venus. Their features suggest that hot, low-density material rose up from the interior and caused the surface to deform and fracture as a result of uplift. Magma was extracted from the plume by partial melting and fed the volcanic edifices and floods that typify many coronae. The second stage involves the flattening of the plume

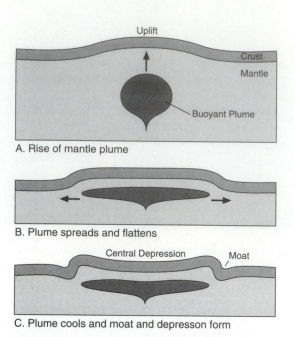

Figure 7.39

Coronae may evolve through a series of steps involving the rise of a hot plume head from the mantle. (A) The heat from the plume and its buoyancy create an uplift, with radial fracturing and volcanism. (B) The large head of the plume hits the base of the strong lithosphere and spreads and flattens, transforming the dome into a circular plateau. Volcanism continues through this stage. (C) The hot plume head cools, removing thermal support of the topography and allowing gravitational relaxation of the topography to form a moat, a rim, and interior depression. Concentric fracturing takes place as the rim stretches while it bends.

head as it is forced against the base of the strong, but relatively thin, lithosphere. The plume head takes on a pancakelike shape, and a broad circular plateau is created. Concentric faults will be produced along the flanks of the plateau and crosscut the earlier radial faults. The concentric fractures around coronae are the result of extension, probably related to downwarping of the moat and upwarping of a raised rim. Depressed moats and raised rims can form through the cooling and gravitational relaxation of a circular plateau that was raised and supported by buoyancy from its high temperature. As the lithosphere bends, grabens can form on the floor of the moat and on up the rim to its crest. Pressure-release melting and volcanism will continue through this stage. The third stage is marked by modification of the dome and includes relaxation, more concentric fracturing, and waning volcanism. Relaxation of a corona involves spreading out under its own weight on the hot surface of Venus, creating compressional belts of sinuous ridges along its outer margins.

Coronae in all stages of development are found on the surface (Figure 7.38), implying that a sequence of ages of coronae exists. However, it is difficult to assign ages to the coronae independently. Only a few have superimposed impact craters. The only general conclusion we can make about their age is that they are about the same age or younger than the average age of the venusian surface.

Calderas

Some large shield volcanoes are so low that it is difficult to see the edifice at all. Their presence, however, is unmistakable because they are marked by large calderas (Figure 7.40). All are thought to have been created by the withdrawal of magma from huge magma chambers by eruption or drainage. These calderas range up to about 200 km across, but most are only 40 to 80 km across. Although not concentrated in the highlands, two of

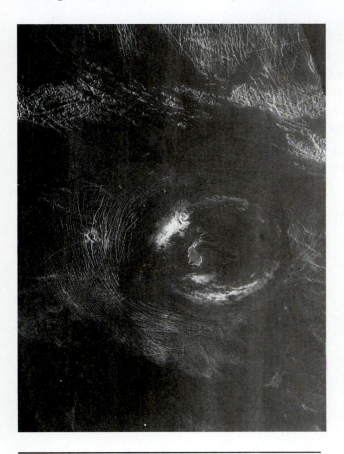

Figure 7.40

Young volcanic calderas like Sacajewea are found on the highlands of Venus. About 100 by 150 km across, Sacajewea lies on the flat plateau of Lakshmi Planum. Volcanic flows several hundred kilometers long radiate away from the faulted rim of the caldera, which is less than 2 km above the surrounding plains.

the best examples of this type of volcano lie on the highland Ishtar Terra. These calderas, Sacajewea (Figure 7.40) and Colette, are similar in shape, if not in composition, to the large calderas on Earth from which thousands of cubic kilometers of rhyolitic ash flows erupted. Examples include the Valles caldera in the Jemez Mountains of New Mexico and the caldera complex in Yellowstone National Park.

Colette caldera is 1 to 3 km deep and is defined by an inner and outer set of arcuate faults that probably formed as the roof of the volcano collapsed when magma within it was withdrawn. The outer ring is about 200 km across in its longest dimension. Surrounding the caldera are long narrow flows of varying brightness that must be volcanic flows of some sort. The flows appear to bury tectonically disrupted areas to the west, but they are older than the folded mountains on the northwest. Another large caldera, Sacajewea, is present in eastern Lakshmi Planum (Figure 7.5). Together these calderas may be the source of the smooth volcanic plains that cover the plateau.

Nothing in the features of these volcanoes proves that silica-rich ash flows did not erupt from them, but explosive eruptions on Venus are less likely than on Earth because of the high atmospheric pressure. In the venusian atmosphere, expanding gas bubbles that power explosive eruptions would have difficulty forming and explosively expanding to form volcanic ash. Moreover, there is still little evidence for silica-rich magmas on Venus.

Lava Domes

Another class of volcanoes has been descriptively called pancake domes because of their distinctive shapes with steep margins and relatively flat tops. Figure 7.41 shows the characteristics of these distinctive landforms. Some have very small pits at their summits that could be small collapse pits or explosion craters. Because of their similarity to lava domes on Earth, we think these features are produced by the extrusion of viscous lavas from central vents. Radial and concentric fractures cut the surfaces of most of these low domes. Fracturing of the cool, solid rind of a flow as more lava is pumped into the interior could easily explain these features. Many of the domes are aligned in rows on the flanks of larger volcanoes and coronae, suggesting that they are fed by dikes. An important difference between terrestrial lava domes and these lies in the large size of the venusian domes. Most of these steep-sided domes are about 500 m high, and they range from less than 10 to more than 100 km across. Their volumes may thus be in

Figure 7.41

Lava domes on Venus look like pancakes. These domes are less than about 60 km across and rise a kilometer above the surface. The domes are cut by radial and concentric fractures that must have formed as the domes were inflated by injection of magma from beneath them. Small summit craters lie near the center of most domes.

excess of 100 km³. Rhyolite domes on Earth are typically less than 1 km across and include only a fraction of 1 km³ of lava. Moreover, the great regularity in shape of the pancake domes is not typical of most rhyolite domes on Earth, especially the larger ones. As shown in Chapter 8, most of the terrestrial domes have short, stubby flows that emanate from one side, destroying the initially circular outline imposed by the pipelike vent.

On Earth, grossly similar lava domes are produced only by viscous silica-rich lava like rhyolite (Chapter 2). Basalt is not known to form such domes. Does this mean that granitic magma is produced on Venus? Has the crystallization of mafic magma led to silica-rich compositions? Has partial melting of pre-existing crust produced viscous silicic magma? Only new space probes armed with sophisticated analytical equipment will be able to answer these questions with certainty, but for now there is a hint that at least small volumes of granitic magma are produced on Venus.

Flood Lavas

Smooth plains are common on Venus, and most of these were probably produced by eruptions of flood-type lavas like those that formed the lunar maria and other large smooth plains provinces on the planets. Over 80 percent of the surface of Venus may be covered by volcanic plains. The formation of small shields and eruption of huge lava floods is the predominant means of producing them. Many of the plains have been subsequently degraded by other processes, including weathering and tectonic processes, so that flow features are not visible on all of them.

Large fields of lava are obvious in many areas of Venus (Figure 7.42). Individual flows are as long as 1000 km. Some show clearly that they erupted from fissure vents, now apparent as grabens and faults parallel to regional zones of rifting. Lava channels occur in the interior of many of these flows. Individual fields typically cover areas of

Figure 7.42

Flood lavas cover large areas of Venus. These flows are about 300 km across and were erupted from vents near the center of the field. The textures on the top of the flows reveal the flow directions. Superposition relationships reveal the sequence of eruption.

20,000 km² or less and would bury any impact crater in their path. Such flows must play a major role in resurfacing the planet. These long flows are probably the result of the eruption of very fluid lavas with high eruption rates. The lavas are probably basaltic in composition; Venera landers found the soils had basaltic compositions in the few areas where they landed.

Lava Channels

One of the surprises of the Magellan mission was the discovery of hundreds of channels shaped by the flow of a fluid (Figure 7.43). Long sinuous channels trace across many of the plains of Venus. Some are clearly related to the emplacement of long lava flows because they lie on radar-bright, lobate flows that are hundreds of kilometers long. Many other channels lack obvious lava flows and resemble the sinuous rilles on the Moon. These channels, like terrestrial rivers and lunar sinuous rilles, are erosional features that were carved into the plains across which they extend. Most of the channels consist of a single channel less than about 2 km wide, but some have complex braided, tributary, or distributary patterns as much as 30 km across. As with sinuous rilles on the Moon, the width of the channels decreases from source to

Figure 7.43

Long, narrow channels like this are probably lava channels carved into the venusian plains. Some channels are thousands of kilometers long, suggesting that high temperatures and high eruption rates kept lava liquid for very long times.

terminus. Deltalike distributaries mark the ends of some channels where they empty onto plains. The most spectacular channel is 6800 km long. On Earth, if a channel that long emanated from Kilaeua volcano on the island Hawaii, it would reach the Mississippi River in the central United States. This huge lava channel is as long as the Nile, Earth's longest river.

The lava channels occur predominantly in the lowlands and on the flanks of Aphrodite Terra.

Where flow directions can be determined, most channels are oriented downslope. Subsequent gentle folding has changed some slopes. Many start in coronae, large irregular volcanic craters, and other volcanic terrain of the uplands and drain into the lowlands. The emplacement of lava from such long channels may be an important mechanism of plains creation. Volcanic vents are rare in the lowland plains.

What type of fluid eroded these channels? Certainly, the high surface temperatures and lack of atmospheric water preclude river erosion. Volcanic fluids as exotic as sulfur, carbonate, and ultramafic silicate magmas are possible. However, arguing only from the abundance of basalt on the terrestrial planets and the rarity of the other materials, basaltic lava is the most likely fluid. Huge eruption rates and the slow cooling at the hot venusian surface would aid in producing long lava channels.

Venusian Volcanic Systems

Magma generation and the distribution of volcanoes on Earth are controlled by the plate tectonic system. Notable for their absence on Venus are chains of stratovolcanoes aligned along deep trenches. Moreover, there are no linear chains of shield volcanoes, like the sea mounts produced on moving lithospheric plates above mantle plumes. These types of volcanic systems are the result of moving plates on Earth, and their absence on Venus is strong evidence against an active system of plate tectonics. Can hot-spot tectonics explain the nature of volcanic systems on Venus, as they explain many of its tectonic features?

The wide distribution, abundance, and sizes of the various volcanoes and volcanic fields on Venus strongly suggest that they are the surface manifestations of mantle plumes or hot spots and that the different types of volcanoes represent variations in the size of the plume, the stage of its evolution, and the structure of the lithosphere (Figure 7.44). Why do shield fields form in some areas and large shield volcanoes in others? Is the phenomenon related to the size of the underlying mantle plume? The progression from small shields to shield fields to large shield volcanoes may form a continuum reflecting the size of the underlying mantle plume. Small plumes may be little more than cylindrical pipes with no enlarged heads. Pressure-release melting in such a rising plume could feed small volcanoes. Large plumes might feed multiple volcanic systems as separate magma bodies separate out from the enlarged head of a starter plume, feeding multiple shield fields or the huge volcanoes that cap lithospheric domes.

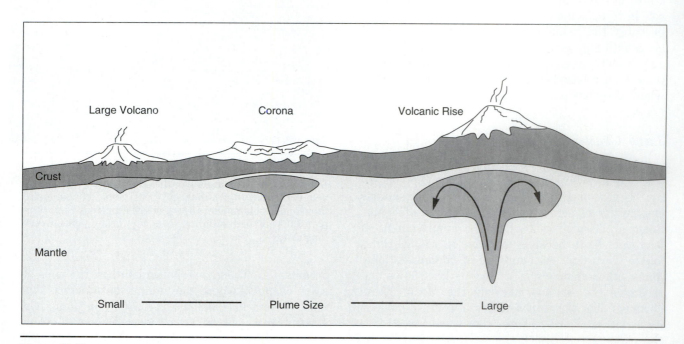

Figure 7.44

Different scales of melting and convection are primarily responsible for the formation of hotspot-related features on Venus. Major shield volcanoes are due to the presence of large bodies of magma, whereas volcanic rises are thought to be primarily due to large-scale upwelling in a mantle plume. Coronae appear to occupy an intermediate position, manifesting both the presence of significant melt and the effects of smaller or shorter-lived upwellings than those responsible for volcanic rises or lithospheric domes.

Volcanic features on Venus are broadly distributed, in contrast to the strong concentration of volcanoes along linear plate boundaries on Earth (Figure 7.45). There is no equivalent of Earth's Ring of Fire on Venus. The distribution of volcanoes on Venus is, therefore, much more like the distribution of hot spots on Earth. Only about 100 hot spots are believed to have been active in the last 70 million years on Earth. Of course, Venus has many more large volcanic centers than this. If Earth dynamics minus plate tectonics tell us anything about Venus dynamics, we might expect about 700 major hot spots on Venus to have developed over the last 500 million years. The lack of crustal recycling and the low rates of erosion on Venus mean that even the oldest of the hot-spot volcanoes would still be visible. Approximately 700 large volcanoes, calderas, lava floods, and coronae and related features have been counted on Venus. Another 500 or so fields of shield volcanoes have also been identified. Thus, the number of hot spots

predicted by comparison to Earth is about the same as the observed number of large hot-spot volcanic centers on Venus.

However, volcanoes are not randomly distributed on Venus. There are few volcanoes in the lowland areas of Venus. Likewise, the major highlands, Aphrodite and Ishtar Terra, have relatively few volcanoes. There is a major concentration of volcanic features in the equatorial region centered on the lithospheric domes at Beta and Atla (Figure 7.45). This region also has many fracture zones, rifts, and several lithospheric domes capped by volcanoes. Estimates of the rate of lava production on Venus are still very crude but range from 1 to 0.5 km^3 per year, averaged over the last 0.5 billion years. This is much lower than the rate of volcanism at Earth's midocean spreading ridges or at convergent plate boundaries, but it is similar to that for intraplate volcanoes related to mantle plumes (0.3 to 0.5 km^3/year). This gives further credence to the idea that much of the volcanism is

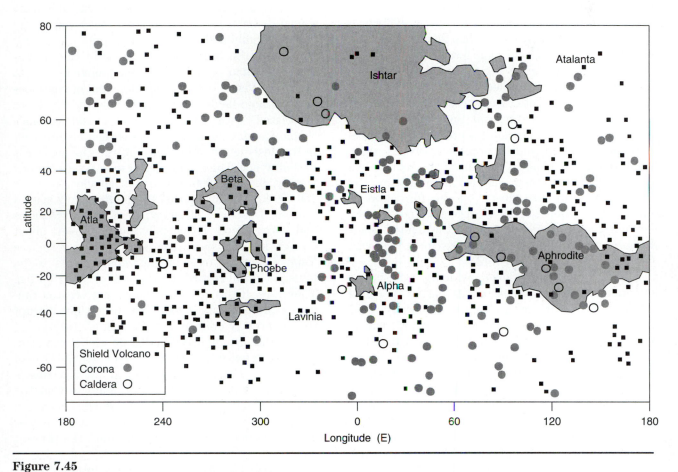

Figure 7.45

Volcanoes are spread across the surface of Venus but they do not occur at random. Each dot on this map shows the location of a volcano (larger than 20 km across), caldera, corona, nova, arachnoid, or field of small shield volcanoes. Volcanoes are more common on the uplands than on lowlands or highlands. An anomalously high concentration of volcanic centers occurs in the western hemisphere in the fractured uplands regions of Beta and Atla Regio.

related to mantle plumes. Most of the volcanic landforms are consistent with basaltic magma compositions. According to the partial analyses made by Soviet landers, the surface soils are quite similar to basalt; that is, they are igneous rocks that are rich in magnesium, iron, and calcium, like those found on the maria of the Moon, the volcanoes of Mars, some asteroids, and in many environments on Earth where partial melts of the mantle can rise to the surface.

Geologic Evolution

The low number of impact craters on Venus shows that it has a young surface. The average age of the entire surface is about 0.5 billion years old. One possible interpretation of the history of Venus calls for a remarkable, perhaps even catastrophic, global resurfacing event that ended 0.5 billion years ago. Little deformation or volcanism has occurred since this time. Purportedly, the volcanic coronae, long lava floods, and highly deformed highlands are remnants of this global event. In essence, we are looking at a fossil surface where little change has gone on for 0.5 billion years. Support for this idea lies in the very uniform distribution of impact craters across the surface. If large areas were resurfaced any time between about 0.5 billion years ago and today, large areas without craters should be apparent. Investigators think that as little as 10 percent of the surface has been resurfaced since this global event by the emplacement of lava flows, by the formation of fault and fold belts, or by weathering and erosion. Of course, earlier global resurfacing events could have occurred, but any evidence for them was destroyed by the latest event.

No other planet shows evidence for similar events in its history. What in the nature of Venus might explain how such an "on or off" history could happen? Some suggest that oscillations of mantle flow set in on a planet such as Venus, which has high surface temperatures and a weak lithosphere. Because of the periodicity of mantle flow, periodic resurfacing events occur on a global scale. This scenario may even account for the lack of a magnetic field in Venus because it predicts that the interior cooled more efficiently and the core solidified during the last billion years. According to the model, about 0.5 billion years ago the vigor of convection diminished and a permanent crust developed. At this stage, the tectonic style changed from recycling of crustal plates to hot-spot volcanism. The model implies that the interior of Venus is completely solid now, except for small partial mol-

ten regions in the upper mantle that feed mantle plumes.

A dramatically different picture emerges from the alternative idea that the surface is being gradually and continuously resurfaced over its long geologic history. In this case, there would be areas with distinct geologic ages. The region of abundant volcanoes and apparent mantle upwelling does indeed appear to have a slightly lower abundance of craters. The complex superposition relationships in the volcanic terrains suggest long periods of volcanism. Variation in the extent of weathering implies a range of ages for lava flows. Moreover, many craters have dark smooth floors, unlike the distinctly fresh craters with parabolic ejecta arcs and bright floors. Perhaps the smooth floors are a sign of degradation and the passage of time.

Those who support this hypothesis conclude that only "small" areas are resurfaced at a time. This process could gradually resurface Venus without making the craters look unevenly distributed. If areas less than 150,000 km^2 were covered in each resurfacing "patch," it would not visibly affect the global distribution of the craters; they would still look randomly and uniformly distributed. Few craters would be buried simply because the craters are so few and far apart. The most obvious way to resurface Venus is by the development of volcanoes and volcanic fields. Many large volcanic fields are associated with shield volcanoes and coronae. Would the development of one corona be apparent in the destruction of impact craters? The answer is probably not. The area covered by a typical corona is only about 50,000 km^2, much smaller than the 150,000 km^2 limit imposed by the randomly spaced craters. Likewise, other types of large volcanoes and large flood lavas individually cover about 130,000 km^2. In short, the likely way to resurface Venus operates on a small enough scale to be invisible with regard to the crater distribution. Therefore, if the volcanic regions are created randomly in space and time, they would not disrupt the apparently uniform distribution of craters.

In this gradualistic model, no global catastrophes are required; no convective overturn is called for. The resulting history implied by this idea is more like that of the Earth or Mars in that the planet was initially hot but cooled slowly because of its large size, retaining enough heat to power active magmatism and tectonism until the present day.

Neither the catastrophic nor the gradual resurfacing hypothesis explains why Venus lacks a system of plate tectonics like Earth, but is instead dominated by mantle plumes. Mars, Venus, and Earth all convect to form mantle plumes, but only Earth recycles its lithosphere through subduction zones. We are forced to conclude that it is Earth that is unique, not Venus.

Judging from what we know about its surface features, the flow of energy on and in Venus takes place in a very different manner than on Earth and produces a different set of geologic features. This occurs, in spite of the planets' similar sizes and densities. A profitable way to examine these differences is in terms of the sources of energy available to drive geologic processes on Venus. One source of energy is the Sun. On Earth, energy from the Sun and the influence of gravity drives the circulation of the hydrosphere and atmosphere, which may be considered the outermost layers of a differentiated planet. Solar energy arriving at Venus drives a similar convective system in its thick atmosphere. Theoretical calculations, laboratory measurements, and radar images show that measured wind velocities are sufficient to move particles. An eolian regime with transportation and deposition of sedimentary particles exists on Venus. These atmospheric movements, driven by an external energy input, along with the weathering of freshly exposed rocks, constitute important, still active, geologic systems on Venus.

Solar energy is also important in other ways. We have seen how heat from the Sun is trapped by the venusian atmosphere to create high surface temperatures. These high temperatures are postulated to have been vital to a mechanism that rid Venus of its water. Once the temperature was sufficiently high, water was in a vapor state and therefore was available for disassociation and loss to space. The lack of water has important consequences for the erosional history of the planet, the rock cycle, and even for the tectonic history of the planet. Less than 1 percent of water in a planet's mantle can lower its melting temperature by 200 to 600 K, increase its fluidity, and stretch the duration of periods of mobile tectonics and volcanism.

Another energy source comes from the heat generated within the planet by radioactive decay or retained from the planet's accretion and differentiation. This energy powers the convection of the mantle. In turn, the vigor, style, and pattern of convection control a planet's volcanic and tectonic systems. Plate tectonics, mantle plumes, and hot spots are important patterns of mantle convection on Earth, Venus, and Mars. In an attempt to tie several disparate observations together, a few scientists have suggested that the high surface temperature, the history of water, the patterns of mantle convection, and the volcanic and tectonic history of Venus are closely related.

Here we need to reconsider some possibilities for the composition of Venus. However, both major possibilities converge on the same final result. Per-

haps Venus has always been poor in water because it formed so close to the hot sun. Alternatively, the early differentiation of Venus may have expelled its water as vapor to the atmosphere. Eventually, as the surface was warmed by the greenhouse effect, the interior and atmosphere of Venus were purged of water. As a result of either of these scenarios, temperatures in the venusian mantle were far below the point of partial melting. A fluid asthenosphere disappeared early or never existed.

Thus, plate tectonics, which requires a shallow fluid asthenosphere to allow thin slabs of lithosphere to slip around as a result of their own weight, died out early or never developed on Venus. In a water-poor Venus, the role of water as a flux for low temperature magma production would not exist. The generation of light granitic magmas may similarly have been prohibited. Some have suggested that the lithosphere of Venus remained buoyant, unable to recycle itself back into the mantle by moving laterally because of the high surface temperatures caused by global greenhouse heating. Instead, continued mantle convection favored the production of many mantle plumes. Swells or domes capped by volcanoes grew above the mantle plumes and were rifted by extension. Rising plumes in a convecting mantle may tell only half the story; complementary downflows, or antiplumes, are necessary too. If Venus lacks a shallow asthenosphere, then the effects of these cylindrical or sheetlike downflowing regions will be coupled by viscous drag to the crust (Figure 7.33). Because the crust is part of a spherical shell, as it is dragged downward, compression ensues. The high temperatures on Venus mean that the compressed crust will be weak and deform ductilely. Without much resistance to these compressive forces, the crust should buckle, bend, fold, and eventually thicken. Deep smooth basins crossed by compressional ridge belts, and with few volcanoes, must have formed above the cold spots. Continued contraction created complexly folded high plateaus. Once a plateau became high enough or when convective downwelling stopped, it relaxed and stretched apart under its own weight. Grabens and fractures would then cut across the already folded regions to create tesserae.

Because our knowledge of age relationships is very limited, comparisons with the other planets are useful in formulating a speculative geologic history of Venus. Essential to understanding planetary history is the development of a geologic time scale based on a knowledge of the sequence of events that produced the various features and rock bodies exposed at the surface. On Earth, Mars, Mercury, and the Moon, the principle of superposition has been used to create a stratigraphic or relative time scale. In some cases, radiometric dates and crater frequencies have allowed a more quantitative assessment of the timing of geologic events. Because of the low crater density and lack of radiometric ages, such evaluations are presently difficult. We can paint the geologic history of Venus only with broad, crude brush strokes, by integrating what we know about its surface and employing analogies with the rest of the inner planets, particularly the Moon and Earth.

Like the other terrestrial planets, Venus probably formed in a fairly short time by the accretion of many smaller planetesimals by gravitational capture. The slow retrograde rotation of Venus may be inherited from this time. A large glancing impact late in its accretion history may have reversed the normal spin of the planet, slowing the rate of rotation as well. The infall of accreting material should have caused high internal temperatures (Figure 7.46).

As a result of this accretionary heating and heat produced by the decay of radioactive elements, the interior of the planet differentiated into layers of different composition and density, probably forming a metallic core and a mantle and crust of silicate minerals. The period of differentiation must have occurred during a period of intense meteoritic bombardment, but no tracts of heavily cratered terrain have been preserved. All of the surface of Venus is younger than the period of accretion and differentiation.

The fluid envelope surrounding the planet was released from its interior during differentiation. A carbon dioxide-rich atmosphere was created. Here an important mystery remains: Did Venus have water that was expelled, or was it always as water-poor as it is today? Water released to the venusian atmosphere may never have condensed to liquid form because greenhouse heating caused temperatures to exceed the boiling point of water. If a venusian ocean formed, it was boiled off by intense greenhouse heating. Eventually, water vapor was broken down to hydrogen and oxygen. The hydrogen may have escaped into space and the oxygen combined with surface materials leaving the present "dry" atmosphere. With continued outgassing, the interior was also purged of water.

The remainder of the geologic history of Venus may have been decided by the rate and manner in which it rid itself of internal heat. The composition and thermal history of a planet are very important to the mechanism and rate of planetary evolution. Calculations imply that Venus, with its Earthlike size, density, and composition, had a thermal history like that of Earth (Figure 7.46). Earth's most important mode of heat loss is presently the forma-

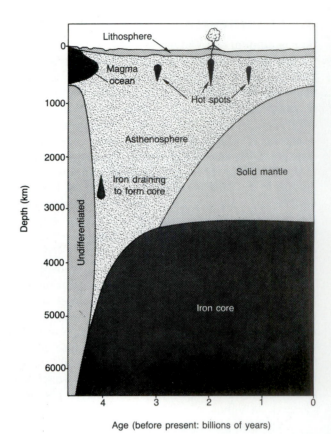

Figure 7.46

The thermal history of Venus may have included (1) an early period of mantle melting and crust formation as a result of accretionary heating, (2) the formation of an iron core diluted with sulfur or oxygen and the consequent development of a thick zone of partial melt in the mantle, and (3) the thickness of the zone of partial melting reduced with time as a result of cooling. Plate tectonics probably did not develop on Venus because it lacks a shallow asthenosphere. The expulsion of heat to the surface may have occurred through a series of mantle plumes that feed volcanoes at rifts or above the hot spots.

tion and cooling of new oceanic lithosphere. Plate tectonics or substantial lateral movement of the lithosphere of Venus is not indicated. No hot spot trails are apparent. No subduction trenches or parallel chains of volcanoes are present.

Venus must be losing its heat by a different mode—through its hot spots. According to this theory, domes, shield volcanoes, and coronae scattered across the planets overlie warm rising plumes (Figure 7.47). Upon development of a rising plume of low density material, for example, heat flows through the crust and is radiated away to space. If partial melts from the plume penetrate the crust, volcanism may be active atop the domes, which also expel heat from inside the planet.

Why would Venus expel its internal heat in such a different manner when compared to Earth?

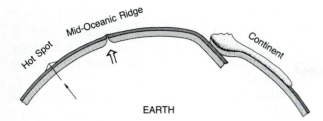

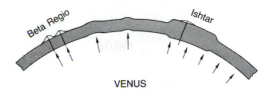

(A) Earth loses most of its heat through the formation of oceanic lithosphere at a midocean ridge and its subsequent cooling. As the terrestrial lithosphere cools and thickens it becomes denser, until it is eventually consumed and recycled back into the mantle at a subduction zone. A much smaller part of Earth's heat is lost through hot-spot volcanoes, such as in Hawaii.

(B) The high temperatures at the surface of Venus may prohibit the formation of the dense lithospheric sinkers that drive Earth's plate tectonics. Instead, Venus may lose its heat through a large number of more uniformly distributed mantle hot spots, which produce volcanoes and surface swells. Internal heat is expelled as the lavas and the swells cool.

Figure 7.47

Tectonic processes on Venus may differ from those on Earth because of different heat-loss mechanisms.

The reason could be related to the high surface temperatures. On Earth, as newly formed oceanic lithosphere moves away from a spreading ridge, cooling and metamorphic transitions form dense minerals in oceanic lithosphere that act as gravitational "sinkers," driving the terrestrial lithosphere back into the mantle. Cooling and metamorphic reactions that form such dense minerals may never occur in the hotter lithosphere of Venus. Instead, the venusian crust may be only partly recycled into the mantle when the crust thickens by crumpling or becomes deeply buried by recurrent volcanism. Gravity could detach slabs of dense metamorphic minerals formed in the crust back into the mantle. Or basal melting and convection could sweep a small fraction of this crustal material to the mantle, but most should return toward the surface as buoyant magma.

Another possible explanation for the absence of plate tectonis is that Venus, in spite of its high surface temperatures, lacks a shallow asthenosphere. Its crust may be tightly bound to a stiff upper mantle that prohibits lateral movements and, simultaneously, allows the effects of convective plumes to be impressed directly on the crust. Why would Venus lack an asthenosphere? It could be because the mantle is water-poor and deforms or melts only at higher temperatures than in the wetter Earth.

This scenario could be summarized as: no water, no asthenosphere; no asthenosphere, no plate tectonics. To investigate the likelihood of this speculative history, it will be important to gain a better understanding of the history of water on Venus. Did Venus ever have water? When and how was it lost? Answers to these basic questions may help us ultimately to resolve how the venusian

mantle convects and how the evolution of convection styles has affected the geologic history exposed on the surface.

Certainly, tectonics and volcanism are not the only processes to shape the evolution of Venus. Its surface features also continue to evolve as the chemically reactive atmosphere combines with surface rocks and fine particles are produced. Eolian transport of the sediments continues to create streaks, dunes, and yardangs.

Conclusions

As the veil of clouds enveloping the surface of Venus is gradually lifted by the development of new technologies and expanded exploration, we may get new perspectives on the geologic history of Venus. These may modify the speculations outlined above. An understanding of the geology of Venus is critically important because of the contrast it presents with Earth and for the clues it will provide about the origin and evolution of the planets. Earth and Venus are similar in size, composition, and distance from the Sun. Yet they are not identical twins and have evolved along remarkably different paths. Venus's thick, carbon dioxide atmosphere and high temperature distinguish it from Earth; these features must have played significant parts in determining the present geology. Perhaps because of its greenhouse climate, Venus lacks water, a hydrologic system, and plate tectonics.

However, Venus is a dynamic planet. Its surface is relatively young; any ancient cratered terrains have been resurfaced by dynamic tectonic and volcanic processes. Tectonic features show that lithospheric extension and compression have shaped the surface. Large and small volcanoes are scattered across the highlands. In some ways Venus is like Earth, with its folded mountain belts, highly deformed, relatively young surface, and abundant volcanism. In other ways, Venus is like the smaller planet Mars, which has a thick lithosphere, crustal domes, and rifts caused by hot spots. We do not yet know for certain how Venus loses heat, which in turn governs the style and vigor of its global tectonic system, but hot-spot convection seems likely and may provide the key to understanding the volcanic and tectonic evolution of Venus.

Important unresolved questions highlight our present understanding, or lack of understanding, of Venus, the planet next door.

How does the bulk composition of Venus compare with that of the Earth or other inner planets? Is it volatile-poor as a result of forming near the Sun? If not, how and when did water escape from the planet?

What style of tectonics shapes the surface of Venus?

Was granite formed on Venus? Are Ishtar Terra or the pancake domes composed of granitic materials?

Is Venus volcanically active today?

Review Questions

1. Why is it so difficult to obtain detailed information about the surface of Venus?
2. Venus and Earth have gravity fields strong enough to retain gases and liquids near their surfaces. Why, then, are there no oceans on Venus?
3. If Venus has an interior structure with a molten core like Earth and Mercury, why does it lack a sizable magnetic field?
4. Why is it important to resolve the question of whether Venus has an asthenosphere or not?
5. Compare the composition and density of the atmosphere of Venus with Earth's.
6. Why is Venus so uniformly hot if its clouds reflect most of the solar energy that reaches the planet? Why are temperatures at the poles only slightly different from those at the equator?
7. List the major types of landforms on Venus. Which are unique to Venus? Which occur on other planets?

8. Do continents like those on Earth exist on Venus? Would you expect continents to develop on Venus?
9. If Venus lacks a system of moving lithospheric plates, how does it lose its internal heat?
10. If Venus has lithospheric spreading centers, how would they differ from those on Earth?
11. Describe the principal tectonic features found so far on Venus. How do they compare with those on Earth?
12. Contrast the effect of mantle plumes and antiplumes on a planet's surface.
13. Outline the major events in the geologic history of Venus.
14. What new information is needed to understand better the history of Venus?
15. If you had the responsibility to choose a landing site on Venus for a remotely controlled mobile robot, what area on Venus would you choose? Why?

Key Terms

Antiplume	Greenhouse Effect	Nova
Arachnoid	Highlands	Tessera
Corona	Lowlands	Uplands

Additional Reading

Bazilevskiy, A. T. 1989. The Planet Next Door. *Sky and Telescope*. Vol. 43, No. 4, pp. 360–368.

Hunt, G. E., and P. Moore. 1982. *The Planet Venus*. London: Faber and Faber.

Hunten, D. M., L. Colin, T. M. Donahue, V. I. Moroz, eds. 1983. *Venus*. Tucson: University of Arizona Press.

Journal of Geophysical Research. 1980. Vol. 85, No. A13. (Complete issue devoted to the findings of Pioneer Venus.)

Journal of Geophysical Research. 1992. Vol. 97, No. E8 and E10. (Two issues devoted to the findings of Magellan at Venus.)

Kasting, J. F., O. B. Toon, and J. B. Pollack. 1988. How Climate Evolved on the Terrestrial Planets. *Scientific American*, Vol. 258, No. 2, pp. 90–97.

Nile River

TABLE 8.1

Physical and Orbital Characteristics of Earth

Mean Distance from the Sun (Earth = 1)	1
Period of Revolution	365.3 d
Period of Rotation	1 d
Inclination of Axis	23° 45 min
Equatorial Diameter	12,756 km
Mass (Earth = 1)	1
Volume (Earth = 1)	1
Density	5.52 g/cm^3
Atmosphere (main components)	N$_2$, O$_2$
Surface Temperature	240 to 320 K
Atmospheric Pressure (at sea level)	1 bar
Surface Gravity (Earth = 1)	1
Magnetic Field (Earth = 1)	1
Surface Area Mass	8.5 × 10^{-11} m^2/kg
Known Satellites	1

CHAPTER 8

Earth

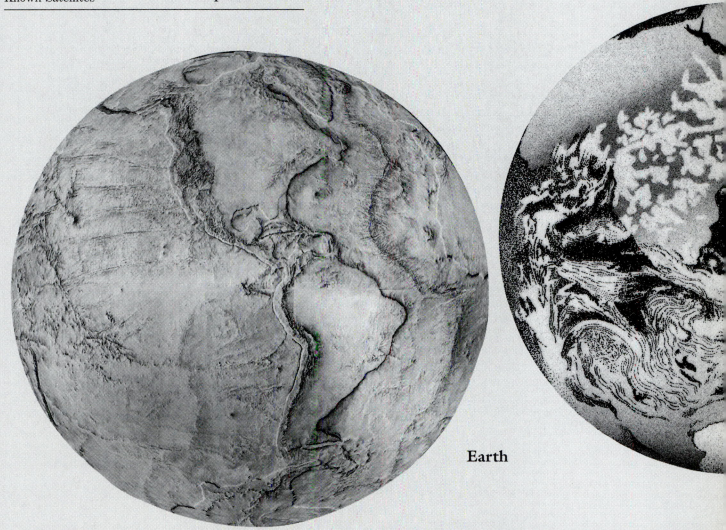

Earth

Introduction

One of the most significant results of space exploration concerns our view of Earth and the dynamics of how the planet changes. Now, "for the first time in all of time, man has seen the Earth from the depths of space—seen it whole and round and beautiful and small." With satellite photography, Earth is seen literally from a new point of view and geology is perceived and understood in a new planetary perspective. Almost without our realizing it, the tools for space studies have brought about a new age of rediscovery of the planet Earth itself.

Viewed from space, Earth presents a fascinating range of surface features (Figure 8.1). By an alien scientist it might be called the Blue Planet, but the deserts, such as the Sahara, are distinctly orange and the lush vegetation gives the tropics a deep greenish hue. Contrasting beautifully with the azure blue are the white streamers of clouds and the massive ice cap of Antarctica. To many, the breathtaking photographs of Earth from space may be simple items of great beauty, but to the student of geology they reveal much more than mere clouds, oceans, and continents. These photographs tell us much about how the planet works as a dynamic system undergoing constant change.

Major Concepts

1. Earth, like the other rocky inner planets, is differentiated with a dense iron-rich core that is still partially molten, a thick mantle of iron- and magnesium-rich silicates, and a crust of silica-rich and magnesium-poor rocks.

2. The present atmosphere contrasts sharply with those around other inner planets because it is poor in carbon dioxide, rich in the residual nitrogen, and also rich in oxygen.

3. Earth's lithosphere is ten to several hundred kilometers thick and is thinner and more mobile than that of the Moon, Mercury, Mars, or Venus.

4. The continents and ocean basins are the principal surface features of Earth. They are created and modified by a distinctive tectonic style called plate tectonics that reflects the style of convection in the upper mantle. Deep mantle convection is driven by mantle plumes like those on other planets.

5. Liquid water is stable at the surface of Earth and forms deep oceans. It is in constant motion, driven by energy from the Sun and Earth's gravity to form an integrated hydrologic system. In dry regions, the wind, also driven by solar energy, creates vast dune fields. In Earth's polar regions and at high altitudes, water is stable as solid ice, which deforms and flows as glaciers.

6. Earth formed 4.6 billion years ago by accretion. It differentiated during this time of heavy meteorite bombardment. Liquid water existed from at least 3.8 billion years ago, the age of the oldest rocks. Life developed in these warm seas by about 3.5 billion years ago. High-standing continents began to form at this time and were not recycled back into the mantle. As the planet cooled, modern-style plate tectonics developed by about 2.5 billion years ago. Repeated cycles of ocean basin formation and closing shaped the present appearance of the planet's surface.

The Planet Earth

Facts about the planet Earth have accumulated for several centuries and the broad outlines of the geology of our planet, from its core to the outer limits of its atmosphere, have been established by painstaking research. With the acceleration of scientific advancements during the last two decades, many of these facts have to be set into a new theoretical framework, one that takes into account the new perspective that Earth is a planet, a member of a diverse community of worlds orbiting the Sun.

Only a small fraction of the results of these investigations is summarized in the table of Physical and Orbital Characteristics (Table 8.1). These terse facts reveal that Earth is nearly the same size as Venus but has a much different atmosphere and surface environment. Moreover, Earth turns rapidly on its axis, and it has a sizable magnetic field and a natural satellite, each a significant contrast with its planetary "twin." Earth shows a much greater variety of geologic features than any other planetary body yet studied. Most large-scale features such as the continents, ocean basins, mountain belts, and volcanoes are the result of Earth's internal heat. Small-scale features, such as fluvial valleys, coastlines, and deserts, are directly or indirectly the products of the atmosphere and energy from the Sun.

All this knowledge about Earth is of utmost importance to planetary studies because it is the conceptual basis by which other planetary bodies are investigated. In turn, whatever new information is obtained about other planetary bodies and how they operate under extremely different conditions will reshape our thinking about Earth itself, its origin, its dynamics, and its history.

The Internal Structure of Earth

The nature of the atmosphere, the oceans, and the surface of the land is known in considerable detail because these can be studied by direct observation, but the internal structures of the planets present some of the most difficult problems faced by geologists. The deepest bore holes on Earth penetrate no deeper than 10 km, and structural deformation and erosion rarely expose rocks that formed more than 20 to 25 km below the surface. Volcanic eruptions provide small samples of material that come from greater depths, possibly as much as 200 km, but aside from these limited data we have no direct knowledge about the nature of Earth's interior. How then are we able to determine the struc-

Figure 8.1

Views of Earth from space are dominated by the ocean and the swirling patterns of clouds, underlining the importance of water in Earth's geologic systems. In this view, a large part of Africa and Antarctica are visible, and climatic zones are clearly delineated. Much of the vast tropical forest of central Africa is seen beneath the discontinuous cloud cover. Several complete cyclonic storms, spiraling over hundreds of square kilometers, can be seen pumping huge quantities of water from the ocean to the atmosphere. Much of this water is precipitated on the continents and erodes the land as it flows back to the sea. A large part of the South Polar ice cap, which covers the continent of Antarctica with a glacier more than 3000 m thick, can also be seen. Of particular interest in this view is the rift system of the Red Sea, a large fracture in the African continent that separates the Arabian Peninsula from the rest of Africa.

ture and composition of Earth or of any planetary body? The evidence comes largely from studies of the physical characteristics of the planet itself—its density, the way in which it transmits seismic (earthquake) waves, the nature of its magnetic field—and from comparisons with meteorites. Although these methods of study do not always provide absolute answers, they indicate what the interior of a planet may be like. Models of the internal structures of other planets, extremely important for understanding planetary differentiation, are strongly influenced by our ideas about Earth.

Evidence from Density Measurements.
A comparison of Earth's average density with the density of the crustal materials provides a very important clue concerning the internal structure of Earth. The overall bulk density of Earth is 5.5 g/cm³ (the density of water is 1.0 g/cm³). Rocks at the surface of Earth, however, are much less dense, averaging 2.0 to 3.5 g/cm³. Since the rocks on Earth's surface are only half as dense as Earth as a whole, there is obviously a mass of greater density in Earth's interior. Significant changes in the density must occur with depth.

Evidence from Meteorites. The evidence that the interior of Earth is composed of material denser than that exposed at the surface is reinforced by the analysis of meteorites. Two major types of meteorites have long been recognized: metallic meteorites composed largely of iron and nickel, and stony meteorites composed mostly of silicate minerals. It is believed that meteorites may be fragments of several small planets or asteroids that broke up from collisions during the early history of the solar system. The metallic meteorites are thought to be fragments of dense cores, and some of the stony meteorites are thought to be fragments of their mantles. If this is the case, Earth's interior could be considered, by analogy, to consist of a mantle of silicate minerals surrounding a core rich in metallic iron. Such a structure fits well with the density measurements. A second argument, based on meteorite studies, holds that a dense iron core must exist inside Earth because the planet's surface rocks are poor in iron compared to the primitive types of meteorites from which Earth probably formed.

Evidence from Earth's Magnetic Field.
Earth possesses a relatively strong magnetic field. At present, Mercury has a measurable field, whereas the Moon, Mars, and Venus are thought to have at most only very weak magnetic fields. The temperature of Earth's interior is far too high for the magnetic field to be produced by a permanent magnet. The magnetic field must be generated electromagnetically by large-scale motion of the material in Earth's interior, combined with rotation on its spin axis. These motions produce a magnetic dynamo. If Earth had a liquid metallic core, convection of the liquid could generate strong electric currents, which in turn would establish a magnetic field. Therefore, many scientists believe that the presence of a magnetic field in a planet is very strong evidence that the planet contains a liquid metallic core that is convecting.

Evidence from Seismic Waves. One of the most important lines of evidence concerning the nature of Earth's interior is the study of seismic waves. As we have seen in the case of the Moon, a network of seismographs enables scientists, in a sense, to x-ray the internal structure of a planet.

In general, the velocities of seismic waves change with the rigidity and density of material through which they are passing. Thus, changes in seismic velocity with depth indicate significant changes in the internal structure of a planetary body. In addition, one type of seismic wave (the S wave) is not transmitted through a liquid, while another type (the P wave) is transmitted, but at greatly reduced velocities.

The simplest way to represent the vast amount of seismic data now available is to draw a graph showing variations in seismic wave velocities with depth (Figure 8.2). This diagram shows the major discontinuities in seismic velocities and contains the basic data for interpreting the internal structure of Earth.

Using the geophysical evidence described above, seismic studies from observations around the world have shown unequivocally that Earth has a concentric structure as idealized in Figure 8.3. Each major unit differs in density, rigidity, thickness, and composition. The main divisions are (1) the core, (2) the mantle, and (3) the crust.

The Core

The core of Earth is a central mass about 7000 km in diameter. Its density increases with depth but averages about 10.78 g/cm³. It is nearly twice as dense as the mantle, and though it constitutes only 16 percent of Earth's volume it accounts for 32 percent of Earth's mass. Most scientists believe that the core is mostly iron and that it consists of two distinct parts—a solid inner core and a liquid outer core. Note in Figure 8.2 that S waves, which will not be transmitted through a liquid, terminate at a depth of 2900 km, and the velocity of P waves is drastically reduced at this level. This is the boundary between a liquid outer core and the surrounding mantle. Heat loss from the core and the rotation of Earth probably causes the liquid core to circulate, and its circulation generates Earth's magnetic field.

The Mantle

The next major structural unit of Earth, the mantle, surrounds or covers the core. This zone constitutes the great bulk of Earth (82 percent of its volume and 68 percent of its mass). The mantle is composed of iron and magnesium silicates. Indeed, fragments of the mantle brought to the surface by volcanic eruptions demonstrate that peridotite, an olivine-rich rock, is an important rock type in the mantle. Within the upper part of the mantle is a weak asthenosphere, which may extend to a depth of as much as 250 km thick, with an upper boundary 10 to 100 km below Earth's surface. Earth's asthenosphere is much closer to the surface of the planet than those presumed to exist for Mars and Mercury. The asthenosphere is distinctive because its temperature and pressure are such that it is near the melting point. In fact, it may be partly molten and structurally weak and thus flow is enhanced. Earth's asthenosphere corresponds to a seismic low-velocity zone (Figure 8.2). The lithosphere is decoupled from the rest of the deeper mantle. Movement of lithospheric plates is allowed by this layer and these movements are in turn responsible for the volcanic activity and crustal deformation observed at the surface of Earth. As discussed in the last chapter, the presence of this shallow asthenosphere may be one of the major differences between Earth and Venus.

The mantle (and overlying crust) above the asthenosphere make up Earth's lithosphere. The lithosphere is defined by its distinctive mechanical behavior. It is a rigid, solid layer about 10 to several hundred kilometers thick, which rests upon the weak partially molten asthenosphere. The boundary between the rigid lithosphere and the soft asthenosphere is gradational. The boundary between the two layers represents a change in

Figure 8.2
The internal structure of Earth as deduced from variations in seismic wave velocities at depth.
(A) **The major subdivisions of the Earth's interior** include its core, mantle, and crust. The asthenosphere is the mechanically weak part of the upper mantle.
(B) **The velocities of both P and S waves** increase to a depth of about 3000 km, where both change abruptly. S waves disappear and do not travel through the central part of Earth, and the velocities of P waves decrease drastically. This is the most striking discontinuity in Earth and is considered to be the boundary between the core and the mantle. Another discontinuity occurs at a depth of 5000 km, indicating an inner core. A low-velocity layer at depths from 100 to 400 km is caused by the asthenosphere.
(C) **Density variations at depth** based on seismic velocities and other geophysical measurements.
(D) **Variations in temperature** with depth in Earth. The dashed line shows the melting point of Earth's materials, and the solid line the temperature. Note that the rocks are near the melting point in the asthenosphere, but due to increases in pressure, they are below their melting point throughout the rest of the mantle. The outer core of iron is above its melting point.

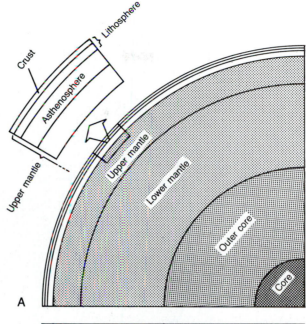

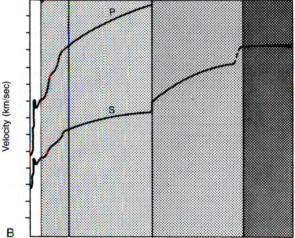

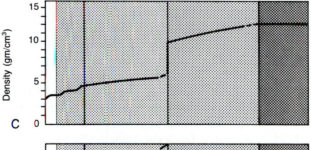

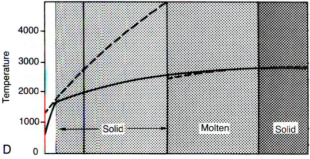

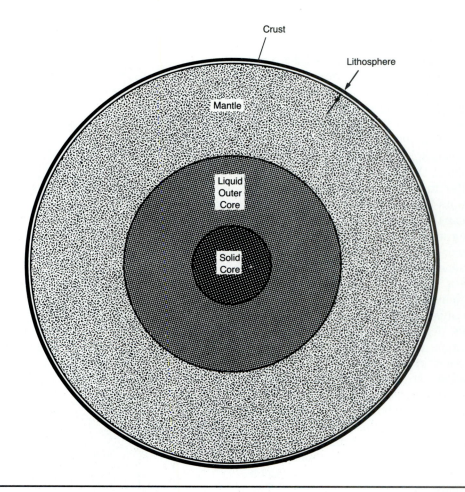

Figure 8.3

Our present understanding of the structure of Earth includes three major components: (1) dense core of metal with a solid inner core and a molten outer core, (2) thick, solid mantle of iron and magnesium silicates, and (3) thin oceanic and continental crust of silicates. The upper mantle contains two mechanically significant layers: (a) asthenosphere, a soft low seismic velocity layer, and (b) lithosphere, a rigid layer that includes the chemically distinctive crust. Surrounding the entire planet is a blanket of surface fluids—water and air.

physical properties as the mantle approaches its melting point.

The Crust

As for other planets, the term crust refers to the chemically distinctive outermost layer of Earth. Today, the term designates the outer layer of Earth extending from the solid surface down to the first major discontinuity in seismic-wave velocity in the lithosphere, which heralds a compositional change. The crust-mantle boundary should not be confused with the lithosphere-asthenosphere boundary. The crust of the **continents** is distinctly different from the crust beneath the **ocean basins**. The continental crust is much thicker (as much as 50 km thick) and is composed of

relatively light granitic (high SiO_2) rock that includes the oldest rock of the crust. By contrast, the oceanic crust is only about 8 km thick and is composed of basalt and gabbro (low SiO_2), igneous rocks having densities much greater than granite. The oceanic crust is young and relatively undeformed by folding. The differences between the continental and oceanic crust, as we shall see, are of fundamental importance in understanding Earth.

The Surface of Earth

The near surface of Earth is composed of four complex geologic components, **the atmosphere, hydrosphere, lithosphere,** and **biosphere.** They are the result of continuing planetary differentiation. Each is a more-or-less continuous envelope or shell, wrapping around and mixing with the others. The atmosphere, hydrosphere, and lithosphere are similar in many ways to those found on some of the other inner planets; however, Earth alone developed a distinctive biosphere, which has had a dramatic effect on the other spheres.

The Atmosphere

Perhaps the most conspicuous features of Earth as seen from space are the brilliant white swirling clouds of the atmosphere (Figure 8.1). Although this envelope of gas constitutes an insignificantly small fraction of the planet (less than 0.01 percent of the mass), it is particularly significant because it moves easily and is highly reactive. It plays a part in the evolution of most features of the landscape and is essential for life. On the scale of the illustration in Figure 8.1, most of the atmosphere would be concentrated in a layer as thin as the ink with which the illustration is printed. As far as we know, all of the life in the solar system exists within this thin film.

It has long been recognized that atmospheres surrounding planetary bodies are highly variable, ranging from the gigantic spheres of gas that constitute the bulk of the outer planets (Jupiter, Saturn, Uranus, and Neptune) to the dense atmosphere of Venus and the thin, tenuous atmosphere of Mars. These variations are so great that it seems certain that atmospheres, regardless of their origins, undergo considerable evolution throughout the history of the solar system.

Earth's atmosphere consists almost entirely of nitrogen (78.08 percent), oxygen (20.95 percent), and argon (0.93 percent). A fourth major component, carbon dioxide, which is essential to all plant life, is present only to the extent of about 0.032 percent. The gases are most dense near sea level and thin rapidly at higher altitudes. At a distance of 5000 km, there is still a trace of an atmosphere, but it is extremely tenuous. The very outer realm of the atmosphere is represented by the **magnetosphere**, a zone of magnetically trapped particles. This part of the atmosphere is a powerful shield from damaging radiation that comes from outer space.

From a geologic point of view, the most significant part of the atmosphere is the lower 13 km. This zone contains about 80 percent of the mass of the atmosphere and practically all of the water vapor, as well as clouds. It is the zone in which evaporation, condensation, and precipitation occur, pressure systems develop, and decay and weathering of the solid surface rock takes place.

The atmosphere is in constant motion; the circulation patterns are clearly seen in Figure 8.1 by the shape and orientation of the clouds. At first glance, the patterns may appear confusing, but upon close examination we find that they are well organized. If we smooth out the details of local weather systems, the global atmospheric circulation becomes apparent. Solar heat, the driving force of atmospheric circulation, is greatest in the equatorial regions and causes water in the oceans to evaporate and the moist air to rise. The warm, humid air forms an equatorial cloud belt, bordered on the north and south by relatively high-pressure zones that are cloud-free in the middle latitudes, where air descends. At higher latitudes, low-pressure systems develop where the warm air from the low latitudes meets the polar fronts. The pattern of circulation is around the resulting low pressure and produces counterclockwise winds in the northern hemisphere (Figure 8.1).

Earth's atmosphere is unique among all others in the solar system in that one of its most important constituents is maintained only by the continuous action of living things. As noted earlier, it is composed primarily of nitrogen and oxygen. These gases are merely traces in the atmospheres of most other planetary bodies (e.g., Mars); if life on Earth were suddenly to cease, significant changes in the atmosphere would certainly occur. For example, most of the oxygen continually produced by plants would soon combine with the iron and carbon in the rocks of the crust. Earth's primeval atmosphere was radically different than the present. It probably formed as a result of planetary differentiation and consisted chiefly of a mixture of carbon dioxide (CO_2), molecular nitrogen (N_2), and water vapor (H_2O), with only small amounts of carbon monoxide (CO) and molecular hydrogen (H_2). These gases were probably extruded from Earth's interior by volcanic processes resulting from the heat produced by accretion and by the decay of radioactive elements. (It is unlikely that Earth's atmosphere was trapped from the solar nebula, as appears to have been the case for the giant outer planets.) However, the composition of this early atmosphere was not static (Figure 8.4). Small quantities of free oxygen (O_2) may have been produced by the decomposition of water in the upper atmosphere, but most oxygen rapidly combined with rock materials or other gases. Some hydrogen probably leaked out into space and a significant amount of carbon dioxide was removed from the atmosphere to form carbonate minerals such as calcite ($CaCO_3$) and dolomite ($CaMg(CO_3)_2$) during weathering of rocks.

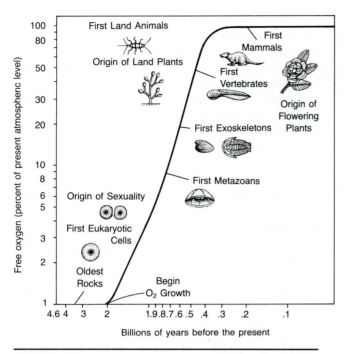

Figure 8.4

The enrichment in oxygen of Earth's atmosphere is traced on this graph. It shows the estimated increase of molecular oxygen (O_2) from a faction of 1 percent to its present level (about 20 percent). The timing of key events in the evolution of the biosphere is also indicated, as the events relate to the evolution of the atmosphere. Land plants of the simplest kind appeared nearly 500 million years ago, but spores that have been interpreted as being those of true vascular plants did not appear until 440 to 430 million years ago. In turn, the oldest organisms generally accepted as being land plants in the modern sense are 420 to 415 million years old. Scorpionlike animals of about the same age are known, but primitive insects did not appear until about 380 million years ago. Banded iron formations are generally 2.2 to 1.6 billion years old and represent the oxidation and precipitation of iron that was once in solution in the oceans.

It seems probable that all of Earth's water was originally in its atmosphere, but as the surface cooled, most of the water vapor eventually condensed as a liquid and collected, forming ancient **oceans**. Within these oceans, the chemical differentiation of Earth continued and the first simple forms of life appeared, probably about 3 billion years ago. Using energy from the Sun, organisms that consumed carbon dioxide and released oxygen (by a process called photosynthesis) eventually evolved, and by about 2 billion years ago they were producing oxygen at a rate faster than it could be removed from the atmosphere by chemical reactions with rocks. Some of the extra oxygen combined with carbon monoxide to form carbon dioxide or with hydrogen to form water. Thus, the atmo-

sphere gradually became more oxidizing. Free molecular oxygen accumulated in the atmosphere and a protective ozone layer developed. This dramatic change in the composition of the atmosphere is reflected in the rock record (Figure 8.4). In the presence of an atmosphere free of molecular oxygen, iron is relatively soluble and must have accumulated in Earth's oceans. However, iron is relatively insoluble in the present ocean because it reacts with oxygen from the atmosphere to form hematite (Fe_2O_3) or other iron minerals that are stable in the presence of oxygen. During the transition from an oxygen-poor to an oxygen-rich atmosphere (about 2.2 to 1.6 billion years ago), large quantities of iron were removed from the oceans to form extensive deposits of delicately banded iron-rich sediments (called banded iron formations). These sedimentary accumulations of iron are the major minable source of iron on Earth. In a sense, Earth's oceans rusted. As the atmosphere remained rich in oxygen, iron was no longer as soluble in terrestrial water. Consequently, iron-rich sediments were rarely produced after this epoch.

An equally important change in the atmosphere–ocean system is reflected by the deposition of carbonate sedimentary rocks (mostly limestones) in the oceans. The formation of these rocks removed a vast amount of carbon dioxide from Earth's atmosphere by biologic and inorganic processes. In fact, if there were no carbonate sedimentary rocks, Earth's present atmosphere would be about as dense as the atmosphere of Venus, which exerts a pressure about 90 times as great. The vast layers of limestones on the continents can be considered as a fossil segment of the early atmosphere. The removal of carbon dioxide increased the proportion of nitrogen, which is nearly inert, in the atmosphere. Thus, the evolution of life created dramatic changes in the composition of the atmosphere by introducing molecular oxygen and extracting carbon dioxide. These processes are reflected in the changing nature of the rock record with time.

The Hydrosphere

If we chose one thing to distinguish Earth from other planetary bodies in the solar system, it would be the great volume of liquid water called the **hydrosphere**. When Earth is viewed from space, the bright blue oceans are one of the most striking features of the planet. Mars, Venus, Jupiter, Saturn and other planetary bodies have clouds, but only Earth has oceans of water and can accurately be called the Blue Planet. Today approximately 71 percent of Earth's surface is covered with water, and during some periods of the geologic past only 15 percent of the planet was dry land.

The existence of liquid water on the planet Earth is made possible by many factors. First, it is important that Earth contained a significant proportion of volatile elements when it formed—implying that some of its constituent planetesimals formed at some distance from the early Sun where temperatures were low enough for volatile compounds to form. Second, these volatile elements needed to be released from the deep interior when the silicate rocks that originally contained them dehydrated as a result of heating during planetary differentiation. Another critical factor is the position of the orbit of Earth around the Sun. Earth is just the right distance from the Sun for liquid water to exist. If it were farther from the Sun, water would be frozen on the surface, the continents would be covered with glaciers like those on Antarctica, and the oceans would be frozen solid. If it were closer to the Sun, the oceans would evaporate and Earth would become an arid inferno like Venus. Also, plant life evolved at a critical time and absorbed some of the original carbon dioxide from the atmosphere, forestalling the runaway greenhouse effect that now exists on Venus. Thus, the original composition, thermal history, and position of the Earth's orbit are in a critical balance that permits water to coexist on the surface as gas, liquid, and solid.

The hydrosphere contains all the free water of the planet; it includes not only the oceans, but all the water in lakes and rivers on land, groundwater beneath the surface, and the water in the glaciers. The total volume is estimated to be 1.5 billion cubic kilometers. If Earth were a smooth sphere, the water in the oceans would cover the entire planet to a depth of approximately 2700 m. The distribution of water on Earth is illustrated in Figure 8.5. About 97.3 percent of all the water in the hydrosphere is in the oceans (liquid salt water and frozen freshwater). The greatest part is in the southern hemisphere. The remaining 2.7 percent of Earth's water is on the continents, mostly in the glaciers of Greenland and Antarctica. The amount of water locked up in polar glaciers is impressively large, totaling 1.8 percent of all the water in the hydrosphere. The distribution of water upon the continents has changed with time. For example, during the recent periods of maximum glaciation, much more water occurred as ice on the land, and sea level was lowered by as much as 140 m. The atmosphere contains only a mere 0.001 percent of the total water in the hydrosphere. Although this is a very small fraction of the total, the influence of this small amount on the evolution of the land-

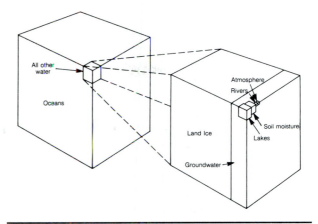

Figure 8.5

The relative amounts of water in the major parts of the hydrologic system can be best appreciated when shown graphically. More than 97 percent of the water on or near Earth's surface is in the oceans. Glaciers contain about 1.9 percent, groundwater 0.5 percent, rivers and lakes 0.02 percent, and the atmosphere 0.001 percent.

scape, climate, and life is far out of proportion to its mass.

One of the most important aspects of the hydrosphere is that its constituents are in constant motion. Water moves in the major ocean basins by systems of currents. It evaporates and becomes part of the general circulation of the atmosphere. It precipitates and moves over the surface of the continents in river systems and percolates through the pore space of the rock and soil as groundwater. This motion makes the surface of Earth one of the most dynamic surfaces of solid planetary bodies.

The Biosphere

The **biosphere** is an interwoven web of life that covers our planet. It includes the forests, prairie grasses, and the familiar animals of the land, together with the numberless creatures that inhabit the sea and atmosphere. Other planetary bodies are chemically differentiated into cores, mantles, crusts, and atmospheres, but only Earth has a biosphere. The biosphere is that part of the differentiated Earth that consists of self-replicating molecules, a property we call life. The biosphere arose and continues to evolve as part of the progressive differentiation of Earth. As a terrestrial envelope, the biosphere is discontinuous and has an irregular shape. It exists within, and reacts with, the atmosphere, hydrosphere, and the upper part of the lithosphere. In vertical dimensions, the biosphere extends from the deep sea trenches of the ocean floor (more than 11,000 m below sea level) to at least 10,000 m above sea level, where spores and bacteria float high in the atmosphere. In lateral dimensions, the biosphere covers the globe, although in the hottest deserts and polar regions and at high elevations only dormant spores and other microorganisms may be found.

If the biosphere were spread out evenly over Earth's surface, it would only form a layer about 10 cm thick, a mere film compared to Earth's mantle, lithosphere, hydrosphere, and atmosphere. However, the biosphere is not distributed evenly over Earth but is concentrated in various environmental niches. Most of the mass in the biosphere, by far, exists in a narrow zone extending from the depth to which sunlight penetrates the oceans (about 200 m) to the snow line in the tropical and subtropical mountain ranges about 6000 m above sea level. It is obvious in space photographs as a green mat of forest in the tropical zones.

The biosphere consists of about one and a half million described species and perhaps more than twice that number yet to be discovered. Each species lives in its own distinctive niche and functions in its own specific way. Yet all living things are linked together with the inanimate part of the planet into the global ecosystem comprising the entire living space of Earth and all the life within it.

Chemically, the biosphere is constructed of only a few elements. Basically, it consists of the relatively volatile elements carbon, hydrogen, and oxygen, with smaller amounts of phosphorus, nitrogen, sulfur, and iron. By far the most abundant single substance in the biosphere is water. With energy drawn from the Sun, the biosphere exchanges chemically within itself and also with the atmosphere, hydrosphere, and lithosphere.

The importance of the biosphere is not limited to its uniqueness. Although the biosphere is relatively small compared to the other major layers of Earth, it has been, and continues to be, a major geologic force operating at the surface. Few people are fully aware of the fact that not only does our atmosphere permit life to flourish, but that the oxygen in the atmosphere was produced by the chemical activity of organisms. The composition of the oceans is likewise affected by the activity of organisms in that most marine organisms extract calcium carbonate from seawater to make their shells and hard parts. When the organisms die, their shells settle to the sea floor and accumulate as beds of limestone. The continued extraction of carbon dioxide to form calcium carbonate in the ocean has a major effect upon the composition of atmosphere as well. In addition, all of the coal, oil, and natural gas (methane) in Earth was formed by the biosphere; so large parts of the near-surface rocks in Earth's crust originated in some way from organic activity.

The record of the evolution of Earth's biosphere is preserved, sometimes in remarkable detail, by fossils, which occur in most sedimentary rocks. Indeed, the numbers of living species represent only about one-tenth of the number of species that may have existed since life first developed on Earth. The oldest sedimentary rocks known on Earth contain chemical hints that life had already made its appearance before these rocks were metamorphosed 3.8 billion years ago. Rocks as much as 3.5 billion years old bear structures (stromatolites) thought to have been formed by primitive forms of life.

An important, but exceedingly difficult question, remains: Are we alone? Did life of any kind arise elsewhere? We have no evidence that life exists anywhere in the solar system or elsewhere in the galaxy, even though it seems reasonable to expect that environments similar to those on Earth exist on other planets revolving about other stars. In fact, recent surveys seeking radio transmissions from advanced civilizations have found none. These searches use radio telescope dishes to look at specific signals beamed toward Earth.

The Lithosphere

In contrast to the inner planets we have studied thus far, Earth's lithosphere is relatively thin. Moreover, it is broken into a number of large fragments or plates, each of which is in motion, and much of the planet's geological activity (earthquakes, crustal deformation, and volcanism) occurs along plate margins.

Earth's lithosphere is divided into two principal regions—continents and basins (Figure 8.6). These major divisions differ markedly, not only in elevation but also in geologic history, rock types, age, density, and chemical composition. Oceanic lithosphere covers 60 percent of Earth and harbors a variety of spectacular landforms, most of which are due to extensive volcanic activity and movements of the lithosphere that continue today. The surface of the continental lithosphere rises above the ocean basins as large platforms. The present shoreline, which is so important geographically and has been mapped in great detail, has fluctuated greatly throughout Earth's history and has no great structural significance. The fact that the continents rise almost 5 km above the ocean basins is much more significant than the position of the shore. The difference in elevation between continents and ocean basins represents a fundamental

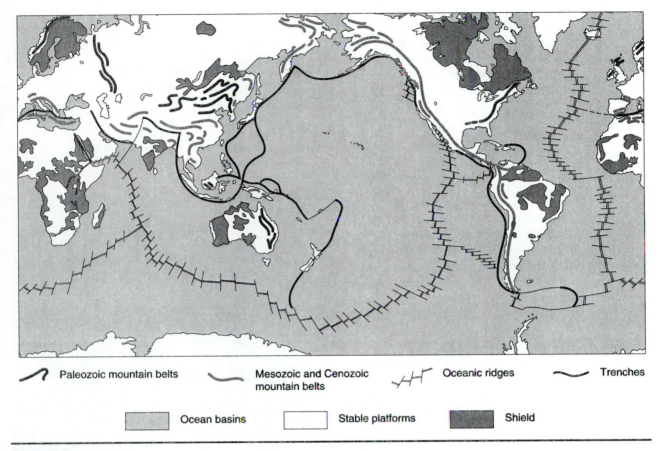

Paleozoic mountain belts   Mesozoic and Cenozoic mountain belts   Oceanic ridges   Trenches

Ocean basins   Stable platforms   Shield

Figure 8.6

A geologic map of Earth shows its major features. Low ocean basins form most of the surface, are dotted with hot spot volcanoes, and are transected by the midocean ridge system. The igneous rocks at the surface of the oceanic crust are mostly basaltic and young (< 0.2 billion years old). The high continental platforms are composed of older (as much as 4 billion years old), more-deformed, and less-dense rocks that have granitic composition.

difference in rock densities in the two types of lithosphere. Rocks of the continents have a lower density than rocks of the ocean basins. It is this density difference that causes the continents to rise and float higher than the denser oceanic crust.

The elevation and area of the continents and ocean basins have been mapped with precision and the data may be summarized in various forms. The data presented graphically in Figure 8.7 show that the continents have a mean elevation of 840 m above sea level, and the ocean floor has a mean depth of about 3700 m below sea level. Only a relatively small percent of Earth's surface rises above the average elevation of the continents or below the average depth of the ocean floor. Figure 8.7 shows that the ocean basins are not only larger than the continental platforms, but also that the average depth of the ocean floor is greater than the average height of the continents.

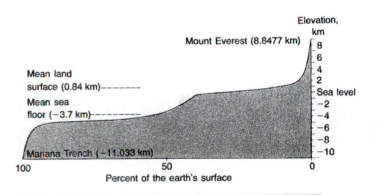

Figure 8.7

A graph of the elevation of the continents and ocean basins shows that the average height of the continents is 840 m above sea level and the average depth of the ocean floor is 3700 m below sea level. Only a small percentage of Earth's surface rises above the average elevation of the continents or drops below the average elevation of the ocean floor.

Major Features of Continents

The broad continental platforms, which rise above the ocean basins, present a great diversity of surface features, with an almost endless variety of hills and valleys, plains and plateaus, and mountains. From a regional perspective, however, the continents are remarkably flat. Most of their surfaces lie within a few hundred meters of sea level. Extensive geologic studies during the past hundred years have revealed several striking facts about the continents:

1. Although each may seem unique, all continents have three basic components: (a) a shield, (b) a stable platform, and (c) folded mountain belts. Geologic differences between continents are mostly in the size, shape, and proportions of these components.
2. Continents consist of granitic rocks, which are less dense than the basaltic rocks that form the ocean basins.
3. The continental rocks are old, some as old as 3.8 billion years.
4. Most of the continental rocks have been extensively deformed by horizontally directed compressive forces.

Shields.

The continental shields are a key to understanding the origin and evolution of continents. One of the most striking characteristics of shields is that they constitute vast expanses of low, relatively flat terrain (Figure 8.8). Throughout an area of thousands of square kilometers, this surface lies within a few hundred meters of sea level. The only features that stand out in relief are resistant rock formations that rise a few tens of meters above the surrounding surface.

A second fundamental characteristic of shields is their structure and composition. Shields are composed of highly deformed sequences of metamorphic rocks and granitic intrusions originally formed under high temperature and high compressive stresses several kilometers below the surface. Their structural complexities are shown by patterns of erosion, the alignment of lakes, and differences in the tones of the photographs. Faults and fractures resulting from crustal deformation are common. They are expressed at the surface by linear depressions, some of which can be traced for hundreds of kilometers. The degree of compression and deformation is illustrated in Figure 8.8. Metamorphic rocks appear in tones of dark gray. Granitic intrusions, which appear in lighter tones, have a more massive texture; some of them cut the regional structural fabric of the metamorphic rock.

(A) The shield, as shown in this block diagram, is composed of complexly deformed crystalline rocks, eroded down to a nearly flat surface near sea level.

(B) Throughout much of the Canadian Shield the topsoil has been removed by glaciers, and different rock bodies are etched out in relief by erosion. The resulting depressions commonly are filled with water and form lakes and bogs, which emphasize the structure of the rock bodies. Dark tones indicate areas of metamorphic rock. Light tones indicate areas of granitic rock.

Figure 8.8

The shield is a fundamental structural component of a continent.

It is apparent from this photograph that the original sedimentary and volcanic rocks that now make up the metamorphic rocks have been intensely deformed by compression and subsequently intruded by granitic magmas. Erosion has since removed the upper cover of the sedimentary and metamorphic terrain, exposing what we now see at the surface.

Rocks of the shield are the oldest rocks exposed in Earth's crust, some having formed more than 3.8 billion years ago. Almost all are more than a billion years old. Yet all evidence of the early period of intense bombardment, which is so prominent on the surface of the Moon, has been erased. At one time, Earth must have been saturated with impact craters like those on the lunar highlands. They have since been completely obliterated by erosion and crustal deformation. This single obvious fact sets Earth apart from most other bodies in the solar system. Earth's crust has been extremely mobile throughout geologic time, being reshaped by both erosion and deposition at the surface and by crustal deformation. Thus, from a planetary perspective, the shields are of great interest because their deformational features demonstrate this tremendous mobility of Earth's lithosphere.

Stable Platforms.   Large areas of igneous and metamorphic rocks, like those exposed in the shields, are covered with a veneer of sedimentary rocks. These areas have been relatively stable throughout the last 600 or 700 million years. That is, they have never been uplifted a great distance above sea level or submerged far below it, hence the term **stable platform**.

Stable platforms form much of the broad, flat lowlands of Earth and are known locally as plains, steppes, and low plateaus (Figure 8.9). Although locally the capping sedimentary rocks appear almost perfectly horizontal, on a regional basis they are warped into broad, shallow domes and basins many kilometers in diameter. One large basin covers practically all of the state of Michigan. Another underlies the state of Illinois.

The relatively flat-lying sedimentary rocks that cover parts of the underlying igneous and metamorphic terrains are predominantly sandstone, shale, and limestone, which were deposited in ancient shallow seas. These flat-lying marine sediments, preserved on all continents, show that large areas of the shields have periodically been flooded by the sea and have then re-emerged as dry land. At present, more than 11 percent of the continental crust is covered with water. At various periods in the past, however, shallow seas spread over a much greater percentage of the land surface, forming platforms covered with sedimentary rocks.

Folded Mountain Belts.   Some of the most revealing features of the planet Earth are the young, folded mountain belts, which typically occur along the margins of continents. Most people think of a mountain as simply a high, more-or-less rugged landform in contrast to flat plains and lowlands. To a geologist, the term **mountain belt** refers to a long, linear zone in Earth's crust where the rocks have been intensely deformed by horizontal stresses and generally intruded by magmas, which cool and crystallize to form **batholiths**. The topog-

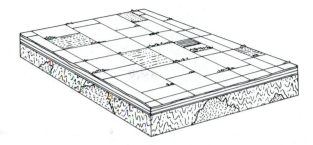

(A) Schematic representation of the stable platform shows the complex metamorphic rocks and a variety of igneous intrusions covered by a relatively thin sequence of horizontal sedimentary layers.

(B) The shield is covered with a veneer of horizontal sedimentary rocks, usually a kilometer of less thick, throughout most of the interior of the United States and on other continents. In some places, erosion has cut through the sedimentary veneer and has exposed the shield below.

Figure 8.9

General characteristics of a stable platform.

raphy can be high and rugged, or it can be worn down to a surface of low relief. It is not the topography of mountain belts that is geologically important but the extent and style of deformation.

Figures 8.10 and 8.11 illustrate some of the characteristics of folded mountains and the extent to which the margins of continents have been deformed. The layers of rock shown in this photograph have been deformed by compression and are folded like wrinkles in a rug. Erosion has removed the upper part of the folds, so the resistant layers from zigzag patterns are similar to those that would be produced if the crests of wrinkles in a rug were cut off.

In many segments of folded mountain belts, the intense compressional forces exceed the strength of

(A) The basic forms of folded, layered rocks are similar to those of a wrinkled rug. In this diagram, the layers are compressed and inclined to the north.

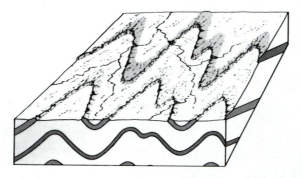

(B) A zigzag pattern at the surface is formed by the traces of the individual layers if the top of the folded sequence of rock is eroded away. Rock units that are resistant to erosion form ridges, and nonresistant layers erode into valleys.

(C) The Appalachian Mountain belt in eastern North America is typical of partially eroded folded mountain belts on Earth. The mountains were formed about 250 million years ago by the collision of two continents at a convergent plate margin. Layers of sedimentary rock, which were originally horizontal, have been compressed into tight folds and later eroded. The internal structure of the mountain belt is revealed by the trace of resistant rock units in this radar image of the eastern United States.

Figure 8.10

The geometry and surface expression of folded mountain belts on Earth.

the rock, and rupture occurs, producing thrust faults in which one block is thrust up and over the other. Movement on a thrust is predominantly horizontal, and displacement can be more than 50 km. Folded mountain belts, like the shields, therefore provide vivid evidence of the great mobility of Earth's lithosphere. Compressive stresses on the Moon, Mercury, and Mars formed the much smaller wrinkle or mare ridges that appear to be the surface expressions of thrust faults. Nonetheless, the lithospheres of these bodies have remained relatively undeformed throughout their histories. This fact is clear because nearly all of their impact craters (regardless of their age) are circular and thus have not been deformed by compressive forces. The lithospheres of these planets, unlike that of the Earth, appear to have been fixed and immovable.

The young, active mountain belts of today are (1) the Cordilleran Belt, which includes the Rockies and the Andes of North and South America, and (2)

the Himalayan–Alpine Belt, which extends across Asia and western Europe. They coincide with zones of intense earthquake and volcanic activity. Older mountain ranges in which deformation ceased long ago include the Appalachian Mountains of the eastern United States, the Great Dividing Range of eastern Australia, and the Ural Mountains of Russia.

The location of young mountains in long narrow belts along the margins of continents is significant because it indicates that Earth's mountain ranges could not result from a uniform planetwide force evenly distributed over Earth, such as one that might be produced from contraction of the planet as it cooled. Mountain ranges must be the result of forces concentrated along the margins of continents. Another important aspect of their restricted location is that many mountain belts extend to the ocean and abruptly terminate at the continental margin. This suggests that some of the older mountain ranges, including the Appalachian Mountains and the mountains of Great Britain, were once connected and have been separated by continental drift and the formation of a younger ocean basin between them.

Continental Rift Systems.

Several continents show evidence that certain zones are under tensional stresses and are being pulled apart. These zones are called rift systems and form part of the major tectonic pattern of Earth. Well-known examples are the East African rift system, the Rhine graben in Germany, the Baikal rift in Russia, and the Basin and Range province in the western United States (Figure 8.12).

Rift systems have a series of long, nearly parallel faults that have large vertical displacements. Typically, an elongate block of the crust subsides between the faults to create a depression called a graben or rift valley. Rift valleys are generally fairly straight for long distances or form a zigzag pattern.

An important feature of rift valleys is that they follow the crest of long, broad upwarps in Earth's crust. The regional upwarping is commonly associated with the stretching and thinning of the lithosphere as hot mantle material rises upward. Young rift zones are commonly associated with active basaltic volcanism.

Rift systems are not restricted to the continents. Some of the continental rift systems seem to be extensions of oceanic ridges. This suggests that there is a planetwide rift system where Earth's crust is under tension and is being pulled apart.

Continental Strike-Slip Fault Zones.

In places, Earth's crust is fractured and displaced horizontally along linear fault zones. These **strike-slip fault zones** are produced when large segments of the lithosphere slide laterally past one rather than colliding (Figure 8.13). The most active are the San Andreas fault zone in the western United States, the Alpine fault in New Zealand, the Dead Sea zone, the Altyn Tagh fault in China, and the Anatolian fault in Turkey. These strike-slip fault zones are characterized by sets of nearly vertical faults with horizontal displacements of as much as several hundred kilometers.

The Major Features of the Ocean Floor

The ocean floor, not the continents, is the typical surface of the solid Earth. It is the ocean floor that holds the key to the evolution of Earth's lithosphere. From a variety of remote-sensing instruments, we have surveyed the ocean floor and have found that the oceanic lithosphere is completely different from the continents.

Among the most significant facts we have learned about the oceanic crust are the following:

1. The oceanic crust is mostly basalt (and its intrusive equivalents), which is erupted at an oceanic ridge.
2. The rocks of the ocean floor are young in a geologic time frame. Most are less than 150 million years old, whereas the ancient rocks of the shields are 700 million to 3.8 billion years old.
3. In general, the rocks of the ocean floor have not been deformed by compression. Their undeformed structure stands out in marked contrast to the complex deformation of rocks in the folded mountains and shields of the continents.

Although most of the topography of the ocean floor can be seen only indirectly, deep-diving research vessels have photographed local areas of the oceanic ridge. Some ocean floor features are visible using satellite measurements (Figure 8.14), but data from seismic profiles such as the examples shown in Figure 8.15 provide more detailed views of the sea floor. From these records we are able to plot accurate depth charts and physiographic or landform maps, and study a variety of geologic features not found on the continents.

The Oceanic Ridge.

The oceanic ridge is the most pronounced tectonic feature on Earth. If the ridge were not covered with water, it certainly would be visible from as far away as the Moon. It is essentially a broad, fractured upwarped segment of the crust, generally more than 1400 km wide, with peaks rising as much as 3000 m above the surrounding ocean floor. The remarkable characteristic of the ridge is that it extends as a nearly continuous feature around the entire globe, like the seam of a baseball. It extends from the Arctic basin, down through the center of the Atlantic into the Indian Ocean and across the South Pacific, terminating in the Gulf of California—a total length of more than 65,000 km. Without question, it is the greatest single structural feature on Earth.

The oceanic ridge is broken by numerous faults, which form linear hills and valleys. The highest and most rugged topography is located along the axis, and a prominent rift valley marks the crest of the ridge throughout most of its length. Oceanic sediments thin rapidly toward the crest.

Throughout most of its length, the oceanic ridge is cut by a series of strike-slip faults, related to **transform faults**, that create steplike offsets of

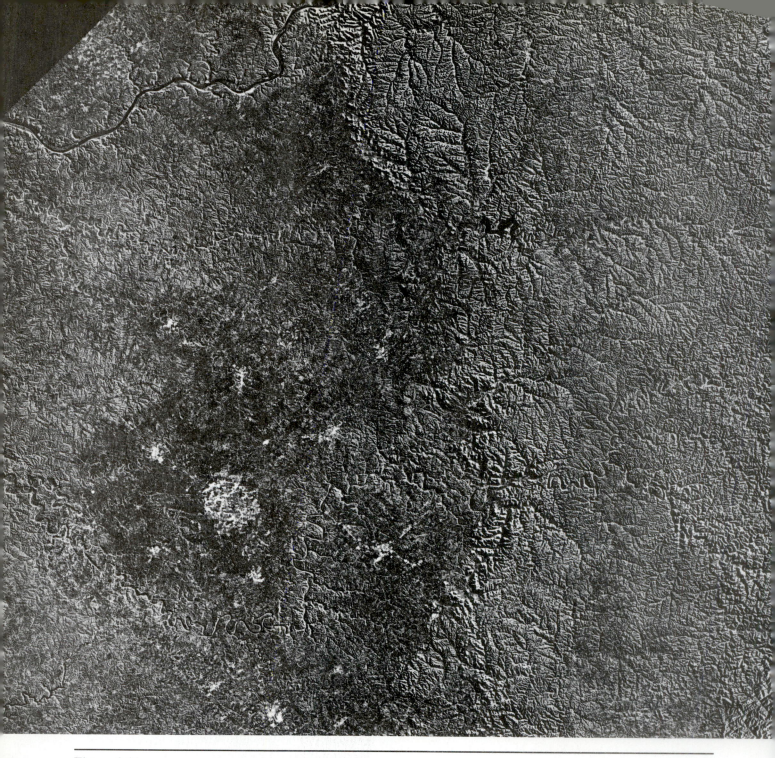

Figure 8.11

A radar image of part of the eastern United States clearly shows the imprint of the major geologic processes operating on the planet Earth. Most striking is the intricate drainage network of the Ohio River and its tributaries, which dissect the surface of the stable platform. The broad domal upwarp of the Cincinnati Arch is near the radar bright spot (towns) and is surrounded by darker patterns in the lower left part of the image. The folded Appalachian Mountains are shown as elongate ridges of resistant sandstone that extend northeastward across the eastern part of the area (compare the style of these folds with those on Venus). The complete absence of impact structures shows that the surface of this area is very young, standing out in marked contrast with the surfaces of most planetary bodies in the solar system.

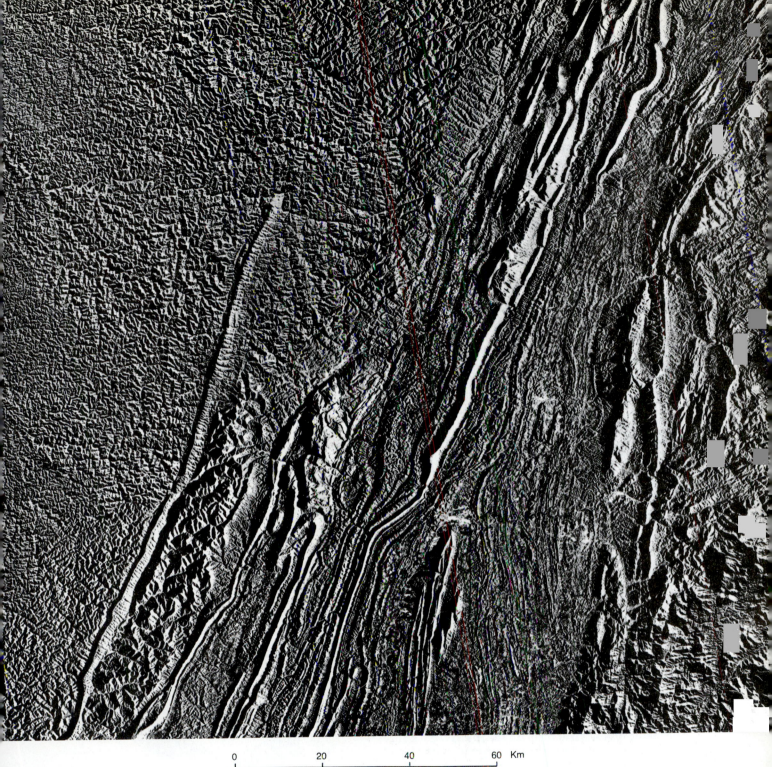

0　　　　20　　　　40　　　　60　Km

Scale

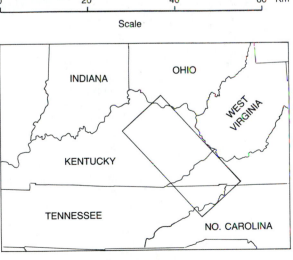

INDIANA　　OHIO

WEST VIRGINIA

KENTUCKY

TENNESSEE

NO. CAROLINA

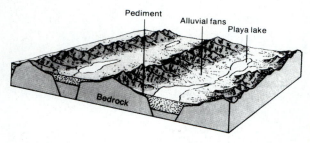

(A) Uplifted blocks called **horsts** form the mountain ranges and are separated by intervening valleys or **grabens** along faults. Sediment shed from the highlands fills the valleys.

(B) This shaded relief map shows the rugged topography of the region centered on Nevada. This type of rift is thought to result from the extension or stretching of the lithosphere.

Figure 8.12

The fault-block structure of the Basin and Range province of the western United States is the result of extensional faulting over a large region.

the ridge (Figure 8.16 and 8.17). Between ridge segments, these faults allow parts of the lithosphere to slip horizontally past one another. The fault zone is expressed by an abrupt, steep cliff, which in places extends even farther as a fracture zone that can be traced for several thousand kilometers.

Detailed studies of the axial zone of the mid-Atlantic ridge were made in 1974 when scientists in deep-diving vessels sampled, observed, and photographed the ridge for the first time. Without doubt, this project made some of the most remarkable submarine discoveries of modern times. The photographs show extensive basaltic lava flows so recent that little or no sediment covers them. Numerous open fissures in the crust were also observed and mapped. In one small area of only 6 km^2, 400 open fissures were mapped, some of which are as wide as 3 m. These are considered conclusive evidence that the oceanic crust is being pulled apart. The eruption of lava from these fractures, which parallel the rift valley, produces the outer layer of the oceanic crust.

Trenches and Volcanic Arcs. Deep-sea **trenches** are 8000 to 11,000 m deep and are the lowest areas on Earth's surface. The most striking examples of trenches occur in the western Pacific, where the trench system extends from the vicinity of New Zealand to Indonesia, to Japan, and then northeastward along the southern flank of the Aleutian Islands. Long trenches also occur along the western coast of Central America and South America, in the Indian Ocean northwest of Australia, in the Atlantic off the tip of South America, and in the Caribbean Sea.

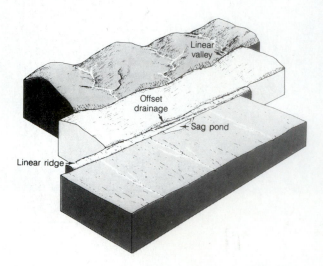

(A) Recent movement along the fault has offset stream drainage patterns on either side and formed sag ponds along the fault. Relative movement between the fault blocks is evident from the direction in which the drainage is offset.

Figure 8.13

The San Andreas Fault, in California, is a major strike-slip fault. It is delineated by prominent, straight ridges and valleys.

(B) This satellite view shows the San Andreas and Garlock strike-slip faults of southern California. It is easy to see why this region is one of the most seismically active in the United States.

Trenches are among the more significant structural features of Earth. Not only are they the lowest parts of Earth's surface, but they are invariably associated with arcuate chains of active volcanoes (called **island arcs**), mountain belts, and zones of intense earthquake activity. Trenches are consistently associated with inclined zones of earthquakes, which dive beneath the volcanic chain.

Islands and Seamounts.   Literally thousands of submarine volcanoes occur on the ocean floor, with the greatest concentration in the eastern Pacific. Some rise above sea level and form islands, but most are submerged and are called seamounts (Figure 8.14 and 8.16). They often occur in groups or chains (e.g., the Hawaiian Islands), with individual volcanoes being as much as 100 km in diameter and 1000 m high. Islands and seamounts testify that volcanic activity is not restricted to the ridges but has occurred in various parts of the ocean basins above mantle plumes or **hot spots** in the mantle.

The surface features of the sea floor result from a combination of tectonic and volcanic processes and provide important clues to the thermal and dynamic processes on Earth. Similar features are not found on the other terrestrial planets, but as we will see, some have speculated that a type of terrain crudely similar to that of the ocean floor is found on Venus.

Earth's Tectonic System

Geologists have long recognized that Earth has a source of internal energy that is manifested repeatedly by earthquakes, volcanism, and mountain building, but it was not until the late 1960s that a unifying theory of Earth's dynamics was developed. This theory, known as **plate tectonics**, provides a master plan that explains the major features of our planet (continents, ocean basins, earthquakes, mountain belts, and volcanism) as a result of the formation, cooling, and destruction of lithosphere. In addition, it explains why Earth is different from the Moon, Mars, and Mercury. The characteristics of the ocean floor, including its surface features as well as its magnetic and seismic properties, show quite clearly that Earth's lithosphere is not stationary but is actually in motion. The continents are carried on moving lithosphere and have repeatedly split, drifted apart, collided, and been sutured together in various patterns. The sea floors, in contrast, are temporary features, opening and closing as the plates move. New oce-

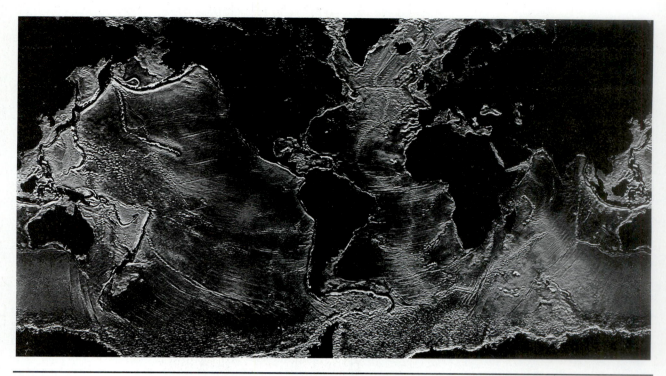

Figure 8.14

A map of the topography of the ocean floor was recently made from satellite measurements of the elevation of the sea surface. This map confirms the data from other methods and reveals many details that were previously unknown. For example, recently discovered seamounts and fractures in the Indian Ocean will help geologists better understand how Africa, India, and Australia drifted northward from Antarctica. The map shows trenches, associated island arcs, and long linear chains of seamounts, which were once volcanic islands. The nature of the midocean ridge, the transform faults, and the fracture zones which transect it are most obvious on the floor of the Atlantic Ocean.

Figure 8.15

Profiles of the ocean floor, produced by seismic studies, provide a wealth of information. A series of profiles placed side by side shows the landforms of a large segment of a midocean ridge. Two segments of the oceanic ridge overlap in the central part of the map.

anic crust is continually being created where the plates move apart and is consumed where the plates converge.

The basic elements of the plate tectonic theory are quite simple and can be understood by studying the diagram in Figure 8.16. The lithosphere of Earth is broken into a limited number of rigid plates, each of which behaves as a separate mechanical unit. The underlying asthenosphere, in contrast, yields to plastic flow. This is possible because the temperature and pressure in the asthenosphere are appropriate to melt some of the minerals, but not the entire rock. Thus, between the solid mineral grains in the asthenosphere is molten rock material, which acts as a lubricant, somewhat like melted water in a slushy snow.

New oceanic lithosphere is constantly being created at oceanic ridges. As it moves laterally away from the ridge, the aging lithosphere cools, thickens, and becomes more dense. In fact, the oceanic lithosphere is simply the cool thermal boundary of the asthenosphere. Ultimately, as cooling oceanic lithosphere slides away from the elevated ridge, it becomes so dense that it sinks back

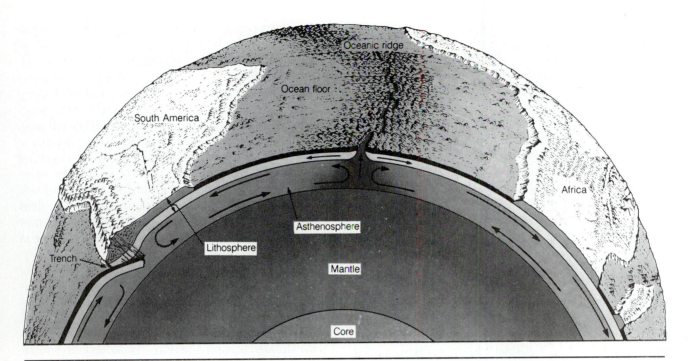

Figure 8.16

Earth's tectonic system is driven by its internal energy. The plate margins are the most active areas of Earth—the sites of the most intense volcanism, seismic activity, and crustal deformation. Where the oceanic lithosphere is pulled apart, upwelling, convecting asthenosphere rises and fills the gap forming the volcanically active midocean ridge. Where converging plates meet, one descends into the mantle and forms a deep-sea trench. These recycled slabs may sink as far as the core-mantle boundary. Some plates contain blocks of continental crust, and plate motion can cause the continents to split and drift apart. Being less dense than the mantle, continental crust cannot sink into the mantle. As a result, where a plate carrying continental crust collides with another plate, the continental margins are compressed and deformed into folded mountain belts. Chains of volcanoes also form above the subduction zone.

into the mantle, pulling the rigid slab with it down a **subduction zone**, where it becomes a part of the mantle again. Back at the ridge, material from the asthenosphere wells upward to fill the void created by the spreading lithosphere. As this rock rises, it partially melts to create basalt, which eventually cools. Then the lithospheric cycle begins again. As a consequence of this large-scale convection, the lithosphere is broken into a series of fragments or plates that are several thousand kilometers across (Figure 8.16).

The continents, formed of relatively light granitic rock and embedded in the denser lower part of the lithosphere, sometimes split and sometimes collide as they are carried about on lithospheric slabs. The continents do not drift through the lower lithosphere. Since Earth is a sphere, the shifting plates are often in collision with each other. As we have seen, oceanic lithosphere is consumed or recycled back into the mantle at oceanic trenches. By contrast, lithospheric plates containing light continental crust cannot sink back into the mantle.

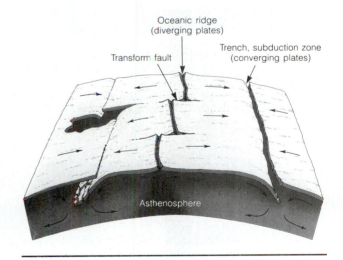

Figure 8.17

Types of plate margins are depicted in this idealized diagram of Earth's sea floor. Divergent plate boundaries occur along the oceanic ridge, where plates move apart. Convergent plate boundaries correspond with deep trenches. Transform plate boundaries coincide with fracture zones that connect two other boundaries; here, transform faults connect oceanic ridges.

Instead, continental margins adjacent to the descending plates are deformed into linear folded mountain belts.

Most plate boundaries are not associated with boundaries between oceans and continents; plate boundaries coincide with ocean ridges and trenches and are characterized by zones of earthquakes and volcanic activity. Three kinds of plate boundaries are recognized, defining three fundamental kinds of deformation and geologic activity (Figure 8.17). These boundaries, defined by the type of motion at the boundary, are called **divergent**, **convergent**, and **transform plate boundaries**.

Processes at Divergent Plate Boundaries

Divergent plate boundaries are characterized by tensional stresses that typically produce long rift zones in the crust, accompanied by fissure eruptions of basaltic lava. As the plate moves apart, decompression induces partial melting in the uprising mantle material, which generates basaltic magma. The magma then rises and is injected into the fissures and fractures of the rift zone; some is extruded as lava on the surface by fissure eruptions and some cools in a magma chamber beneath the surface. As the material cools, it becomes part of the separating plates. This ribbon of new material gradually splits as the plates continue to separate. Thus, new oceanic crust continually develops along divergent plate margins. About 4 km^3 of new basaltic crust are generated each year along the rift zone of the oceanic ridge. The cooling of this magma is one of Earth's most important means of expelling internal heat. Where the zone of spreading intersects a continent, rifting occurs and the continent splits. The separate continental fragments drift apart with the separating plates, creating a new and continually enlarging ocean basin in the site of the initial rift zone.

Several examples showing various stages of continental rifting and the development of new ocean basins can be cited (Figure 8.18). The initial stage is represented by the system of great rift valleys in East Africa. The long, linear valleys (which are occupied partly by lakes) are huge, down-dropped fault blocks that result from the initial tensional stress. Volcanism along the rift zone, including the great volcanoes of Mount Kenya and Mount Kilimanjaro, occurs as magma is injected into the rift zone. A more advanced stage of

(A) Continental rifting begins when the crust is uparched and stretched so that faulting occurs. Continental sediment accumulates in the depressions of the down-faulted blocks and basaltic magma is injected into the rift system. Flood basalt can be extruded over large areas of the rift zone during this phase.

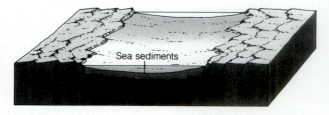

(B) Rifting continues and the continents separate enough for a narrow arm of the ocean to invade the rift zone. The injection of basaltic magma continues and begins to develop new oceanic crust. Remnants of continental sediment can be preserved in the down-dropped blocks of the new continental margins, but most of the sea floor is covered with marine sediment.

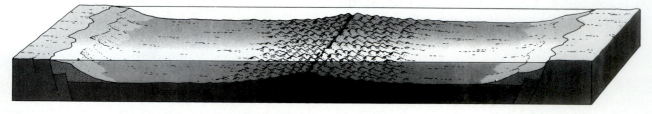

(C) The ocean basin grows larger as spreading continues. The continents move off from the up-arched spreading zone, and parts of the continental crust can be covered by the ocean.

Figure 8.18

Stages of continental rifting are shown in this series of diagrams. The major geologic processes at divergent plate boundaries are tensional stress, block faulting, and basaltic volcanism.

rifting is exemplified by the Red Sea (Figure 8.19), where the Arabian Peninsula has been completely separated from Africa and a new, linear ocean basin is just beginning to develop. The Atlantic Ocean represents a still more advanced stage of continental drifting and sea-floor spreading, where the American continents have been separated from Africa and Europe by thousands of kilometers. The mid-Atlantic ridge is the boundary between the diverging plates, with the American plates moving westward relative to Africa and Europe. The mid-Atlantic ridge has been intensely scrutinized in Iceland, where it rises above sea-level.

Processes at Convergent Plate Boundaries

The boundary between converging plates is a zone of complicated geologic processes, which include igneous activity, crustal deformation, and mountain building. The geologic processes acting in this area depend upon the nature of the converging plates.

When both plates at the convergent boundary are oceanic, one dives beneath the margin of the other and descends into the asthenosphere, where it is heated and ultimately absorbed into the mantle. The great system of deep-sea trenches in the Pacific marks the zone where the Pacific plate descends down into the mantle. Earthquakes and volcanic activity also mark this plate boundary where the lithosphere is being destroyed.

When one of two colliding plates contains a continent, the lighter continental crust resists subsidence and always overrides the oceanic plate. The Rockies and Andes mountain chains result from the encounter of the American and eastern Pacific plates. Earthquakes that consistently rock Chile,

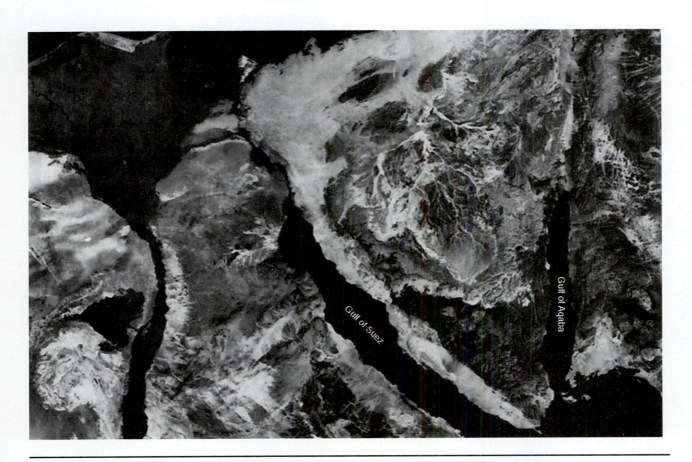

Figure 8.19

The Red Sea rift, which separates Africa from Asia, is a major fracture in Earth's crust. This feature, an important part of Earth's rift system, represents the incipient stages of crustal fragmentation and the movement of continental plates. The rift extends up the Red Sea and splits at its northern end, with one branch forming the Gulf of Suez and the other extending up the Gulf of Aqaba, into the Dead Sea, and up the Jordan valley, on the right of the photo. Movement of the Arabian plate in a northeasterly direction, away from Africa, has created a new ocean basin, which is in the initial stage of its formation. As the rift widens, the edges of the continental block break off, forming a series of steps leading down toward the depression. These steplike blocks can be seen along the Gulf of Suez as distinct parallel lines in the bedrock.

Peru, and Central America result from the encounter of the Pacific and American plates. The active Cascade volcanoes of the northwestern United States are also related to the subduction of a small oceanic plate beneath the North American continent.

When continental crust exists on both converging plates, neither can subside very far into the mantle and both plates are subjected to compression. The continents are ultimately fused or welded together into a single continental block, and a mountain range with folds and thrust faults marks the line of suture. The deeper roots of the mountain belt become metamorphic rocks as they recrystallize under new and usually higher temperature and pressure. The Alps and Himalaya mountain systems resulted from the collision of the African and Indian plates with Eurasia, which also produces the volcanism and earthquakes that torment the Mediterranean and Near East. The nature of deformation resulting from plate collision can be clearly seen where mountain ranges have been eroded so that the resistant layers stand out in prominent relief. The radar image of part of the Appalachian Mountains in Pennsylvania is a classic example (Figure 8.10). The sequence of sedimentary rocks originally deposited in horizontal layers is now tightly folded as a result of a plate collision (between North America and Africa) more than 200 million years ago. The folds are huge flexures, which can be traced across a large part of the state by a series of zigzagging ridges. Deformation of this magnitude is produced wherever continents collide and is one of the most dramatic expressions of the tectonic system. Such folds are commonly accompanied by the development of thrust faults, similar but with more displacement than those associated with wrinkle ridges on the Moon and Mars.

The major geologic processes characteristic of converging plate margins are shown in Figure 8.20. The **subduction zone** (or zone of underthrusting) usually is marked by a deep-sea trench, and the movement of the descending plate generates an inclined zone of seismic activity. As the plate moves down into the hot asthenosphere, partial melting of the oceanic crust or the overlying wedge of mantle generates magma, which (being less dense than the surrounding material) moves upward. Some magma is extruded at the surface as lava flows or domes or as ash flows. The magmas range in composition from basalt to andesite; rhyolite is important in continental volcanic arcs. These produce composite volcanoes, which form a volcanic island arc or a chain of volcanoes in the mountain belt of the overriding plate. Commonly, a large part

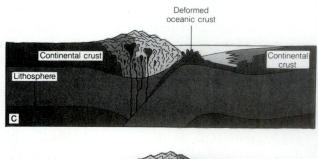

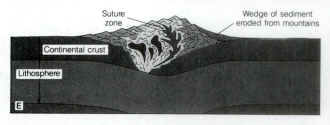

Figure 8.20

The major geologic processes at convergent plate boundaries are illustrated in these cross sections, which show the collision of two continents. Initially partial melting of the mantle above the descending plate produces andesitic volcanism and plutonism on the margin of the continent. When the continents meet, oceanic crust and continental margin sediments are deformed by compression to produce mountain belts; rocks in the crust of both plates become metamorphosed due to high temperatures and high pressures in the mountain roots.

of the magma is intruded in the deformed mountain belt to form batholiths. In both cases, new, low-density material which is difficult to subduct, is added. It appears that continents grow by accretion of these marginal belts. This is an important mechanism in the differentiation of the outer part of Earth.

Because continents are never consumed into the mantle, they preserve records of plate movements in the early history of Earth—records in the forms of ancient faults, old mountain belts, granitic batholiths, and sediments deposited along ancient continental margins.

Processes at Transform Fault Boundaries

The third type of plate boundary occurs where the plates slide horizontally past each other along a special type of fault called a **transform fault**.

Transform faults allow two plates to slip past one another. These nearly vertical faults can join converging and diverging plate boundaries together in various combinations.

Where transform faults connect two diverging plate margins, they create a fracture zone. Fracture zones, however, are not what they might seem to be at first sight, and one must keep in mind the motion of plates produced at the spreading ridge. The apparent offset of the oceanic ridge suggests a simple (but extremely long) strike-slip fault. Careful study of Figure 8.17, however, shows that the relative motion between plates occurs in the area between ridges—the only place where the fault forms a boundary between plates. Even though the cliff, or fracture zone, persists beyond this point, the plates on either side of the fracture are moving in the same direction and at the same rate and are actually linked together (Figure 8.17).

Transform faults also can join ridges to trenches and trenches to trenches. In all cases, transform faults are parallel to the direction of relative plate motion, so there is neither divergence nor convergence along this type of boundary. As a result, plates are neither enlarged nor destroyed. The plates slide passively along the fracture system, producing only fracturing and seismic activity.

Earth's Structure and Plate Tectonics

Let us consider the present structural features of Earth and how they fit into the plate tectonic theory. The boundaries of the plates are delineated with dramatic clarity by the belts of active earthquakes and volcanoes (Figure 8.21). Seven major lithospheric plates are recognized, together with several smaller ones. The oceanic ridge, where the lithosphere is pulled apart, extends from the Arctic down through the central Atlantic into the Indian and Pacific oceans. Movement of the plates is away from the crest of the ridge. For example, the North and South American plates are moving in a westward direction and are interacting with the eastern

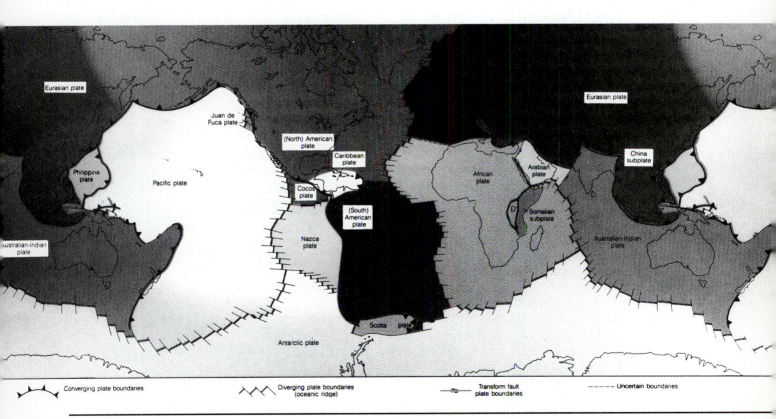

Figure 8.21

Maps showing the seven major plates of the lithosphere and the geologic activity associated with their margins. Plates of oceanic lithosphere form at the oceanic ridges and move toward the trenches. The plate margins are outlined with remarkable fidelity by zones of earthquake activity. Earthquakes occur along the crests of the oceanic ridges where plates are being pulled apart and along the deep-sea trenches and mountain belts where plates are colliding.

Pacific, Cocos, and Nazca plates along the west coast of the Americas. The Pacific plate consists only of oceanic crust and is moving from the ocean ridge northwestward to the system of deep trenches in the western Pacific basin.

The great central theme of the plate tectonic theory is that Earth is a dynamic planet. Although from a personal perspective the surface of the Earth appears to be stable and unchanging from year to year, or even throughout a lifetime, it is continually being modified as heat and mass are redistributed within the planet. The footprints left on the Moon by the astronauts will remain unaltered for eons unless disturbed by a chance impact by a meteorite. The same imprints on Earth would be erased in a few hours to a few years at most, being modified by wind, running water, and overgrowth of plants, or deformed and covered by Earth's movement, volcanism, or a variety of other geologic processes. Change is constant and often dramatic on Earth, in contrast to the static condition that exists on the Moon, Mercury, and most of the icy planetary bodies in the outer solar system.

Other Forces Causing Crustal Deformation

In addition to the plate tectonic system, there are other forces that operate upon Earth and that are capable of deforming its crust. Foremost among these are (1) isostasy, (2) tidal forces, and (3) the impact of meteorites. Their influence compared to plate tectonics is minor, but some of these forces have been more significant in the past and are of paramount importance on other planetary bodies.

Isostasy. Earth's crust is continually responding to the force of gravity in an effort to reach gravitational balance or isostasy (from the Greek *isos*, "equal," and *stasis*, "standing"). Isostasy occurs because the lithospheric plates are buoyed up on the plastic asthenosphere beneath, with each portion of the crust displacing the mantle according to its bulk and density. Denser crustal material sinks deeper into the mantle than crustal material of low density.

Isostatic adjustment in Earth's crust involves the surface layers floating on denser material like an iceberg on the ocean. Suppose that the ice is melted by heat from the Sun only at the surface. As the upper part of the ice melts, the submerged part rises to maintain a floating balance. The same processes may be involved in Earth's crust. When weight is added in some area, such as the delta at the mouth of a river, the area subsides, displacing

the subcrustal material. Similarly, as a mountain is eroded and the weight of the upper rocks is removed, the area rises to compensate for the removed material.

As a result of isostatic adjustments, high mountains and plateaus usually have roots that extend more deeply into the mantle than do the roots of areas of low elevation (Figure 8.22). Any change in an area of the crust, such as removal of material by erosion or addition of material by sedimentation, volcanic extrusion, or accumulations of large continental glaciers, will cause an isostatic adjustment.

The concept of isostasy is therefore fundamental to studies of major features of the crust of all planetary bodies (not only continents, ocean basins, and mountain ranges on Earth, but rebound of crater floors and emplacement of lava flows on other planets).

Tidal Forces. Sensitive instruments capable of measuring small changes in the tilt of Earth's crust indicate that the solid lithosphere moves up and down in response to the gravitational attraction of the Moon and the Sun. During a tidal cycle, Earth's crust may move up and down as much as 30 cm. The amplitude of solid or earth tides, as they are called, is a function not only of the tide-raising force but also of the elastic properties

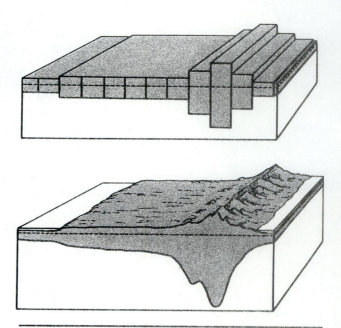

Figure 8.22

Isostasy is the tendency of segments of a planet's crust to establish a condition of gravitational balance. Differences in both density and thickness can cause isostatic adjustments in the crust of a planet.

of the lithosphere. The rise and fall of the lithosphere in response to tidal forces periodically stresses the planet but generates very little thermal energy.

Meteorite Impact. The impact of meteorites is a fundamental and universal process in planetary development. Earth, like all of the other solid planetary bodies in the solar system, was once pockmarked with thousands of craters. The period of intense bombardment of the planets by meteorites was an early event, and the number of impacts has decreased exponentially during the last 4 billion years of planetary history. The dynamics of Earth's lithosphere and modification of its surface by erosional processes have removed most impact structures formed at that time, but over 100 craters have been identified (Figure 8.23). Most impact craters that have been found are highly degraded and lack many of the features found on pristine lunar craters (Figure 8.24). Nonetheless, even during its most recent history, Earth's surface has been modified by rare and generally small impact events.

The youngest large impact crater to be identified on Earth lies on the northern tip of Mexico's Yucatan Peninsula (Figure 8.25). Chicxulub Crater is largely buried by young sedimentary rocks now, but small variations in the gravity field over the crater have allowed it to be detected. The crater appears to be about 180 km in diameter, but a vague outer ring is interpreted by some to show that the crater is 300 km in diameter. If the latter estimate is accurate, Chicxulub would be larger than Meade, the largest crater on Venus. This crater is even more interesting when its age of formation is taken into account. Recent radiometric dates place it at 65 million years old, the exact boundary between the Cenozoic Era and the Mesozoic Era, the two most recent eras of geologic time. This boundary also marks the demise of some of Earth's most famous creatures, the dinosaurs. Many scientists have concluded that the impact led to the extinction of these large reptiles and to the extinction of many other forms of plant and animal life in what has been called a **mass extinction**. Vast amounts of dust are expelled high into the atmosphere of a planet by an impact and it is conceivable that the dust could temporarily block incoming sunlight and cool the planet. Acids are also generated from the nitrogen in the air during its heating by a meteor streaking through the atmosphere. These acids may have changed the environment for many marine animals and plants. Others think that large fires may have been ignited by the impact, also adding to material in the atmo-

sphere and blocking out solar radiation. However, the connection between impact and extinction is still being debated. Perhaps the impact was coincident with an extinction already in progress driven by changing climates and shorelines. Proponents of this idea point out that dinosaurs were showing signs of demise millions of years before the impact. Others point out that some life forms became extinct shortly before and others shortly after the impact event.

Volcanism on Earth

There is no more dramatic proof that Earth has an internal heat engine than the eruptions of molten rock that produce volcanoes. Volcanism is a product of the tectonic activity of Earth. Volcanic features are among the most significant landforms to be studied in planetary geology because they provide a window into the planet's interior and give tangible evidence of the processes operating far below the surface. From the studies of volcanic features, we can learn about thermal conditions below the surface, the structure of the lithosphere, and the thermal history of a planet through time. An understanding of volcanism in its broadest context is essential to understanding the origin and evolution of planetary bodies.

Earth is, and has been, far more active volcanically than one might suppose. More than 500 active volcanoes (those that have erupted at least once within recorded history) are exposed above sea level and hundreds more are considered to be inactive. In addition, many thousands of extinct volcanoes have cones and craters sufficiently well preserved that they must have been active only a few thousand years ago. But these spectacular landforms tell only part of the story. Most of Earth's present volcanic activity is invisible, hidden beneath the sea. Moreover, much of the past volcanic activity on the continents is masked by a cover of sedimentary rocks or has been obliterated by erosion.

Volcanic Processes

Volcanic processes require the generation of magma, its rise to the surface, and its eruption. There is a remarkable diversity in the nature of these three steps, which lead to the variety of volcanic landforms on the planets.

Magma generation requires that temperatures exceed the melting temperature of rock (for silicates 1900 to 900 K, depending on rock composition and pressure). An important reservoir of partially

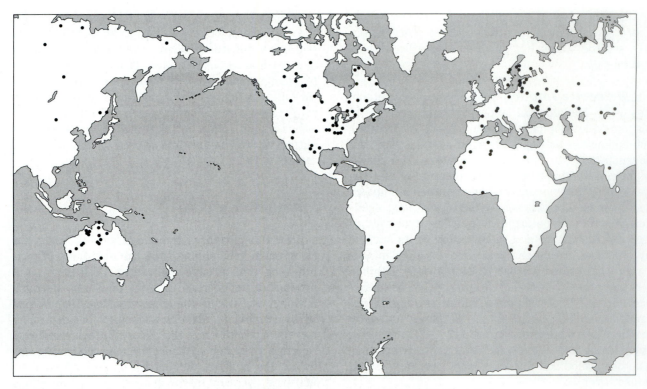

(A) Craters have been found on almost all of the continents but are rare, indicating that Earth's surface has been resurfaced by tectonic, volcanic, and sedimentary processes.

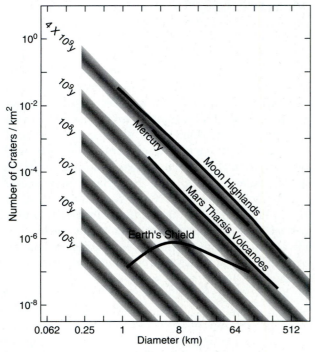

(B) The crater frequency on Earth is the lowest on any of the inner planets. Unlike the Moon, superposition relationships and radiometric dating are much more useful for establishing ages than crater densities.

Figure 8.23

Impact craters are not common on Earth, but several dozen have been identified.

Figure 8.24

This eroded meteorite impact crater is evident only as a circular water-filled cavity. It is about 30 km across and formed in the Canadian shield. Although many impact features must have formed on Earth during its history before 3.8 billion years ago, these have all been obliterated by recycling of the crust back into the mantle. Craters formed after that time have also been destroyed or strongly modified like this example by Earth's surface fluids or its active tectonic system.

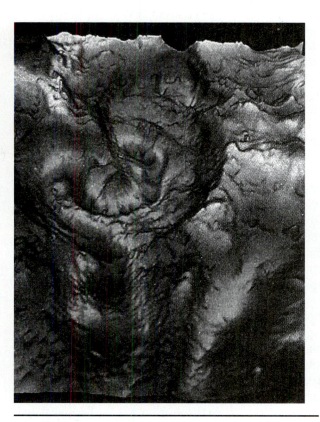

Figure 8.25

Chicxulub Crater is the largest impact feature yet found on Earth. It is at least 180 km in diameter, as shown on this gravity map. This crater formed 65 million years ago and is now largely buried by younger sedimentary rocks. This impact has been blamed for the mass extinctions, at the end of the Mesozoic Era, that included the demise of the dinosaurs.

molten rock is the asthenosphere. In general, this portion of the mantle gives rise to basaltic magma. If the fraction of melting becomes large enough, pools or pods of melt may accumulate as the liquid separates from residual crystals. Magma may escape from this residue completely if a sufficient mass of low density melt forms. Magma may rise as inverted teardrop-shaped blobs or, if it intersects a fracture, as thin streams or sheets of melt.

These magmas from the asthenosphere may in turn be heat sources for melting at shallower levels. For example, hot basaltic magma, generated in the asthenosphere, may rise into the lithosphere (the crust or upper mantle) and produce melts that may rise along separate paths to the surface or become mixed with the original magma.

Eruptions at the surface are driven by a variety of physical and chemical processes. One common method is the result of the expansion of volatile gases (mostly water or carbon dioxide) released from the magma in a process analogous to that of opening a bottle of pop to release the dissolved gas. As magmas approach the surface, the drop in pressure allows gas to separate from the liquid and to expand, sometimes explosively, fragmenting the melt and sending a shower of volcanic debris onto the surface. Other times, a rising magma body encounters water near the surface (groundwater, lakes, oceans, or even glaciers), producing steam instantaneously. The conversion from liquid to steam involves a tremendous

expansion and may also explosively fragment a magma and expel it from the vent.

Basaltic Volcanism: Flood Basalts, Fissure Eruptions, Basaltic Plains

Floods of basalts extruded as fissure eruptions are the most extensive type of volcanic activity on Earth's continents (and on the other inner planets as well). Fissure eruptions emit large volumes of very fluid basaltic lava, which fill depressions and rapidly cover broad areas with flat-lying layers of basalt flows. These volcanic extrusions do not build up high mountains, and when viewed from space they may not even appear to have a volcanic origin. Repeated eruptions flood the landscape and build up extensive plains so that an area is completely resurfaced. Subsequent uplift and erosion causes these plains to be dissected into a series of plateaus, and these regions on Earth are referred to as basaltic plateaus.

One of the most impressive aspects of flood basalts is their colossal dimensions. A lava field may cover an area of 100,000 to a million km². For example, in southern Brazil, more than 1 million km³ of basalt was extruded in a relatively short period of geologic time (10 million years). Similar floods have occurred in the Deccan Plateau of India, the Ethiopian Plateau of Africa, and large areas of Siberia, Greenland, Antarctica, and northern Ireland. Much older flood basalts are found in northern Michigan and the Piedmont region of the eastern United States. Many of these floods of basaltic lavas are related to upwelling mantle plumes. Partial melting of rising mantle results from the low pressures. Flood lavas probably record the rise of a large plume head through the mantle; the tail of the plume feeds much smaller volumes of basalt to the surface in the many hot-spot tracks found on Earth.

Excellent examples of flood basalts are in the Columbia Plateau of eastern Washington and Oregon and western Idaho (Figure 8.26). The Columbia River basalts cover an area of about 200,000 km² with a total thickness of 1 to 2 km. This great accumulation of lava flowed to the surface through numerous fissures over 100 km long. Vast dike swarms now mark these fissures. Individual flows are commonly more than 100 km long and 10 to 30 m thick. Many ponded to form large lava lakes,

burying the underlying topography. In some cases, a ghost or imprint of the former landscape is expressed in the surface of the flood basalts, possibly a result of differential subsidence of the lava over underlying ridges.

Where the rate of eruption is relatively high but the total volume of lava per eruption is low, a broad shield volcano is produced (Figure 8.27). Although fissure vents are common on such volcanoes, many eruptions are centered on pipelike conduits, which bring magma from some depth. Many shield volcanoes have summit craters or calderas produced by collapse, and small cinder and spatter cones are commonly developed on the volcanoes' flanks. The Hawaiian Islands are examples of large, basaltic, shield volcanoes. The shields are commonly several kilometers high and tens of kilometers wide. Volcanic islands like Hawaii are composed of stacks of overlapping shields—five are exposed at the surface. Many other shield volcanoes dot the floors of the oceans, but large basaltic shields are less common on the continents.

Other accumulations of basaltic lavas combine features of both flood basalts (with fissure eruptions) and shield volcanoes (with central vents). These basaltic plains are typified by the Snake River Plain in southern Idaho (Figure 8.28) where small, 2 to 15 km across, and low shields, less than 100 m high, are interspersed with fissure-fed lavas. Over 300 shields are on this plain and the volume of lava extruded from each is relatively low—on the order of 1 to 5 km³. In marked contrast to flood lavas, the lava flows are usually less than 10 m thick and 30 km long. Low shield volcanoes may coalesce and form the major internal structure of a basaltic plain. Small cinder cones, as well as spatter cones, which are built entirely from spatter accumulating at the base of a lava fountain, may also be produced over some fissure eruptions. The individual flow units have a number of characteristic flow features. A typical basaltic flow will develop lobate flow fronts so that although the flow is elongate its margins are highly irregular (Figure 8.28). In many areas, lava flows are confined to channels and flow as open rivers of lava. This prevents heat loss and enables the flow to remain mobile for a longer period of time. The frozen sides of a lava channel will build up from small overflow surges and spatter to form a lava levee. Commonly, the top and margins of a lava flow solidify first, leaving the interior liquid. Beneath this solid upper crust the lava will remain mobile, and if the liquid breaks through the lower end of the crust and runs out, a long lava tunnel or lava tube will form. The roof of a lava tube may subsequently collapse to form a long open

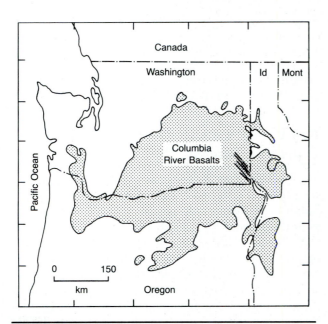

Figure 8.26
Huge eruptions of flood basalt buried this part of the northwestern United States about 14 to 17 million years ago, creating the Columbia River basalts. Individual lava flows, erupted from large fissures, can be traced from near the Idaho-Oregon border to the Pacific Ocean—a distance of almost 700 km.

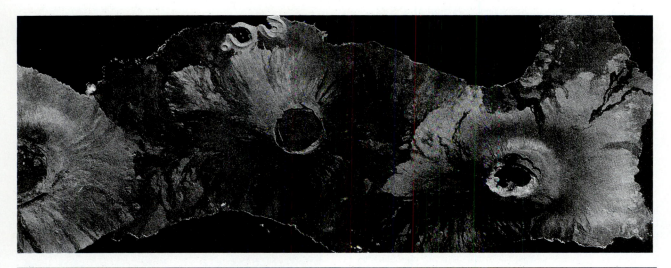

Figure 8.27

Shield volcanoes related to mantle plumes are the largest volcanoes on Earth. They are composed of innumerable thin basaltic lava flows with relatively little ash. Some of the best examples are the islands and seamounts of the Pacific, like these in the Galapagos Islands. The island of Hawaii, rising approximately 10,000 m above the sea floor, is the largest volcano on Earth—but is not as large as several of the shields found on Mars or Venus.

trench or a series of depressions aligned in a sinuous pattern. Lava channels and lava tubes help to preserve the heat and fluid condition of a flow and thus permit it to be transported a long distance from its source. Many tube systems on Earth are over 20 km in length; on the Moon and Mars tube systems exceed 100 km in length.

Silicic Volcanism: Domes, Ash-Flow Shields, and Composite Volcanoes

The type of volcanism associated with young mountain belts and island arcs is quite different from the basaltic fissure eruptions and basaltic plains. The silicic magma is relatively cool, richer in volatile components (like water) and more viscous than basaltic magma. As a result, the mechanism of eruptions and the volcanic products are different from those associated with basaltic eruptions.

Many silicic lava flows are so thick and viscous that small volumes hardly flow at all, but form massive plugs or bulbous domes over the volcanic vent (Figure 8.29). These eruptions are generally small in volume (2 or 3 km^3), but the domes may coalesce to form larger fields. Small eruptions of volcanic ash generally occur just prior to the appearance of pasty lava at the vent. Silicic lavas erupted on the continent are commonly accompanied by the formation of basaltic shields or cinder cones. In some cases, the basalt seems to have added heat to the continental crust, inducing the formation of silicic magma by partial melting of already silica-rich crust (Figure 8.30).

Eruptions of large volumes of silicic magma as ash, instead of lava, form broad **ash-flow shields** (Figure 8.31). These shields have dimensions similar to basaltic lava shields—50 to several hundred kilometers across and a few kilometers high—but represent a dramatically different eruption style. As silicic magma works its way to the surface, confining pressure is released and trapped gas bubbles rapidly expand. Near the surface, the magma explodes, ripped apart by the pressure of the rapidly growing bubbles. Towering columns of ash are propelled to heights of 20 to 30 km by expanding gases released from the magma and by the convective movements of the heated atmosphere. Showers of particles falling out of these clouds produce **ash-fall** blankets several meters thick near the vent. If the density of the eruption cloud exceeds the ability of the atmosphere to support it, vast quantities of still incandescent ash collapse to the ground and rush outward as **ash flows**. Once on the ground, these ash flows behave like fluid lavas but consist instead of jostling bits of ash lubricated by gas swallowed at the flow's front. These streams sweep over hills and travel distances of hundreds of kilometers at tremendous

Figure 8.28

Basaltic plains like those identified on Mars and the Moon are also found on Earth. These volcanic provinces are developed by relatively small eruptions of lava from many central vents, forming groups of low shield volcanoes only a few kilometers across. Fissure vents, like the one shown here, cut across some of the shields as well. The Snake River plains of southern Idaho were formed by small eruptions of this sort, and contrast sharply with the vast floods of lava erupted from fissures in the adjacent Columbia River basalts.

Figure 8.29

Domes of silicic lava form because silica-rich lava is so viscous that it resists flow. It piles up over the vent to form large, bulbous domes and short, thick lava flows.

velocities (approaching 100 m/s or 360 km/h). Some deposits are more than 100 m thick and cover thousands of square kilometers. Although most are much smaller, some prehistoric eruptions expelled over 3000 km^3 of magma in one episode. Repeated eruptions build up very large shields and eventually drain huge magma chambers. Circular calderas (large volcanic craters) form when the unsupported roof of the magma chamber collapses after the expulsion of the magma. The calderas may be up to 100 km across; some contain younger volcanic domes and thick accumulations of ash.

   Composite volcanoes have high, steep-sided cones around their vents (Figure 8.32). They are composed of interlayered ash and viscous lava flows, combining features of lava domes and ash-flow shields. This is probably the most familiar form of continental volcano, with such famous examples as Mount St. Helens, Fuji, Vesuvius, and Etna. These cones may be 1 to 2 km high and have diameters of about 5 to 10 km. Usually, a crater at the summit marks the position of the vent. Subsidiary vents occur on the flanks and on the surrounding platforms. These vents may be marked by pluglike domes, lavas, or even cinder cones (Figure 8.33). Explosive eruptions, similar to but smaller than those that produce ash-flow shields, may cause composite volcanoes to collapse and produce calderas. Crater Lake, Oregon (Figure 8.34), formed when a volcano's summit collapsed after its last major eruption and the resulting caldera filled with water. Composite volcanoes emit many types of magma but andesite is common. Moreover, this type of volcano is typical of island arcs and volcanic arcs on the continents.

   The size, shape, type of magma, and eruption style of volcanoes are governed by many factors. Three of the most important are (1) the composition of the magma, (2) the source and amount of thermal energy, and (3) the nature of stress in the

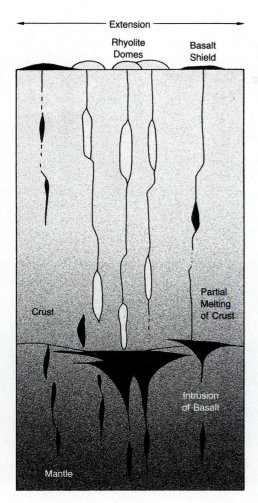

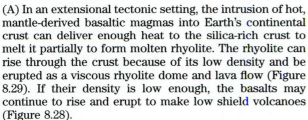

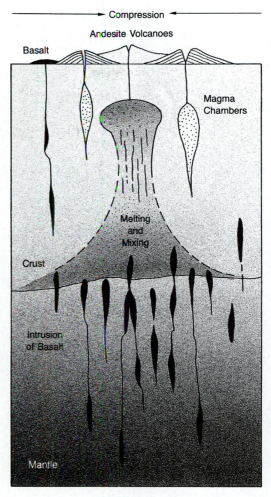

(A) In an extensional tectonic setting, the intrusion of hot, mantle-derived basaltic magmas into Earth's continental crust can deliver enough heat to the silica-rich crust to melt it partially to form molten rhyolite. The rhyolite can rise through the crust because of its low density and be erupted as a viscous rhyolite dome and lava flow (Figure 8.29). If their density is low enough, the basalts may continue to rise and erupt to make low shield volcanoes (Figure 8.28).

(B) In a compressional tectonic setting, the intrusion of basalt into crust may lead to the development of an andesitic composite volcano and associated vents. In this setting basaltic magma may mix with silica-rich crustal melts to form andesite, for example. The amount of heat, the composition of the magmas, and the nature of the stress field are all controlled to a large extent by the tectonic setting of the volcanic system. Thus, it is no surprise that the style and composition of volcanic systems are related to their plate tectonic setting.

Figure 8.30

Cross-sections of volcanic systems on the Earth as inferred from geological studies.

lithosphere. Obviously, the composition of magma may determine whether it flows as a lava or explodes to form a gas-rich column of ash. The thermal energy in a volcanic system determines the amount of magma produced, its temperature, and the length of time the system will be active. The state of stress, extensional or compressional, is important in determining if long fissures or central conduits are formed. Extension of the lithosphere favors the production of magma-filled fractures and shortens the length of time a magma requires to

reach the surface. Compression inhibits the upward movement of magma, perhaps enhancing the mixing of different magmas. These factors are summarized diagrammatically in Figure 8.30, which also serves as a reminder of the three parts of all volcanic systems—a magma source, an ascent route, and eventual eruption through a vent.

Many of the factors that govern volcanic processes on Earth are controlled by its tectonic system. The flow of energy and material in this system determines how much heat can be delivered where

Figure 8.31

Large deposits of volcanic ash are distributed around a large collapse crater or caldera, such as this one in northern New Mexico. These ash-flow shields develop on the continents of Earth above large magma chambers containing silica-rich magma. Although the shields have very gentle slopes, individual eruptions may release as much as 3000 km³ of magma. Repeated eruptions and collapse of the roofs of the magma chambers mark the development of these shields (Figure 8.34). Eruptions of lava may fill the central caldera.

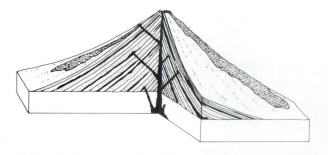

Figure 8.32

Composite volcanoes are built up of alternating layers of ash and lava flows, which characteristically form high, steep-sided cones. Mount Fuji and Mount St. Helens are examples of this type of volcano.

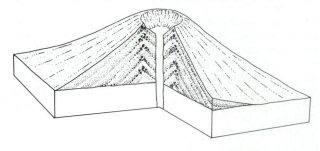

(A) Cinder cones are built up almost exclusively of pyroclastic material thrown from a central vent.

(B) A cinder cone's internal structure consists of layers of ash inclined away from the summit crater.

Figure 8.33

A cinder cone is a distinctive type of small volcano.

and also the state of stress experienced by a region. Therefore, it should come as no surprise that the different types of volcanism can be generally related to certain tectonic environments. Rifting and the production of numerous fractures is associated with the production of basaltic plains or flood basalts (depending on the volume of individual eruptions). In continental settings, silicic dome clusters or ash-flow shields are typical expressions of the accumulation of silicic melts produced by the heat

from these basaltic magmas as they pass through or stall at the base of the crust. Lithospheric subduction, accompanied by compressional stress, is usually associated with the formation of composite volcanoes in island arc chains or on the continents. Magmatism associated with mantle plumes may produce huge basaltic shields in the ocean basins, or it may produce ash-flow shields and basaltic cinder cones and low shields on continental crust. The more silica-rich magmas may come from partial melting of the continental crust.

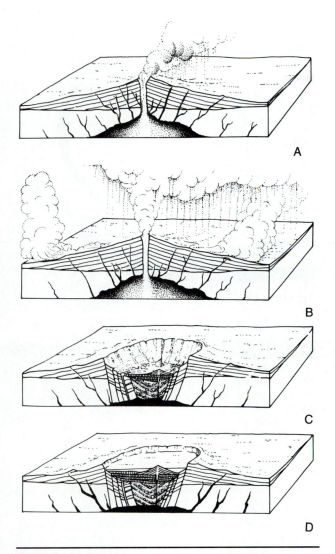

Figure 8.34

The evolution of the caldera at Crater Lake, Oregon, involved a series of great eruptions followed by the collapse of the summit into the magma chamber.

(A) Early eruptions formed the prehistoric composite volcano, Mount Mazama.

(B) Explosive eruptions of ash flows emptied the magma chamber so that the top of the volcano was not supported.

(C) Collapse of the summit of the volcano into the magma chamber formed the caldera. Renewed eruptions of lava formed a small volcano in the caldera.

(D) A lake formed in the caldera, and minor eruptions within the caldera continued to feed small volcanoes in the lake.

Earth's Hydrologic System

We have seen in the previous sections of this chapter that the major features of Earth (continents, ocean basins, mountain belts, volcanoes, etc.) are produced by the tectonic system. The details of sculpturing and shaping the surface of the planet, however, are caused by the hydrologic system. The hydrologic system operates on a global scale extending over the entire Earth. In the broadest sense it includes all possible paths of motion of Earth's hydrosphere (Figure 8.35). The system operates as heat from the Sun evaporates water from the oceans, the principal reservoir for Earth's water. Most of the water returns directly to the oceans as rain. Atmospheric circulation carries the rest over the continents, where it is precipitated as rain or snow. Water that falls on the land can take a variety of paths back to the oceans. The greatest quantity returns to the atmosphere by evaporation, but the most obvious return is by surface runoff in river systems that funnel water back to the oceans. Some water also seeps into the ground and moves slowly through the pore spaces of the rocks. Plants use part of the groundwater and then expel it into the atmosphere by transpiration. Much of the groundwater slowly seeps into streams and lakes or migrates through the subsurface back to the oceans. In polar regions, water can be temporarily trapped upon a continent as glacial ice, but the ice in glaciers moves from cold centers of accumulation into warmer areas, where it ultimately melts, returning to the system as surface runoff.

Water in the hydrologic system—moving as surface runoff, groundwater, glaciers, and waves and currents—erodes and transports surface rock material and ultimately deposits it as deltas, beaches, and other types of sedimentary formations. In this way, the surface material is in constant motion, generally moving under the action of gravity from highlands toward lower regions. The movement of material by the hydrologic system is constantly resurfacing the planet. New surfaces are produced by erosion and by deposition of sediments, which cover older surfaces.

Another way to gain an accurate conception of the magnitude of the hydrologic system is to study Figure 8.1, which permits a view of the system in operation on a global scale. A traveler arriving from space would observe that Earth's surface is predominantly water. The movement of water from the oceans to the atmosphere is expressed in the flow patterns of the clouds. This motion is one of the most distinctive features of Earth as viewed from space, and it stands out in marked contrast to conditions on Moon, Mars, and Mercury.

We can also gain insight into the magnitude of the hydrologic system by considering the volume of water involved. From extensive measurements of precipitation and stream runoff, it has been estimated that if the hydrologic system were interrupted and water did not return to the oceans (by precipitation into the oceans and by surface runoff from the continents), sea level would drop 1 m per year. All of the ocean basins would be completely

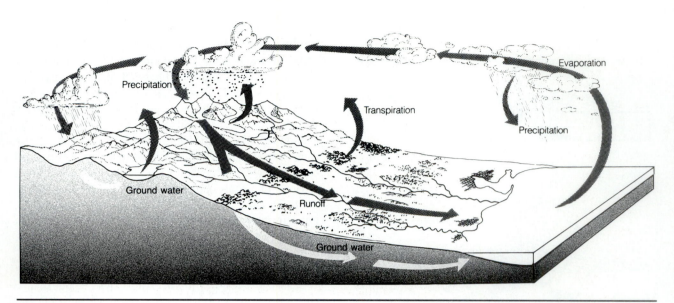

Figure 8.35

The circulation of water in the hydrologic system is driven by solar energy. Water evaporates from the oceans, circulates around the globe with the atmosphere, and is eventually precipitated on the surface as rain or snow. The water that falls on the land returns to the ocean by surface runoff and groundwater seepage. Variations in the major flow patterns of the system include the temporary storage of water in lakes and glaciers. Within this major system are many smaller cycles, shortcuts, and two-way paths. Some occur in just minutes, as when rain evaporates before it falls to the surface of Earth. Others endure for millions of years, as in the case of water locked up in minerals or deposits of sediments.

dry within about 4000 years. Stated another way, Earth's continents are washed by the equivalent of the whole ocean every 4000 years. The recent glacial epoch demonstrates this point clearly. The hydrologic system was partly interrupted as much of the water that fell upon the northern hemisphere froze and formed huge continental glaciers, which prevented the water from flowing immediately back to the oceans as surface runoff. Consequently, during this **ice age**, which ended about 10,000 years ago, sea level dropped more than 100 m.

In the following sections of this chapter we consider some of the details of how the hydrologic system operates and the surface features it produces.

Weathering

The interface between the atmosphere, hydrosphere, and lithosphere is an environment of constant change for minerals and rocks. Solid rock exposed at the surface rapidly breaks down and is decomposed by **weathering**. Most rocks in the lithosphere were formed several kilometers below the surface and were in equilibrium at the high temperatures and pressures that exist deep within Earth's crust. Consequently, when they are ex-

posed to the much lower temperature and pressure at the surface, to the gases in the atmosphere, and to liquid water, they become unstable, experiencing chemical changes and mechanical stresses. As a result, new minerals that are stable at the surface are formed, and solid rock is transformed into small, decomposed fragments that can be removed by agents of erosion. Water is of prime importance in chemical weathering because it takes part directly in the chemical reactions and carries elements of the atmosphere into contact with the minerals of the rocks. In addition, water removes the products of weathering, thus exposing fresh rock. The rate and depth of chemical weathering, therefore, is greatly influenced by the amount of precipitation.

The products of weathering can be seen over the entire surface of the land, from the driest deserts and frozen wastelands to the warm, humid tropics. The major products include a blanket of loose, decayed rock debris, the regolith, which forms a discontinuous cover over solid bedrock. Earth's regolith thus has a very different origin than the impact-generated lunar regolith. The thickness of the regolith ranges from a few centimeters to many meters, depending on the climate, the type of rock, and the length of time that uninterrupted weathering has proceeded. Other major products of weathering are the soluble compounds carried away by streams and groundwater. The salts in ocean water have accumulated from these weathering products.

River Systems and Fluvial Processes

Stream valleys are the dominant landform on Earth's continents; scarcely a landscape exists that does not have some feature produced by the action of running water in the fluvial system. An attempt to appreciate the significance of streams and stream valleys in the regional landscape of Earth, however, is a problem of perspective, much like trying to appreciate the abundance of craters on the Moon from viewpoints on its surface. To an astronaut on the Moon, the surface appears to be an irregular landscape of rolling hills cluttered with rock debris. Indeed, crater rims may appear only as rounded hills. Viewed from the ground, Earth's stream valleys also appear rather nondescript and may be considered only as relatively insignificant, irregular depressions between rolling hills, mountains, and broad plains. But viewed from space, stream valleys can be seen to dominate the landscape of most of Earth (Figure 8.36).

The dominance of stream valleys and associated stream deposits on Earth's continents is an expression of the importance of the hydrologic system and its control in shaping the landforms of Earth. No other planet in the solar system has a hydrosphere like Earth, and no other planet has a surface continually modified by running water.

Major Features of River and Stream Valleys

A **river** system is a network of connecting channels through which water is collected and drains back to the ocean. Rivers also transport large volumes of suspended silt and mud, as is shown in Figure 8.37. The amount of the sediment load carried by a river is controlled by the velocity of the water, size of the drainage basin, elevation of the land, and climate. Maximum sediment load occurs in large rivers that drain a mountainous topography in a humid climate.

A river consists of a main channel and all of the tributaries that flow into it. Each river system is bounded by a divide (ridge), beyond which water is drained by another river system. The drainage system acts as a funneling mechanism for removing precipitation and weathered rock debris. A typical river system can be divided into two subsystems: a collecting system and a dispersing system.

(A) A Skylab photograph of an area in the arid southwestern United States shows the regional patterns of a river system and its valleys.

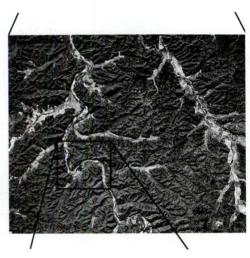

(B) A high-altitude aerial photograph of a portion of the area shown in A reveals an intricate network of streams and valleys within the tributary regions of the larger streams.

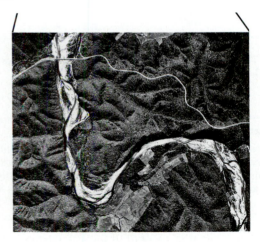

(C) A low altitude aerial photograph of part of the area shown in B reveals many smaller streams and valleys in the drainage system.

Figure 8.36

Erosion by running water is the dominant process in the formation of the landscape.

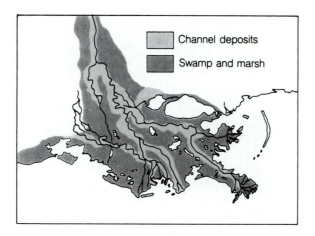

(A) The Mississippi Delta is dominated by fluvial processes, which produce bird-foot extensions. The Mississippi River pours more than a million metric tons of sediment into the ocean each day. Eventually the river finds a shorter route to the ocean and abandons its active distributary channel for a shorter course. The abandoned distributary ceases to grow and is eroded back by wave action. Abandoned river channels and inactive subdeltas have developed on each side of the present river.

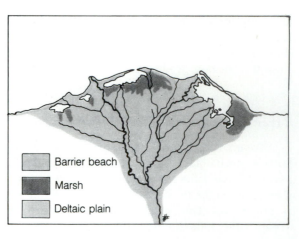

(B) The Nile Delta is dominated by wave action that produces an arcuate delta front.

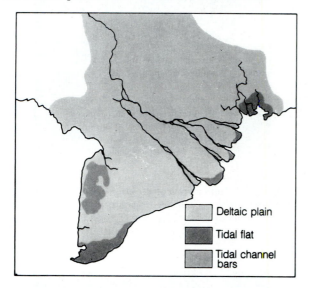

(C) The Mekong Delta is dominated by tidal forces that produce wide distributary channels.

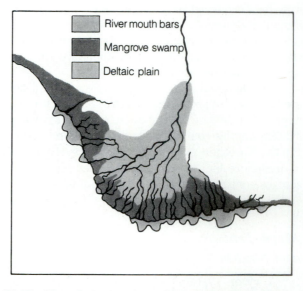

(D) The Niger Delta has formed where stream deposition, wave action, and tidal forces are about equal. An arcuate delta front and wide distributary channels are thus produced.

Figure 8.37

Deltas form when sediment is deposited at the mouth of a river. The deltas of major rivers are records of the transport of vast quantities of sediment by rivers. Sediment eroded from the land is transported by a river system and deposited in the sea. This material is deposited as banks of mud, sand, and clay at the edge of the continent.

The collecting system of a river, consisting of a network of tributaries in the headwater region, collects and channels water and sediment to some base level, usually the ocean. The streams commonly form dendritic (treelike) patterns with intricate systems of branches, which extend upslope toward the divide. By plotting all of the tributaries longer than 1 km in the drainage system shown in Figure 8.38, we obtain the intricate pattern shown in the enlargement. Moreover, each of the smallest tributaries shown in Figure 8.38 has its own system of smaller and smaller tributaries, so that the

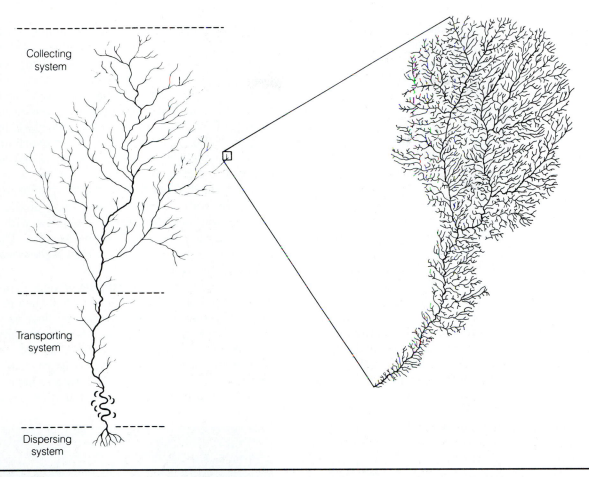

Collecting
system

Transporting
system

Dispersing
system

Figure 8.38

The major parts of a river system are characterized by different geologic processes. The tributaries in the headwaters constitute a subsystem that collects water and sediment and funnels them into a main trunk stream. Erosion is dominant in this area. The main trunk stream is a transporting subsystem. Both erosion and deposition can occur in this area. The lower end of the river is a dispersing subsystem, where most sediment is deposited in a delta or an alluvial fan and water is dispersed into the ocean. Deposition is the dominant process in this part of the river.

total number becomes astronomical. The network of tributaries is important because it is a clear expression of the fact that the main source of water in Earth's rivers is from rain and surface runoff. If the main source of water were from groundwater seepage and springs (as on Mars) or from melting of ice, integrated tributary networks would not develop. Thus, the patterns of Earth's drainage networks are a result of the nature of its hydrologic system in which there is a permanent reservoir of water (oceans) and a constant cycle of water from the oceans to the atmosphere, precipitation on the surface, and surface runoff.

The dispersing system of a river consists of a network of distributaries at the mouth of a river, where sediment and water are dispersed into the ocean, a lake, or a dry basin. The major processes in the dispersing system are the deposition of the coarse sediment load and the dispersal of

fine-grained material and river waters into the basin. Huge accumulations of sediment, called **deltas**, are commonly built at the mouths of major rivers where they enter the ocean.

Processes of Stream Erosion

Removal of regolith or soil is one of the most important processes of stream erosion. Loose, weathered rock debris (sand and mud) falls under the action of gravity or is easily washed downslope into the drainage system and is transported as sediment load in streams and rivers. In addition, a considerable amount of soluble material is carried in solution. As the regolith is removed, it is continually being regenerated by weathering.

Downcutting of stream channels is a second basic process of erosion in all stream channels. Hard silicate minerals carried by rivers are capable of abrading and eroding channel floors. The process of abrasion in rivers is similar to that by which the wire saws used in quarries cut large blocks of stone. An abrasive such as quartz, when dragged across a rock by a wire, can cut through a stone block with remarkable speed. So it is with the abrasive action of a river system.

Headward erosion is the third fundamental process involved in the erosion of a stream valley, and every tributary is involved in the process. Above the head of a valley, water flows down the slope as a sheet, but it converges to a point where a definite stream channel begins. As the water is concentrated into a channel, its velocity and erosive power increase far beyond that of the slower-moving sheet of water on the surrounding, ungullied surface. Thus, the head of the valley is eroded much more rapidly than are the valley walls, and the valley is extended upslope. In addition, groundwater moves toward the valley so that the head of the valley is a favorable location for the development of springs and seeps. These in turn help to undercut overlying resistant rock and cause headward erosion to occur much faster than does the retreat of the valley walls.

In continental highlands, erosion can occur at a surprising rate, and as soil and loose rock material is washed away by the system of streams and rivers, a new surface is created. On the basis of extensive sampling of sediment transported by rivers and radiometric dates of landform changes, it is possible to estimate the rate of erosion over large regions. Mountainous areas are being reduced at an average rate of 50 cm per 1000 years, whereas moderately sloping surfaces (shields and platforms) are being reduced at a rate of less than 5 cm per 1000 years. An average of 15 cm of material over the entire surface of a continent is removed every 1000 years. Of course, erosion of the continents is interrupted periodically by deposition of sedimentary rocks as sea level rises or continents drop. Thus, any surface feature is a temporary thing; Earth is constantly being resurfaced by deposition and by erosion. This is certainly not the case for most planetary bodies, such as the Moon, Mercury, and the icy satellites in the outer solar system. Their landforms are mostly old fossils of the period of intense bombardment, with only parts of their surfaces being covered or eroded subsequently.

Stream Deposition

Erosion is the dominant process in the high headwaters of a drainage system; but where the stream gradient is very low, a river system is unable to transport all of its sediment load and significant deposition occurs. Sediment deposition is usually caused by a drop in velocity of the flowing water. For example, as water velocity drops along the inside of a meandering river bend, some sediment is deposited; erosion occurs on the outside of the bend, where water velocity is higher (Figure 8.38). Likewise, if a river overflows its banks during flood stage, the water is no longer confined to a channel but flows over the surface as a broad sheet. This flow pattern significantly reduces the water's velocity, and some of the sediment settles out. Natural levees grow with each flood and soon form high embankments. Thus, a river actually can build its channel higher than the surrounding area.

Another cause of stream deposition involves a large supply of sediment. A river will then deposit the excess materials on the channel floor as sandbars and gravel bars, forcing a stream to split into two or more channels, and form an interlacing network of channels and islands (Figure 8.39). Such streams are called **braided rivers** and bear many similarities to the dry channels found on Mars.

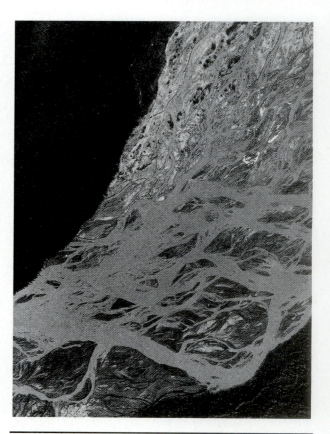

Figure 8.39

A braided stream pattern commonly results if a river is supplied with more sediment than it can carry. Deposition occurs in the middle of the channel, causing the river to develop new channels.

Alluvial fans are stream deposits that accumulate in dry basins at the bases of mountain fronts. An arcuate, fan-shaped deposit (Figure 8.40) results from the sudden decrease in velocity as a stream emerges from steep upland slopes and enters an adjacent basin with a gentle gradient. Alluvial fans form mostly in arid regions, where streams flow intermittently. As several fans build basin-ward at the mouths of adjacent canyons, they eventually merge to form a broad slope of alluvium at the base of the mountain range.

The major areas of sediment deposition, however, are the areas where rivers enter lakes or the ocean. Here the velocity suddenly diminishes, and most of the river's sediment load is deposited to form a delta. The manner in which a delta grows depends on such factors as the rate of sediment supply, the rate of subsidence, and the removal of sediment by waves and tides. If stream deposition dominates, the delta is extended seaward. If waves dominate, the sediment delivered by the river is transported along the coast and deposited as beaches and bars. If tides are strong, sediment is transported up and down the distributary channels, forming linear bars in wide distributary channels.

The great deltas of Earth are deposited by major rivers that have maximum discharge—in general, those that drain the tropical regions (Figure 8.37)—so large deltas, and the rivers that feed them, are partially controlled by climate. Some deltas cover many thousands of square kilometers and have remarkably flat surfaces, only a few meters above sea level. Indeed, large parts of most major deltas are below sea level. The rocks formed from the sediment accumulated in deltas are commonly fine-grained shales and are the most common type of sedimentary rock on Earth.

Drainage Patterns

On a regional basis, the drainage patterns of the major rivers of Earth reflect the regional structure and tectonic history of the continents they drain. They develop on a high mountain belt on the leading edge of a continent and flow across the shield or stable platform toward the trailing or inactive margin of the continent. This results in a distinctive asymmetry, so that the drainage pattern of an idealized major river would consist of a large collecting system in a mountain belt with a trunk stream flowing across the stable platform or shield and emptying into the sea along the trailing, or tectonically inactive, edge of the continent. The Amazon River in South America and the major north-flowing rivers in Russia are examples of this basic pattern.

Although the orientation of major river systems depends mostly on the tectonic history of the continental mass, other factors can significantly modify and alter the drainage pattern. Continental rifting will effectively behead or dismember a previously established river system. Changes in sea level can markedly alter and reduce the size of river systems and eliminate many smaller ones. Continental glaciation will completely obliterate the drainage system beneath it and force the major rivers to establish a new course along the margins of the ice. After the continental ice sheet retreats, a new and commonly complex pattern is reestablished through a system of overflowing ponds and lakes. Outpouring of flood basalts can also completely disrupt and destroy the drainage system over which they flow, and a new pattern has to be established on or adjacent to the volcanic surface. Also, as a continent drifts into an arid zone, desert sand can obliterate large parts of the previously established drainage system, as in the case of the Sahara in North Africa. The Niger River, for example, which is diverted by the Sahara, makes almost a complete circle.

Figure 8.40

Alluvial fans form in arid regions where a stream enters a dry basin and deposits its load of sediment. The drop in the velocity of the stream results in the inability of the stream to carry sediment.

Origin and Evolution of River Systems

Two important questions concerning a river system involve its age and its history, but the answers are extremely complicated. Unlike impact craters, rivers do not form instantaneously. Very few rivers (and certainly no major ones) begin or end without some relationship to the drainage systems that preceded them. Instead, a drainage system continually evolves by headward erosion and stream capture, adjustment to structure, and adjustments and modifications to marine transgressions, continental glaciation, desert sand, and continental rifting. As a drainage system evolves, each period or generation inherits something from its ancestors. The reason for continual evolution of a river system is that the hydrologic system is continuous.

Groundwater and Karst Landforms

The movement of water in the pore spaces of rocks beneath Earth's surface is a geologic process that is not easily observed and therefore not readily appreciated. Yet groundwater is an integral part of Earth's hydrologic system. It is distributed everywhere beneath the surface and occurs not only in humid areas but also beneath desert regions, under the frozen Arctic, and in high mountain ranges. In many areas, the amount of water seeping into the ground equals the surface runoff.

Like other parts of the hydrologic system (rivers and glaciers), the groundwater system is open, with inputs, a means of transport, and places for discharge of water. However, movement in groundwater systems is generally very slow compared to the flow of water in surface streams.

Groundwater systems exist because solid bedrock, like loose soil and gravel, contains pore spaces. The pores, or voids, within a rock can be spaced between mineral grains, cracks, solution cavities, or vesicles. Water seeps into the ground, pulled downward by gravity; below a certain level, all of the openings of the rock are completely filled with water.

Groundwater returns to the surface at springs and enlarges and extends some stream valleys, as is thought to have happened in some martian valley networks. But erosion caused by terrestrial springs is not as spectacular as the great outflow channels that surround Chryse basin on Mars.

In many parts of Earth, underlain by thick layers of limestone, solution activity of groundwater has produced a spectacular and distinctive landscape called **karst topography**. Instead of an integrated network of stream valleys, karst topography is made of features produced by solution activity of groundwater and collapse. The most common and widespread features in karst areas are depressions called **sinkholes**, which develop over caves. Groundwater dissolves the rock in small cavities and fractures and enlarges them into caves and caverns. Ultimately, the roofs of these growing caves collapse, creating small craterlike depressions. In the limestone regions of Kentucky and Illinois, for example, are an estimated 600,000 sinkholes, which vary in size from one meter to more than 30 m in depth. Few through-flowing streams traverse karst areas. What streams do occur disappear quickly by flowing into sinkholes. Most valleys, therefore, are short and end abruptly as the streams disappear into the subsurface.

Limestone is the most common rock type susceptible to solution in water, but dolomite, rock salt, and gypsum can also develop karst features. The great karst regions of Earth are the limestone terrains in tropical and semitropical regions where rainfall is abundant. These are shown in Figure 8.41. Regions of low rainfall are much less susceptible to karst development because there is simply not enough water to accomplish the necessary solution activity. In extremely cold climates, full development of karst topography is inhibited by the presence of permafrost. The permanently frozen groundwater does not circulate, so solution activity cannot occur. Likewise, karst terrains do not develop in the nearly insoluble metamorphic and igneous rocks of the shields. The most favorable areas for karst development are on the gently dipping limestones of the stable platforms in temperate climatic zones. China has the largest area of karst on Earth; the spectacular landscape (Figure 8.42) is well known from the classic art of southern China.

Shorelines

As seen from satellite photography, shorelines are one of the most obvious surface features of Earth because of the striking contrast between the reddish land and the blue water. One might therefore suspect that the shoreline is a fundamental structural feature of Earth, but this is certainly not the case. Shorelines are temporary geographic features, changing constantly as a result of tectonic movements, wave erosion, coastal deposition, and numerous factors that cause sea level to rise and fall. Shorelines are sculptured by waves and currents into many shapes and forms—ranging from steep rocky cliffs to low beaches, quiet bays, tidal flats, and marshes.

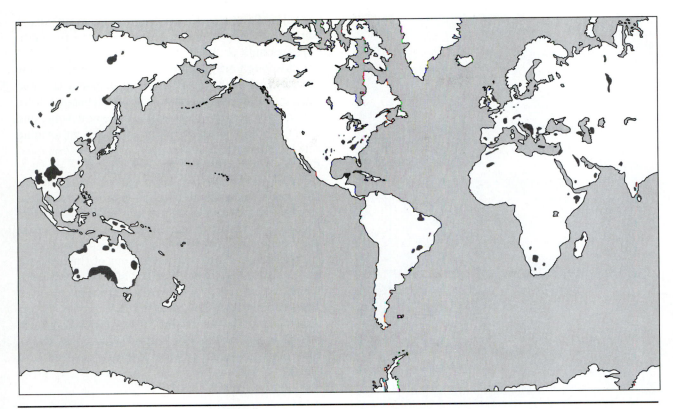

Figure 8.41

The major areas of karst topography on Earth are restricted to areas where outcrops of limestone occur in humid climatic conditions. Limestone outcrops also occur in many other areas, but the climate is in most cases too arid to develop a typical karst topography.

Throughout the history of Earth, shorelines have been very temporary features, constantly changing as a result of the rise and fall of sea level. You will remember from previous descriptions of the major features of continents that the shields and stable platforms are remarkably flat surfaces eroded down to within a few hundred meters of sea level. Thus, any significant change in sea level will result in a major change in the coastline. For example, nearly all coasts on the planet today have been profoundly affected by the rise and fall of sea level caused by the ice age. During periods of glaciation, water was locked upon the land in the form of vast ice sheets. As a consequence, sea level dropped several hundred meters. When the glaciers melted, sea level rose and flooded large parts of the continents. In some places the sea has expanded more than 400 km inland from its former position only a few thousand years ago.

Indeed, one of the most significant surface processes on Earth has been the expansion and contraction of seas over nearly flat continental platforms. The result of this action has caused the platforms to be resurfaced many times, as they were covered with sediment deposited in these shallow seas. What causes sea-level change and the transgression and regression of water over the flat continental platforms? As noted above, one major cause is probably climate change driven by orbital factors (the tilt of Earth, its wobble, and the evolution of the shape of its orbit) as well as by the composition of the atmosphere (a slight increase in carbon dioxide content in the atmosphere will cause the global temperature to rise several degrees). During cold periods sea level is low, and water is trapped on the continents as ice; during warm periods the continents are flooded as the glaciers disappear. Aside from climate change, sea level changes because of the mobility of Earth's lithosphere. For example, an increase in the rate of sea-floor spreading would warm and inflate the midoceanic ridges, reducing the volume of the ocean basins. This would cause the sea to expand over low continental areas. Slow rates of convection would allow the oceanic ridge to subside, increasing the volume of the ocean basins and causing the sea to withdraw from the continents. Hot spots in the mantle beneath the continents could also cause upwarping and regression of the seas.

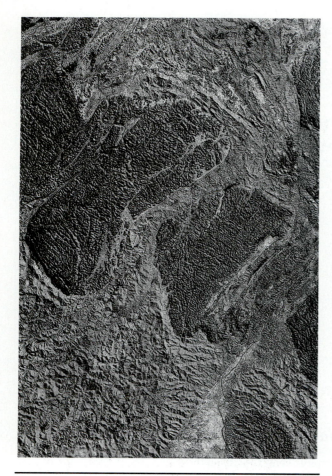

Figure 8.42

The tower karst topography of central China forms the dark lineated terrains in this satellite photo. These towers, often depicted in Chinese art, are some of the most spectacular landforms on Earth.

As a result of sea-level change, Earth's continents have been resurfaced hundreds of times as sedimentary rocks (shale and limestone) deposited in the shallow seas buried continental margins. Unlike other planetary bodies in the solar system, Earth has continually been resurfaced since it originated as a planet.

Deserts and Eolian Systems

Although the liquid water of the hydrosphere is nearly pervasive on Earth, more than a third of its present land surface is arid desert in which there is little rainfall. The largest deserts are the Sahara, the Arabian, the Kalahari, and the great desert of Australia (Figure 8.43). All of these deserts lie between 10 and 30 degrees north or south of the equator. Other smaller deserts occur in the rain shadows of high mountain ranges. The most hostile of Earth's arid regions seemingly exist without a trace of precipitation and have temperatures as high as 350 K (150 F).

Although arid regions are characterized by low rainfall (generally less than 15 cm per year) and high rates of evaporation, climatic changes have repeatedly occurred throughout geologic time, and running water is still a dominant factor in forming much of the desert landscape. The drainage in most deserts rarely reaches the sea (the Nile and the Colorado rivers being major exceptions). That is, the rainfall is so small and the rate of evaporation so great that water does not flow out of the desert region. The result of this internal drainage is that the products of weathering (sand, silt, and loose rock fragments) are not carried out of the area but are deposited to form a variety of distinctive landforms. Sand and gravel washed down from highlands during flash floods accumulate as fan-shaped deposits in adjacent basins. Temporary lakes form in low areas during wet seasons, but most of the time these areas are dry and typically covered with a crust of salt.

Eolian Systems

Sand dunes, products of the circulation of the atmosphere, are the spectacular landform in the great deserts of Earth. Between 25 and 35 percent of Earth's deserts are covered with active sand dunes, which form great **sand seas**. Outside the deserts, the activity of the wind is obscured by the much more effective actions of running water.

In the solar system, only Earth, Mars, Venus, and Titan have substantial atmospheres moving over solid surfaces; each planet is subject to the same basic eolian processes, but each atmosphere is strikingly different. The other planetary bodies, such as the Moon, Mercury, and the satellites of the outer planets, do not have thick atmospheres and therefore lack wind activity. Furthermore, there is no evidence to suggest that there was any eolian activity on these planets in the past. The outer planets, except Pluto, have enormously thick atmospheres, but their atmospheres do not interact with solid surfaces.

Wind Erosion.   Wind erosion in the deserts of Earth has resulted in the excavation of large shallow depressions called **deflation basins** or **blowouts**. They commonly develop where calcium carbonate in the bedrock is dissolved by groundwater, so that loose material is left to be picked up and transported by the wind. In the Great Plains area of the United States, tens of thousands of small deflation basins dot the landscape (Figure

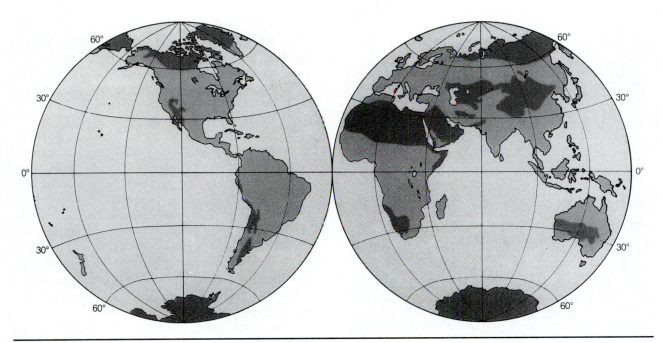

Figure 8.43

The major deserts of Earth, such as the Sahara, the Arabian, the Kalahari, and the great desert of Australia, all lie between 10 and 30 degrees north or south of the equator. These areas are under almost constant high atmospheric pressure and are characterized by subsiding dry air and low humidity. Desert and near-desert areas cover nearly one-third of Earth's continents.

8.44). They can be shallow depressions several meters in diameter or large basins more than 50 m deep and several kilometers across. The largest deflation basins, covering areas of several hundred square kilometers, are associated with the great desert areas of Earth. Commonly, deflation produces no distinctive landform that can be recognized from orbit.

Generally, winds on Earth can move only sand and dust-sized particles, so deflation leaves concentrations of coarser material called **lag deposits** or desert pavements. Such residues may ultimately accumulate into sheets of coarse angular fragments that essentially cover the surfaces, protecting the finer material beneath from further erosion (Figure 8.45), and look like the surface of Mars.

In areas where soft, poorly consolidated rock is exposed, small-scale features produced by wind erosion can be both spectacular and distinctive. Wind abrasion is essentially the same process as the artificial sandblasting used to clean building stone. The most common large-scale features eroded by the wind are elongate yardangs like those found on Mars. On Earth, yardangs are usually less than 10 m high and are aligned parallel to the prevailing wind. They tend to form in groups in which the individual ridges are separated by smooth, wind-eroded troughs (Figure 8.46).

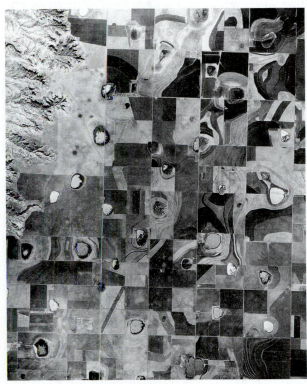

Figure 8.44

Deflation basins in the Great Plains of the central United States are produced where solution activity has dissolved the cement that binds the sand grains together in the horizontal rocks. The loose sand is removed by the wind to form a basin. Water trapped in the basin further dissolves cement, and the basin is enlarged.

Figure 8.45

Wind selectively transports sand and fine sediment, leaving the coarser gravel to form a lag deposit called desert pavement. The protective cover of lag gravel limits future erosion.

Figure 8.46

Wind-eroded ridges, or yardangs, from Iran are similar to those discovered on Mars. These yardangs are developed in easily eroded fine-grained lake sediments in the Middle East. The ridges are up to 150 m high.

Large landforms produced by eolian erosion are rare on Earth. The most common features formed by wind abrasion are pebbles shaped and polished by the wind. Such pebbles are commonly distinguished by two or more smooth, polished surfaces that meet in a sharp ridge; these, too, are found on Mars.

Wind Deposition. Samples taken from many parts of Earth show that large quantities of small dust particles are transported great distances by wind. For example, some dust falls in the eastern United States were composed of particles that originated 3200 km to the west, in Texas and Oklahoma. Dust storms are major processes in Earth's deserts, although not of global scale like the martian storms. Great terrestrial dust storms sometimes reach elevations of 2500 m and advance at speeds of up to 200 m per second.

On Earth, great quantities of wind-blown dust transported from deserts and glaciated areas are eventually deposited as vast sheets that may mask pre-existing landforms. Such **loess** deposits may exceed hundreds of meters in thickness, but most blankets of loess are much thinner. Loess covers about one-tenth of Earth's land surface. An estimated 500 million tons of windblown dust are carried from Earth's deserts each year. (This is only slightly less than the amount of sediment deposited each year by the Mississippi River). Some dust is deposited on the continents downwind from deserts, such as in China (Figure 8.47), but dust from the Sahara, Australia, and South America is carried out to sea. Loess deposits are particularly widespread in semiarid regions along the margins of great deserts (Figure 8.48). The equatorial tropics and the areas formerly covered by continental glaciers are free of loess. Moreover, loess sheets are short-lived because they are easily stripped away by the relentless activity of running water.

As we saw on Mars, wind commonly deposits sand in the form of dunes, which generally migrate downwind. The existence of sand dunes on a planetary surface indicates that eolian processes operating on the planet have been or are capable of redistributing large quantities of sediment and resurfacing large regions of a planet.

The great sand seas of Earth are an effective resurfacing agent and vividly illustrate the dynamics of Earth's atmosphere. The dunes of terrestrial sand seas assume a variety of fascinating shapes and patterns, such as those shown in Figure 8.49. *Longitudinal dunes* are long parallel ridges of sand, elongate in the direction of prevailing wind. They occupy vast areas of central Australia, called the Sand Ridge Desert, and are well developed in the Sahara and Arabian regions as well. *Transverse dunes* have a wavelike form with the crust perpendicular to the prevailing wind. They typically develop where there is a large supply of sand and a constant wind direction. Where sand supply is limited, crescent-shaped dunes called *barchan dunes* tend to develop. *Star dunes* are mounds of sand having a high, central point from which three or four arms or ridges radiate in all directions. Strong winds and limited sand supply develop elongate sheets or stringers.

Polar Regions and the Role of Ice

In Earth's polar regions, most of the water is in a solid state as ice and is not free to move rapidly in river systems. As a result, the sculpting action of running water is subdued and ice takes over as the major geologic agent. If deposits of ice become thick enough, they begin to deform and flow. As ice moves, it leaves its imprint in distinctive landforms not seen in the temperate, desert, or tropical regions of Earth. Viewed from space the importance of ice in Earth's geology is clear. During the winter in the northern hemisphere more than 50 percent of the land and up to 30 percent of the oceans may be covered by a blanket of snow and ice. Moreover, even in the present warm interglacial time, about one-tenth of the land surface (Antarctica and Greenland) is covered with glaciers. At times in the very recent geologic past (the Ice Age), 30 percent of the surface of the land was covered with glaciers.

In the polar regions, ice occurs principally as (1) glaciers, (2) ground ice, and (3) sea ice. Each of these accumulations of ice originates in a different way and constitutes a separate dynamic system. Together they replace the river system as the major process operating on the surface in the polar regions.

Glacial Systems

Glaciers are flowing masses of solid ice that are found in the cold mountains and polar regions of Earth. No other inner planet has glaciers, but the ice-covered surfaces of the satellites of Jupiter, Saturn, and Uranus show evidence of glacial flow. Thus, it is important to understand the dynamics of ice and how ice responds to gravitational, impact, and isostatic stresses in the crusts of the planets.

Continental Glaciers. Two enormous glaciers of continental proportions exist today in the polar regions, covering nearly all of the land masses of Greenland and Antarctica. The Greenland glacier covers an area of about 2 million km², nearly 80 percent of the island. Only a narrow fringe along the margins of Greenland is free from ice (Figure 8.49). The general shape of the enormous mass of ice is like that of a flattened drop of water on a table. The upper surface is typically smooth and featureless and the base is relatively

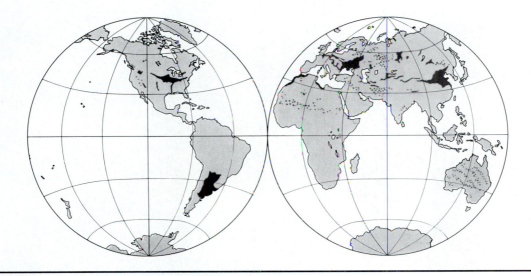

Figure 8.47

Major loess deposits cover large areas of Earth. They are particularly widespread in semiarid regions that are adjacent to deserts, as in China and South America. The loess deposits of Europe and the United States are derived from sediments deposited along the margins of continental glaciers. The tropics and areas formerly covered by glaciers are free from loess.

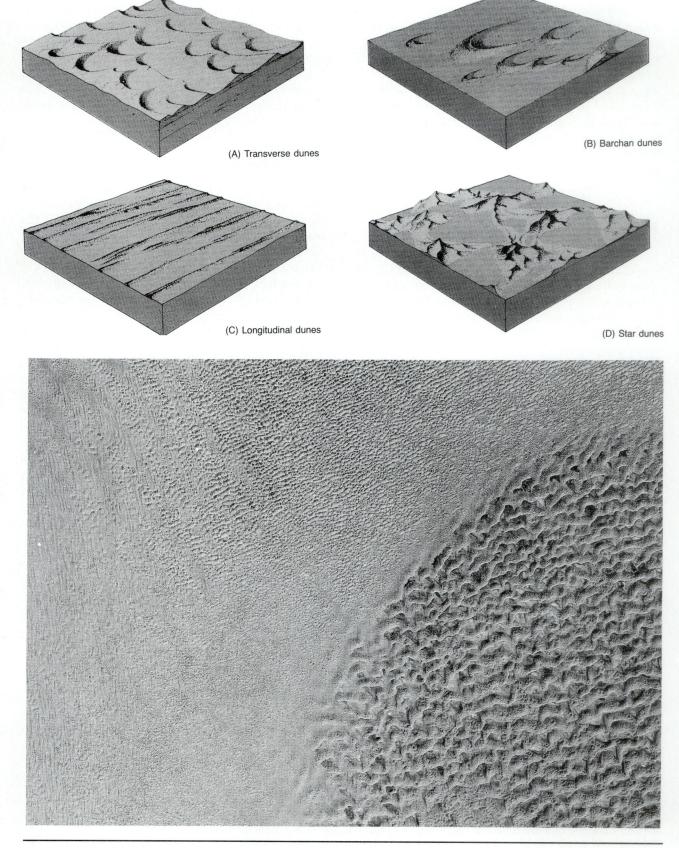

(A) Transverse dunes

(B) Barchan dunes

(C) Longitudinal dunes

(D) Star dunes

Figure 8.48
Satellite view of one of Earth's sand seas showing some of the major dune types found on Earth

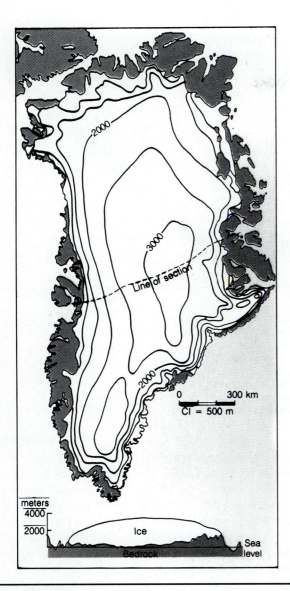

Figure 8.49

The Greenland ice sheet covers nearly 80 percent of the island. In this diagram, the thickness of the glacier is shown by contour lines. Note from the cross section that the central part of Greenland has been depressed below sea level by the weight of the ice.

flat. The Greenland glacier is over 3000 m thick in its central part, but it thins toward the margins. The zone of accumulation is in the central part of the island, where the ice sheet is nourished by snowstorms moving from west to east. The snowline lies from 50 to 250 km inland so that an area of melting constitutes only a narrow belt along the glacial margins.

The glacier of Antarctica is much larger than that of Greenland and contains more than 90 percent of Earth's ice. Much of the glacier is over 3000 m thick, and the weight of this huge mass of ice has depressed a large part of the continent's surface

below sea level. The thick ice covers nearly the entire continent and only small isolated areas of rock are exposed along the coast. The general contour of the Antarctic glacier is that of a very broad, low dome, so the surface is essentially an ice plateau sloping gently for about 4 km from the highest elevations near the center toward the coast.

Antarctica, like Greenland, is surrounded by water, so there is an ample source of moisture to feed the glacier continually. Evidence indicates that Antarctica has been completely covered with ice for more than 30 million years, the time since Antarctica drifted to its polar position.

The glaciers in Antarctica and Greenland provide enough data to construct conceptual models of how a glacial system functions. The basic elements of a continental glacier system and the landforms it produces are shown in Figure 8.50. Water enters the system as snow in a zone of accumulation and is transformed into ice by compaction. Adequate precipitation is essential for the growth and maintenance of a glacier. Cold temperature alone is not sufficient. A number of areas in Siberia are cold enough to support glaciers but have insufficient snowfall; hence glaciers do not develop. Under its own weight, the ice flows out from the zone of accumulation toward the margins, where it leaves the system through melting and evaporation. If more snow is added in the zone of accumulation than is lost by melting or evaporating at the end of the glacier, the ice mass increases in size, and the glacial system expands. If the accumulation of ice is less than is lost by melting and evaporation, there is a net loss of mass, and the size of the glacial system is reduced. If accumulation and loss are in balance, the mass of ice remains constant and the terminus of the ice remains stationary. Ice within the glacier, however, continually flows toward the terminal margins, whether these are advancing, retreating, or stationary.

As shown in Figure 8.50, a continental glacier is a roughly circular or elliptical plate of ice, rarely more than 3000 m thick. Ice does not have the strength to support the weight of an appreciably thicker accumulation. If more ice is added by increased precipitation, the glacier simply flows out from the centers of accumulation more quickly. Because of the low gravity on Mars, much thicker accumulations of ice are necessary to initiate flow; such great thicknesses are not found even at the martian poles.

As glacial ice flows, it erodes, transports, and deposits vast amounts of rock material and greatly modifies the pre-existing landscape. Glaciers erode bedrock (1) by glacial plucking and (2) by abrasion.

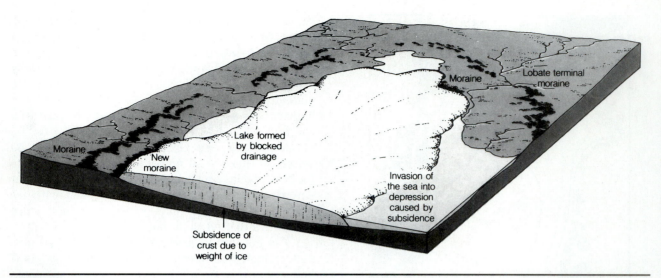

Figure 8.50

A continental glacier system covers a part of a continent and causes a number of significant changes in the regional physical setting. The weight of the ice depresses the continent, so the land slopes toward the glacier. Consequently, glacial lakes form along the ice margins. Alternatively, an arm of the ocean can invade the depression. The original drainage system is greatly modified, as some streams flow toward the ice margins and form lakes. The glacier advances more rapidly into lowlands so that the margins typically are lobate. As the system expands and contracts, ridges of sediment are deposited along its margins, and a variety of erosional and depositional landforms develop beneath the ice.

Glacial **plucking** is the lifting out and removal of fragments of bedrock by a glacier. Beneath the glacier, meltwater seeps into fractures, where it freezes and expands, wedging blocks of rock loose. The loosened blocks freeze to the bottom of the glacier and are plucked or quarried from the bedrock, becoming incorporated into the moving ice. **Abrasion** is essentially a filing process. The angular blocks plucked by the moving ice freeze firmly into the glacier, and they act as tools that grind and scrape the bedrock. Aided by the pressure of the overlying ice, the angular blocks are very effective agents of erosion. The rock fragments are then carried suspended in the moving ice with large blocks transported side by side with small grains, without sorting or separation of the material according to size. As a result, the deposits of a glacier are unsorted and unstratified and thus differ markedly from stream deposits. Most of the particles transported by a glacier are deposited near the terminus, where melting dominates.

Glacial Landforms and the Ice Age.
There is a wealth of undisputable evidence that only 15,000 to 20,000 years ago glaciers covered more than 30 percent of the land. Throughout most of Canada, Scandinavia, and parts of Russia, the surface is dominated by glacial landforms sculpted during the last ice age. During this time, the normal hydrologic system was interrupted

throughout large areas of the planet and was considerably modified elsewhere. The glaciated areas of Earth are among the youngest regional surfaces on the planet and record one of the most significant recent events in the history of Earth.

On a regional basis, the landforms produced by glaciation are spectacular. The moving ice eroded away several meters of soil and solid bedrock so that now many thousands of square kilometers of North America and northern Europe have little or no soil cover, and the effects of glaciation are seen everywhere in the polished and grooved surfaces of bare bedrock. The huge volume of soil and rock eroded by the glaciers was transported outward to the ice margins where it was deposited to form a series of ridges called terminal moraines, which mark the former position of the ice sheet (Figure 8.50). In places, the eroded material was shaped by the moving ice into swarms of stream-lined hills called drumlins (Figure 8.51). Much of the glacial sediment was reworked by meltwater and deposited in streams and lakes. These deposits range up to 300 m thick, so they commonly blanket the landscape upon which they rest.

If a glacier expands beyond the shores of a continent, the ice will continue to move out to the sea as great plates of floating ice, known as **ice shelves**, but will still be attached to the land. Ice shelves are the largest floating masses of ice on Earth, averaging about 300 m thick. They com-

Figure 8.51

Drumlin fields in the Canadian shield are stream-lined hills of glacial sediment that were shaped by the movement of a glacier and show the direction in which the ice flowed. The dark smooth patches are lakes.

monly protrude 50 to 75 m above the water, and their surfaces are remarkably smooth and flat. The margins of the ice shelves typically form a line of vertical cliffs. Movement of the ice within the shelf is very rapid, on the order of 1000 to 1200 m per year. The shelf ice of Antarctica breaks off into enormous tabular icebergs characterized by flat tops and bases and nearly vertical sides. A single iceberg may be more than 15 km across and more than 600 m thick.

The presence of so much ice upon the continents had a profound effect on Earth's crust and its hydrologic system. Major isostatic adjustment of Earth's crust resulted from the weight of the ice, which depressed large areas of the continents. If the margin of the glacier is near the coast, an arm of the sea may invade the depression. In Canada, the entire region around Hudson Bay was depressed below sea level. In Europe, a similar depression was the area around the Baltic Sea. Ever since the ice melted, the land in these areas has been rebounding from these depressions and the former sea floor has risen almost 300 m. It is still rising at a rate of about 2 cm per year.

One of the most important effects of the Ice Age was the repeated worldwide fall and rise of sea level, corresponding to the repeated advance and retreat of glaciers. During a glacial period, water

that normally returned to the ocean by runoff became locked upon the land as ice, and sea level was lowered. As the glaciers melted, sea level rose again. These fluctuations caused the Atlantic shoreline to recede between 100 and 200 km so that vast areas of the continental shelf were exposed.

Another major effect of glaciation was the modification of regional drainage patterns. Prior to glaciation, the landscape of North America was eroded mainly by running water. Much of North America was drained by rivers flowing northward across Canada and into the Atlantic Ocean because the regional slope throughout the north central part of the continent is and was to the northeast. As the glaciers spread over the northern part of the continent, they effectively buried the trunk streams of the major drainage systems, damming up the northward-flowing tributaries along the ice front. This damming created a series of lakes along the glacial margins. As the lakes overflowed, the water drained along the ice front and established the present course of the Missouri and Ohio rivers.

Another significant and obvious effect of glaciation was the formation of a myriad of lakes. Indeed, glaciation created more lakes than those produced by all other geologic processes combined. The reason is obvious if we recall that a continental glacier completely disrupts the preglacial drainage system.

The surface over which the glacier moved was scoured and eroded by the ice, which left innumerable closed, undrained depressions in the bedrock. These depressions filled with water and became lakes.

Glacial Dams and Catastrophic Flooding: The Channeled Scablands.

The continental glacier in western North America moved southward from Canada only a short distance into the state of Washington, but it played an important role in producing a complex of interlaced deep channels found there. These Channeled Scablands cover much of eastern Washington and consist of a network of braided channels from 15 to 30 m deep. The term *scabland* is appropriately descriptive because, viewed from space, the surface of the area has the appearance of great wounds or scars (Figure 8.52). Many of the channels have steep walls and dry waterfalls or cataracts. In addition, there are deposits with giant ripple marks and huge bars of sand and gravel. These features attest to an exceptional degree of erosion by running water, one that would be considered catastrophic by normal standards. Yet, today, there is not enough rainfall in the area to maintain a single permanent stream.

Briefly, the scablands were eroded by the following process. A large lobe of ice advanced southward across the Columbia Plateau and temporarily blocked the Clark Fork River, one of the major northward-flowing tributaries of the Columbia River (Figure 8.53). The impounded water backed up to form a long, narrow lake extending diagonally across part of western Montana called Lake Missoula. As the glacier receded, the ice dam failed, releasing a tremendous flood over the Columbia Plateau. The enormous discharge spread over the basalt surface, scouring out channels and forming giant ripple marks, bars, and other sediment deposits. Estimates suggest that during the flood as much as 40 km^3 of water per hour might have discharged from Lake Missoula. Since glaciers advanced several times into the region, catastrophic flooding occurred many times.

Ground Ice

In the polar regions, groundwater is permanently frozen to a depth ranging from a meter or so to several hundred meters below the surface. This **ground ice** exists in the pore space between soil and rock particles as distinct lenses, veins, or wedges. The expansion, contraction, and partial

Figure 8.52

The Channeled Scablands of Washington consist of a complex of deep channels cut into the basalt bedrock. The scabland topography is completely different from that produced by a normal drainage system. It is believed to have been produced by catastrophic flooding.

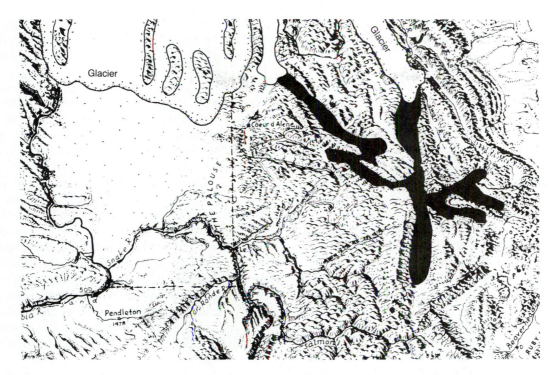

(A) The ice sheet in northern Washington blocked the drainage of the northward-flowing Clark Fork River to form Lake Missoula, a long, deep lake in northern Idaho and western Montana.

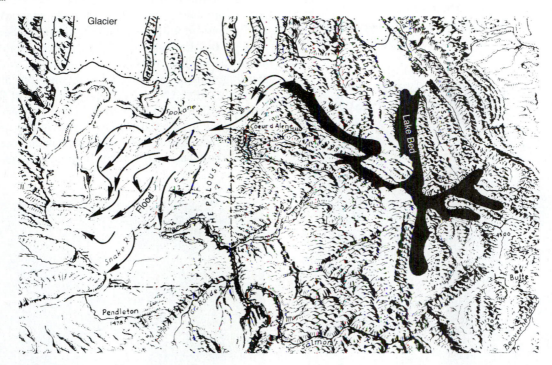

(B) As the glacier receded, the ice dam that formed Lake Missoula failed catastrophically, and water from the lake quickly drained across the scablands, eroding deep channels. Repeated advance and retreat of the glacier probably produced several ice dams that failed as the ice melted, and each caused catastrophic flooding.

Figure 8.53

The origin of the Channeled Scablands is attributed to catastrophic flooding caused by breaking of an ice dam.

melting of the ground ice produces some unique and spectacular landforms in the subarctic and arctic regions of Earth.

The permanently frozen ground is usually referred to as **permafrost**. It develops where the average annual temperature is 0°C (273°K) or below. When this happens, frozen groundwater penetrates deeper than the summer thaw does. The thickness of the permafrost zone, however, is limited because the temperature within Earth rises on the average of 1 K for each 30 m; ultimately, temperatures above freezing are reached. Ground ice may, however, extend to formidable depths, from between 250 and 350 m in North America to a maximum of approximately 1500 m in Siberia. At the present time, ground ice underlies approximately 26 percent of Earth's land surface, so it is a significant factor in developing the features of the polar regions.

A variety of landforms are produced in permafrost areas, depending upon the surface material, slope, and the thermal conditions in the near-surface layer. The surfaces of permafrost areas often display distinct geometric shapes collectively referred to as **patterned ground** (Figure 8.54). One distinctive type of patterned ground, which looks like a smaller version of the polygonal cracks found on Mars, is marked by polygons that may range from a few centimeters to more than 100 m across. Although their origin is complex, cracking of the surface is instrumental in their development. During exceptionally cold winters, the ground ice contracts because of thermal shrinking. Cracks are thus formed that divide the ground surface into roughly equidimensional cells, much like the shrinking of drying mud produces mud cracks. During the summer thaw, water seeps into the cracks, subsequently freezes, and forms vertical veins or wedges of ice. Repetition of the process over many centuries produces a network of ice wedges and an unmistakable ground pattern.

Melting of the ground ice and the migration of the meltwater commonly produces shallow irregular depressions called **thermokarst**. These features are similar in many respects to the sinkholes in normal karst regions, but the fundamental process is melting of ground ice rather than solution of rock material. Features interpreted to result from thermokarst are also abundant on Mars.

Sea Ice

In addition to glaciers and permafrost, the polar regions are unique in that the seas are presently covered with **sea ice** (Figure 8.55). Sea ice is formed by the freezing of seawater and is distinct from the ice in ice shelves and icebergs, which are floating masses of glacial ice formed on land and transported to the ocean. The thickness of floating sea ice is limited to only about 3 to 4 m because the ice acts as an insulating layer, preventing the loss of heat from the water below that is necessary for the deeper water to freeze. Beyond a given thick-

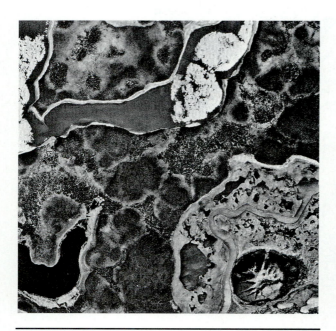

Figure 8.54
Patterned ground is common in Earth's polar regions and is formed by the alternate freezing and thawing of water in the ground's near-surface layers. The mound near the bottom left corner of the photo is an ice-cored hill called a pingo. Patterned ground and similar mounds may exist on Mars.

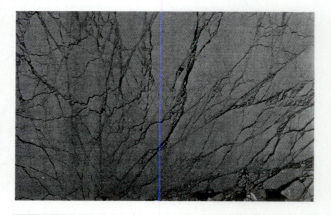

Figure 8.55
Ice covers much of the sea surface in Earth's polar regions. The ice is commonly fractured forming intricate patterns of dark lines seen in this photo. The ice is not thick (only 3 or 4 m) because as it thickens it insulates the underlying seawater from the frigid air above.

ness, the rate at which ice forms at the base of the plate of sea ice equals the rate at which ice is lost by melting and evaporation at the upper surface. Compared with the huge volumes of ice in the glaciers of Antarctica and Greenland, the thin veneer of sea ice may seem trivial, but its large areal extent has great importance on the heat exchange between the oceans and the atmosphere. Sea ice constitutes our best terrestrial example of the ice dynamics that operate on the icy satellites of the outer planets.

The sea ice is dynamic because it is subject to the force of waves, winds, currents, and tides and breaks up into fragments that move about. The individual fragments may be forced together with such great pressure that the ice margins buckle to form pressure ridges and rough hummocky surfaces; they are commonly subjected to tensional forces that cause them to split apart along fracture systems. Water rapidly moves up into the fracture to freeze and form a younger generation of ice. This movement is reflected in the complex fracture systems and multiple generations of ice that fill the cracks (Figure 8.55).

Geologic History of Earth

The rocks that make up Earth are a record of events. They record the interaction of energy and matter with the passage of time and tell us of the major processes that shaped the planet. From the principles of superposition, radiometric dating, and the fact that the terrestrial biosphere changes with time, it has been possible to establish a relative, as well as an absolute, time scale for Earth. Most of the original time scale was pieced together from sequences of rocks and fossils studied in Europe during the mid-nineteenth century.

Major units of geologic time were generally named after a distinctive characteristic of the time period they represent or of a geographic area where rocks of that age are well exposed. The standard *geologic column* for Earth is shown in Figure 8.56. The column provides a framework for understanding the great tapestry of events that is the history of Earth.

Four main subdivisions of Earth history can be defined:

1. Accretion (4.6 billion years ago) and internal differentiation accompanied a period of intense meteorite bombardment, perhaps including a giant impact with a body the size of Mars (Figure 8.57). No rocks are preserved unmodified from this stage (4.5 to 3.8 billion years ago).

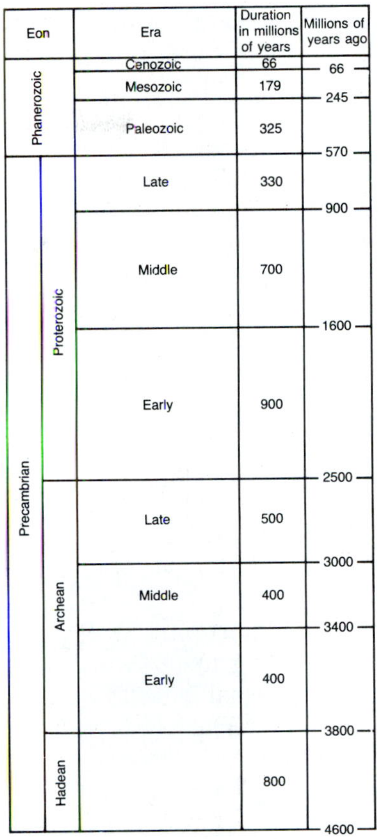

Eon	Era	Duration in millions of years	Millions of years ago
Phanerozoic	Cenozoic	66	66
Phanerozoic	Mesozoic	179	245
Phanerozoic	Paleozoic	325	570
Proterozoic	Late	330	900
Proterozoic	Middle	700	1600
Proterozoic	Early	900	2500
Archean	Late	500	3000
Archean	Middle	400	3400
Archean	Early	400	3800
Hadean		800	4600

Figure 8.56

The geologic time scale for Earth was originated in Europe during the midnineteenth century on the basis of the principles of superposition and the progressive changes that occurred in the fossil record. Later, radiometric dates provided a scale of absolute time for Earth.

2. Development of continental nuclei and the origin of life (3.8 to 2.5 billion years ago).
3. Stabilization of continental platforms and the oxidation of the atmosphere (2.5 to 0.6 billion years ago).
4. The modern plate tectonic period, typified by the fragmentation and construction of continents and the growth and destruction of ocean basins, as well as the expansion of life on land (0.6 billion years ago to present).

The evolution of Earth is summarized in Figure 8.58, which shows the thermal history of Earth. Figure 8.59 shows how the surface of Earth may have looked during the periods of time described below.

The Hadean Eon (4.6 to 3.8 Billion Years Ago). Earth, like the other inner planets, formed by the gravitational accretion of solid planetesimals that condensed from a gaseous nebula centered on the Sun. The accretion of millions of

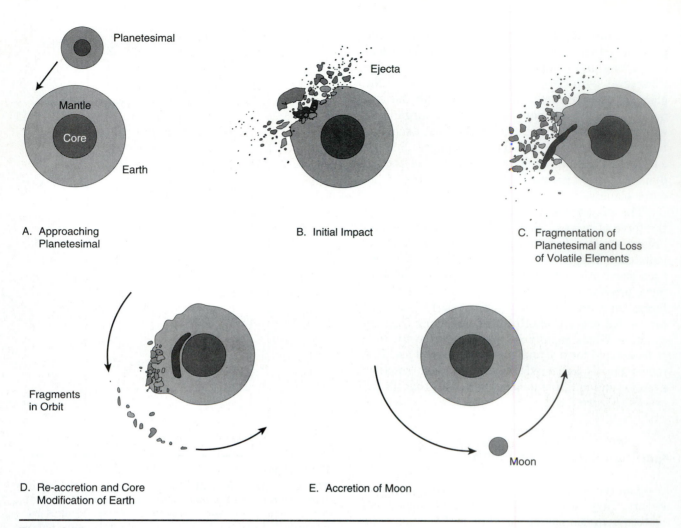

Figure 8.57

A giant impact during Earth's accretion may have ejected material to form the Moon and melted much of its outer portion to create a magma ocean and blow-off its atmosphere.

planetesimals a kilometer or so across occurred in a surprisingly short period by geologic standards, probably less than 100 million years.

The young Earth was probably relatively homogeneous; however, it was much hotter than it is today because of accretionary and radioactive heating. High internal temperatures, which had a profound effect on the future of Earth, resulted. By the time the growing Earth was about half its present diameter, it must have become hot enough for iron compounds to melt and form dense blobs that percolated down toward the interior of the planet to create a spherical metallic core. As a result, less dense silicates were displaced upward.

This rearrangement or differentiation of the interior also released further heat, adding to the effects of accretion and radioactive decay. By this time, much of the outer part of Earth must have been partially molten, perhaps forming a globe-encircling ocean of magma. Differentiation of Earth

continued as gases escaped from the interior to form an atmosphere of carbon dioxide, water, and nitrogen, similar in composition, if not bulk, to that of Mars and Venus. Eventually a cool crust formed on the upper surface of the magma ocean. Its composition was probably basaltic—that is, rich in iron and magnesium silicates and plagioclase feldspar. Simultaneously, the lower part of the magma ocean crystallized from its base upwards, forming a mantle rich in olivine or other magnesium silicates.

Sometime during this early accretion-differentiation episode Earth may have collided with a Mars-sized object (Figure 8.57). Upon impact, the temperature, as high as 6000 K, created by the conversion of kinetic energy to heat, would have vaporized much of Earth's crust and mantle. Moreover, a large crater penetrating nearly to the core may have formed. Most of the excavated materials temporarily thrown into space would fall back into

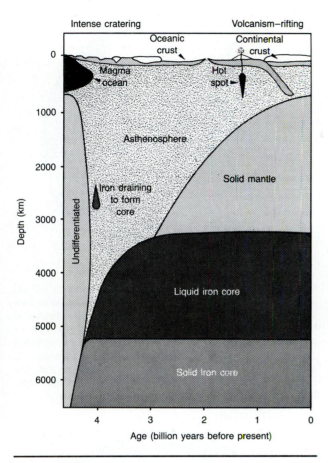

Figure 8.58

The thermal history of Earth is similar in many respects to that of Venus in that it formed by accretion and consequently became hot enough for internal differentiation to occur very early in its history. Following differentiation the planet has cooled, but its internal temperatures have been moderated by radiogenic heat. Because of its larger size, Earth cooled more slowly than Mars, and its interior is still hot, its lithosphere is thin, and volcanism is common. Following a very active period of volcanism (perhaps with the formation of a magma ocean), primitive crust formation, and recycling, Earth's lithosphere became thicker and stronger, and plates of lithosphere developed that were periodically pulled back into the mantle. The movements of these slabs of lithosphere carry the less dense continents with them. Plate tectonics may be unique to Earth.

the crater within a few minutes, but much of the vaporized material may have recondensed in orbit to form volatile-poor solids. Eventually, the Moon may have formed from this debris by a secondary accretion process. Iron from the core of the impacting body may have plunged rapidly through Earth's mantle, adding to the mass of the growing core. If a giant impact occurred, most of the outer part of the Earth must have melted to form a magma ocean, even if one had not formed earlier. Consequently, if any atmosphere remained after the impact it must have boiled away at this point.

Earth's early crust was probably very different from modern continental crust, but no trace of this primordial crust has been preserved. In fact, the oldest rocks so far found on Earth are "only" 3.8 billion years old. A crust of anorthosite like that which formed on the Moon probably did not form on Earth because plagioclase will not float in basaltic magma that contains a small amount of water. Magmas on the more volatile-rich Earth almost certainly were "wet" as compared to those on the Moon.

This period of time unrecorded by any existing rocks is called the Hadean Eon after the mythical fiery home of the dead—Hades—an appropriate name for this the hottest stage of Earth's evolution. Earth must have experienced a period of intense meteoritic bombardment during the first half-billion years of its history. This epoch of meteorite impact is clearly recorded on the surfaces of the Moon, Mercury, and Mars. In all probability, the surface of the young Earth appeared much like that of the Moon (Figure 8.59). The impact of small and large bodies probably created a densely cratered terrain over the entire planet.

This cratered surface was probably short-lived, however. Earth's early lithosphere was probably thin and its interior vigorously convecting so that any piece of solid crust formed was shortly swallowed back into the mantle. Large impacts, like those that formed the basins on the Moon, almost certainly penetrated Earth's thin crust and promoted this recycling and churning process. It may be no coincidence that the oldest rocks preserved at the surface of Earth (3.8 billion years old) formed shortly after the rapid decline of the bombardment rate discovered on the Moon (3.9 billion years ago). Before the decline in bombardment, the crust could not become stabilized at the surface.

The Archean Eon (3.8 to 2.5 Billion Years Ago). One of the most significant features of Earth is the presence of distinctive continental highlands whose origins can be traced back to about 3.8 billion years ago. Only small fragments of deformed and metamorphosed rocks are this old, but they demonstrate that light silicates developed and began to accumulate to form the continental crust by at least 700 million years after the accretion of Earth. The oldest rocks on Earth include metamorphosed sedimentary rocks, demonstrating that Earth had cooled enough so that liquid water existed on the surface of the planet by at least 3.8 billion years ago; in fact, the condensation of liquid water may have been much earlier than this. The oceans that now cover so much of Earth were derived from water vapor in the atmosphere after

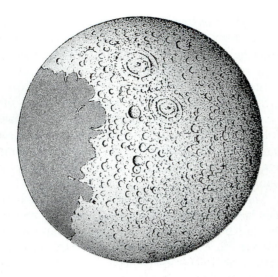

(A) Intense bombardment continued after formation of the planet by accretion of planetesimals. The planet differentiated, a solid crust formed, and an atmosphere and hydrosphere were outgassed. Meteorite bombardment and vigorous convection in the mantle recycled the primitive crust of Earth that probably formed over a global magma ocean, back into the mantle. No rocks from this or early ages are preserved on Earth.

(B) Embryonic continents developed, probably by partial melting of basaltic crustal rocks over hot spots, as shown in Figure 8.60, or in primitive island arcs. The continental blocks became stabilized at the surface by intrusion of low-density granitic rocks, so they could not be recycled back into the mantle. These highlands were immersed in an ocean of liquid water that was cycled across the continents as rain and in rivers, which eroded the highlands. Simple forms of life evolved in the oceans.

(C) Continental platforms developed and sedimentary rocks were deposited in shallow seas along their margins. Riftlike troughs cut the continents, suggesting that some type of plate tectonics began in this stage. Life expanded, producing oxygen in the atmosphere. Carbon dioxide was removed to form carbonate sedimentary rocks, reducing the pressure of the atmosphere dramatically.

(D) Plate tectonics persisted as lithospheric slabs participated in the convection of the mantle. The continents grew slowly, but rifting, mountain building, hot-spot activity, and climate change produced oscillations in the level of the sea, yielding cycles of marine deposition on the continents. Pangaea, a large supercontinent, developed by collision of smaller blocks of continental crust. Life rapidly diversified in the seas and eventually on land as well.

Figure 8.59

The history of Earth is summarized in these diagrams.

(E) The breakup of Pangaea about 300 million years ago followed as a result of rifting and the growth of a new ocean basin. The continents took on their present outlines.

it had cooled sufficiently to allow water to condense and rain to fall. The removal of water vapor from the atmosphere greatly reduced the pressure of the atmosphere, as well as its capacity to cause greenhouse warming. Over time, depressions (probably impact basins) filled with water, the first oceans began to take form, and a dynamic hydrologic system of rivers and oceans was fully established.

In these oceans, life evolved by about 3.5 billion years ago. Evidence of simple one-celled organisms is found nearly from the beginning of the rock record. Oxygen was produced by these primitive plants and as a result a protective ozone (O_3) layer began to form by the action of sunlight on oxygen in the upper atmosphere.

The terrains preserved from this period consist of metamorphosed accumulations of mafic lavas (greenstones) pierced and deformed by slightly younger granite plutons. Distinctive members of these ancient terrains are magnesium-rich lavas erupted at very high temperatures—an indication of the higher temperature of Earth during its early history. Earth has cooled by about 300 K during the last 3 billion years of its history (Figure 8.59). These ancient granite-greenstone terrain may have become stabilized at the surface by the presence of the low-density granites. However, most crust formed during the Archean Eon was recycled back into the mantle by vertical sinking or by incipient plate tectonics.

Perhaps the most important development in this stage of Earth history was the tremendous growth of continental crust during the latter part of this epoch. Between 3.0 and 2.6 billion years ago,

50 to 90 percent of the present volume of continental crust formed. Rocks of this age form Earth's shields and underlie the thin sedimentary veneer of the platforms. The mechanisms by which this low-density silicon- and aluminum-rich crust developed are still uncertain. The granitic portion of the continental crust may have formed as silica-rich magma was produced from basaltic crust. It is speculated that the primitive basaltic crust may have warped down into the warmer interior of Earth by the loading of younger volcanic rocks on its surface above a mantle plume or by down buckling above a mantle cold spot (Figures 8.59 and 8.60). The deep basaltic crust would have heated as it was forced to greater depths and perhaps melted to form granite that rose and intruded overlying greenstones creating the composite granite-greenstone terrains. Such a sequence is similar to the hypothesis developed for the origin of highland plateaus presently found on Venus. Others suggest that the silicic magmas that created the ancient continental crust were produced in modern-style subduction zones as lithospheric slabs were thrust into the planet's warm mantle. In any case, tremendous volumes of light, more silica-rich rocks were produced by reworking of older crustal material and accumulated to form protocontinents. These continents and the lithosphere that contained them were probably thin compared to present-day continents.

The presence of ancient metamorphosed sedimentary rocks clearly indicates that the temperature range at the surface of Earth from 3.8 billion years ago to the present was roughly constant. The temperature permitted water to exist in a liquid state. Earth was not hot enough for its oceans to evaporate (like Venus) or cold enough for them to freeze (like the satellites in the outer solar system). This nearly constant temperature range on Earth's surface is indeed a remarkable fact.

The great amount of geologic activity during this period, driven by the heat in the mantle, distinguishes Earth from the other planets. For example, the Moon's crust had long before become stabilized and its lithosphere thick; only the mare lavas were erupted during this period.

The Proterozoic Eon (2.5 to 0.6 Billion Years Ago).
A dramatic change in the character and behavior of the continents occurred about 2.5 billion years ago, shortly after the period of major crustal growth. Unstable, short-lived, mafic Archean crust was transformed into the stable, long-lived, more granitic crust typical of the Proterozoic and subsequent ages. By the beginning of this stage, the continents, although not in their

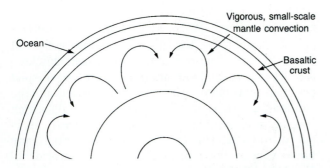

(A) A primitive basaltic crust develops over a convecting mantle.

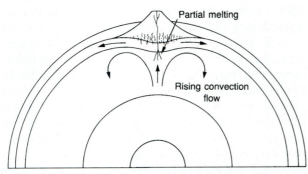

(B) In a rising mantle plume, partial melting of the mantle might have produced voluminous basalt. As the weight of the overlying volcanic rocks pressed the basaltic crust downward, it became progressively hotter, until the melting point of the crust was reached. Partial melting of the basalt may have produced low density granite, parental to continental crust.

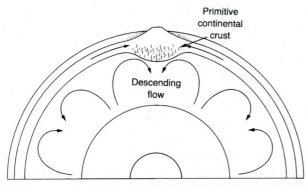

(C) Alternatively, partial melting of basaltic crust above a descending flow produced more silica-rich magmas with lower densities. These granitic plutons made the crust difficult to assimilate back into the mantle. Small landmasses emerging above the sea were weathered and eroded, producing sediments adjacent to the islands. In this way small islands of metamorphosed igneous rocks, and sediment derived from them, formed small continental embryos, which were later swept together to form the continents.

Figure 8.60

The origin of the first continents on Earth is a matter of conjecture, but two different hypotheses are shown here.

present shape or distribution, had roughly their present volume, thickness, and stability against recycling back into the mantle.

This transition is marked by a significant increase in the abundance of carbonate—sandstone—shale sedimentary rocks that are typical of modern continental margins covered by shallow seas. Most modern platforms are veneered by similar, but undeformed, rocks. Accumulations of sedimentary rocks of this age range up to about 10 km thick; similar thicknesses are found in modern environments, as for example on the rifted margins of continents.

The first well-defined continental rifts also appear to have developed at this time and indicate that the crust was brittle, relatively thick, and probably cooler than in the earlier Archean Eon. During the Proterozoic Eon, large fault-bounded rifts formed and filled with sediments and volcanic rocks. These troughs are similar in many respects to the failed arms of rifts found on the present continents. These rifts may be key features indicating the existence of modern-style plate tectonics by about 2 billion years ago. Many abortive attempts at continental rifting occurred and left vast dike swarms and linear belts of distinctive complexes of mafic and granitic rocks.

These important changes may reflect differences in the energy content of Earth as it cooled and may be related to a thicker and stronger lithosphere, as well as a more coherent layer of continental crust. In spite of a near constant atmospheric temperature, the interior of Earth appears to have cooled about 100 K for each billion years of its early history. Lithospheric thickness probably changed in a complementary fashion, with the Hadean lithosphere perhaps only 10 to 20 km in thickness, increasing to perhaps 25 to 40 km thick by the early Proterozoic. As Earth's lithosphere became thicker and more rigid over large distances, tectonic plates became defined. Subduction of these plates back into the mantle must have begun when old, cold lithosphere became dense enough to fragment and sink back into the mantle, pulling the rest of the plate with it down a newly defined subduction zone and corresponding trench. The density difference between thin, young lithosphere and old, thick lithosphere is an important driving force for plate tectonics, one amplified by metamorphic changes that create denser minerals in the subducting slab. Segments of the lithosphere rich in low-density rocks like granite were not only high standing but also were nearly impervious to subduction. Thus, the transition from the Archean to the Proterozoic marked a fundamental threshold in Earth's history and led to the development of modern crustal conditions and plate tectonics.

Earth's atmosphere was created by volcanic activity but changed in response to the evolution of the biosphere and changes on the continents. Vast amounts of carbon dioxide were probably removed during this or an earlier time and trapped in the limestones preserved on the continents. If all the carbon dioxide in carbonate rocks were released to the atmosphere, it would exert a pressure about 50 times greater than the pressure of today's atmosphere. Increasingly complex forms of oxygen-producing life developed during the Proterozoic Eon, and consequently the level of free oxygen in the atmosphere increased. As a result of the buildup of oxygen, the iron dissolved in the oceans precipitated in shallow basins on the margins of continents to make delicately banded deposits of oxidized iron.

The Phanerozoic Eon (0.6 Billion Years to Present).

The tectonic and hydrologic systems continued to operate, resulting in slowly enlarging continental mass and the growth and destruction of ocean basins. Each continental shield was eroded to near sea level and was in gravitational equilibrium. As the continents drifted with the lithospheric plates, slight changes in sea level permitted shallow seas to expand over the shields, and extensive layers of horizontal sedimentary rocks were deposited on the continents to form the stable platforms. These rocks contain abundant fossils, which record a dramatic diversification of life after about 600 million years ago. All major groups of animals are represented in these rocks, although marine invertebrates dominate. These fossils also reveal changes in local temperatures and geographic settings that correspond to plate movements through the various climate zones on Earth. By about 300 million years ago, plate tectonics developed a huge supercontinent, called *Pangaea*, or "one earth" (Figure 8.61). The southern continents were grouped near the South Pole, where glaciation occurred. North America and Europe were in the equatorial zones where lush swamps flourished that later developed into extensive coal deposits.

The history of Earth during the last 300 million years is known in much greater detail than that of the earlier Earth; by combining a variety of geologic evidence from various continents and ocean basins, we are able to reconstruct the sequence of events that resulted in the breakup of Pangaea and the development of the present continents and ocean basins. The results of this synthesis are

100 million years ago

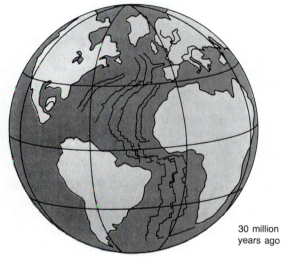

30 million years ago

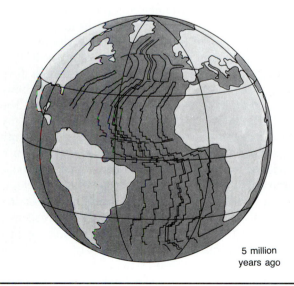

5 million years ago

Figure 8.61

The breakup of Pangaea and the evolution of the Atlantic Ocean.

shown in Figure 8.61. The initial event was the splitting of the continents along a north–south rift above a mantle plume. The fractures that formed enlarged to form a rift that became an oceanic ridge in the Atlantic Ocean. Africa and India moved northward and collided with Europe and Asia to form the Alps and the Himalaya Mountains. The westward drift of North and South America over the Pacific plate and resulting collisions with small island arcs and other far-traveled fragments resulted in the growth of the continents' western portions and produced the Sierra Nevada, Rocky, and Andes mountain chains, which have strongly deformed sedimentary and metamorphic rocks intruded by igneous rocks.

During this stage of Earth's history, significant changes occurred in the animal and plant communities. The periodic extinctions of many forms of land and marine life that occurred over relatively short periods of time have been used to subdivide Earth's geologic history. For example, dinosaurs flourished from about 200 million years ago and became extinct about 65 million years ago. Mammals then became the dominant animals on land. As noted earlier, some geologists relate the demise of dinosaurs and other forms of plant and animal life to rapid changes in climate induced by a meteorite impact known to have occurred at this time. However, many others think the impact and the extinctions are merely contemporaneous events and not related in a cause-and-effect fashion. Evidence of meteorite or comet impact has not been found at all horizons in the rock record where mass extinctions are known to have occurred.

While the continents drifted toward their present positions, they set the stage for the most recent major event in geologic history, the Ice Age. Antarctica drifted over the South Pole possibly as long as 30 million years ago and has been in a deep freeze ever since. The northern continents, however, form a patchwork of land and sea around the North Pole, and climatic conditions controlled by orbital parameters and changing oceanic currents and atmospheric composition fluctuated to produce a series of four major glacial and interglacial epochs (Figure 8.62). During the Ice Age, the hydrologic system was interrupted significantly. Much of the precipitation on the land, which normally returned directly to the sea, became trapped on the continents as glacial ice. Consequently, sea level dropped as much as 100 m over the entire globe. The moving ice scoured the landscape, forced the crust to subside under its weight, and created a myriad of lakes. The area beneath the ice was completely depopulated of all life, and both plant

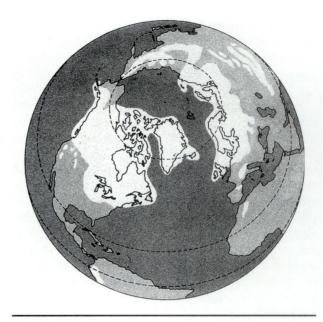

Figure 8.62

The amount of ice on Earth's continents has fluctuated greatly as the result of cyclical changes in climate. The advance of these continental glaciers greatly changed the face of the northern hemisphere. The extent of ice sheets (white) from the last ice age is shown on this map.

and animal communities were forced to migrate before the changing margins of the ice sheets.

Conclusions

The grand central theme of Earth's history is change—constant and sometimes rapid change resulting from the great energy sources of internal heat and solar radiation. Driven by the planet's attempt to rid itself of the heat within, the tectonic system of Earth has continued uninterrupted from the beginning. The style of lithospheric recycling may have changed, but the system has continued to operate. Unlike the Moon, Mercury, and Mars, Earth has been resurfaced millions of times; ocean basins have been created and destroyed; continents have grown, drifted together, and rifted apart. Erosion, accomplished by the gravity-driven flow of surface fluids (water, wind, and ice), constantly renewed Earth's surface, and volcanism has been vigorous and long-lived. These factors make Earth unique among the planets, but much more similar to Venus than to any of the others. The Moon, Mercury, Mars, and the satellites of the outer planets have experienced only the initial stages of a tectonic system; throughout most of their histories they have remained much as they were soon after the creation of the solar system. The larger size of

Earth and its substantially slower cooling rate produced many of its distinctive characteristics. Geologic processes on these other objects do not approach the dynamic activity of Earth's volcano-tectonic system. No other planet supports a liquid ocean of water or an atmosphere rich in oxygen. Perhaps most important, no other planet is known to harbor life.

Review Questions

1. In light of Earth's vigorous interior and high internal temperatures, why are the volcanoes on Mars so much larger than those found on Earth?

2. Where did the free oxygen in Earth's atmosphere come from? Why are the atmospheres of Mars and Venus carbon-dioxide-rich and oxygen-poor? What happened to the carbon dioxide that once resided in Earth's atmosphere?

3. What are the major differences between the crust of Earth and the crust of the Moon? What events or processes caused this dramatic difference in crustal properties?

4. What are the major differences between the crust of the continents and the crust of the ocean basins?

5. Compare Earth's hydrologic cycle (using diagrams) with that found on Mars. What is the major source of energy for this system?

6. How did the oceans form on Earth? What changes have occurred in the ocean during Earth's history?

7. How does a river system grow longer?

8. Describe the important elements of a typical terrestrial river system. How do these compare with the channels found on Mars?

9. How do springs originate? Are they important erosion agents on Earth's continents?

10. Compare the glacial ice caps found on Earth with the ice caps of Mars in terms of their composition, age, erosive power, and variability.

11. Has sea level remained constant through the history of Earth? What causes sea level changes? How does a continent change with sea level rise, with sea level drop?

12. Give an example of a flowing solid at the surface of Earth. Do you think the movement of this material may be important on any other worlds in the solar system?

13. What controls the distribution of the major deserts of Earth?

14. What evidence is there that Earth's crust and lithosphere are in motion?

15. Describe with sketches the fundamentals of Earth's tectonic system. What is the major source of energy for this system?

16. How is the type of volcanism on Earth related to the tectonic setting of the volcano?

17. What is the origin of the linear chains of volcanic islands found on the ocean floor? How would volcanic activity differ if a mantle hot spot were active on Mars?

18. Explain convection in Earth's interior. Is the mantle necessarily liquid?

19. In what ways are the rocks of Earth's oceanic crust similar to those of the lunar maria?

20. Why are the oldest rocks on Earth found on its continents?

21. Why did continents develop on Earth and not on the Moon, Mercury, or Mars?

22. Explain why the oldest rocks found on the Moon are about 700 million years older than those found on Earth.

23. Draw a sketch of the interior structure of Earth and briefly discuss the core, mantle, asthenosphere, lithosphere, and crust. At what stage of Earth's history did its layered structure develop?

24. Is it plausible to think that meteorite impact has shaped not only the physical evolution of Earth but its biological evolution as well?

Key Terms

Abrasion	Granite	Plucking
Alluvial fan	Ground Ice	River
Ash Fall	Hot Spots	Sand Seas
Ash Flow	Ice Age	Sea Ice
Ash-Flow Shield	Ice Shelves	Seamount
Batholith	Island Arc	Shield
Blowouts	Isostasy	Shoreline
Braided River	Karst	Sinkhole
Composite Volcano	Lag Deposits	Stable Platform
Continental Glacier	Loess	Stream Valley
Continental Rift	Magnetosphere	Strike-Slip Fault
Continents	Mass Extinction	Subduction Zone
Convergent Boundary	Mountain Belt	Thermokarst
Deflation Basins	Ocean	Tidal Force
Delta	Ocean Basins	Transform Boundary
Desert	Oceanic Ridge	Transform Fault
Divergent Boundary	Pangaea	Trench
Flood Plain	Patterned Ground	Valley Glacier
Folded Mountains	Permafrost	Weathering
Glacial Dam	Plate Tectonics	

Additional Reading

Hamblin, W. K., and Christiansen, E. H. 1995. *Earth's Dynamic Systems*, 7th Edition. New York: Prentice Hall.

Cloud, P. 1987. *Oasis in Space*. New York: W. W. Norton and Company.

TABLE 9.1

Physical and Orbital Characteristics of Jupiter

Mean Distance from Sun (Earth = 1)	5.20
Period of Revolution	11.86 y
Period of Rotation	9.8 h
Inclination of Axis	3.08°
Equatorial Diameter	143,800 km
Mass (Earth = 1)	317.8
Volume (Earth = 1)	1,335
Density	1.3 g/cm^3
Atmosphere (main components)	H_2, He
Temperature (at the cloud tops)	125 K
Gravity (Earth = 1)	2.34
Magnetic Field (Earth = 1)	20,000
Known Satellites	16

Io Eruption Plumes

© TB 93

CHAPTER 9

The Jupiter System

Io

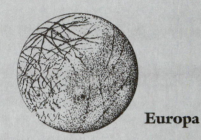

Europa

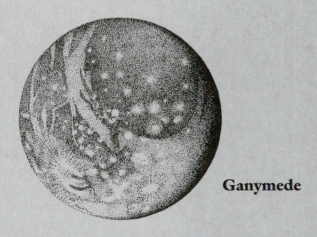

Ganymede

Earth

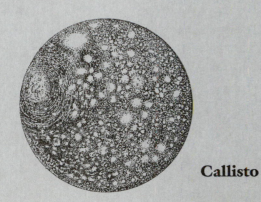

Callisto

Introduction

Jupiter and its moons form a planetary group of incredible beauty (Figure 9.1).

The giant planet has a volume 1300 times greater than the Earth, but its near-surface layers are composed mostly of gases swirling in complex patterns. Orbiting the Sun beyond the asteroid belt, Jupiter with its set of at least 16 orbiting satellites is the center of a small planetary system. Jupiter is the largest planet in the solar system; if it had been about 100 times more massive, Jupiter might have evolved into a star and our solar system would have had two suns, like many stellar systems in our galaxy that contain two or even more stars. As it is, Jupiter radiates more energy than it receives from the Sun. The heat emitted from it may have been sufficient to alter drastically the composition of its satellites as they formed from the condensing solar nebula.

Jupiter and the rest of the giant planets are fundamentally different from the rocky inner planets. Besides being larger, they are composed predominantly of gas and have no solid surfaces at all. A vast amount of data has been returned to Earth by several flyby spacecraft, including Pioneers 10 and 11 (1973 and 1974) and Voyagers 1 and 2 (1979). As planetary bodies they are intriguing, together with their large systems of moons and rings.

Major Concepts

1. Jupiter, a giant gas- and ice-rich planet of the outer solar system, is the center of a system of at least 16 ice and rock satellites and a narrow ring of much smaller particles. Part of the reason Jupiter became so much larger than the rocky inner planets is that it formed in a cool region of the ancient solar nebula where water ice was stable.

2. Jupiter consists principally of hydrogen and helium; perhaps a core of silicates and ice is embedded deep within the planet. The brightly colored atmosphere is banded and has large semipermanent storm systems.

3. Io, the innermost Galilean satellite, has a very young surface and is presently volcanically active. Tidal flexing of this moon provides a continued input of energy to melt parts of the interior of this rocky planet.

4. Europa has a relatively smooth, but fractured, icy surface. No vestiges of the intense bombardment remain on Europa; only a few small craters have been identified. Resurfacing by the outpouring of watery lavas and the flow of solid ice have shaped the surface features. Tidal flexing of Europa has kept the interior warm.

5. Ganymede, the largest Galilean satellite, has a varied surface dominated by heavily cratered terrain and large swathes of bright younger, intensely grooved terrain, apparently formed by foundering of old crust and flooding with volcanic flows of liquid water.

6. Callisto, the outermost of the large satellites, is dominated by heavily cratered terrain. This moon did not expand during its late history.

7. The satellites of Jupiter appear to have condensed in a thermal gradient centered on Jupiter. Therefore, the inner moons are refractory, silicate-rich, and ice-poor, whereas the outer moons are ice-rich.

Jupiter and Its Satellites

The characteristics of Jupiter, which orbits the Sun at a distance five times that of Earth, exemplify many features of the **outer planets** (those that lie beyond the orbit of Mars). Like the other outer planets, Jupiter is huge. It has a diameter of 143,800 km, more than ten times that of Earth. Its mass is about 318 times that of the Earth, but its volume is more than 1300 times Earth's. Therefore, Jupiter's density is very low—1.33 g/cm³, just slightly more than water. (By contrast, the Earth has a density of 5.5 g/cm³.) Obviously, Jupiter's composition must differ greatly from Earth's or from that of any other inner planet. It is composed mainly of hydrogen and helium, the two lightest elements, and has only a small core of rock and ice, in sharp contrast with the Moon and Mercury, which have no atmospheres and are just small rocky bodies. Jupiter's composition is like that of the Sun and to that of the original nebula from which the planets formed.

Other significant facts about Jupiter include its system of at least 16 sizeable moons and its rapid period of rotation. One day at the top of the cloud layer is only about 10 hours long. The rotational velocity at this level is therefore much greater than it is at the surface of the much smaller Earth. This rapid rotation helps to produce the slightly flattened shape of the planet and its huge magnetic field, as well as the tremendous storm systems seen in the atmosphere.

The Interior of Jupiter

The dramatic difference in the bulk composition of Jupiter from the inner planets we have already examined is expressed in the nature of its

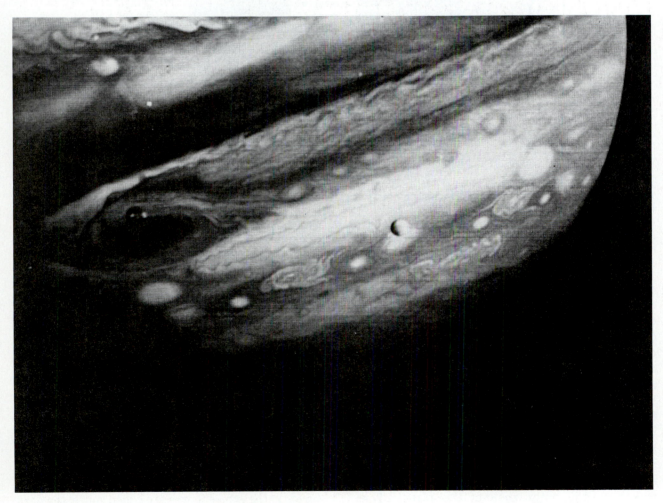

Figure 9.1

Jupiter and two of its moons as seen from Voyager 1 on February 13, 1979, at a distance of about 20 million km. Jupiter is huge by terrestrial standards. The Red Spot (in the lower left) is larger than Earth. Io (to the left) and Europa (to the right) are each the size of Earth's Moon.

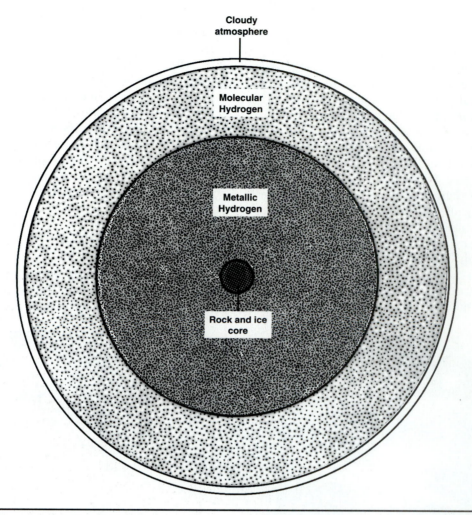

Figure 9.2

A model of the internal structure of Jupiter is based on measurements by Pioneer 10 and the Voyager spacecraft. Jupiter consists largely of liquid hydrogen, with a small solid core of rock and ice and an extremely thick atmosphere of hydrogen and helium.

churning interior. The internal structure of Jupiter may appear as depicted in Figure 9.2. A small, solid core, composed of water ice and rocky materials (mostly silicate minerals and iron) is shown at the center. This core is probably several times larger in diameter than our own planet. The core is surrounded by a 30,000- to 40,000-km-thick layer of hydrogen in a liquid **metallic** state (H⁺) because of the high pressures. When combined with Jupiter's rotation, convective motions in this layer of ionized hydrogen may create the tremendously powerful magnetic field around the planet. Another layer of molecular hydrogen (H$_2$) 14,000 to 20,000 km thick makes up the outer shell of the planet. Thus, Jupiter really has no solid surface—only a gradual transition zone between liquid and gaseous hydrogen. The brightly banded "surface" seen from Earth and photographed by passing spacecraft is actually a thick cloud layer, composed mostly of crystals of ammonia ice. Although hydrogen is the

most abundant gas (over 80 percent by mass), helium (about 18 percent) and small amounts of water, methane, ammonia, and various carbon-hydrogen compounds have also been detected in the atmosphere. The processes on such a planet are drastically different from those previously considered. No surficial geologic structures exist; since there is no crust, no impact craters, no volcanoes, or other geologic features can be produced.

Observable features are dramatic if ephemeral, and the images returned of Jupiter are among the most beautiful of any taken during the exploration of space (Figure 9.1). The light, delicately colored bands in the atmospheres are called zones; the dark bands are called belts. These semipermanent structures are similar to terrestrial cyclones and anticyclones, masses of rising and falling air produced by the circulation of Earth's atmosphere. On Jupiter these cloud patterns are stretched into thin belts by high-velocity winds. Other visible features include loops, streaks, plumes, and (most notably) spots; these are the result of shearing between the opposing wind directions of each band. The best known is called the Great Red Spot; it varies in appearance but seems to be a relatively permanent group of storms, a cyclonic disturbance like a terrestrial hurricane. There is at least one other such feature that is invisible from Earth. It is much smaller than the Red Spot and lies in the northern hemisphere.

The Rings

Although the 1979 Voyager mission had many highlights, one of its more important results was the discovery of a **ring** around Jupiter that is similar to but much smaller than the familiar rings of Saturn (Figure 9.3). The ring lies in Jupiter's equatorial plane and appears to be almost 6000 km wide and about 30 km thick and may be banded like Saturn's. Since it is so thin and dark, it was not discovered through telescopes on Earth. The ring probably consists of a swarm of micron-sized particles of silicate dust or dirty ice. The outer margin is sharp, whereas the inner margin grades imperceptibly toward the planet's atmosphere. Apparently, some of the material from the rings spirals inward to the planet because of changes in the magnetic field or the slight pressure of sunlight.

The ring particles may be formed by collisions on the two small satellites Adrastea and Metis, which lie near the outer edge of the ring. These satellites are nearly **co-orbital,** meaning that they share the same orbit. Both satellites are dark and small, with diameters of a few tens of kilometers. Collisions and sputtering at the surface of these bodies may send showers of small particles spiraling inward to Jupiter's atmosphere. (Sputtering refers to the ejection of atoms from rocks or regolith exposed at the surface of an airless planet. Ejection is caused by the impact of solar wind ions on the surface.) If such is the case, then the ring is being continually or episodically regenerated by new surges of material that gradually drain into Jupiter's gravitational well. The origin, evolution, and implications of planetary ring systems are discussed in Chapter 10.

The Galilean Satellites

The photographs of Jupiter and its moons sent back to Earth from the Voyager spacecraft in March and July of 1979 reveal a wide range of differences between the planetary bodies of the

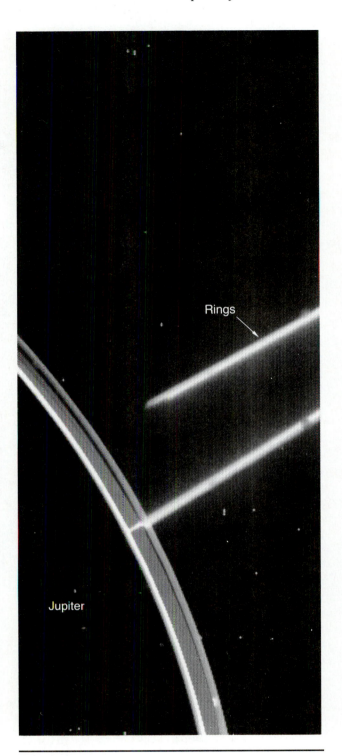

Figure 9.3

The rings of Jupiter were discovered by Voyager 1 and excellent photographs were later taken by Voyager 2. The individual rings are probably made up of very small dust-size fragments of rock or ice that orbit Jupiter. If collected together to form a single body, they would form an object less than twice that of Jupiter's tiny moon, Amalthea, which is 270 km by 155 km in size.

outer and inner solar system and unmask four new worlds for geologic exploration. The Galilean moons (the four large moons of Jupiter discovered by Galileo in 1610), previously known as little more than specks of light in a telescopic field of view, have now been photographed with resolution sufficient to show objects as small as 600 m across. The Voyager cameras have found that these moons are not only complex but are also strikingly different, with their surfaces varying greatly in age, composition, appearance, and history (Figure 9.4). In order outward from Jupiter, the Galilean moons are Io, Europa, Ganymede, and Callisto. We will focus our attention on these four fascinating bodies, each unique and totally different from the planetary bodies of the inner solar system. Table 9.2 compares the four moons in terms of their basic physical properties. Their sizes are similar to terrestrial planets, but their formation in the icy outer solar system and their relationship to Jupiter have produced decidedly nonterrestrial geologic features and unique, fascinating histories. No samples have been returned from the satellites of Jupiter, their surfaces are incompletely photographed, and the best of the images have resolutions of only a few hundred meters. It is thus not surprising that geologists have not yet agreed upon interpretations of many features observed on these moons or upon their histories. In most cases, we have tried to present what appears to be the most reasonable explanation at the moment. Alternative explanations are discussed as well, demonstrating that the progress of science involves gathering observations and formulating hypotheses, sometimes multiple ones, followed by further testing with new observations. It seems clear that exploration of the moons of Jupiter will not only provide an exciting new understanding about the Jupiter system but, through comparative planetology, new insight into our own planet.

Io

Io, the innermost of the Galilean satellites, has intrigued astronomers for years because of its perplexing physical properties. Its yellowish color was reported near the turn of the century. Occasionally it shows brightness variations visible with telescopes. Spectral studies suggest that its surface is covered by sulfur compounds and that water ice is extremely rare. The satellite is surrounded by a very thin nebula, not really an atmosphere, which has a yellow glow characteristic of sodium. It also contains the volatile elements sulfur, oxygen, and potassium, apparently ejected from the surface of Io by the solar wind. All of these peculiar properties increased the desire of many to see pictures of its surface and attempt to decipher these puzzles. The trajectory of Voyager 1 permitted a very close encounter with Io, and numerous images with resolution of about 600 m were obtained. As it turns out, the exploration of Io has been one of the most spectacular and significant results of the Voyager mission.

As can be seen in the global views of Io (Figs. 9.5 and 9.6), the planet is unlike anything yet examined in the solar system, even though it is similar in size and density to the Moon. The surface displays a variety of pastel shades of yellow with hints of orange, mottled with irregular patches of white, gray, and black. The first impression one receives is that the surface color is a mass of confused splotches, but upon closer examination one finds some degree of order. The equatorial region is bright and has large irregular patches of white, whereas the midlatitude regions are darker, and the poles seem to be covered with a bluish frost. Numerous small dark spots are scattered across the surface, giving the planet its distinctive character. At first, the dark spots were thought to be impact craters, but images of higher resolution revealed that the dark pits and markings bear little

TABLE 9.2
Characteristics of the Major Satellites

Satellite	Diameter (km)	Density (g/cm³)	Surface Composition
Io	3,630	3.57	Silicates and sulfur
Europa	3,138	2.97	Water ice
Ganymede	5,262	1.94	Water ice
Callisto	4,800	1.86	Water ice
Earth's Moon	3,500	3.30	Silicates

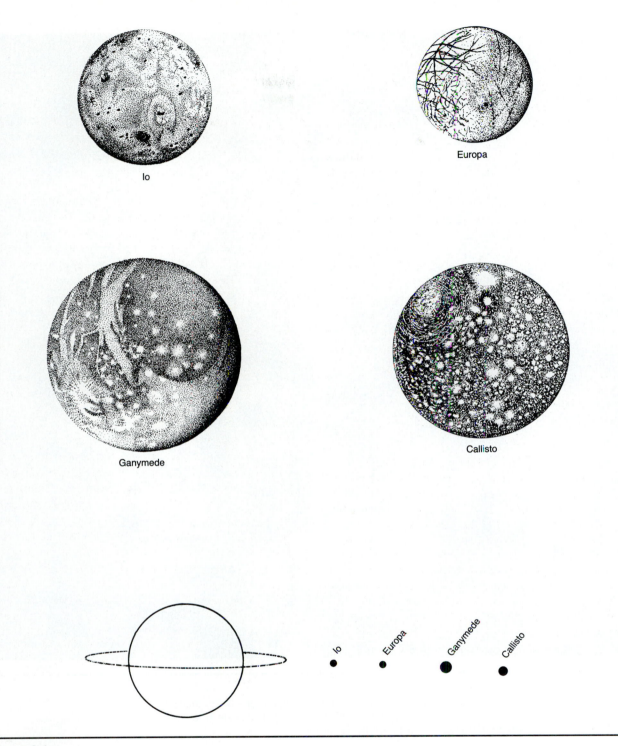

Figure 9.4

The four large, or Galilean, moons of Jupiter are shown here to scale. They are roughly the size of the Moon and are found to be as diverse and fascinating as the terrestrial planets. The orbital positions of the moons are shown in the lower sketch.

resemblance to impact craters. Unlike the Moon, Mercury, Mars, or even the Earth, no impact craters have been identified on Io. The complete absence of impact craters indicates that Io is being continually resurfaced by erosion and or deposition and that all of the features we now see are probably less than one million years old.

Volcanic Features. Close inspection of high-resolution images shows that the surface of Io is dominated by volcanic features. Indeed, Io is believed to be the most volcanically active body in the solar system. The volcanic features identified include volcanic pit craters, larger calderas, lava flows, low domes, a variety of shield volcanoes,

Figure 9.5

Io is perhaps the most spectacular of all the moons of Jupiter. Its surface is dominated by recent volcanic features. The surface is covered by sulfur and sulfur dioxide frost, which create the striking orange and yellow coloration of the moon.

volcanic ash plains, plus deposits of sulfur and other volatile materials. The principal volcanic features include over 200 calderas. These irregular pits may be as large as 200 km across and are in many ways similar to the large calderas related to ash-flow shields on Earth's continents. Most calderas do not appear to cap high volcanic structures as they do on the lava shields of Mars, and they may represent a fundamentally different type of eruption mechanism. Commonly, the calderas are surrounded by diffuse halos or by a variety of bright and dark patterns. Some calderas do have complex radiating volcanic flows that extend out from the vent area in a sinuous pattern. The lava flows that are associated with about half of the calderas are long, narrow flows with a variety of colors ranging from black to yellow and brown (Figure 9.7). In general, they appear to be similar to some of the basaltic flows on Hawaii. By terrestrial standards, however, the flows are very large with lengths of several hundred kilometers and widths of several tens of kilometers being very common.

One of the most spectacular results of the Voyager mission was the discovery of active volcanic eruptions on Io. At least nine volcanoes in the throes of violent eruption were photographed during the brief encounters. These were identified by plumes of eruptive material seen against the black sky on Io's horizon and by large brilliant flares seen in the twilight zone (Figure 9.8). If a spacecraft were to pass Earth, even with all of its volcanoes, it is unlikely that even one would be experiencing an eruption of the magnitude of those seen on Io. In addition to the nine actively erupting volcanoes photographed by Voyager, 11 volcanoes showed signs of very recent eruptions in the form of a series of diffuse rings surrounding the dark centers of eruption. All indications are that Io is the most thermally active body in the solar system and that it must have many large subsurface magma chambers.

The volcanic eruptions on Io appeared as enormous umbrella-shaped plumes spraying material approximately 300 km above the surface and measuring 1000 km across. These are enormous eruptions indeed. The internal structure in the ejecta plume was revealed by ultraviolet light, in which zones of ejected gases and fine-grained pyroclastic material can be seen extending far beyond a central region of coarse fragmented material (Figure 9.9). The morphology of the sprays of ejected material from the various volcanoes is quite variable. In addition to the umbrella-shaped plumes, some ejecta is thrown out in jetlike streaks erupting from fissures up to 225 km long. Other eruptions occurred in a diffuse manner with no well-defined central fountain. It is interesting to note that significant changes in one volcanic spray were detected within a 5 1/2-hour period, suggesting fluctuation in the eruptive energy during that time.

The geometry of the ejecta plumes appears to be related to the type of eruption. The jetlike streaks are probably fragmented material erupted from fissures. The umbrellalike plumes develop from central vents, with the outer plumes, which are detected only by ultraviolet light, being volatile material. The gas may consist predominantly of sulfur or sulfur compounds such as H_2S and SO_4 (sulfur compounds are common in terrestrial volcanic gases as well). On reaching the surface these volatile gases become so cold that a fraction rapidly condenses into fine, solid particles, which fall back to the surface as a film that mantles older volcanic deposits. Prominent bright rings surround some of the calderas, and it has been suggested that they consist of a thin coating of sulfur and other salts produced in this fashion. So much sulfur is released from Io's interior by this process that it has created an extremely thin atmosphere that eventually escapes into space and creates a doughnut-shaped ring, or torus, that encircles Jupiter, through which Io orbits. It is very possible that Io's surface is composed predominantly of sulfur and sulfur compounds ejected from the volcanoes. Even its magmas might be liquid sulfur, not the compounds of silicon and oxygen typical of terrestrial and even lunar rocks.

Large, explosive eruptions like those observed on Io often produce ash flows that consist of a dense cloud of volcanic ash or dust that rushes away from the center of eruption. These ash flows are very mobile and produce thin, sheetlike deposits in an apron around the volcano. Individual flows may be hundreds of kilometers long and over 10 km wide. It may be that the colored halos (Figure 9.9) around some of the calderas are ash-flow deposits, in contrast to the lava flows erupted from quieter volcanoes like those in the Hawaiian Islands and from volcanoes on several of the other planets. Calderas often result from explosive ash eruptions, as large magma chambers are rapidly emptied. As a result, the roofs collapse, filling the voids created by eruption. The surrounding plains may also consist of successions of ash flows.

The vast tracts of smooth plains between the volcanoes are called **intervent plains**. These plains are composed of visible layers, the margins of which are marked by a variety of irregular scarps and unusually uniform relief. Some scarps appear to be flow fronts; others are irregular and jagged, suggesting scarp retreat as a result of some erosional process. Others are obviously fault scarps, which mark the margins of relatively long, straight, narrow grabens.

Unlike the other Galilean satellites, Io has no water, but studies of Io's **spectral reflectance** (color) suggest a variety of sulfur-rich materials at the surface. Sulfur is a complex substance occurring in numerous vividly colored forms including reds, yellows, browns, and blacks; these probably account for Io's dramatic coloration.

Tectonic Features.   The undeniable presence of active volcanism on Io demonstrates that high temperatures exist within the planet—its heat engine is still alive. This thermal energy may drive an active tectonic system. Geologic evidence that this is the case is found in the photographs of its surface (Figure 9.10). Many of the plains units are

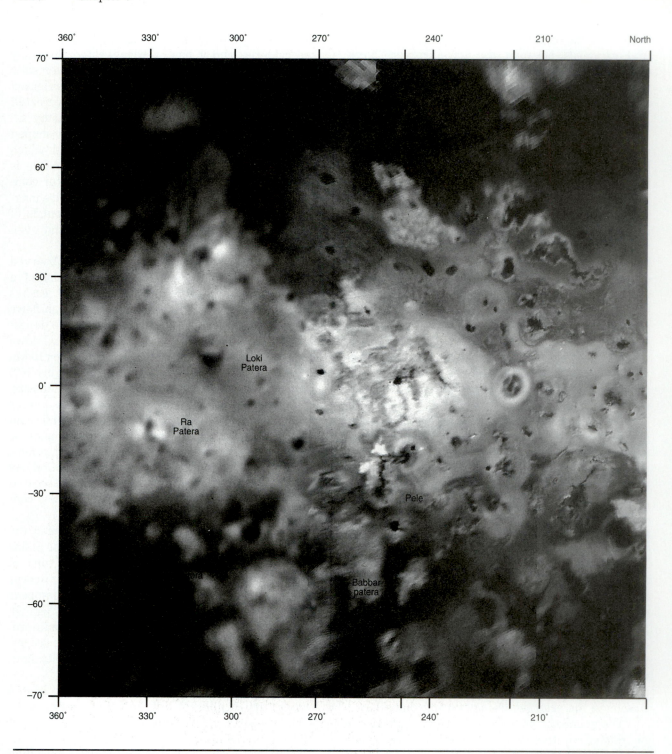

Figure 9.6

This shaded relief map of Io shows a unique surface dominated by volcanic features. The pockmarks are mostly volcanic calderas as much as 200 km in diameter. The light patches are plains covered with volcanic ash and sulfurous frost. Mountains and plateaus exist near the poles, some rising 8 km or more above the surrounding plains.

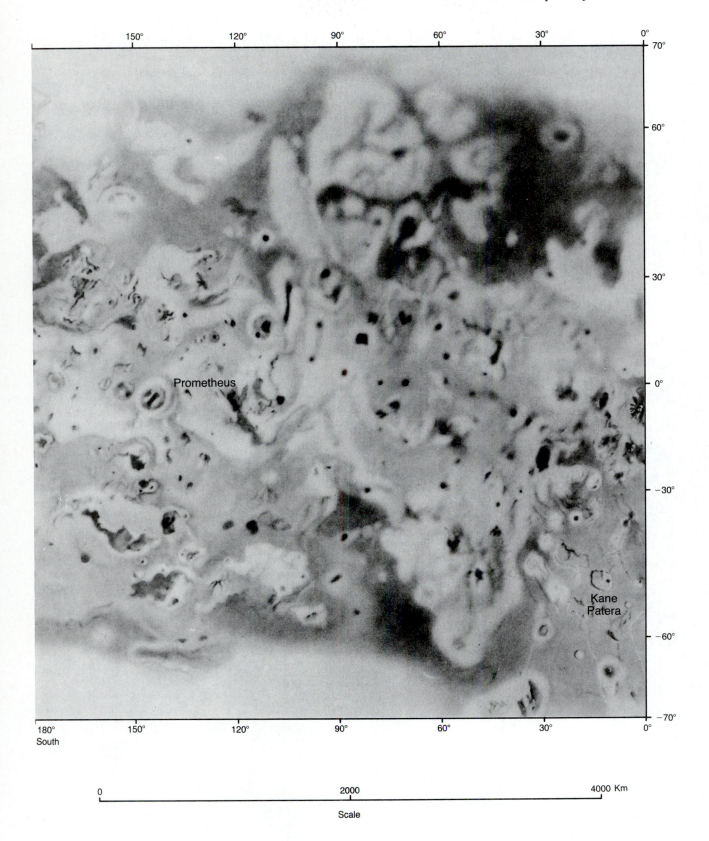

Scale

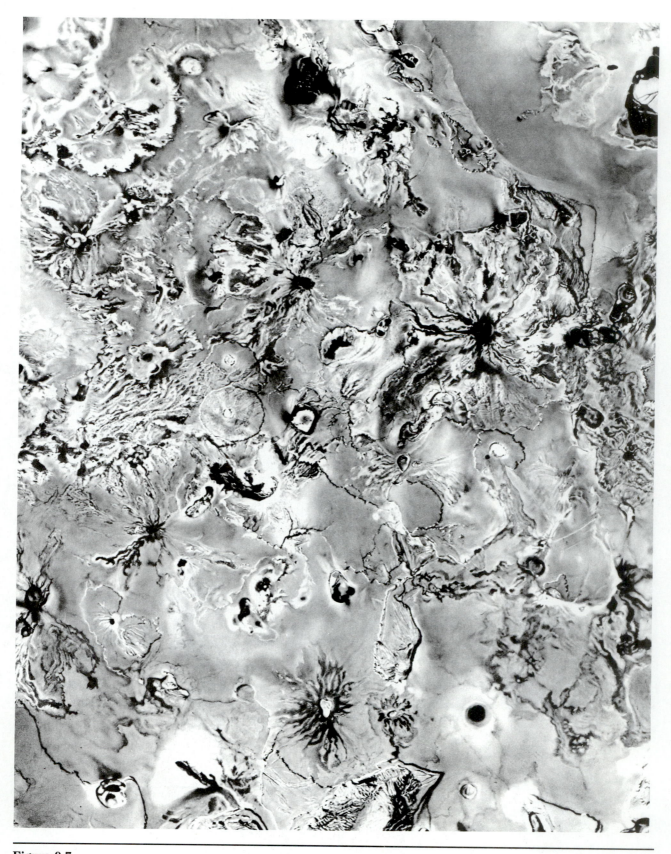

Figure 9.7

Volcanic features on Io include semicircular calderas from which radiate long, narrow lava flows; some of which may be covered by a thin layer of sulfur. Smooth volcanic plains are covered with volcanic ash. The complete absence of craters indicates that the entire surface of Io is very young.

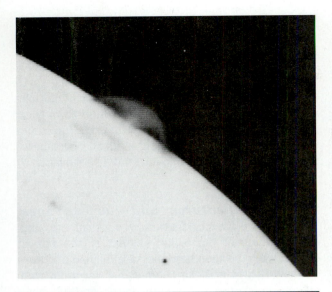

Figure 9.8

Many volcanic eruptions on Io are extremely explosive, with ejection velocities of more than 3000 km per hour. This is more violent than the explosions of Mt. St. Helens on Earth. In this image, an enormous volcanic plume is silhouetted against the dark sky. On Io, an eruption sprays volcanic ash upward like water from a fountain. The ash is deposited in a ring around the vent (see Figure 9.9).

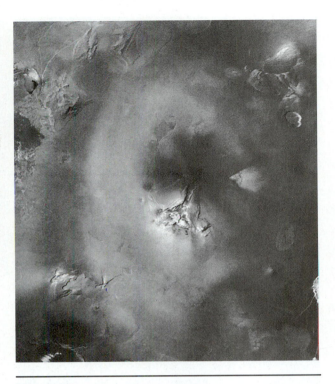

Figure 9.9

The great erupting volcanoes on Io, like Pele shown here, produce distinctive surface markings. The huge heart-shaped ring surrounding the vent was deposited from an umbrellalike fountain. The dark areas, some of which are surrounded by light rings, are calderas.

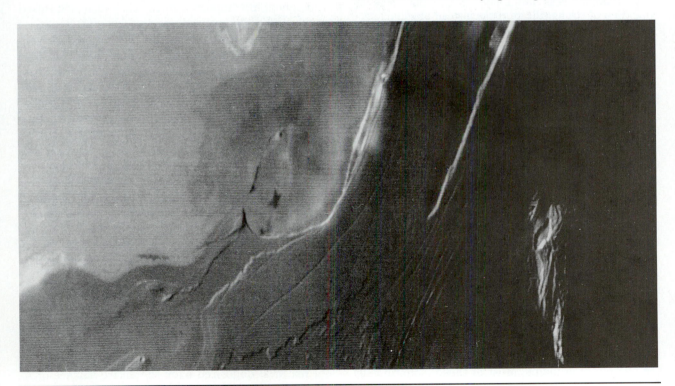

Figure 9.10

Structural deformation on Io is indicated by long linear fault scarps in the plains and plateaus near the southern polar region. The serrated edges of the plateau margins on the bottom of the photo indicate some kind of erosional processes operate on Io. Rugged mountains, up to 2 km high, are visible on the right.

Figure 9.11

Mountainous areas are found on Io. They may be composed of strong silicate rocks instead of sulfur.

cut by scarps that resemble faults produced by tension in the lithosphere. Although there are no apparent linear chains of volcanoes or fold belts, there are irregular isolated mountains in the polar regions and large plateaus bounded by faults—suggesting that the outer rigid shell of Io is fracturing in response to motions within a fluid substratum. Most movements appear to be vertical with little or no lateral movement occurring. From these observations it is inferred that although there is active tectonism, there is no system of integrated plate tectonics on Io that could form folded mountain belts or subduction zones like those on Earth. Like the Earth's Moon, Io appears to have only one planetwide lithospheric plate.

Rugged mountainous areas are also scattered across the surface of Io (Figure 9.11). These isolated massifs attain heights of almost 10 km above the surrounding plains. Most of these mountainous regions are disrupted along numerous linear traces, presumably faults, and have sinuous spines whose flanks almost appear to be gullied. In a few instances, faulted, sloping segments of the intervent plains form ramparts around the mountains. Some

mountains are also the sites of volcanic pit craters. The presence of high mountains on Io indicates the presence of strong silicate rocks within its crust. Sulfur is not strong enough to support topographic variations of this magnitude; on Io, it begins to deform under its own weight when accumulations are only about 2 km thick. Thus, aside from their importance as indicators of tectonism, the mountains of Io provide convincing evidence for silicate materials and silicate volcanism at Io's surface.

Impact Features and Surface Ages. The principal surface feature on most of the planetary bodies of our solar system is the impact crater. We have seen how they are formed and have observed their endless geomorphic variation from planet to planet. Although most craters were formed billions of years ago, even Earth with its active geologic systems retains evidence of past encounters with large comets and meteorites. Io, however, does not. This stands out in marked contrast to all other planetary bodies in the solar system and clearly indicates that Io has an extremely young and dynamic surface. Although in distant views Io's volcanic calderas resemble impact craters, in detail they are very different. The calderas have irregular outlines; few are circular or polygonal like impact craters. They also lack raised rims and the textured ejecta blankets typical of lunar craters, and none of Io's volcanic craters have raised central peaks. At the resolution of some pictures, impact craters as small as 1 or 2 km across would be visible if they were present, but none have been found. From comparisons of crater production rates on the more extensively studied inner planets, it seems likely that a crater larger than 10 km in diameter would be produced on Io every few million years. Apparently any craters that form are rapidly erased or modified beyond recognition. Impact craters that formed on Io have probably been buried by the extensive volcanic deposits obvious in the photographs. Calculations based on the impact rates on other planetary bodies in the solar system indicate that Io is being resurfaced at the rate of 1 m per 1000 years.

It is also possible to calculate the rate of resurfacing from the number and size of observed eruptions. Each eruptive center is extruding about 1000 tons of material a second, or 100 billion tons per year. This would amount to an average of 10 m of material deposited over the entire surface of Io in one million years. This figure is comparable to the erosional stripping rate calculated for the Earth—60 m per million years. The surface of Io is thus one of the youngest and most dynamic in the solar system.

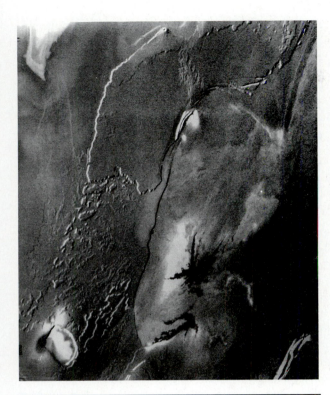

Figure 9.12
Eroded plateau margins in the south polar region are reminiscent of those of Earth. They also resemble the etched terrain on Mars and may be the result of various volatile materials escaping from pore spaces in the rocks.

Although there is no liquid water or even winds generated in a thick atmosphere, there is an erosive process on Io; a process unfamiliar to us operates over the surface, stripping away some craters as well as volcanic deposits and creating an exotic etched terrain. In the central part of Figure 9.12, a long mesa appears to have been produced by scarp retreat. The sublimation of various volatile species of sulfur may cause this, sculpting a terrain like the thermokarst terrains on Mars. In the lower right corner of the photograph, an isolated mountain is surrounded at its base by an apron of what may be erosional debris shed from the upper slopes. Perhaps it is the intense radiation associated with Jupiter and its magnetic field that induces the escape of some of the molecules that make up the surface. In any case, Io is being actively resurfaced, most intensely by a veneer of young volcanic deposits.

Dynamics and Internal Structure.
The geologic development of a planet is largely determined by its composition, its size, and the energy available within it. Io has a density of about 3.6 g/cm³, almost the same as the Moon. This density is consistent with the speculation that Io

consists mostly of silicate minerals, like the Moon. Io is also similar to Earth's Moon in its size (Table 9.2), and yet, 4.6 billion years after its formation, when other bodies of comparable size have cooled and become geologically dormant, it still has enough internal energy to melt and eject tons of material through numerous volcanoes on its surface. To keep a silicate-rich body the size of Io warm enough to remain partially molten for this length of time would require an unreasonably large proportion of radioactive elements as an internal energy source. It is more likely that the energy that heats Io is from an external source. Earlier, we discussed how tides can be raised in a satellite's lithosphere by the gravitational attraction of its primary. This effect has caused Io and the other Galilean satellites to rotate synchronously with their orbital periods; they thus keep the same hemisphere pointed toward Jupiter (as the Moon does toward Earth). If this were the whole story, any tidal bulge created by Io's attraction to Jupiter should always be in the same place, not allowed to relax and heat the satellite. However, Io's orbit is eccentric (not perfectly circular); thus, variations occur in the Jupiter–Io distance and consequently in the tidal strain. The eccentricity is caused by gravitational interactions in the satellite system. Io, Europa, and Ganymede have achieved **orbital resonance**—that is, their orbital periods are simple multiples of one another. In this case, for each orbit Ganymede makes around Jupiter, Europa makes two, and Io four. This resonance forces Io to have a slightly eccentric orbit, causing the distance between Jupiter and Io to change and thereby changing the height of the tidal bulge. As the energy from these changes is dissipated, heat is produced, which apparently melts the inner regions of Io and provides abundant magma. This external heating mechanism is a relatively permanent energy source for such a small body and may keep Io thermally active long after the Earth, with its radioactive heat engine, has cooled and become geologically dormant.

Because of this unusual energy source, the internal structure of Io may be unique. When the effects of the extraordinary mechanical heat engine and heating from its own radioactive elements are compounded, it is very likely that Io is extremely differentiated and has experienced extensive outgassing of volatile elements. After its accretion, whatever water (H_2O) and carbon dioxide (CO_2) existed in the planet were probably rapidly released as the planet heated and have since escaped because of Io's small mass and low gravitational field, leaving a volatile-poor planet. Sulfur, which is also fairly volatile, may condense rapidly from these emissions and return to accumulate at the surface.

Since Io is continuously heated, a thick lithosphere has never formed. It may be as thin as 20 km and rest atop an ocean of silicate or sulfur magma. One idea for Io's structure is shown in Figure 9.13. A thin, solid crust of sulfur and sulfur compounds is underlain by an asthenosphere of molten sulfur. This, in turn, may be underlain by a solid layer of silicate rock, which may locally penetrate the sulfur crust to produce the rugged mountains. Beneath the silicate layer, a much thicker body of partially molten or convecting silicates probably exists that may extend to the core. From its assumed composition, density, and high internal temperatures, Io's core probably is composed of molten iron and should produce a magnetic field. If this is the case, there may be silicate volcanoes where the silicate crust reaches near the surface. Intrusions of silicate magmas onto the base of the zone of molten sulfur may produce violent explosions above it that penetrate the crust and produce sulfur volcanoes. It seems likely that silicate magmas are also erupted to the surface.

The high internal temperatures that have existed inside Io for billions of years may have allowed extensive differentiation into shells of distinctive density and composition. Lateral density differences may not be pronounced at present; as a result, lithospheric plates do not appear to be consumed by subduction. Rather, a continuous vertical recycling of lithospheric material may exist. As magmas are poured onto the surface, building a new crustal layer, the underlying layer is depressed, melts at the base through a type of thermal erosion, and begins a new path toward the surface. This style of vertical cycling is reminiscent of that proposed by some for Venus and the early Earth. Such a process active for so long would have allowed very extensive chemical differentiation of the satellite.

From the facts in hand, we can speculate about Io's geologic development. Io probably accreted from a cloud of debris circling Jupiter about 4.5 billion years ago, incorporating its own complement of radioactive elements, water, carbon dioxide, and sulfur. Early melting of the small moon probably occurred just as on Earth's own Moon but was driven in part by tidal heat from its interaction with Jupiter, as well as from radiogenic and accretionary heat. This melting was probably more extensive and occurred more rapidly on Io than on the Moon because of the extra energy source and resulted in a molten core (sulfur- and iron-rich), a magnesium-silicate mantle, and a lighter silicate crust. (It has been suggested that Io's core is composed of iron oxides instead of iron or iron sulfides. Sulfur not extracted to the core would become enriched in its outer layers and available for volcanic outgassing.) By analogy with the Moon and taking into account Io's larger heat production, differentiation probably occurred during the period of intense bombardment, and many craters that formed during this period were probably short-lived. Although Io degassed large amounts of volatiles, nitrogen, water, and carbon dioxide were rapidly lost to space because of Io's intrinsically low gravitational attraction and their volcanic ejection to high altitudes, where the gas interacts with Jupiter's **magnetosphere**. Gradually, a sulfur concentration was built up on the surface as it condensed rapidly in cold space, while the interior of the planet became volatile-depleted.

Continuously heated from an external source, volcanism has been a persistent process. Repeated melting and differentiation at the base of the silicate lithosphere may have eliminated all lateral density contrasts in this thin, rigid zone, allowing vertical movements and continuous recycling of light silicate materials to dominate the recycling of lithospheric material. Of course the sulfur-rich veneer thickened and may have become tens of kilometers thick (depending on how much sulfur Io had originally and how much was sequestered in its core). Since sulfur has a lower melting point than most silicate minerals, once it became thick enough, it could form an "ocean" of molten sulfur above the silicate lithosphere with its own cap of solid sulfur-rich materials. In some ways this condition is analo-

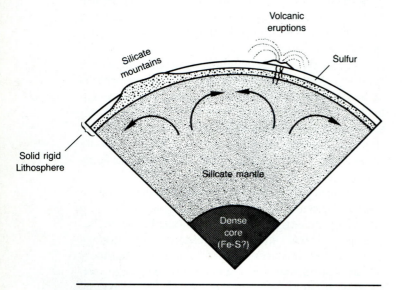

Figure 9.13

A model of the interior of Io shows a thin crust of solid sulfur and solid sulfur dioxide perhaps including molten sulfur. This layer surrounds a thicker crust of silicates that may protrude as mountains. A mantle of silicates probably envelopes a denser metallic core at the center of the planet.

gous to the situation at Earth's North Pole, where ice floats atop liquid water and covers a silicate-rich lithosphere. The shell of sulfur-rich materials, if it really exists, could be called a **thiosphere** (*thio* in Greek means "sulfur"), analogous to the *hydro*spheres (water-rich), *cryo*spheres (ice-rich), and *litho*spheres (rock-rich) of other planetary bodies that are not as extensively differentiated. In this regard, Io may be one of the most chemically fascinating bodies in the solar system.

Europa

Europa is the next satellite out from Io. Europa is only slightly smaller and less dense (2.97 g/cm³) than Io. Models of its condensation history and chemical composition suggest that Europa accreted from material richer in water, making it less dense than Io. This water appears to have been released during differentiation to form an icy mantle surrounding a silicate core. Telescopic observations of Europa's spectral characteristics confirm that water ice is exposed on its surface.

The surface of Europa resembles that of a fractured ice pack in the Arctic Ocean (Figure 9.14). The maximum relief on Europa is on the order of 1000 m, in contrast to features that rise from 4000 to 9000 m on Io. This low relief suggests that the subsurface material is partly liquid and slushy so that isostatic adjustment of the ice prohibits the development of high landforms. It is uncertain if silicate materials are exposed at the surface at all.

There are two major terrains on Europa: plains and mottled terrain (Figure 9.15). The bright plains appear to be the older of the two terrains and are crossed by a distinctive set of narrow bands or stripes. The mottled terrain, also cut by these features, has a much rougher, pitted topography and is darker than the plains. Craters are rare, but a variety of fractures and ridges distinguish Europa from its Jovian companions.

Craters. Like Io, the surface of Europa is also distinctive in that its surface is essentially uncratered. Photos of neighboring Callisto and Ganymede at a resolution similar to that in Figure 9.15 would show numerous craters in the range of 50 to 100 km in diameter, whereas on Europa there are none. The surface of Europa is completely younger than the period of intense bombardment and must also be younger than the grooved terrain on Ganymede (described in a later section). The key question is how much younger. The lack of numerous craters perhaps indicates that geologic

processes in Europa's crust are still modifying its surface. Based on impact crater production rates on the inner planets, the bright plains may be only several hundred million years old.

Careful study of all the photographs of Europa taken by the Voyager spacecraft have only found five fresh craters of 10 to 30 km in diameter. One is a bowl-shaped crater with a sharp rim. Another is shallow and surrounded by a system of dark rays. One has a bright central peak (Figure 9.16). There is also a small multiring crater with several concentric rims and troughs. Another is a bright crater with a dark halo.

A second class of probable impact craters includes large, flat, circular, brown spots 100 km or more in diameter. The best example of this type of crater is shown in Figure 9.17. This feature, named Tyre Macula, is probably a **crater palimpsest**, a feature deemed to have an impact origin but lacking a topographic basin. *Palimpsest* is derived from Greek and refers to parchment that has been scraped clean to allow further writing; usually the erasure is not complete and the earlier writing can still be seen. Tyre Macula consists of vague multiple concentric rings with no detectable surface relief. It appears as a dark circular spot near the center of some of the large regional fracture patterns. Other similar dark patches with nearly radial lineation patterns can be seen on the global mosaic (see Figure 9.14) and are suspected to be crater scars. If this is the case, impacts occurring sporadically across Europa's surface may have developed zones of weakness in the crust, which influenced structural patterns of fractures and ridges.

Tectonic Features. The most spectacular feature on Europa is the intricate network of streaks that extends across the surface in various directions. These are thought to be fractures in Europa's crust that have been filled by ice and slush from below to form dikelike structures—a type of icy-fissure eruption. Some fractures range up to several thousand km long and 70 km wide. Although some have been called dark and appear brown on the color-enhanced images, they are in fact gray against the white of Europa's plains surface.

The fracture systems of Europa may be subdivided into sets that apparently have different structural origins. Some are global in extent, whereas others are restricted to local areas. Some are straight, others curved. Some form thin, sharp lines, whereas others are faint broad swaths. All, however, appear to be fractures and present a pattern similar to pack ice in the Earth's polar regions that has been fractured by repeated breakups.

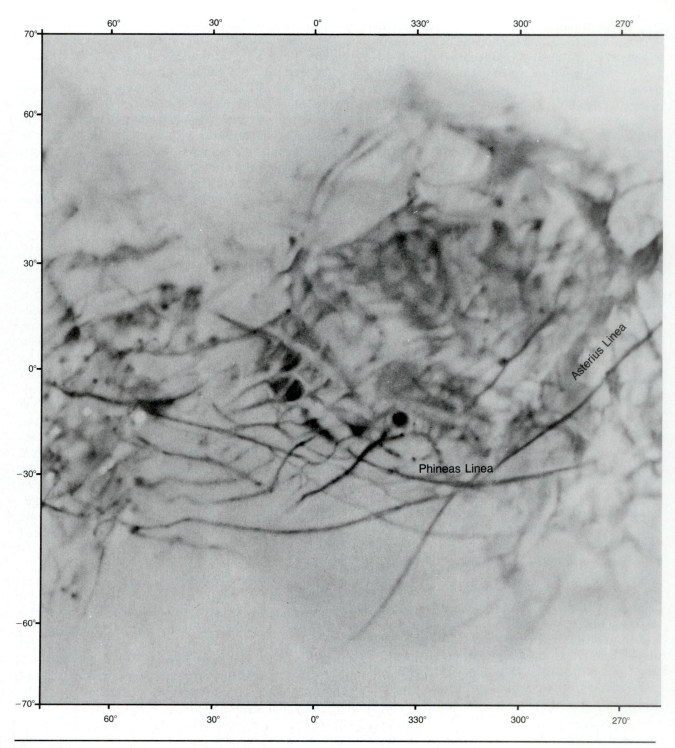

Figure 9.14

A shaded relief map of Europa was compiled from the images obtained from the Voyager missions. It shows an icy terrain with complex systems of fractures unlike anything else in the solar system. The almost complete absence of impact craters indicates that the surface is young.

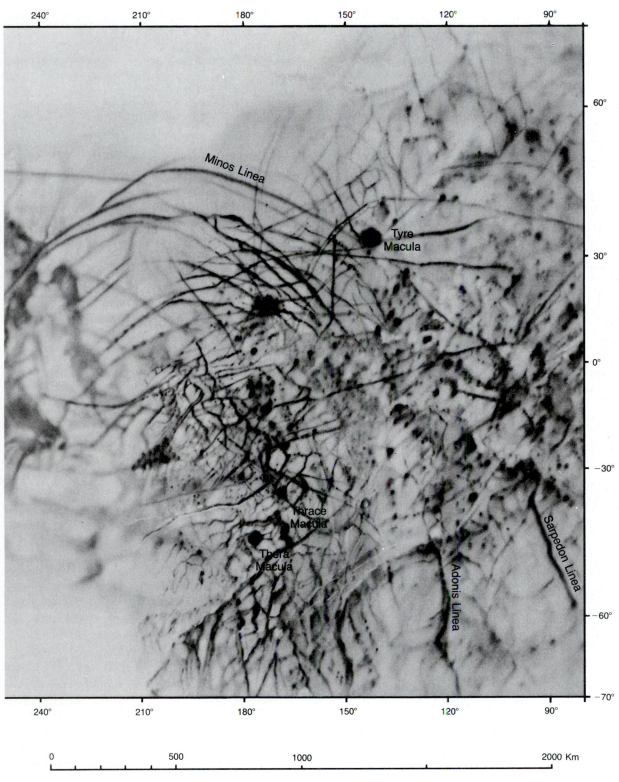

Scale at 0° Latitude

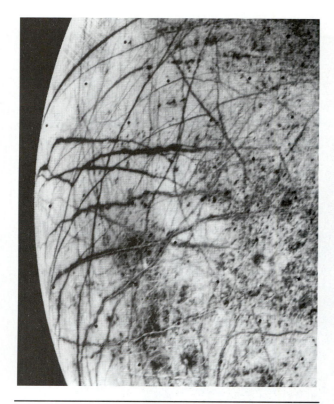

Figure 9.15

A complex fracture system dominates the surface of Europa. The fractures cut two different types of terrain: (1) bright plains, which appear to be older, and (2) dark, mottled terrain, which is pitted and rougher.

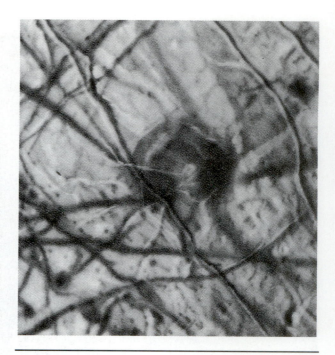

Figure 9.17

Some impact structures on Europa are vague and lack significant relief. The dark circular patch in the middle of this map is thought to have formed by impact. It probably lost its relief because the weak ice in Europa's lithosphere flowed.

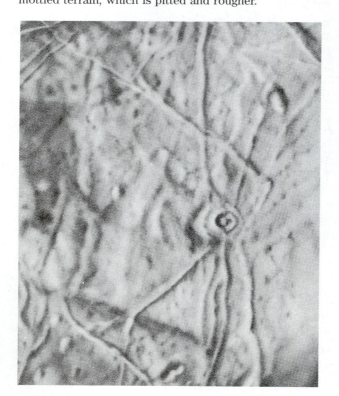

Figure 9.16

Craters on Europa are rare, but several small multiring impact structures, such as this one, have been detected.

Dark Wedge-Shaped Bands. Dark wedge-shaped bands form the most conspicuous fracture systems on Europa. These coarse, dark fractures extend across the equatorial region in a general southeasterly direction (Figure 9.18). They are among the widest fractures on Europa but taper rapidly and ultimately pinch out. The margins are sharp, and the material filling a fracture is relatively dark, so the wedge shape is quite distinct. This wedge shape and the apparent displacement of some crossing lineations suggest that expansion and rotation of blocks of the pre-existing crust was involved in their formation. This suggests that some form of incipient plate tectonics and strike-slip faulting may be responsible for Europa's crustal deformations and that slight rotation of the crustal plates opened gaps into which the warmer icy slush mixed with some rocky material was injected to form a small stripe of new crustal material.

Triple Bands. Triple bands constitute the second most conspicuous fracture type on Europa. These consist of fractures filled with dark material along their margins, with narrow stripes of bright material running down the center (Figure 9.19). Locally, the central stripe is a ridge. The triple bands commonly merge with a single brown streak, suggesting that many faint lineations may be unre-

Figure 9.18

Fracture systems on Europa occur in various shapes and forms. The most conspicuous are the dark, sharply delineated, wedge-shaped bands shown in this map.

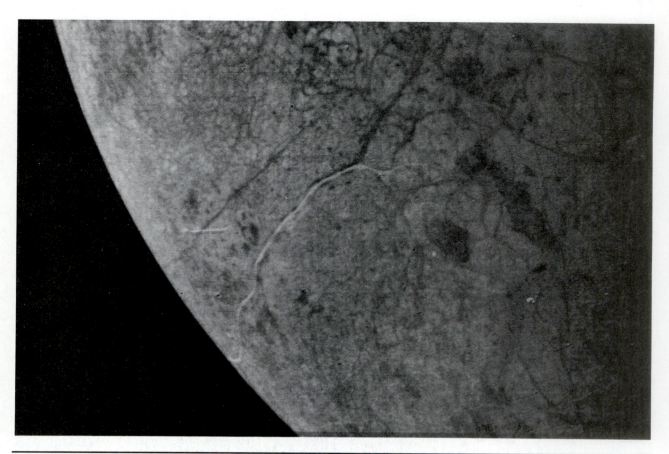

Figure 9.19

Triple bands, like the one that extends across the center of this image, are probably fractures that have been filled with clean ice so that the dark fracture zone has a bright central stripe.

solvable triple bands. This fracture system is global in extent and contains some of the longest fractures in Europa. Cross-cutting relations show that parts of some triple bands are older than some fracture systems, whereas other parts are younger, thus suggesting that the stress fields forming the global system existed over an extended period of time and must have had diverse orientations.

Generally, the triple bands disappear when traced into the brown, mottled terrain, suggesting that they are older and were obliterated when the mottled terrain was formed or that the physical properties of the mottled terrain are sufficiently different from the plains to prevent fractures of this type from forming.

The triple band fracture system is believed to have formed as a result of global expansion. Zones of weakness induced by tidal stress may have guided the bands' orientations. The resulting fractures were filled by slush and water from below, perhaps containing rocky material. This intrusion formed a dark dike of new crust. Subsequent stress reopened the fracture, and clean water formed the bright central stripe.

Gray Bands.   Gray bands occur in the southern part of Europa (Figure 9.20) and are distinctive in that they are the widest fractures on the planet. They are cut by other structural features and are thus considered to be relatively old. Their restriction to a local area suggests that they are the result of local tectonics.

Ridges.   Ridges are clearly visible only near the terminator and near the south polar region, where the low Sun angle enhances features with low topographic relief. In all probability, they are present over much of Europa's surface. The most striking examples are the curved white ridges in the southern hemisphere, which can be traced near the terminator as a series of broad, graceful scallops (Figure 9.20). They are 5 to 10 km wide and rise at most only a few hundred meters above the surrounding surface. The ridges are the youngest tectonic features on Europa, as is evidenced by the fact that they cut all other features. Some ridges, however, merge with triple bands. It is probable that the ridges were formed by intrusion of clean ice into fractures, much like triple bands.

Figure 9.20

Graceful, arcuate, scalloped ridges are some of the most striking features in the southern plains of Europa. These are apparently the youngest tectonic features on Europa. The gray bands are part of Europa's global fracture system.

Brown Spots.   In addition to the major terrain types (plains and mottled terrain) and the fracture system, irregular patches of brown material occur in various isolated areas. This material is similar in texture, color, and albedo to the material filling dark fractures and is especially abundant where major structural systems intersect. This material is probably the result of intrusion and extrusion of contaminated water and slush from below.

Internal Structure and Dynamics.

One idea of Europa's internal structure is shown in Figure 9.21. A water-rich lithosphere (or cryosphere), possibly 100 km thick, may have formed from a vast ocean that covered the entire planet shortly after its accretion, initial differentiation, and outgassing. An alternative to this thick ice model for Europa's cryosphere is that the icy outer shell is only a few tens of kilometers thick. If such is the case, then the brown spots and mottled terrains may represent the peaks of silicate massifs nearly protruding through the icy shell. Excavation of these materials by impact or magmatism may create the dark coloration of some areas of Europa. This outer icy crust may be underlain by a watery asthenosphere over which the cryosphere may deform. Because of the absence of lateral density differences in the icy outer layers of Europa, most movements are probably vertical; subduction is not

apparent. It is not known if Europa has a molten core or a detectable magnetic field.

Because of Europa's small size, rapid radiative cooling might be expected to have quickly produced a thick, rigid lithosphere that could preserve many impact craters. The absence of many craters and the apparent youth of Europa's surface suggests that a small external energy input has slowed its cooling substantially and allowed thermally driven geologic processes to continue to the present, or nearly so. The tidal heating mechanism that heats Io has apparently also heated the interior of Europa, allowing repeated eruptions of watery lavas to obscure its early history. Moreover, a warm, thin layer of ice deforms much more easily than most silicate materials, by slow flowage or glacierlike creep. Hence, even if they are not buried by volcanism, large craters may have been relatively short-lived features, as relaxation (the slow upward flow of deep materials) tends to obliterate any depressions on the surface.

As to what created the cracks, there is no simple explanation. It is hard to imagine that they are the result of simple planetary expansion unless Europa, too, has some external energy source to keep it warm and crack its youthful surface. Possibly the tidal heating mechanism may keep Europa slightly warm, as it does Io. Alternatively, the cracks may be the result of deformation arising from within as deeper, warmer layers of ice or silicate material convect and disrupt the surface ice. Another suggestion holds that the fractures may be the result of tidal disruption of the icy crust from the constant gravitational tug-of-war going on between the satellites and Jupiter.

Ganymede

Just as the planets of the solar system show a regular decrease in density outward from the Sun, so do the Galilean satellites of the Jupiter system. Ganymede, the largest satellite in the solar system, is over twice as massive as Earth's Moon and has a diameter about the same as Mercury's; yet its density is only 1.94 g/cm^3. The low density suggests that a substantial portion of its bulk, perhaps half, consists of water in the form of a liquid or a solid. Thus, Ganymede, together with Callisto, represents a class of large, low-density planetary objects common in the outer solar system.

Most models of accretion and radioactive warming of these ice-rock mixtures suggest that they became differentiated as ice was melted and formed a predominantly icy mantle, or cryosphere, a liquid water asthenosphere, and a rocky core. Ganymede's outer icy shell is probably much thicker than Europa's.

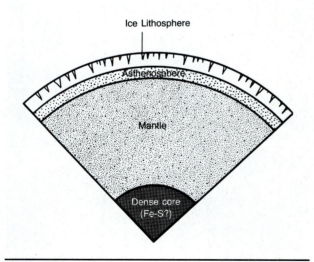

Ice Lithosphere

Asthenosphere

Mantle

Dense core
(Fe-S?)

Figure 9.21

A model of the internal structure of Europa may be constructed, based on measurements of density, composition, and the nature of the surface terrains. Europa has an icy lithosphere about 100 km thick, which has been repeatedly fractured. The lithosphere may surround an asthenosphere of water and silicate minerals (perhaps forming a muddy mixture), which in turn surrounds a denser core of silicates.

Surface Features. Viewed from a distance, the surface of Ganymede almost seems familiar and resembles in many ways the airless planetary bodies of the inner solar system—Mercury and the Moon. There are large areas with significant color and brightness contrasts (Figure 9.22). Bright, rayed impact craters as well as highly degraded rayless craters pockmark its surface and attest to its ancient age. Close inspection reveals that most of the bright regions are striated by innumerable crisscrossing grooves and have distinctly fewer craters than the darker irregular patches. Two distinct terrain types are thus distinguishable—the dark, cratered terrain and the bright, usually grooved terrain (Figure 9.23).

The Cratered Terrain. The cratered terrain is dark and nearly saturated with craters a few tens of kilometers in diameter, a condition similar to that found in the ancient lunar highlands, which date back to the period of heavy bombardment that ended almost 4 billion years ago. If the rate of

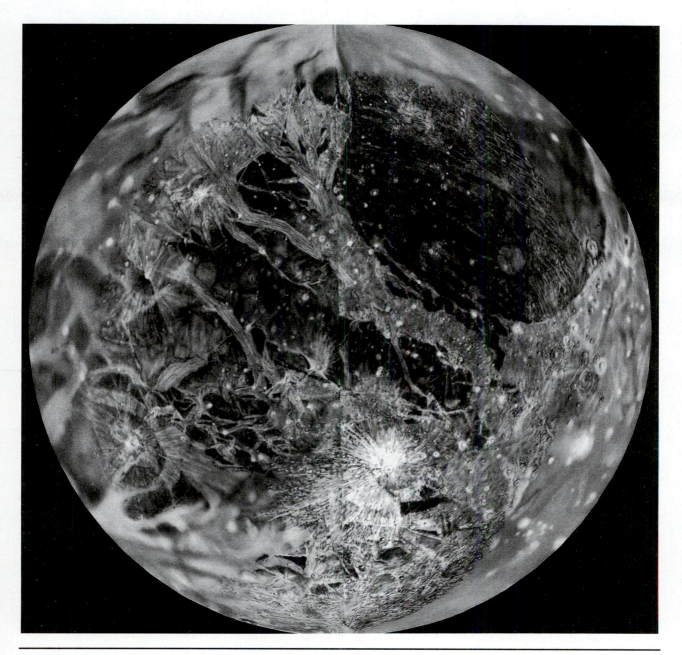

Figure 9.22

Ganymede is the largest of Jupiter's Galilean moons. It has a bulk density of only 1.94 g/cm^3 (about half that of the Moon) and is therefore probably composed of rock and ice. The various tones in this image represent different types and ages of terrains. The long white streaks are rays associated with impact structures.

Figure 9.23

A shaded relief map of Ganymede, compiled from the Voyager images, shows a variety of fascinating surface features, which represent a complex geologic history. Two major terrain types are distinguished: (1) an old, dark, cratered terrain that appears to have been fragmented and (2) a younger, bright terrain grooved by complex ridges and valleys formed by tectonic activity.

meteorite impact was approximately the same on Ganymede as on the Moon, then these regions are about the same age and record the events of this cataclysmic era. Regardless of their absolute age,

the dark, cratered terrains are the oldest surfaces recognized on Ganymede. They appear to be composed of older ice, darkened by mixing with silicates from meteoritic projectiles. Spectral studies

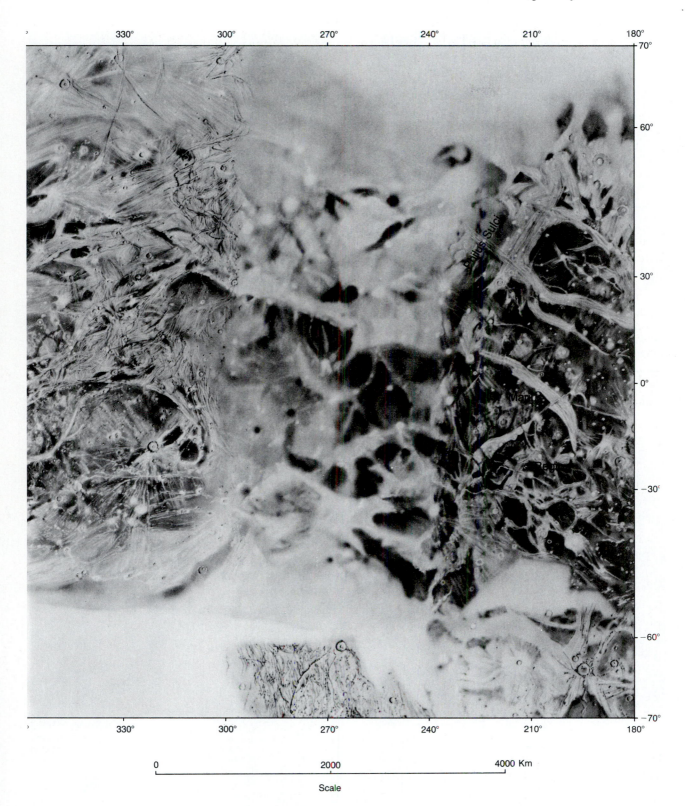

show that silicates are present at the surface of Ganymede in greater abundance than on Europa.

Most of Ganymede's impact craters are shallow, and some have broad central uplifts or small inner rings and pits (Figure 9.24). Although most of the craters are degraded and old, the fresher craters display crisp outlines and prominent ray systems. In Figure 9.25 craters with light and dark rays are prominent. Larger craters have subtle outer rings. This is especially obvious on the younger craters, which contain obvious white rings. Many areas have large circular bright spots, which may be very

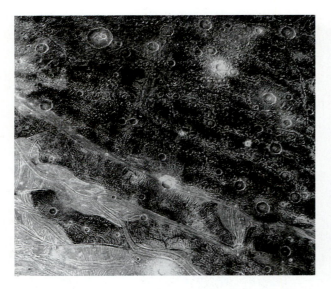

(A) Some craters are well defined and are marked by light rims and central peaks. Most are shallow and are roughly the same size. Light-colored palimpsests are common on Ganymede's dark-cratered regions.

(B) The youngest craters have bright ejecta blankets and systems of bright rays produced as the impact sprayed clean ice from below the surface.

Figure 9.24
Impact craters on Ganymede are shown in the images above.

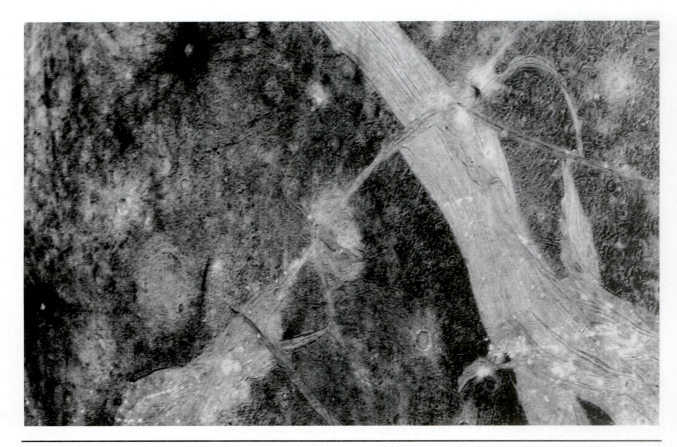

Figure 9.25 Craters on the ancient, dark terrain of Ganymede show a variety of impact features. Most of the younger impact structures have thrown out bright, clean ice from below the surface; yet some craters have systems of dark rays. Note the abrupt change in width of the grooved terrain near the top of the images, suggesting differential growth or horizontal displacement.

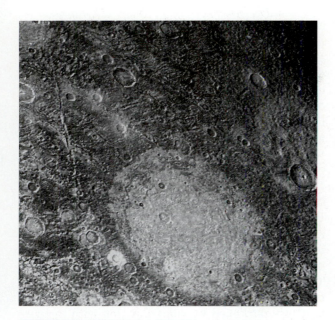

Figure 9.26

Palimpsests are common impact structures on Ganymede. They are vestiges of original craters that have been subdued by ice flowage and isostatic adjustments. The original topography is almost completely erased so all that remains is a "ghost" crater marked by its light color.

ancient craters that were almost erased by later events (Figure 9.26). Their topography might have been reduced by slow, viscous relaxation of the icy crust during Ganymede's early history. An alternate view, worthy of further consideration, suggests that ancient craters collapsed immediately after impact, this because of the low melting point of ice and the relative ease with which it deforms, compared to the silicate lithospheres that dominate the inner planets. The lack of mountainous features and large basins also suggests that viscous flow in the cryosphere may have occurred. However, remnants of large multiring impact structures are partially preserved on the old cratered terrain—fragments similar to the complete multiring structure discovered on Callisto. One, located in the Galileo Regio area (Figure 9.23), consists of an extensive system of rings that are expressed as narrow, regularly spaced furrows, which appear to be grabens. The furrows cover an area larger than the United States but are centered on a partially preserved palimpsest only 500 km in diameter. This ancient crater may have formed when Ganymede's icy lithosphere was very thin indeed—perhaps only about 10 km thick and barely able to preserve a crater at all—and attests to the dramatic differences in crater morphology that are caused by lithospheres of different thicknesses and target materials of different strengths.

Bright and Grooved Terrains. Dividing large patches of dark, cratered terrain are broad swathes of bright, apparently younger, terrains. Much of these bright terrains are crisscrossed by narrow grooves. The **grooved terrain** is the most peculiar structure on Ganymede. Nothing quite like it has ever been observed on other planets of the solar system. In any attempt to analyze and understand it, we must remember that Ganymede has an outer shell of ice, and we may not be able to think in terms of surface and tectonic processes with which we are familiar on the terrestrial planets. As shown in Figure 9.27, the bright terrain occurs in a series of complex systems of stripes or bands, which divide the older cratered terrain into a series of isolated islands several hundred to as much as 1000 km across. A few small patches of bright, relatively smooth plains lacking grooves are found within this terrain or at its margins. Individual grooves are only a few hundred meters deep but may be hundreds of kilometers long, similar in many respects to graben. Generally they run parallel to the edge of the cratered terrain; occasionally they terminate against one another and are sinuous; some are offset in faultlike fashion (Figure 9.28). Each segment of grooved terrain is a mosaic of groups or bundles of grooves, which terminate abruptly at the boundary with an adjacent system. Some systems cut across others but apparently do not displace or offset them. Some of the sets of grooves, however, are offset laterally as though by strike-slip faulting. They typically form curved or arcuate patterns.

The crater frequency on Ganymede's grooved terrain varies from place to place, indicating that it developed over some period of time, but on average the crater frequency is approximately one-tenth that of the older cratered terrain. The crater frequency of the youngest sections of the grooved terrain is about the same as that of the oldest lunar maria and the oldest martian plains, suggesting that formation of the grooved terrain ceased about 3 billion years ago, at about the same time as the extrusion of the oldest mare basalts on the Moon. Impact craters in the grooved terrain are generally deeper than their counterparts of similar size in the ancient cratered terrain. There are also few palimpsests developed in grooved terrain. These facts suggest that Ganymede cooled substantially by the time the grooved terrain formed. As a result, the icy lithosphere was thicker, stiffer, and more able to support substantial topographic relief over millions of years.

The origin of the grooved terrain presents some perplexing problems. These stripes of bright terrain are probably not the result of impact pro-

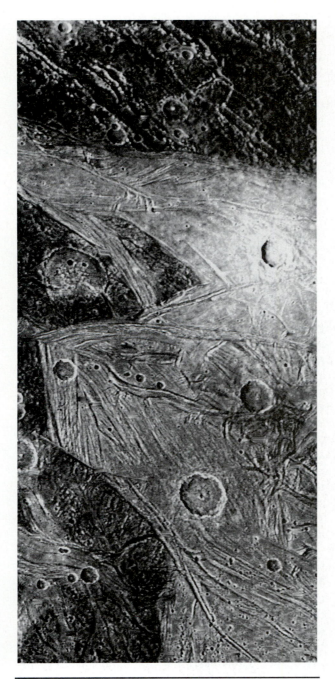

Figure 9.27 The grooved terrain is one of the most fascinating landscapes on Ganymede. It consists of bright regions with complex bands of low ridges. The ridges are about the same size as the ridges in the Appalachian Mountains on Earth. Multiple sets of grooves intersect at various angles, others are offset by faults, and others butt into each other, forming crazy-quilt patterns. The grooved terrain and the manner in which it fragments the ancient, dark, cratered terrain indicates breaking, faulting, and spreading of the crust. This type of tectonics is extremely exciting to geologists because it was the first time that lateral motion of crustal blocks and the creation of linear bands of new crust was seen on a planetary body other than the Earth.

cesses but must involve some type of deformation of Ganymede's crust that accompanied total obliteration of older terrains. More than half of the ancient cratered surface appears to have been disrupted by grooved terrain—obviously the result of some process acting on a global scale. Nothing like it is seen on the terrestrial planets, and until now geologists had given little, if any, thought to the tectonic processes involving a lithosphere composed of ice. Being relatively younger and brighter than the cratered terrain, the grooved terrain is believed to be composed of relatively clean ice. Present theories consider the bright grooved terrain to have been formed by volcanic resurfacing, forming the bright terrains, and then fracturing to make down-dropped blocks, forming the grooves. The terrain is bright because clean ice was extruded from beneath the older cratered surface that was darkened by contamination with meteoritic debris. Volcanic features similar to those found on the terrestrial planets and Io, such as shields, cones, fissures, or flow fronts, have not yet been identified on Ganymede, but it is almost certain that volcanic processes helped to modify the crust. Smooth plains that occur as patches within the grooved terrain were produced by the flow, ponding, and solidification of an extremely fluid volcanic liquid, probably water (Figure 9.29). But how do these watery magmas reach the surface? Since the form of ice found at the surface of planets is less dense than the liquid, watery magmas would only rise partway to the surface and then freeze as subsurface intrusions. However, if the water contained bubbles of water vapor or other gases, liquid water "magmas" would be less dense than the surrounding ice and could rise completely to the surface and extrude as lavas. The rifting that formed these terrains may have been in response to global expansion, which resulted in the formation of dominantly normal faults that acted as avenues for the volcanic liquids.

Internal Structure and Dynamics.
Ganymede's low density (less than 2 g/cm^3) indicates that it may contain no more than about 50 percent by mass of dense silicate materials. Most of its bulk is made up of frozen water (Figure 9.30). During accretion, these ices and the silicates were probably well mixed, but an early differentiation event probably resulted from a combination of accretionary, radiogenic, and perhaps tidal heating. (Tidal stresses may have been much greater early in its history, before its synchronous rotation evolved.) The icy part of the planet may have melted quite rapidly, leaving the silicates unmelted. The heavy metallic and rocky materials could have

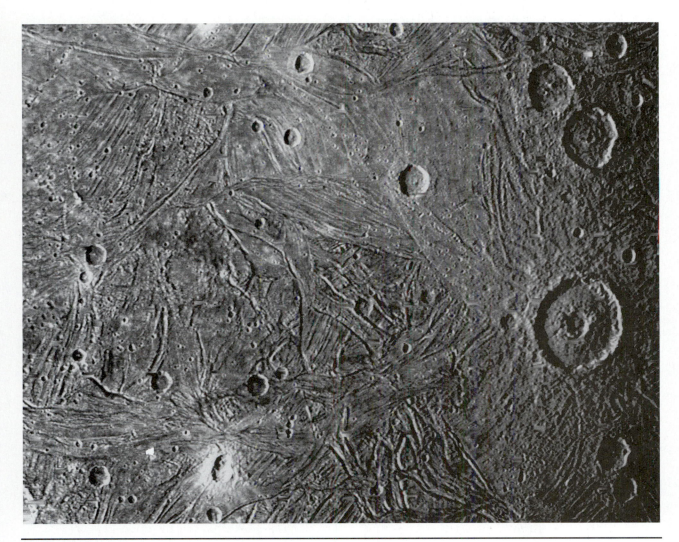

Figure 9.28

Complex geologic relations in the grooved terrain indicate that Ganymede had a complex tectonic history in which crustal fragments were moved about and new crust generated in the fracture system; perhaps the closest thing to plate tectonics seen on a planetary body other than Earth.

sunk to the center of the satellite, forming a core of rock, an intermediate zone of mixed rock and ice. The outer portion of the planet could consist almost entirely of liquid water. This liquid ocean would have rapidly frozen at the low surface temperature (almost 120 K), thickening from the top down. This ancient crust may have survived to the present as the heavily cratered terrain, dark because of meteoritic contamination. Because several forms of ice expand (rather than contract like most materials) as they freeze and cool, thickening of the icy outer rind may have resulted in slight expansion at Ganymede's surface, fracturing and faulting the cratered terrain to produce large tracts of grooved terrain. Their lighter color and fewer craters may be attributed to the eruption of younger, clean ice through the fractures. Others suggest that

Ganymede expanded as a result of heating by radioactive decay in its silicate core. When this expansion occurred is difficult to determine, but it apparently is not a recent event because of the number of craters superimposed on the grooves; some estimates place its formation at about 3.5 billion years ago. Thermal models indicate that icy Ganymede may have remained in this expanded condition without contracting as Mercury and the Moon did. The originally thin, icy lithosphere is probably very thick by now and may be completely solid. Theoretically, present-day convection in a solid, icy mantle seems possible, but there is no obvious geologic evidence of young tectonic features. Aside from meteorite impact, the reduction of topographic relief by slow, viscous creep seems to be the only process active for a very long time.

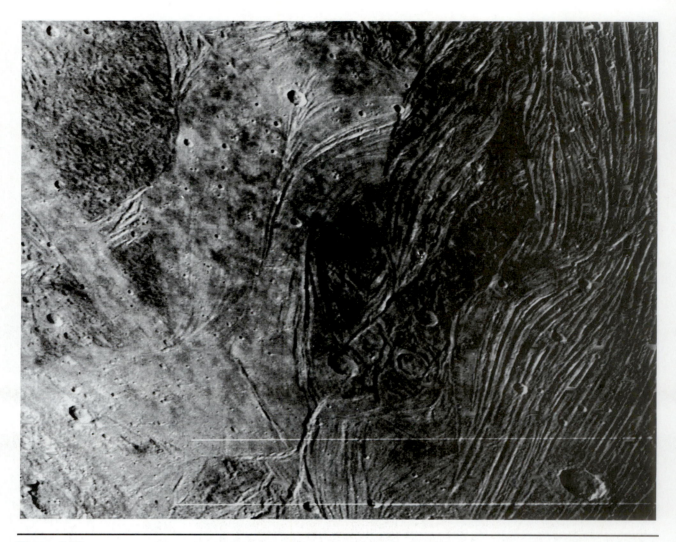

Figure 9.29

Plains in the grooved terrain of Ganymede are believed to have been produced by volcanic eruptions of slushy ice extruded as lava to bury parts of the grooved terrain.

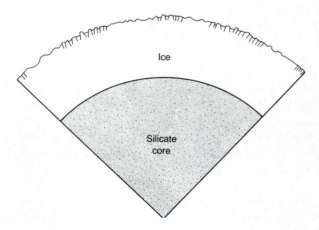

Ice

Silicate
core

Figure 9.30

A model of the internal structure of Ganymede may be constructed based on its size, density, surface composition, and external features. The icy mantle may include a watery, icy slush zone or some silicate materials. A large core of silicate rock is postulated to account for the planet's density (1.94 g/cm^3), which is intermediate between water ice (1 g/cm^3 and silicate rock 3 g/cm^3).

In summary, the first closeup observations of Ganymede revealed a new type of icy world, different in composition and history than any of the rocky inner planets. Ganymede is probably differentiated into a silicate core and icy crust. Subsequent to a period of heavy bombardment like that experienced by the inner planets, ancient impact basins and large craters, which normally produce high topographic relief, were reduced, but not completely eliminated, by viscous flow in the icy lithosphere. Later, the crust of Ganymede experienced a period of even more dramatic change, when sections of the older cratered terrain were replaced by bright volcanic terrains and extensional faults. The development of this terrain appears to be correlated with the thickening of the icy lithosphere.

Callisto

The outermost Galilean satellite is Callisto (Figure 9.31), a low-density body (1.86 g/cm^3) slightly smaller (4800 km in diameter) than Ganymede. Callisto is the most heavily cratered planetary body in our solar system (Figure 9.32). Because of its densely cratered surface, it resembles the Moon, but it lacks regions of major brightness contrast like that between expansive lunar maria and the highlands. Spectral studies of Callisto show that its surface is composed of water

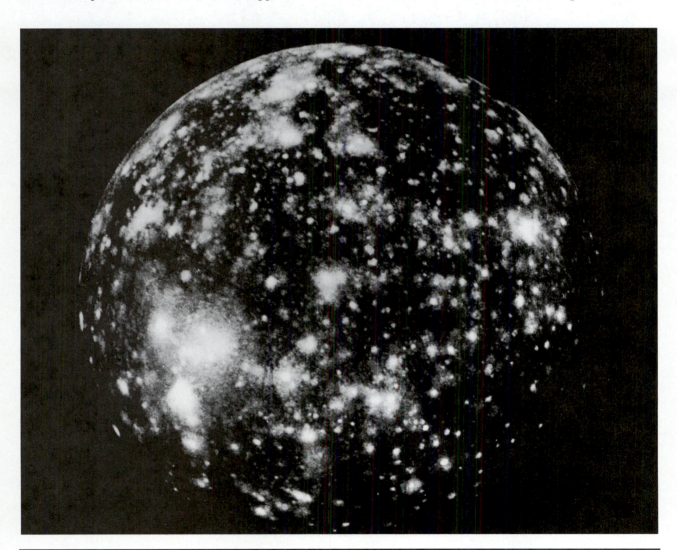

Figure 9.31

Callisto may be the most heavily cratered planetary body in our solar system. Its surface is composed of ice that is relatively dark. Many hundreds of moderate-sized craters are visible in the mosaic of Voyager 2 images. Many craters have bright rays of clean ice ejected from below the surface by the impact process. There are few craters larger than 150 km in diameter, indicating that scars left by large impact structures may not survive on the icy crust of Callisto but are subdued or obliterated by ice flowage.

Figure 9.32

The shaded relief map of Callisto was compiled from Voyager images by the U.S. Geological Survey. It shows the extent of the densely cratered terrain and two large multiring structures. The dominance of impact craters and lack of other surface features indicate that Callisto has the oldest surface in this part of the solar system, having changed little since the period of intense bombardment.

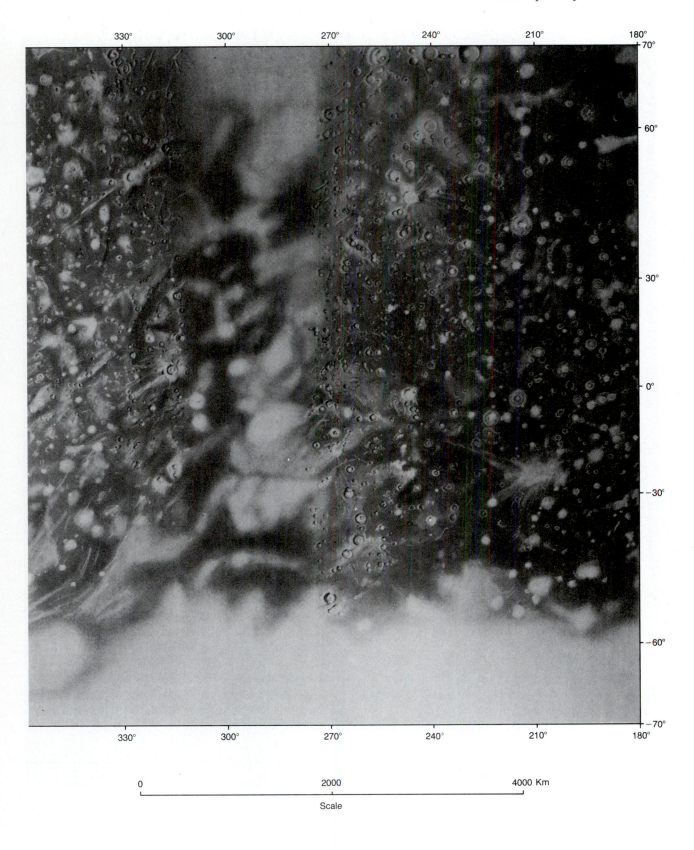

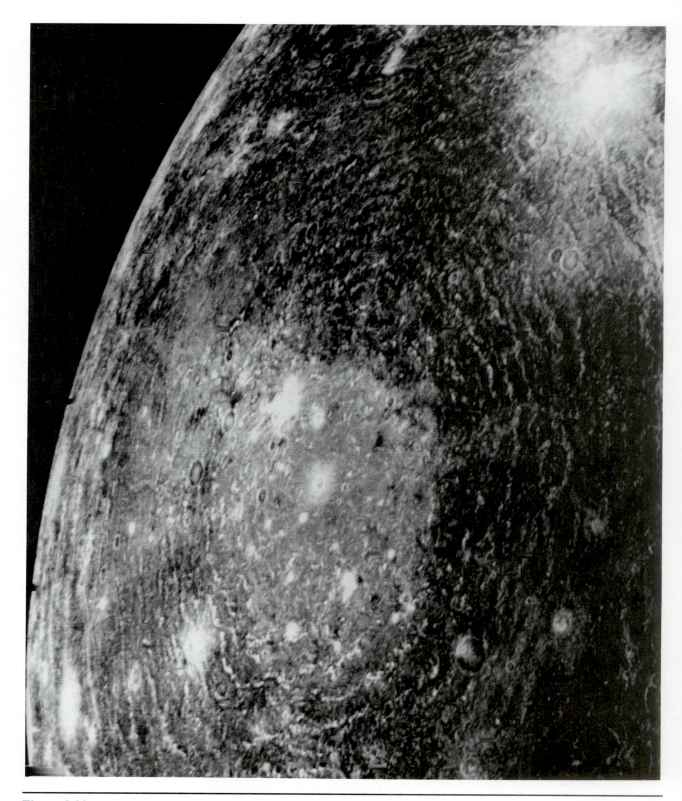

Figure 9.33

The multiring structure known as Valhalla is the most prominent impact feature on Callisto. It is similar in many respects to the multiring basins on the Moon and Mercury but is distinctive in that it has many more rings. This is probably due to the difference in the way an icy crust responds to large impacts as compared to the silicate crusts of the inner planets. The bright central area of Valhalla is about 300 km across, and the series of discontinuous rings extends out from the center to a distance of about 1500 km. Note the decrease in crater frequency toward the center of the multiring structure. This results because older craters were destroyed by the large impact that formed the interior of the ringed structure.

ice. Like Ganymede, the internal structure of Callisto is believed to consist of a central silicate core surrounded by a thick cryosphere or mantle of ice. This conclusion is based on its low density, the presence of exposed water, and its inferred accretionary history. Nonetheless, the surface of Callisto is relatively dark, being about half as reflective as Ganymede (but still twice as bright as the Moon) suggesting that the surface of Callisto is dirty ice. Temperatures at the equator are a frigid 120 K.

Impact Craters. Images sent back from Voyager 1 show that although Callisto is densely cratered, somewhat like the highland regions of Earth's Moon, it is different from the Moon in several important ways. There is a distinct paucity of craters larger than 50 km in diameter, as compared to the inner planets. The reason is not immediately obvious, but the anomalous distribution in crater size may be due to the peculiarity of the population of impacting particles in the outer solar system, or it may be due to the response of ice to impact (large craters may have been preferentially destroyed).

A glance at Figure 9.31 reveals that many craters have bright areas in their centers that are reminiscent of the central peaks in craters on the terrestrial planets. These bright areas are probably the result of clean ice that was melted during impact and refrozen at the surface. Many of the craters have central pits. The bright rims, rays, and ejecta are also believed to represent cleaner ice from the substratum, brought to the surface during the impact.

The densely cratered terrain indicates that most of the surface of Callisto is very old, probably dating back to the period of intense bombardment ending about 4 billion years ago. The extreme range in the degree of preservation of the ejecta and rays indicates that the cratering process continued throughout geologic time, with the rate of cratering decreasing exponentially with time similar to the cratering histories of the terrestrial planets.

Multiring Structures. Aside from the densely cratered terrain the most striking features on Callisto are several large multiring structures (Figure 9.33). Although these features are similar in some respects to the large multiring impact basins on the Moon, Mercury, and Mars, there are several important differences. The most obvious difference is that the rings are more numerous and are much closer together. Just as important is the fact that there is no central basin. Concentric rings on Callisto encircle Valhalla (Figure 9.33), a bright central

palimpsest about 300 km in diameter, and extend another 1000 km outward from the bright interior, like gigantic ripples. At least 25 separate rings exist; segments of some rings are continuous for about a thousand kilometers (Figure 9.34). Other, smaller multiring structures are shown in Figure 9.35. Most investigators have concluded that these ancient ring structures are the result of large impacts early in Callisto's history. Nonetheless, these are in striking contrast to the multiring impact basins on the terrestrial planets, which generally have less than five well-defined rings. In addition, the multiring basins on Callisto lack the radially textured ejecta blankets that surround basins on other bodies. Another difference on Callisto lies in the regularity of the spacing of the rings. On the inner planets the spacing increases outward from the center of the basin.

The difference between the multiring structure on Callisto and those on the terrestrial planets is probably due to the way ice responds to the impact of a large asteroid-sized body. Viscous flow would be expected in an icy crust during or subsequent to impact and would be expected to reduce the topographic relief of the basin and ringed structures, leaving only low, closely spaced ridges and a central area of high reflectivity to mark the former position of the impact basin.

Beyond Valhalla, the surface of Callisto is nearly saturated with small craters, but the central area is far less heavily cratered, as can be seen in Figure 9.33. The crater frequency increases from the center outward to the outer rings, where it reaches the average frequency of the surface of the satellite. These relationships imply that Valhalla was formed relatively late in the period of intense bombardment.

Internal Structure and Dynamics. Callisto's surface features indicate that its structure and evolution were significantly different from its near neighbor Ganymede. Figure 9.36 shows a possible arrangement for Callisto's interior. Since it has a lower density than Ganymede, it probably has a larger fraction of ice and a smaller rocky core. Callisto shows no evidence of the global tectonism that resulted in the production of Ganymede's grooved terrain. It has also retained more craters and large multiring features than has Ganymede. Both of these observations imply that after the heavy bombardment Callisto's lithosphere deformed less than Ganymede's. Cold, thick, rigid lithospheres are less likely to deform than thin ones, so it appears that Callisto cooled rapidly (after a presumed differentiation epoch), forming a thick, relatively immobile lithosphere before the

Figure 9.34

Details of the multiring structure Valhalla show that the rings are discontinuous. The presence of superposed craters on the ring system shows that the multiring basin formed early in Callisto's history.

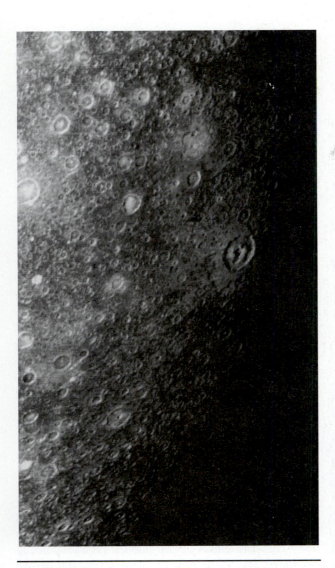

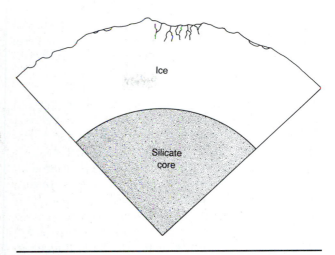

Figure 9.36

The internal structure of Callisto is probably similar to that of Ganymede; yet the tectonic history of the two planetary bodies was vastly different. Perhaps Callisto's crust became thicker and stronger more rapidly than Ganymede's and was unable to respond to movement of material in the asthenosphere below.

planet's silicate rocks). Additional small differences in the amount of heat inherited from accretion, heat generated by differentiation, or past tidal interactions with Jupiter may explain why Callisto cooled more rapidly than Ganymede and retained an ancient cratered surface.

The Other Moons of Jupiter

Although substantial geologic information is only available for the four Galilean satellites, we know a little about Jupiter's other satellites. Inside the orbit of Io lie four relatively small satellites. Two, Adrastea and Metis, are nearly co-orbital and appear to control the rings of Jupiter. Both are small (a few tens of kilometers in diameter) and dark (suggesting they lack ice). Amalthea, the next moon out from Jupiter, is large by comparison. The images obtained by Voyager 1 have an effective resolution of 8 km, so we know some details about Amalthea's overall shape, size, and color, but many important questions concerning its surface features remain. It is a small, elongate, and irregularly shaped body about 270 km long and 170 km wide. In a general way, Amalthea resembles Phobos, the larger moon of Mars, but it is ten times larger and is therefore the first intermediate-sized planetary object observed in detail. It is in synchronous rotation—its long axis is pointed toward Jupiter at all times—and whizzes around the planet every 12 hours, only 1.55 Jupiter radii above the top of the clouds. Amalthea is a dark object, much

Figure 9.35

Small multiring structures are also found on Callisto; most of them are less than 250 km in diameter.

end of the period of heavy bombardment, which probably occurred about 4 billion years ago. Similar rapid lithospheric thickening also occurred on the Moon and Mercury, and evidence of any earlier deformation of their thin lithospheres was destroyed by numerous impact events.

Small differences in the amount of internal energy available in Callisto and Ganymede may account for the significantly different surface features on these satellites. The principal difference lies in the apparent absence of a period of expansion, rifting, and watery volcanism on Callisto, at least after the heavy bombardment. Why didn't Callisto expand like Ganymede? Compared to Ganymede, Callisto's slightly smaller size and smaller silicate fraction may have resulted in more rapid heat loss and less radiogenic heat production (radioactive elements condense with and reside in a darker than the Galilean satellites, and is similar to some of the very dark asteroids. There are several bright spots on its surface, but its overall reflectivity is less than 10 percent. Amalthea is also reddish, probably because of a coating or alteration of its surface material, although it is possible that the red color is characteristic of the bulk of Amalthea. Some suggest that sulfur spewed from the volcanoes of Io is collecting on Amalthea's surface.

The irregular surface of Amalthea is scarred by a number of craters formed by meteorite impact. The largest has a diameter of about 90 km and a depth of about 8 km. It is much deeper than craters of similar diameter on the Moon (less than 5 km deep). Large indentations on Amalthea appear to be impact craters that have light markings on their walls. These bright patches may represent brighter, deeper material stripped of its sulphur-tainted regolith by slumping on steep slopes. The irregular shape of Amalthea probably results from its small size and from a long history of impact cratering that fragmented it. The orbits of these small satellites are all nearly circular, are **prograde**, and lie in the plane of Jupiter's equator.

Thebe is a small satellite, about 100 km across, that orbits Jupiter beyond Amalthea. No good images of its surface or details about its density or composition have been acquired yet.

The eight satellites of Jupiter that lie beyond Callisto can be divided into two groups of four, each of which have substantially different orbital properties and perhaps origins as well. The inner group travels in prograde but inclined orbits at 11 to 12 million km from Jupiter. They range in diameter from about 10 to about 180 km. Although the spectral data are meager, the satellites appear to have surface compositions like some primitive (undifferentiated) asteroids. The outermost group consists of four satellites with diameters ranging from 20 to 40 km. They all have irregular, highly inclined orbits and travel around Jupiter in the direction opposite to all the other satellites, at a distance of about 22 million km. When compared to Callisto, these satellites have substantially different sizes, orbital characteristics, and perhaps compositions as well. Some scientists speculate that the two outer groups represent the shattered remains of two asteroids captured by Jupiter. This hypothesis will remain in the realm of conjecture until more definitive data about their compositions and surface features can be obtained.

Conclusions

Jupiter is dramatically different from any of the other planets we have examined so far. It is huge by comparison to the inner planets. Its gaseous constitution bears little resemblance to the rocky inner planets. These are not quirks that Jupiter developed over its 4.5 billion year history. Instead, many of these fundamental differences were inherited from the creation of the planet itself. Consider the scenario for the origin of the planets outlined in Chapter 2. Jupiter lies farther from the Sun than any of the inner planets. In fact, it is reasonable to conclude that the temperature in this part of the ancient solar nebula was only about 100 K. At this temperature, solid particles condensing from the gases of the solar nebula consisted not only of metals, oxides, and silicates, but also included huge quantities of water ice. During this nebular blizzard, a large portion of the gaseous nebula became solid and accreted to form icy planetesimals. In fact, a **snow line** of sorts must have existed in the ancient nebula at the outer limits of the asteroid belt. Beyond this distance from the Sun, ice is stable, just as snow is stable on many cold mountain peaks but unstable on the warmer, lower slopes. Moreover, the snow line acted as a filter, solidifying and thereby removing water vapor from the nebular gas being swept out of the inner solar system during the Sun's early, violent T-Tauri phase, the first 1 million years or so of its existence. Computer simulations suggest that accretion of these icy planetesimals, and lesser proportions of silicates, occurred in several hundred thousand years to form a protoplanet about ten times as massive as Earth. So huge was this icy body that the gases of the nebula catastrophically collapsed onto it by gravitational attraction before they could be swept out of the solar system by T-Tauri winds. Jupiter, a massive ball of hydrogen and helium diluted by traces of other constituents, was born. The tremendous gravitational force exerted by this titan perturbed the orbits of planetesimals in the adjacent asteroid belt so much that no large planet ever formed there. In this zone, impact caused fragmentation not accretion. Some planetesimals were accelerated so much that they were tossed out of the solar system. Moreover, Jupiter stole much of the mass from this zone to feed its own rapid growth. Perhaps even the relatively small size of Mars can be blamed on Jupiter. Its gravitational forces may have ejected much of the mass from the feeding zone for Mars.

One of the challenges facing planetary scientists studying the Jovian satellites is to explain why there is such a great range of physical properties and evolutionary histories among objects that presumably formed very near to one another from similar nebular materials. Obviously the great size differences between the Galilean satellites and Jupiter's other moons is one anomaly, but what causes the differences among the individual Galilean satellites? Perhaps the answer lies in the nature of the early history of the Jupiter system. Earlier we saw how the planets forming close to the protosun accreted from volatile-poor substances (silicates) because low-temperature compounds (ices) could not condense at the prevailing temperature. Thus, the inner planets are small and volatile-poor because the pool of materials that condensed at high temperatures was also small. At distances from the Sun that allowed ices to condense, much larger planets formed because water was a substantial part of the condensable material in the nebula. It is plausible that Jupiter was also surrounded by its own miniature disk of nebular gas and dust.

Furthermore, calculations of Jupiter's early history suggest that during its first several million years of development it may have radiated about one ten-thousandth as much energy as the Sun does today. This central heat source would have established a localized thermal gradient in the condensing nebular cloud. Following this hypothesis, it appears that Amalthea and the small inner satellites accreted from refractory materials condensed at high temperatures (600 to 1200 K) and are small because of the restricted quantities of materials that could condense at these temperatures. The condensates that went into Io may have formed at temperatures low enough to allow the crystallization of hydrous silicates. However, Io appears to have efficiently expelled this water during volcanism caused by its unusual tidal heat source. The particles that accreted to form Europa may have condensed at even lower temperatures, but this moon appears to have retained the water released during differentiation to form a relatively thin cryosphere. The materials of the two outer Galilean satellites, Ganymede and Callisto, probably formed at temperatures low enough (85 to 160 K) for substantial quantities of icy condensates to crystallize and accrete. Thus, they are larger and less dense than their companions Io and Europa, having incorporated a larger fraction of low-density volatiles from the Jovian nebula. As mentioned earlier, the outermost eight satellites may not have formed from this Jupiter-centered nebula; if they did, we would expect them to be little more than balls of ice.

An important result of the study of the Jupiter system is that the Sun and its system of planets are analogous in many ways to Jupiter and its system of moons. Both systems apparently formed in very similar ways and may have been dominated by a central energy source that established a temperature gradient in the ancient nebula, resulting in substantial chemical differences in the bodies that formed around these islands of warmth.

The differences in the geologic histories of the Galilean satellites are partly a result of dramatic chemical differences but are compounded by the operation of tidal heating of satellites within the system. Although the time at which orbital resonance was achieved is not known, the resulting tidal heating has, to varying degrees, reshaped the surfaces of the satellites. Io, which experiences the largest energy input, is still volcanically alive; its surfaces are probably all less than a few million years old. Europa, with a modest tidal input, has a youthful surface as well, but there are some impact craters. Surfaces on Europa may be as young as several hundred million years old. Ganymede, affected little by tidal heating, nonetheless has surfaces that postdate its intense bombardment. Perhaps a larger endowment of radioactive elements allowed it to expand and remain geologically active until after about 3.5 billion years when its distinctive grooved terrain remolded its surface appearance. Callisto, with the most accelerated thermal history of all the Galilean satellites, has intensely cratered surfaces that date from 4.0 billion years ago. It appears to be untouched by tidal heating. Perhaps nowhere in the solar system are the fundamental natures of planetary composition and thermal history so strikingly portrayed. These elements—the interaction of heat with matter over the course of time—create planetary histories and produce the endless variation we see in the surface features of planetary objects.

Review Questions

1. Contrast the composition of Jupiter's atmosphere with that of Mars. Suggest some reasons why they are so different.
2. Do features like the Giant Red Spot occur in the atmospheres of the Earth?
3. In what characteristics does the atmosphere of Venus contrast with the atmosphere of Jupiter? How do the atmospheres contrast in origin?
4. Does Jupiter have a surface?
5. Why do the rings of Jupiter lie in its equatorial plane?
6. What indicates that the outer satellites of Jupiter (those that are beyond the Galilean satellites) might have formed elsewhere and then been captured?

7. Contrast the characteristics of the crust of Io and Callisto.

8. What heat-producing mechanism keeps Io volcanically active? How does this differ from the way in which the Earth is heated? Would you expect the mechanism that keeps Io warm to heat satellites in any other planetary systems?

9. What causes some volcanic eruptions to be explosive, with tall eruption plumes, and others to be quiet eruptions of lava?

10. Compare the composition of the materials found at the surface of Io and Europa.

11. How do we know that the surface of Europa is relatively young?

12. What might the broad stripes that cut across the surface of Europa reveal about its tectonic history?

13. Would it be difficult for liquid water to erupt as a "lava" onto the surface of a planet with an icy crust like Europa? Why?

14. Why are large old craters on Ganymede and Callisto much shallower and more subdued than those found on the Earth's Moon?

15. Do any of the inner planets have terrains similar to the grooved terrain on Ganymede?

16. What is the evidence that Ganymede and Callisto have thick outer layers composed of water ice?

17. Are there significant differences between the multiring impact features found on Callisto and those found on the inner planets? Why?

18. Contrast the geological histories of Ganymede and Callisto. What accounts for the lack of grooved terrain on Callisto?

19. Prepare a table that compares the diameters, densities, surface compositions, ages, and features of the Galilean satellites.

20. What events or processes created the differences between the rocky, ice-poor Galilean satellites (Io and Europa) and the ice-rich satellites (Ganymede and Callisto)?

Key Terms

Co-Orbital

Grooved Terrain

Intervent Plains

Magnetosphere

Metallic Hydrogen

Orbital Resonance

Outer Planets

Prograde

Palimpsest

Ring

Snow Line

Spectral Reflectance

Thiosphere

Additional Reading

Morrison, D., ed., 1982. *Satellites of Jupiter*. Tucson: University of Arizona Press.

Science, 1979. Vol. 204, pp. 945–1008. (Reports on Voyager 1 Encounter with the Jovian System).

Voyager to Jupiter. 1980. NASA-SP 439. Washington, DC.

TABLE 10.1

Physical and Orbital Characteristics of Saturn

Mean Distance from Sun (Earth = 1)	9.54
Period of Revolution	29.5 y
Period of Rotation	10.2h
Inclination of Axis	29°
Equatorial Diameter	120,660 km
Mass (Earth = 1)	95
Volume (Earth = 1)	760
Density (g/cm^3)	0.69 g/cm^3
Atmosphere (main components)	H_2, He, CH_4, NH_3
Temperature (at 1 bar)	120 to 160 K
Magnetic Field (Earth = 1)	0.4
Gravity (Earth = 1)	1.16
Known Satellites	17

Saturn and Satellites

Mimas

Enceladus

Tethys

Dione

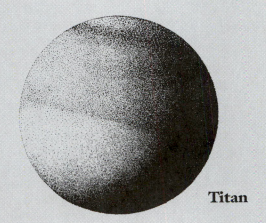

Rhea

Titan

Lapetus

Earth

CHAPTER 10

The Saturn System

Introduction

Saturn is considered by many to be the most beautiful object in the known universe. Anyone who has seen the planet through a powerful telescope will never forget the sight. A pastel-toned, yellowish globe, noticeably flattened at the poles, appears to float in the center of a delicate ring system as a simple but elegant jewel (Figure 10.1). In reality, Saturn is mostly a gigantic ball of gaseous hydrogen and helium surrounding a zone of liquid hydrogen and possibly a small rocky core in the interior. Saturn contains more than 100 times the volume of Earth but has a density less than that of water.

Saturn is the center of a large system of satellites. With its elaborate system of rings and multiplicity of satellites, Saturn resembles a miniature solar system. Thus, study of Saturn adds new insights to those gained from the examination of the Jupiter system and may ultimately reveal more clues about the origin and evolution of the Sun's system of planets. Although Saturn was visited by Pioneer 11 in 1979, our first detailed glimpses of the satellites were provided by the Voyager spacecraft as they flew through the system in November 1980 and August 1981.

Major Concepts

1. Saturn, somewhat smaller than Jupiter, is a large gas- and ice-rich planet well known for its ring system. It also has at least 17 satellites, most of which are icy.

2. Saturn consists principally of hydrogen and helium but appears to have an ice-and-rock core. The atmosphere has cloudy bands parallel to the equator and contains storms like Jupiter's but is not as colorful.

3. Saturn's ring system consists of a myriad of small ice particles in orbit around the planet. The rings are defined by small shepherd satellites and by resonances with larger satellites. The ice particles may be material that never accreted to form a moon or may be the result of collisions in the satellite system.

4. Saturn has seven moons with diameters greater than about 400 km. The larger satellites are farther from Saturn.

5. Titan is larger than the planet Mercury and has an atmosphere rich in nitrogen. Its surface is obscured by haze but may be covered by hydrocarbons precipitated from the atmosphere. Methane (natural gas) or ethane (another hydrocarbon) liquids may exist at the surface.

6. Six icy satellites, ranging from 400 to 1500 km in diameter, lack atmospheres and reveal surfaces modified by impact, tectonic, and volcanic processes. Enceladus, Tethys, Dione, and perhaps Rhea had extended thermal histories, indicated by young plains. The period over which their interiors remained warm may have been extended by tidal heating, particularly in the case of Enceladus. In addition, each of these bodies has large fractures or rifts, suggesting that each body cracked, perhaps while cooling. Mimas and Iapetus are impact-dominated bodies. It is possible that some satellites of Saturn were disrupted during collisions with incoming projectiles and then reaccreted.

Saturn and Its Satellites

The table of physical and orbital characteristics (Table 10.1) summarizes some numerical information about Saturn, the sixth planet from the Sun. Saturn shares many general characteristics with the other three large planets of the outer solar system. It is the second-largest planet, with a diameter about 80 percent that of Jupiter and ten times that of Earth. Because of its composition (rich in hydrogen and helium), Saturn has the lowest density of all the planets, even less than that of liquid water. Like Jupiter, Saturn has a large magnetic field, generated deep within its interior, and a rapid rotation rate. Its rapid spin (a day is only about 10 hours long) causes the fluid planet to bulge at the equator and flatten at the poles. The diameter at Saturn's equator is 13,000 km greater than the diameter measured through its spin axis. Saturn lies twice as far from the Sun as Jupiter and actually radiates more energy from its interior than it receives from the far distant Sun.

Saturn has a marvelous system of at least 23 orbiting moons and rings (Table 10.2). In fact, Saturn has more moons greater than 100 km in diameter than any other planet. The satellites range from moons the size of Mercury to ring particles smaller than sand grains. Most of the moons consist mainly of water ice, with lesser amounts of silicate rocks and even smaller, but significant, quantities of methane and ammonia.

The Interior of Saturn

Saturn has about the same composition as the Sun and the solar nebula from which the solar system formed some 4.6 billion years ago. Hydrogen is the most abundant element (nearly 80 percent), and helium is believed to constitute nearly 18 percent of the planet. Other elements—oxygen, iron, neon, nitrogen, and silicon—make up most of the remaining 2 percent. There is no particular reason to believe that a large planet like Saturn has selectively lost any of the gases that were originally available to form it. Its internal structure must therefore be dominated by different forms of hydrogen, forms that are stable at the high temperatures and pressures in the deep interior.

Most models of the internal structure of Saturn are similar to that shown in Figure 10.2. The cloud zone forms a thin envelope that is only 0.1 to 0.3 percent of the planet's radius. Beneath this is a clear atmosphere of hydrogen, which extends downward and gradually increases in density. There is no solid surface, just gas that increases in density and temperature down to a level where hydrogen changes from a gas to a liquid. This zone beneath the cloud cover is a zone of molecular hydrogen.

TABLE 10.2
Characteristics of Saturn's Major Satellites

Satellite	Diameter (km)	Density (g/cm³)	Notes
Atlas	40×20		A-ring shepherd
Prometheus	140×74		F-ring shepherd
Pandora	110×66		F-ring shepherd
Janus	220×160	0.67	Co-orbital, near rings
Epimethus	140×100	0.64	Co-orbital, near rings
Mimas	394	1.43	Heavily cratered
Enceladus	502	1.20	Extensive resurfacing and fractures
Tethys	1060	1.25	Some resurfacing and fractures
Telesto	30×16		Irregular shape
Calypso	24×22		Irregular shape
Dione	1120	1.43	Some resurfacing and fractures
Helene	34×30		Irregular shape
Rhea	1530	1.33	Heavily cratered
Titan	5150	1.88	Nitrogen-rich atmosphere
Hyperion	380×228		Irregular shape
Iapetus	1436	1.2	Dark hemisphere, heavily cratered
Phoebe	220		Retrograde orbit, silicate composition
Moon	3500	3.30	

Figure 10.1

Saturn is unique among the planets in that it has a bright complex system of rings. This image was constructed by Voyager 2 on August 4, 1981, from a distance of 20 million km. Three of Saturn's small icy moons can be seen. From left to right (in order of their distance from the planet) they are Tethys, Dione, and Rhea. The shadow of Tethys appears as a prominent black dot on the southern hemisphere. The atmosphere of Saturn is characterized by soft pastel yellow hues and shows many contrasting light and dark bands.

A significant transition occurs at a depth of about 30,000 km (half the radius of the planet), where molecular liquid hydrogen (H_2) becomes metallic liquid hydrogen (H^+, in which the molecules have been stripped of their outer electrons). The metallic hydrogen zone in Saturn is much smaller than that in Jupiter because of the lower pressure in Saturn. As for Jupiter, the magnetic field of Saturn presumably is caused by convection in the metallic hydrogen zone.

A second abrupt transition marks the change between the metallic hydrogen zone and the ice-silicate core, which is about 17 times as massive as Earth. It is assumed that most of the iron, silicon, and other heavy elements together with much of the water, ammonia, and methane ices are separated because of their density and are concentrated in a central core.

Saturn radiates 2.8 times as much heat as it receives from the Sun. It was first assumed that this heat was primordial, like that emitted from Jupiter, a vestige from the period of accretion. But Saturn is smaller than Jupiter and should have lost much of its original heat. Consequently, Saturn should now be radiating less heat than Jupiter. Some process within Saturn must be generating this excess heat. Perhaps, the separation of helium from liquid metallic hydrogen occurs in the interior,

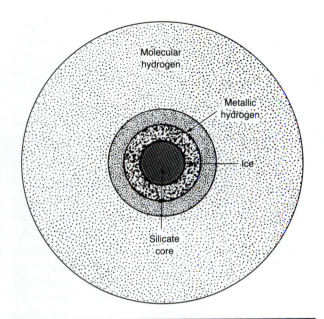

Figure 10.2

The internal structure of Saturn is believed to be similar to that of Jupiter and the other gaseous planets. A thin outer envelope of clouds covers a thick layer of hydrogen, which increases in density downward. A significant boundary is believed to exist at a depth of 30,000 km, where molecular liquid hydrogen converts to metallic liquid hydrogen. The core of Saturn may consist of ice and silicate rock at such a high temperature that it may be in a liquid state. Saturn does not have a well-defined solid surface.

releasing energy in the process. If part of the helium forms droplets and rains down to deeper levels, it would release enough heat by its fall and loss of gravitational potential energy to produce the excess heat. Apparently because of this sort of internal differentiation, the upper atmosphere of Saturn is depleted in helium compared to Jupiter or the Sun. Saturn's upper atmosphere has a measured concentration of about 6 percent helium (by volume) compared to 11 percent in Jupiter's upper atmosphere. Why isn't Jupiter experiencing this same sort of chemical differentiation? One possibility is that the interior temperature of Jupiter is slightly higher than Saturn's, allowing helium to remain in solution with the voluminous metallic hydrogen.

The Atmosphere

Saturn's atmosphere is roughly similar to Jupiter's but the patterns of circulation are more obscured and muted because of a lack of contrast and a layer of overlying haze. This gives Saturn a soft, velvety appearance, subduing the color in its atmospheric patterns. At close range the haze appears

more tenuous, however, and with computer enhancement of the Voyager photographs, a wealth of detail can be seen (Figure 10.3). The most prominent features on both Saturn and Jupiter are narrow, contrasting bands that circle the planets parallel to their equators. On Saturn, the velocity of the winds in these bands is enormous, moving up to 1500 km per hour (15 miles per min or 900 miles per hour). They are traveling about four times as fast as the equatorial clouds of Jupiter, which makes Saturn's clouds the fastest in any planetary system.

At higher latitudes, where wind speeds are lower, warm air rising from the interior can break up to form large circular storm systems with complex eddies (Figure 10.3). These appear on the photographs as a variety of oval spots ranging in color from white to brown and red. The smaller eddies may be short-lived and last only a few days, but the larger features may persist for a year or more. The largest storms on Saturn are the oval spots; they are substantially smaller than those on Jupiter. In addition, there are ribbonlike jet streams (Figure 10.3).

What could cause such powerful winds and create the swirling storm and jet streams? Unlike Earth, Mars, Venus, and probably Titan, whose atmospheres are driven by solar energy, the rapid winds and atmospheric motion on Saturn are

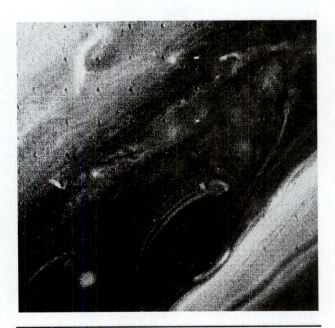

Figure 10.3

Saturn's atmosphere consists of contrasting bands that circle the planet parallel to the equator. The ribbonlike structures are believed to be large-scale waves lying in a rapidly moving jet stream. Cyclonic storms, complex eddies, and convective oval clouds are all observed in various bands.

caused when heat from the planet's interior is transferred to the atmosphere.

The Rings

Since the rings of Saturn were discovered by telescopic observations almost 400 years ago, they have amazed and intrigued astronomers, who strained at the eyepieces of their telescopes to glimpse fine details of these magnificent structures. Now that we have seen them up close through the eyes of the Pioneer and Voyager spacecraft, they are even more astonishing and we are perhaps as surprised as the first people who saw them. Even with the explosion of our knowledge about them, the rings remain not only one of the most beautiful features in the heavens, but also one of the most scientifically challenging.

The rings of Saturn consist of billions of particles, ranging in size from the finest dust grains to large boulders 10 meters or more in diameter. Judging from their brightness and densities, the particles are probably made of water ice. Each particle moves in its own independent orbit and at its own independent speed. Each can be considered as a separate satellite, but the particles are concentrated in a thin region in Saturn's equatorial plane, giving the appearance of a solid disc (Figure 10.4).

The main ring system extends from about 7000 km above Saturn's atmosphere out to the F ring, a total span of 74,000 km. It lies much closer to the planet than any of the major satellites. In fact, the bulk of the rings lies less than one planetary radius away from the surface of Saturn. Measurements made by instruments on the Voyager spacecraft indicate that the main rings of Saturn are generally no more than a few hundred meters thick. (If scaled

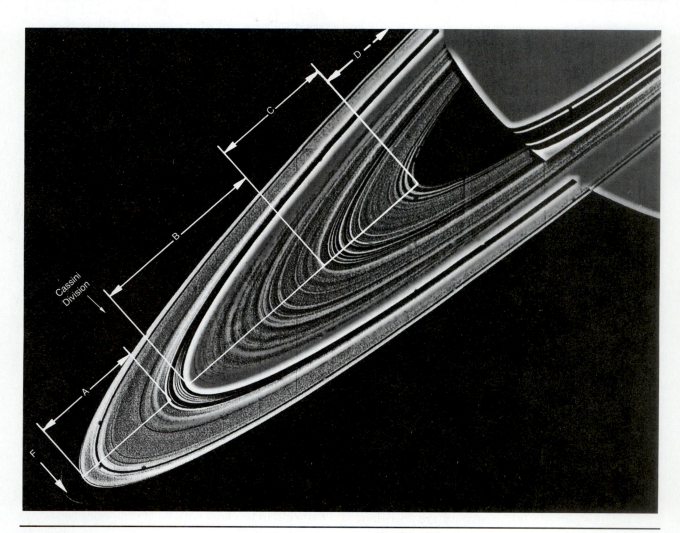

Figure 10.4

The rings of Saturn are extraordinarily complex. The classical divisions of the rings have been labeled A, B, C, and D, but photographs and measurements made by the Voyager spacecraft indicate that there are thousands of separate rings. In the photograph shown here, over 95 concentric structures can be counted.

down to the size of a football field, the rings would be thinner than a razor's edge.) This is truly an incredible geometry and is responsible for the delicate beauty of the ring system. In terms of mass, the material in the rings is roughly equivalent to a satellite of ice 500 km in diameter.

Although ring material is organized into an extremely complex system of tens of thousands of separate ringlets, there are seven major sections. Each section is separated by a gap that is relatively devoid of particles or by an abrupt change in the spacing or sizes of the particles. The rings of Saturn were labeled in order of their discovery, and so the labels have nothing to do with the rings' relative positions. Let us consider them in order outward from Saturn (Figure 10.4).

The ring particles are reddish, tan, or brownish, and there are definite color differences between the rings. The color of pure water ice is white, so it is believed that the ring colors come either from impurities, such as iron oxides, or from structural damage to the ice crystals, caused by prolonged exposure to energetic atomic particles or by ultraviolet radiation from the Sun.

The D Ring

The innermost ring that we now know to orbit Saturn is the D ring. It extends from within 7000 km of Saturn's atmosphere to the inner edge of the C ring but is not visible in telescopic view nor in normal Voyager photos. Its existence was suspected from Earth-based photos but was never confirmed until long-exposure photographs from Voyager detected it.

The D ring consists of many very thin ringlets, which appear to have as intricate a structure as the more dense regions in the C and B rings. The material of the D ring is probably very fine grained, possibly derived from the C ring leaking past a hypothetical satellite resonance that defines the inner edge of the C ring. It is probable that D-ring material spirals rapidly into Saturn's atmosphere because of the drag force on it (Figure 10.5).

The C Ring

The C ring is the innermost of the classical rings of Saturn and can be faintly seen from Earth, even through a relatively small telescope. The C ring is distinguished by a regular ordering of light and dark bands (Figure 10.6). Two prominent gaps 200 to 300 km wide occur and add structural definition to the ring. Various measurements suggest that the particles in the C ring have a mean diameter of 1 m, and that the particles are spaced about 20 m apart on the average. Probably a few objects with

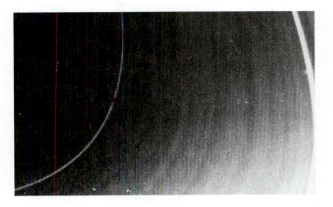

Figure 10.5

The D Ring is very faint inside the C ring, but even in this tenuous region the ring clearly possesses a great deal of fine structure.

C-Ring

Figure 10.6

The C Ring is 19,000 km wide and displays a regular ordering of light and dark bands. Its inner edge is sharp and well defined whereas the outer boundary is slightly less distinct.

diameters of tens of meters to a kilometer or so are present. The C ring is deficient in very small particles when compared to the B and A rings.

The B Ring

The B ring is the large central body of the ring system. It is separated from the A ring by the Cassini Division (see below) and from the C ring by a rapid transition to less dense, nearly transparent material. It is the brightest, widest, thickest, and most opaque of Saturn's rings and probably contains most of the ring mass. Various measurements indicate that the B ring is no more than 2 km thick, the thickest in the main ring system. Data from the Voyager spacecraft and from radar observations further show that the most abundant particles are about 10 cm in diameter. The largest common particles are about 10 meters across. Dust-sized particles also are a significant fraction of the B ring.

Voyager pictures show that the B ring displays an astonishingly detailed small-scale structure, consisting of thousands of fine bright and dark ringlets distributed without apparent regularity. These appear similar to grooves on a phonograph record and range in width down to the limits of the resolution of the Voyager camera (Figure 10.7).

Unexpected features of the ring system are radial features, or spokes (Figure 10.8). Most of the spokes appear in the central and most opaque part of the B ring. The spokes commonly are 10,000 km long, with widths ranging from 100 to 1000 km. Sequences of pictures taken of the spokes allowed scientists to develop a time-lapse movie showing growth and development of the spokes as the rings rotate. The inner section moves faster so that the

Figure 10.7

The B Ring is the dominant structure in Saturn's ring system. It is brighter than all other rings combined and presumably contains most of the ring mass. It also has the most extensive and uniform small-scale structure. Thousands of ringlets are discernible. This photograph shows a segment of the B ring 6000 km wide. The thinnest ringlets discernible are about 10 km across.

Figure 10.8

Radial spokes in the B Ring of Saturn were discovered by the Voyager cameras, and time-lapse photography has permitted scientists to study their motion. The spokes typically have a length of about 10,000 km and widths ranging from 100 to 1000 km. They are found pointing radially outward from Saturn in the densest part of the B ring.

spoke becomes tilted and sheared with time and eventually disappears. Once formed, each spoke is visible for most of the 10 hours it takes for a particle to complete one orbital revolution. Meanwhile, new spokes form sporadically at new locations in the ring. Although there are many unanswered questions, the spokes are believed to be composed of a cloud of dust levitated above the ring plane by electrostatic forces similar to static electricity that makes dust stand up on a record album (Figure 10.9).

The Cassini Division

The Cassini Division is the major gap separating the B and A rings. It was first noted in 1675 by Jean Cassini, a French astronomer. The inner edge (B-ring edge) is a sharp discontinuity at a 2 to 1 orbital resonance with the moon Mimas (where a ring particle orbits Saturn exactly twice for each orbit of Mimas). The outer edge is in resonance with Iapetus.

The Cassini Division is about 4000 km wide. The interior of the Cassini Division was once thought to be devoid of matter. However, it is seen in Voyager images to contain a great deal of structure (Figure 10.10). It seems likely that the gaps within the Cassini Division were cleared of ring particles by small satellites 10 to 30 km in diameter embedded in the rings. However, no such objects were found by the Voyagers. Another suggestion calls on the repeated gravitational tug of Mimas on these particles to pull and push them out of the Cassini Division.

The A Ring

The A ring, as seen on the Voyager images, is much more uniform than the B ring and appears relatively free of fine ringlets. The inner edge is sharp and opaque like the B ring, but most of the A

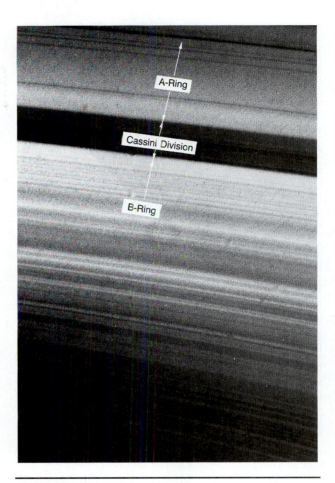

Figure 10.10

The Cassini Division was once thought to be nearly devoid of matter but, as shown on this detailed view, it has fine, bright rings, each of which contains additional small-scale structures.

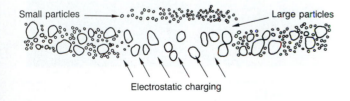

Figure 10.9

The origin of the spokes remains a mystery. The most plausible theory is that clouds of fine particles within the ring are levitated by electrostatic forces, perhaps related to the huge electrical discharges from Saturn detected by instruments on both Voyager spacecraft. **Large particles are too massive to be lifted.**

ring is quite transparent. The outer edge is also sharp and well defined, probably because of gravitational interactions with a small **shepherd** satellite. Particle sizes in the A ring span a broad range—from about 10 m in diameter down to micrometer-sized dust grains, which abound. In the A ring are several narrow ringlets due to orbital resonance as well as the prominent Encke Division.

The F Ring

The F ring consists of a narrow band of material located some 4000 km beyond the outer edge of the A ring (main ring system) (Figure 10.11). The total width is only 700 km, so that on many Voyager photographs this ring appears only as a thin line. The narrow restriction of the F ring is thought to result from the gravitational focusing or shepherding of two small satellites, Pandora and Prometheus, on either side of the ring. Both the ring and the satellites are slightly eccentric in their

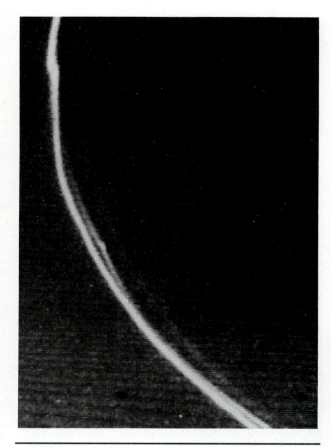

Figure 10.11

The F Ring is like a thin line circumscribing the classical rings of Saturn. It was discovered by Pioneer Saturn. At moderate resolution the F ring appears to clump, forming a spectacular variety of kinks, strands, and braids.

orbits, and at times the inner satellite can approach within grazing distance of the ring. The F ring is thought to be composed of fine-grained particles.

The photographs of the F ring provided many surprises. At moderate resolution, they revealed that the F ring is not uniformly bright all the way around. This may be due to variations in concentrations of ring material along the ring's orbit. At higher resolution, the Voyager 1 photographs showed kinks, warps, and knots in the F ring; the outer two strands appeared to be braided, or at least to intersect (Figure 10.11). But surprisingly, when Voyager 2 photographed five different areas of the ring, braids were found in only one small area. It thus appears that whatever created the braidlike structure is temporary or confined to small regions.

The origin of the clumps, kinks, and braids in the F ring remains one of the mysteries of the Saturnian system. The same kinds of electrostatic and magnetic forces that deflect the spokes in the B ring from strictly gravitational orbits may move finer particles in the F ring away from the larger

fragments. It has also been suggested that the strangely shaped forms of the F ring result from complex gravitational pull from the eccentric shepherding satellites. The data on the F ring, however, are insufficient to provide a clear explanation of this surprising ring structure.

The F ring is also remarkable in that it has an intricate fine structure of many ringlets (Figure 10.12).

The E Ring

Beyond the traditional system lies the E ring, which is scarcely more than a slight concentration of debris in the satellite orbital plane. It is visible only when the ring system is viewed approximately edge on. Observations made in 1980 show that

Figure 10.12

The internal structure of the F Ring is shown in this computer generated image of the bright F ring. This synthesized picture was made from data obtained from occultation measurements. The bright areas indicate denser spacing of material than in the dark zones. The highest resolution of the Voyager imaging system is seldom better than 10 km, but with the occultation measurements made by a different instrument, structures only hundreds of meters across can be discriminated.

material extends out to at least eight planetary radii, thickening toward the outer edge. The density of the particles may peak near the orbit of Enceladus. Voyager detected that only the E rings scattered light significantly, so many of its particles must be micron-sized, like those in Jupiter's ring.

Origin of the Rings

The recent discoveries of rings around Jupiter, Uranus, and Neptune suggest that ring formation may be the rule, not the exception, for large planets. The presence of so many ring systems implies that rings developed as part of the normal history of the outer planets. Three major hypotheses have been proposed to explain the origin of planetary rings: (1) tidal disruption of a satellite or comet, (2) large impact on a satellite, and (3) halted accretion.

The idea of tidal disruption was first proposed by a French mathematician, Edouard Albert Roche, in 1848. According to this hypothesis, Saturn's rings were generated when a single large body that came too close to the planet was fragmented into a myriad of pieces. The body may have been a comet that suffered a chance gravitational encounter. Calculations suggest that during the history of the solar system 10 to 100 large icy planetesimals or comets should have passed close enough to Saturn to be torn apart. As an example, in 1992, comet Shoemaker-Levy 9 was ripped into at least a dozen fragments when it passed close to Jupiter. If its orbit had been slightly different, a new ring could have been born. Instead, the fragments of the disrupted comet crashed into Jupiter in 1994. Alternatively, a small satellite that formed in the planet's gravitational envelope could have been broken apart. In any event, fragmentation would have resulted from tidal disruption (a step further than the tidal massaging suffered by Io). Roche calculated that these tidal forces could exceed the cohesive gravitational forces of a liquid satellite if the satellite came closer than about 2.5 times the radius of Saturn. This disruption threshold, or **Roche limit**, lies close to the outer edge of Saturn's main ring system. For rocky or icy satellites larger than about 40 km in diameter, the Roche limit is about 1.5 times the radius of the primary.

The second hypothesis for ring formation is that a moon, or moons, in the ring system was fragmented by collision with a large comet. The multitude of small particles went into orbit as a cloud of debris. The particles eventually fell into a disk around the planet's equator. Collisions of particles in the rings made the particles even smaller. The multiring basins on Callisto and Ganymede and the giant craters on Tethys and Mimas show that impact by large bodies in the outer solar system did indeed occur. It is not strictly by chance that catastrophic collisions should occur in the area occupied by the rings. The satellites of Saturn tend to become smaller at distances closer to the planet and a smaller moon would be more likely to fragment, for a given energy of collision, than would a larger one. In addition, the gravitational field of Saturn focuses the trajectories of meteorites and comets, so the number of meteorites passing a satellite close to the planet is significantly greater than for satellites at greater distances. Studies of the variations in the rings suggest that they are changing on a short time scale. Some scientists conclude from this that the rings could not have survived for billions of years. The particles slowly spiral into the atmosphere of Saturn or toward a ring moon. According to this model, the present ring system is quite young (between 10 and 100 million years old). Thus, the rings of Saturn would be the result of ongoing processes of creation by collisional disruption of small moons and eventual dispersal of the ring particles. The present small moons of Saturn may be the sources of new rings in the future. The principal problem with an impact origin for the rings is that it is difficult to form a moon as close to Saturn as the rings are today. Ring particles presumably would not accrete inside the Roche limit because of strong tidal forces.

The third hypothesis holds that the rings formed from the primordial circumplanetary nebula out of which the planet's satellites grew. Accretion began with cooling of the nebula and condensation of gaseous matter into solid grains, but (according to this hypothesis) the ring material did not grow into a large satellite. Accretion may have halted because of the gravitational effects of orbital resonances with small nearby satellites, which prevented ring material from accreting. Accordingly, the detailed structure of the rings of Saturn made visible by Voyager could be a fossil record of an intermediate stage in the accretion of planetary bodies. According to this hypothesis, the rings are long-lived features.

In short, the origin and evolution of planetary ring systems remain one of the enigmas in the solar system.

The Satellites

The Saturnian moons are a diverse and remarkable group of icy worlds. Unlike the Jovian moons, there are no regular variations in size and composition with distance from Saturn. The satel-

lite's dimensions range from the size of small asteroids to Titan, which is larger than the planet
Mercury. Seventeen satellites have now been identified. Nine of the largest satellites are shown in
Figure 10.13. Fifteen are **regular satellites** (those
that have nearly circular prograde orbits in Saturn's equatorial plane). The two other satellites,
Iapetus and Phoebe, do not orbit in the equatorial
plane. Iapetus is in a prograde orbit inclined at 15
degrees to the plane of the system, and Phoebe is in
a **retrograde orbit** (meaning it moves in a direction opposite that of most satellites) with an inclination of about 150 degrees (Figure 10.14). Except
Titan, which is in a class by itself, the six larger
satellites of Saturn (those larger than 400 km in
diameter) have much in common. All are intermediate in size (400 to 1500 km in diameter) and all are
composed primarily of water ice. The ten smaller
satellites are irregular objects and may include
captured asteroids and fragments from intersatel-
lite collisions.

The six larger satellites fall into convenient
pairs that almost match their order outward from
Saturn—Mimas and Enceladus (400 to 500 km in
diameter), Tethys and Dione (approximately 1000
km in diameter), and Rhea and Iapetus (approximately 1500 km in diameter). Hyperion and Phoebe
are smaller (irregular bodies with long dimensions
of 250 and 220 km).

Various measurements, both from Earth-based
observations and from spacecraft, indicate that the
satellites are composed predominantly of water ice.
All of the satellites for which we have estimates
have densities of less than 2 g/cm^3. The densities of
Phoebe and Hyperion, as well as those of the
smaller satellites, are unknown. For most of the
satellites, a composition of approximately 30 to 40
percent rock and 60 to 70 percent water ice would
match the estimated densities.

As noted before, there is good reason for ice to
be a dominant substance in the outer solar system.
When a gas whose composition is similar to that of
the Sun is cooled under conditions believed to have
existed in the early solar system, some oxygen in
the gas will combine with silicon to form silicate
rocks at relatively high temperatures. When the
silicon is exhausted, a substantial amount of oxygen
will remain. As temperatures continue to drop, the
remaining oxygen will combine with hydrogen, the
most abundant element in the gas, and will form
water. At even lower temperatures, ices of methane and ammonia should also form. Also, the tendency for ice to lose vapor to space depends
strongly on temperature. At a distance from the
Sun less than that of the asteroid belt, a mass of
water ice will evaporate in a relatively short time.

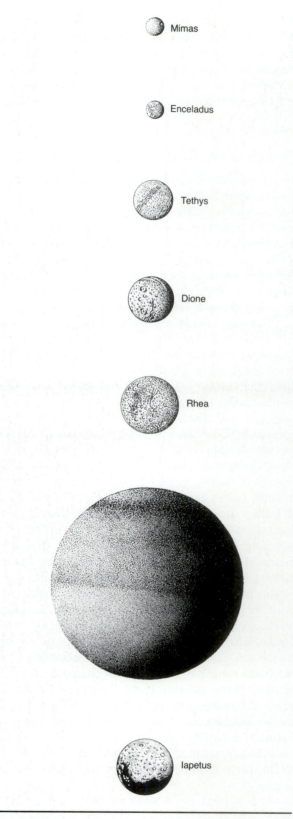

Figure 10.13

The major satellites of Saturn are small icy bodies
each of which records a fascinating history unknown
before the Voyager encounters. The six larger satellites
have much in common. Titan has an atmosphere and is in
a class by itself. The eight small satellites (not shown) are
irregular in shape.

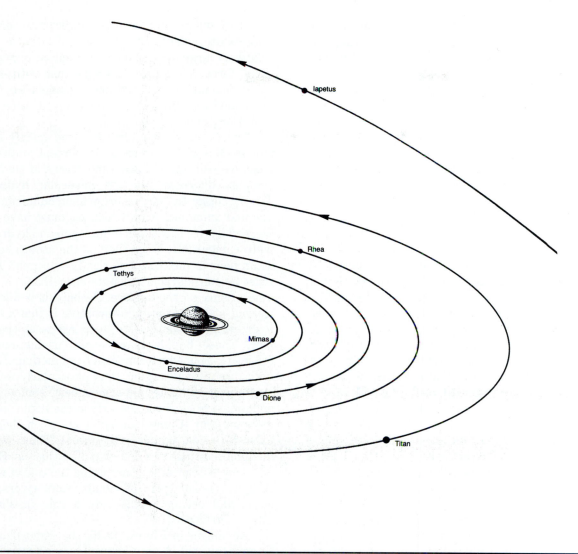

Figure 10.14

The moons of Saturn form a miniature planetary system. All but two of the satellites lie within the equatorial plane (and the plane of the planet's rings). Iapetus has an orbit inclined at 15 degrees and Phoebe at 150 degrees. Only the orbits of the seven largest moons are shown. At least five small satellites lie inside the orbit of Mimas.

At greater distances (Jupiter and beyond) a body of ice is stable for billions of years. Thus, ice is expected to be a major constituent in all bodies that accreted at low temperatures in the outer solar system.

One might anticipate that the small icy moons of Saturn would be of little geologic interest because of their cold origin, their primitive compositional character, and their apparent lack of internal sources of heat. Ice, however, has a melting point far lower than that of rock, so relatively little heat is needed to melt the interior of an icy moon. Therefore, the moons of Saturn could develop some planetary dynamics with relatively little internal energy. In addition, these moons probably contain

small amounts of substances such as ammonia hydrates and compounds of methane and water. The interior of a body containing such material melts at a lower temperature than does one consisting of rock and ice alone. Although this was known long ago, the new information about the small icy satellites of Saturn obtained from the Voyager missions surpassed all expectations. Tethys, Enceladus, Dione, and perhaps Rhea show a surprising diversity of evidence for a long-lived source of internal heat that has been responsible for a complex history of tectonic and volcanic activity. In addition, observations suggest that there may be many combinations of energy sources (tidal heating, short-lived radionuclides, accretion of impact heating, and long-lived radionuclides) and compositions (low-melting-point condensate) that can result in internal melting and prolonged geologic activity in small, predominantly icy planetary bodies.

As the primary bodies in Saturn's miniature planetary system, these satellites provide the most important information about the origin and evolution of the Saturnian system. We discuss Titan first because of its unique properties and follow this by accounts of eight other satellites, in the order they occur outward from Saturn (Figure 10.14).

Titan

Saturn's giant moon Titan is considered by some planetary scientists to be the most intriguing moon in the solar system. With a diameter of 5150 km, it is one of the largest satellites in the solar system. In addition, Titan has long held a special allure as the only satellite known to have a substantial atmosphere. Its pressure and nitrogen-rich composition are similar to that of Earth. Voyager 1 was therefore programmed to fly past Titan at a distance of 2500 km in an effort to study its atmosphere and bulk properties and possibly to see through an opening in its cloud cover and photograph its surface. Unfortunately, Voyager 1 could not see through the impenetrable smog and opaque clouds, so the results of the mission offer little for geologic study. The data obtained about Titan's atmosphere, however, makes up for this deficiency.

As revealed by Voyager, Titan is slightly smaller than Jupiter's Ganymede and slightly larger than Callisto. In many respects it could be considered their near twin. Their sizes are within 5 percent of each other and their densities are all between 1.8 and 1.9 g/cm^3. (Titan is the densest of the moons of Saturn.) The similarities in size and density suggest that the bulk composition and structure of the three planetary bodies are probably similar, being composed of about 45 percent water ice and 55 percent silicate rock. Because of probable heating during accretion, some planetary differentiation seems very likely, so Titan's interior probably consists of a rocky core surrounded by a mantle of ice that may once have been liquid. However, Titan is very different from Ganymede and Callisto in that it has an atmosphere. Why? The answer is probably that Titan formed in a colder part of the solar nebula than Ganymede and Callisto and managed to incorporate ices of methane (CH_4) and ammonia (NH_3) into its bulk composition. These ices probably did not condense near Jupiter because it was too warm. With planetary differentiation, a fraction of these compounds es-caped by outgassing from the interior to form an atmosphere, which Titan could hold because of its relatively large mass and consequent large gravitational force. Therefore, it seems that with only a small amount of methane and ammonia ice, Titan developed an atmosphere and evolved differently than Ganymede and Callisto.

Titan's atmosphere is composed mainly of nitrogen (N_2) with as much as 10 percent methane—what we call natural gas—and traces of the more complex hydrocarbons and molecular hydrogen. Theory suggests that the original amounts of methane and ammonia, from which N_2 must have been derived by disassociation of nitrogen and hydrogen, were comparable. Why, then, is the atmosphere of Titan poor in methane but rich in nitrogen? There must be considerable amounts of methane in some kind of sink on the surface. Perhaps it is methane ice, but a more exciting alternative is that a sea of hydrocarbons about 1 km deep exists at the surface. These liquids may consist of methane or ethane (C_2H_6), which can form from methane, along with dissolved nitrogen. It is particularly intriguing to note that the mid-atmospheric temperature and pressure lie almost exactly at the triple point of methane and ethane (the point at which they can exist as solids, liquids, and gases). Thus, we can imagine a situation somewhat like that on Earth, with methane or ethane assuming the role of water. A hydrocarbon cycle of clouds, rain, rivers, glaciers, and oceans might play a role in shaping Titan's surface.

As shown in Figure 10.15, the atmosphere of Titan is layered. A multilayered haze merges into a darker hood over the north pole, but such a mantle is not present at the south pole. In addition, the atmosphere in the southern hemisphere is light, with a well-defined boundary at the satellite's equator. The northern polar region is darker than the neighboring areas.

A theoretical cross section of Titan's atmosphere is shown in Figure 10.16. Above 400 km is a layer transparent to visible light in which ultraviolet radiation is absorbed. Next are several thin, high, haze layers; below them is a thick smog layer between 200 and 300 km above the surface. This smog consists of tiny aerosol particles composed of hydrogen and carbon. It is presumed that these particles have been aggregating into larger particles and falling to the surface over the history of Titan. They may have accumulated into a gooey deposit hundreds of meters thick. Methane clouds and methane rain may occur below an altitude of 40 km but are unconfirmed. The temperature at the surface is estimated to be about 94 K, while the pressure is 1.5 times that of Earth. Nonetheless,

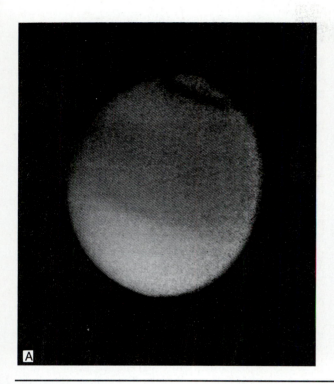

Figure 10.15

The atmosphere of Titan. (A) Titan's atmosphere is dense and opaque and completely obscures the surface of the satellite. (B) This closeup shows several layers of high haze above the opaque clouds.

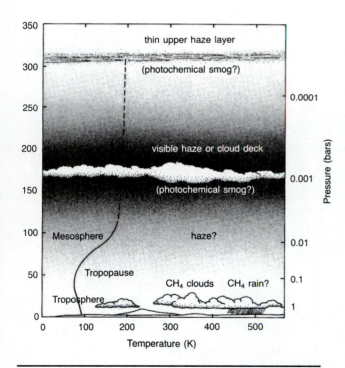

Figure 10.16

Titan's nitrogen-rich atmosphere extends up to 600 km above the surface. It consists of several thin layers of high haze, a main, thick smog layer between 150 and 200 km above the surface, and methane ice clouds less than 40 km in altitude. The curved line shows how temperature and pressure change with altitude.

the mass of the gas above each square meter of surface on Titan is about 10 times more than on Earth because it takes more gas to exert similar pressures at the low gravity of this satellite.

The dynamics of the atmosphere could involve the continuous formation of a multitude of large, complex organic molecules. When methane in the atmosphere is exposed to ultraviolet sunlight, its molecules lose one or more of their hydrogen atoms. These molecule fragments combine with other carbon compounds to form the more complex hydrocarbon aerosols. In the process, some hydrogen is released and may escape into space. The hydrocarbons ultimately precipitate and accumulate on the surface. Atmospheric methane is renewed by evaporation from surface liquids or frosts.

Studies of Titan's fascinating atmosphere may reveal a few clues about the formation of complex hydrocarbons on planetary surfaces, such as those that led to the evolution of life on Earth. On both worlds, several organic molecules were produced from the interaction of sunlight with volatile fluids. On Earth, these molecules became complex and developed the ability to reproduce themselves. The large quantities of organic materials on Titan make it a most interesting place to study organic chemistry.

Mimas

With a diameter of only 392 km, Mimas is the smallest and innermost of the major satellites that orbit Saturn but was known from telescopic observation before the twentieth century. This small, icy body has only a thousandth the mass of Earth's Moon and a hundred-thousandth the mass of Earth. Its small size is significant because of what it might tell us concerning planetary evolution. Before the exploration of the outer solar system, it was generally believed that the main source of heat in a planetary body was the decay of radioactive elements in the body's interior. Thus, the dynamic history of a planetary body, which is governed by the generation of heat, depends on the quantity of radioactive elements balanced against the loss of heat, which in turn is governed by the surface area to mass ratio. One would therefore assume that larger bodies (especially those made of silicate rock) would generate more heat but lose it at a slower rate than smaller bodies (especially those made of ice). They would thus be more likely to melt and remain hot longer and to develop an active tectonic system. For example, Earth's Moon and Mercury are each less than half the diameter of Earth and have long been geologically inactive. Mars is over half the diameter of Earth and has been moderately active volcanically throughout its history. These ideas suggest that a tiny icy moon such as Mimas would be an inactive sphere of ice that would not have changed (because of internal heat) since it formed.

This generalization holds for Mimas. It is less than 400 km in diameter and composed mostly of ice (density, 1.4 g/cm^3). Its surface is heavily cratered, and there is no evidence that volcanism or resurfacing of any type has occurred since the early period of intense bombardment (Figure 10.17). Many impact craters are old and severely degraded. Rayed craters have not been recognized, possibly because of the high albedo, or brightness, of the satellite's surface and the ease with which ejecta could escape into orbit about Saturn. Most of the craters are bowl shaped and are sharper and much deeper than craters of comparable size on the Moon or on the large icy satellites of Jupiter. This is probably a natural result of the extremely low gravity field of this small body. The unique depth of craters on Mimas could also be influenced by the fact that Mimas is cold enough to prevent ice flow

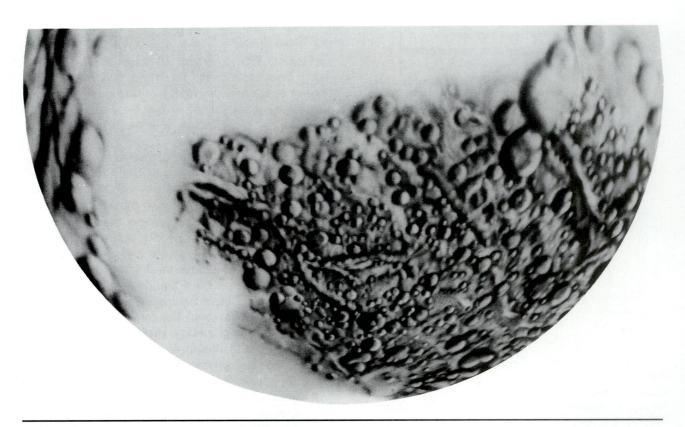

Figure 10.17

The surface of Mimas, as shown in this shaded relief map, is saturated with small impact craters, many of which are old and degraded. Linear troughs, which are probably fractures, cut across the cratered terrain. This small, icy satellite is only 400 km in diameter.

and isostatic rebound. Craters greater than 20 km in diameter have rudimentary central peaks.

Because it rotates once for each revolution it makes about Saturn, Mimas always keeps the same face toward Saturn. This sort of spin-orbit coupling has evolved for all of Saturn's large satellites, with the possible exception of Titan. As a result, Mimas and the other tidally locked satellites have leading faces (the side facing forward with respect to orbital motion) and trailing faces or hemispheres.

The most striking feature on Mimas is a giant crater (130 km in diameter), nearly centered on the leading hemisphere. The image of this crater, dubbed Herschel, first appeared on the television screen as an enormous cyclopean eye staring out into space, giving Mimas a bizarre appearance (Figure 10.18). (Yet there was also a sense of déjà vu. The resemblance to the Death Star spaceship from the movie *Star Wars* is uncanny.) The very size of the crater seems improbable. How could Mimas receive such an impact and survive? The crater rim is steep and sharp, appears to be little modified by superposed craters, and is evidently younger than most of the rest of the surface. The crater walls have an estimated average height of 5 km and parts rise more than 10 km above the crater floor. An enormous central peak 20 to 30 km in diameter at its base rises to a height of 6 km above the crater floor. This crater, which is about one-third the diameter of the satellite, must have been produced by the impact of a body approximately 10 km across. For a body the size of Mimas, which is weakly bound by its own gravity, this is very nearly as large an impact as could occur without breaking Mimas apart. A larger projectile would have shattered the cold and brittle satellite. If an earlier Mimas had been blown apart by one such impact, the debris would probably remain in a narrow band where once the orbit of the satellite had been. Gradually the debris should re-accrete by mutual gravitational attractions to form a new moon. Comparisons with the size of craters on neighboring moons suggest to a few scientists that disruption may actually have happened—possibly several times.

In addition to craters, the surface of Mimas is scarred by a number of long narrow troughs (Figure 10.19). The most conspicuous troughs trend northwest and are as much as 90 km long, 10 km wide, and 1 to 2 km deep. Some are straight and may represent deep-seated fracture systems. Others are less regular and may consist of chains of coalesced craters. Geologists speculate that Mimas was nearly broken apart by the impact, which produced the large crater on the opposite side of the satellite, and that the shock wave that traveled through the satellite produced the large fractures.

With its small size, Mimas is expected to be frozen all the way through; with no source of internal energy, it has probably remained essentially unchanged since the giant crater was formed billions of years ago. In fact, Mimas is so small that it may never have become internally differentiated. Accretion may have only raised its internal temperature less than 100 K, well below the point at which water ice melts. The interior of Mimas may thus consist of an intimate mixture of ice and rock.

Enceladus

In terms of size, Mimas and Enceladus are in the same class (392 km versus 500 km in diameter); the next largest pair, Tethys and Dione, are an order of magnitude more massive. Before the Voyager encounter, scientists therefore expected Mimas and Enceladus to be similar—both being primitive, small, inactive spheres of ice. They could hardly have been more wrong. Voyager 2 came within 100,000 km of Enceladus and provided excellent images with resolution of about 2 km. As it turns out, Enceladus is remarkably different from Mimas. It is without question the most geologically active satellite of Saturn, and its surface contains a wide diversity of crater-free terrains, probably

Figure 10.18

The enormous impact crater on Mimas is about one-third the diameter of the satellite itself and records an impact that very nearly broke Mimas apart. The giant crater gives Mimas a surprising resemblance to the Death Star spacecraft of *Star Wars*.

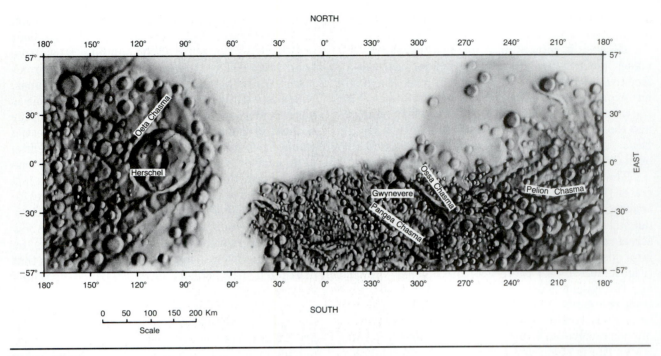

NORTH

180° 150° 120° 90° 60° 30° 0° 330° 300° 270° 240° 210° 180°

57° 57°

30° 30°

Oeta Chasma

Herschel

0° 0° EAST

Ossa Chasma

Gwynevere

Pelion Chasma

-30° -30°

Pangea Chasma

-57° -57°

180° 150° 120° 90° 60° 30° 0° 330° 300° 270° 240° 210° 180°

0 50 100 150 200 Km

Scale

SOUTH

Figure 10.19

The shaded relief map of Mimas was prepared by the U.S. Geological Survey from images obtained by the Voyager spacecraft. In addition to its densely cratered terrain and the large impact crater, the surface of Mimas is scarred by long narrow troughs trending to the northwest. As a result of its small size and absence of internal heat, Mimas has remained nearly unchanged since the formation of the giant crater billions of years ago.

implying that resurfacing occurred within the last 100 million years or so.

As seen in Figures 10.20 and 10.21, the surface of Enceladus can be divided into several provinces, each characterized by different landforms and different geologic histories. At least five terrain types are recognized based on crater populations and other surface features.

Cratered Terrain. The oldest surface on Enceladus is the cratered terrain, which covers most of the northern hemisphere. The craters are typically bowl shaped, with sharp rims, and many larger ones have central peaks. From inspection of Figure 10.20, it is obvious that the crater population is not nearly as dense as that on Mimas (or Dione, Rhea, Tethys, or Iapetus). Impact craters up to 35 km in diameter are abundant, but there are no larger craters and the craters are not densely packed together. Based on this crater frequency, even the most densely cratered terrain may be younger than the period of intense bombardment.

Two types of cratered terrain have been identified on Enceladus. One, located near the terminator in Figure 10.20, is characterized by craters 10 to 20 km in diameter that are highly flattened and subdued. The other has craters of the same size,

but the original topography is preserved. The craters are deep, and the bowl-shaped rims rise high above the crater floors. Apparently the thermal histories of the two regions have been different. A thermal event is postulated for the younger cratered terrain in which the icy crust was plastic or even liquid within 10 to 20 km of the surface, causing the craters to be altered by plastic flow and relaxation much as we have seen on other icy bodies.

Cratered Plains. South of the cratered terrain, several regions are plains on which craters 5 to 10 km in diameter are sparsely distributed. In this area, the crater population is only one-third that of the older cratered terrain. Between the craters are large areas that lack impact structures larger than 2 km in diameter.

Smooth Plains. A third major province, an area in which the dominant landform consists of an open pattern of rectilinear grooves (Figure 10.21), is called the smooth plains. Two types of smooth plains are recognized: (1) the area to the southwest, in which craters 2 to 5 km in diameter are very sparsely distributed over the surface, and

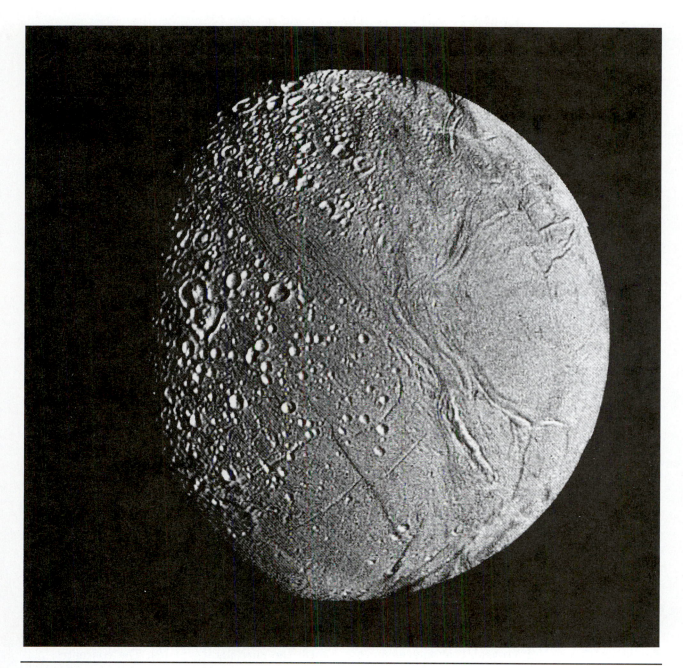

Figure 10.20

The surface of Enceladus preserves a surprising record of geologic activity. The smooth, uncratered terrain is geologically young, which suggests that Enceladus has experienced a period of relatively recent internal melting and resurfacing by an exotic form of volcanic activity in which water and icy slush were extruded. Linear sets of grooves tens of kilometers long are probably faults resulting from crustal deformation.

(2) the area to the northeast, which is essentially free from craters and must be younger.

Ridged Plains. The ridged-plains province is the youngest terrain on Enceladus; it is distinguished by a complex of subparallel ridges, especially along its contacts with older terrain. Some ridges have a relief of more than 1 km and appear to be small-scale versions of the grooved terrain on Ganymede. A long corridor of the ridged-plains province extends northward from the equator along the 330 degrees longitude line almost to the north pole. It is along the margins of this corridor that we find the best evidence of crustal separation (where two large craters are bisected by the ridged terrain as though they were split by crustal spreading), but perhaps the truncated craters indicate down-dropping and flooding of grabens. Several

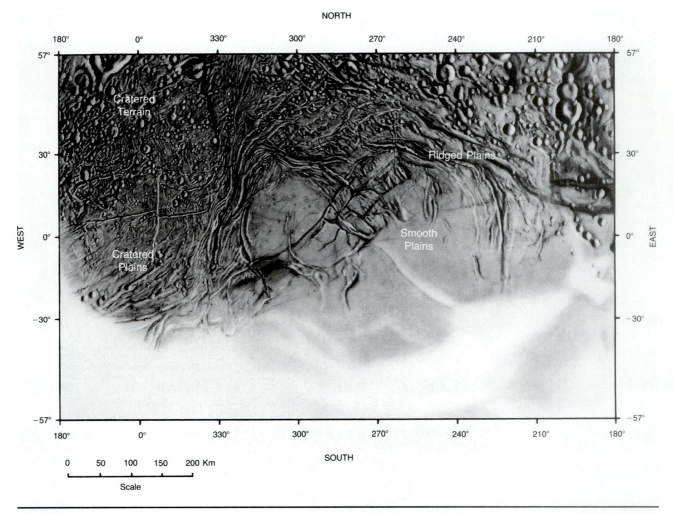

Figure 10.21

The shaded relief map of Enceladus prepared by the U.S. Geological Survey shows the relationships between the variety of terrains on this small, icy body.

stages in the development of the cratered terrain are also seen in this area.

Close study of the computer-enhanced image of Enceladus (Figure 10.20) reveals that an even more detailed subdivision of terrain types is possible and reinforces the fact that Enceladus has had a complex history during which there has been an involved sequence of thermal events and multiple stages of resurfacing parts of the planetary body.

Tectonic Features. Two styles of tectonic features are seen on Enceladus: (1) linear grooves that occur in the smooth plains are probably graben and indicate crustal extension and brittle fracturing; (2) the curvilinear valleys and ridges of the ridged plains resemble the grooved terrain on Ganymede, especially in the north polar region. On Enceladus, there are double ridges with intervening valleys, simple grabens, and one-sided valleys (possibly fault scarps).

The grabens suggest normal faulting of a thin, brittle lithosphere, accompanied by extrusion of water or ice slush as an exotic form of volcanism to form the surrounding plains. The concentric patterns of ridges near the border of the smooth-plains province possibly resulted from convective upwelling. New crust formed in the center of the unit and caused compression along its margins. Voyager 2 did not reveal any ice volcanoes erupting and there are no obvious volcanic cones or calderas.

However, fissure eruptions of water and slush must have occurred many times on Enceladus because the variety of landforms clearly indicates a continuing tectonic system, not an isolated event such as the impact of a large comet.

The big question concerns what heat source drives the geologic activity in Enceladus and has kept the interior molten and active throughout much of its history. Primordial heat inherited from accretion and differentiation cannot explain the geologic activity because tiny Enceladus would have cooled and frozen solid within a few hundred million years after its formation. Radioactive heating also seems implausible because such a small, icy body would have little uranium or other radioactive elements. The best explanation for a heat source might be heat from tidal stress induced by an orbital resonance with Dione. Enceladus is in a 2:1 resonance lock with Dione. That is, Enceladus orbits Saturn exactly twice for each orbit of Dione. Thus, Enceladus is forced into an eccentric orbit. This causes tidal bulges on Enceladus to expand and contract alternately, a process that generates heat. Such a process occurs on two of Jupiter's moons, Io and Europa, so that Io is continually being heated and is the most volcanically active planetary body in the solar system, and Europa has a young but cracked icy shell. Recent theoretical estimates of the amount of tidal heating experienced by Enceladus suggest that tidal heating may be adequate to soften the interior and perhaps melt water ice. In addition, if small amounts of ammonia are mixed with the water ice of Enceladus, the melting point of this mixture (173 K) would be about 100 K lower than for pure ice and could reduce the amount of heat needed to mobilize the icy crust.

Clearly, Enceladus joins Io in showing that radioactivity does not always dominate the thermal and geologic histories of planetary bodies in the solar system. Enceladus has only a hundred-thousandth the mass of Earth and yet it, nonetheless, may still be active geologically.

Tethys

The best global view of Tethys, with a density of about 1.2 g/cm^3 and a diameter of 1060 km, is shown in Figure 10.22. It is clear from this photo that Tethys, like several other Saturnian satellites, exhibits two major terrain types: (1) a heavily cratered terrain at the top of the photo (western hemisphere), which is extremely rough and hilly as a result of many large, highly degraded craters, and (2) a relatively smooth surface at the bottom of the photo (east, the trailing hemisphere) on which the crater population is clearly smaller. A few craters on the smooth plains are large, but most of the craters are quite small. The boundary between the two terrain types is not marked by a sharp change in color or tone, but the textures of the two surfaces are clearly distinct.

The plains to the east were probably produced by volcanic processes that generated slushy ice, which was extruded by fissure eruptions and flooded the older, densely cratered terrain, creating a new, smooth surface. Subsequent impact upon

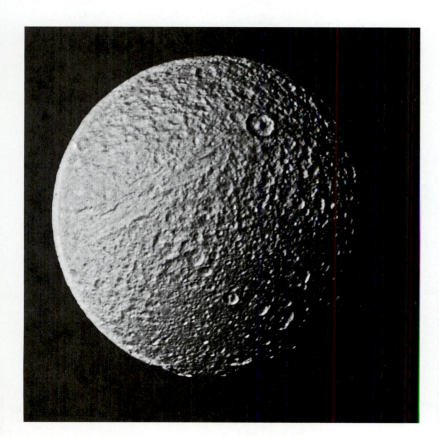

Figure 10.22

The surface of Tethys shows two terrain types: (1) heavily cratered terrain (top and right) and (2) a more lightly cratered terrain (bottom right). The relatively smooth, lightly cratered surface probably resulted from the extrusion of water and slush, which flooded the older cratered surface and provides evidence of internal heating early in the history of Tethys.

this new surface was relatively light and was dominated by smaller projectiles.

The images of Tethys transmitted by Voyager 2 reveal an enormous impact scar, named Odysseus, on the leading hemisphere, centered on 120 degrees longitude (Figure 10.23). The rim of Odysseus has a diameter that is 40 percent of the diameter (400 km) of Tethys itself. This structure looks like the large crater on Mimas, but the whole of little Mimas would fit into the great scar of Tethys, the largest and deepest crater in the Saturnian system. Odysseus was so large that it partly destroyed the curvature of the satellite. The original crater must have been very deep and its rim and central peak must have been high. Structural rebound of the crust has occurred, however, so that the shape of the floor now matches the spherical shape of the satellite. Both the rim and central peak have collapsed and now appear subdued. Apparently the interior of Tethys was sufficiently warm during the early history of the satellite to allow such gravitational rearrangement (isostatic adjustment) of its surface features. The same adjustments have been made on large craters on Ganymede and Callisto but apparently not on much smaller Mimas.

Tethys is also scarred with an enormous fracture system (Ithaca Chasma) centered on the Saturn-facing hemisphere (0 degrees longitude) (Figure 10.24). This huge trench extends as a continuous feature for 2000 km, almost three-quarters of the way around the satellite. The width of the valley complex is about 100 km and its depth ranges from 3 to 5 km. It is thus comparable in size (but not in detailed form) to the great Valles Marineris on Mars. The fracture system represents about 5 to 10 percent of the surface of Tethys. Its southern reaches are bordered by relatively smooth, uncratered plains. Some geologists suggest that the fracture system was produced because the surface of Tethys, which is largely water ice, cooled and contracted quickly compared to the still-warm interior. This contraction may have pulled the crust apart, forming a major rift zone. It is unclear why the vast stretch mark would take the form of a single regular girdle rather than a more complex fracture pattern. Others believe that the fracture system, which is situated on a circle that passes through the crater Odysseus, may have been caused by the focus of violent shock waves that accompanied the large impact or alternatively by stresses built up during viscous relaxation of the crater.

One other feature of Tethys deserves mention. As can be seen in Figure 10.24, parts of the surface are notably darker than others. For example, a dark zone trends essentially north–south near 270

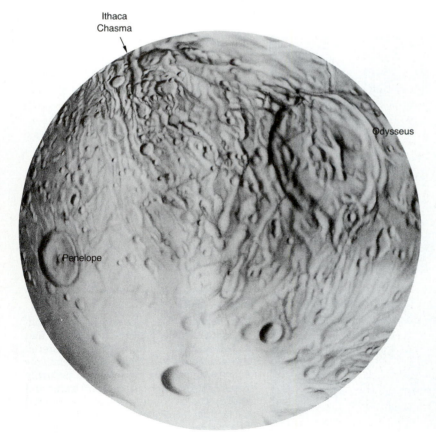

Ithaca Chasma

Odysseus

Penelope

Figure 10.23

The giant crater on Tethys is Odysseus. It is the largest impact structure in the Saturnian system. It is about 400 km in diameter but it is relatively shallow. Isostatic adjustment and the flow of soft ice has flattened the floor and subdued the rim and central peak, modifying the deep bowl shape that characterizes fresh craters.

degrees longitude. Boundaries are relatively sharp, and at first glance the dark marking appears to have been painted on the surface from an external source. This enigmatic feature has so far escaped adequate explanation.

Dione

Dione is about the same size as Tethys, with a diameter of 1120 km, but it has a distinctly higher density (1.4 g/cm³). Presumably Dione has a higher fraction of silicate rocky material and a correspondingly greater source of radiogenic heat. The value of 1.4 g/cm³ is consistent with a rocky core that comprises about one-third of the satellite's mass. In addition, Dione has a different global appearance. As seen in Figure 10.25, Dione shows some of the largest variations in surface brightness in the Saturnian system, second only to that of Iapetus. The trailing hemisphere is characterized by a dark surface covered by a network of broad, wispy, bright markings. This type of surface feature had never been seen before anywhere else in the solar system. The strange bright markings, tentatively called wispy terrain, do not form a pattern typical of that produced by crater rays but may be associated with regional fractures or faults. These markings are brighter than the brightest features seen on Jupiter's moons and are believed to be composed of water ice or frost, emitted from the interior of Dione along the fracture systems—a unique type of volcanic ejecta. Perhaps the frost was carried upward by more volatile substances such as methane. Closeup views of Dione show that craters 50 to 100 km in diameter are crossed by the bright streaks, indicating that the streaks formed after the period of intense bombardment.

High-resolution photography shows that the leading hemisphere of Dione is intensely cratered, with huge fractures extending across the surface for hundreds of kilometers (Figure 10.26). Nongeologists may become a bit bored to find another ball of ice scarred with craters and cut by deep canyons, but the specialist marvels at the diversity of craters and the events in the Saturnian system they record. The surface of the leading side of Dione consists of at least two terrains of contrasting topography (Figure 10.27). One, evidently older, is heavily cratered with both large and small impact structures. The craters range from small features near the limits of resolution to nearly 100 km in diameter. The large craters have well-developed central peaks and raised rims. There are also a few craters with bright radiating patterns probably representing clean ice thrown out as rays from

impact. The other terrain is cratered with small impact structures but lacks larger craters. Why? The difference in the crater population of the two terrains may represent distinct events in the satellite's history. The first period of Dione's history may have been dominated by impact from large bodies, probably material left over from the accretion of the solar system. Then parts of Dione were resurfaced; that is, a thermal event occurred in which internal melting produced liquid water and icy slush, which was extruded as fissure eruptions and which buried parts of the already cratered surface. This layer of new material was thick enough to mantle and obscure the pre-existing cratered terrain. Dione was then subjected to another period of bombardment from smaller bodies, which modified the newly formed plains with small craters. The second generation of small projectiles could have resulted from collisions near Saturn between bodies left over from the period of accretion. About this same time, fractures formed over parts of the surface of Dione and were partly filled with extrusions from the interior. This could imply that Dione generated enough heat to produce a continual outpouring of material onto its surface over a sufficient period of time for the cratering population to vary markedly. Perhaps the source of internal heat resulted from concentrations of radiogenic nuclides, which would be consistent with Dione's higher density and greater amounts of rock material.

The absence of high-contrast boundaries between the older and younger terrains may be because the fresh and older icy surfaces are about the same color. A similar blurring of the boundary between the lunar highlands and maria would result if the basalts of the maria were the same color and tone as the highlands. The fundamental distinction between the two terrain types is in crater populations, not color or tone.

You can see from Figure 10.26 that Dione, like other icy satellites observed at close range, is nearly devoid of rayed craters. This is probably because the icy satellites are generally bright, which would obscure fresh ray patterns.

Another major topographic feature occurs in the polar regions (Figure 10.27). In the north, a long, linear valley resembling a rift valley extends for nearly 500 km near latitude 70 degrees north. The trough is part of a string of features that radiate from the rim of a large degraded crater. Much of the rest of the northern hemisphere is crisscrossed by a complex, branching valley system that extends toward the equator. These features indicate a significant period of crustal evolution and tectonic modification for Dione, possibly similar to that on Tethys.

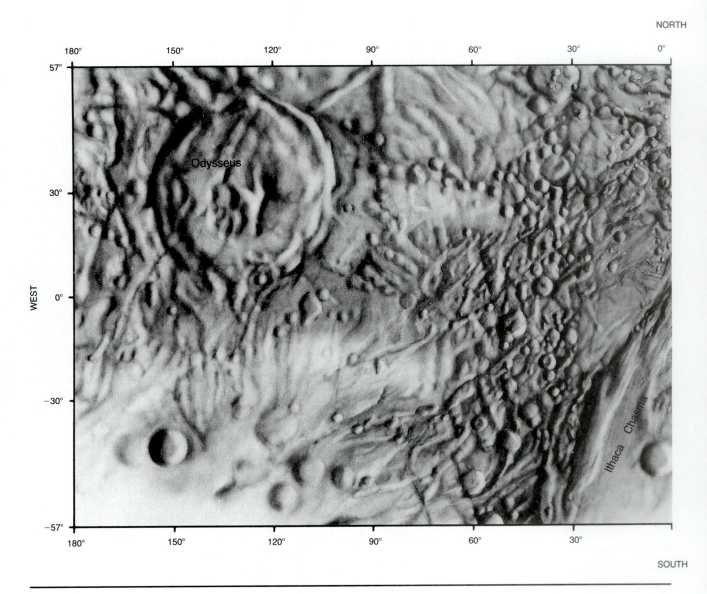

Figure 10.24
The shaded relief map of Tethys was prepared by the U.S. Geological Survey from images made by Voyagers 1 and 2. Two features dominate the landscape. The most prominent is the great canyon extending from the north pole a third of the way around the satellite. The other is the giant impact structure, Odysseus.

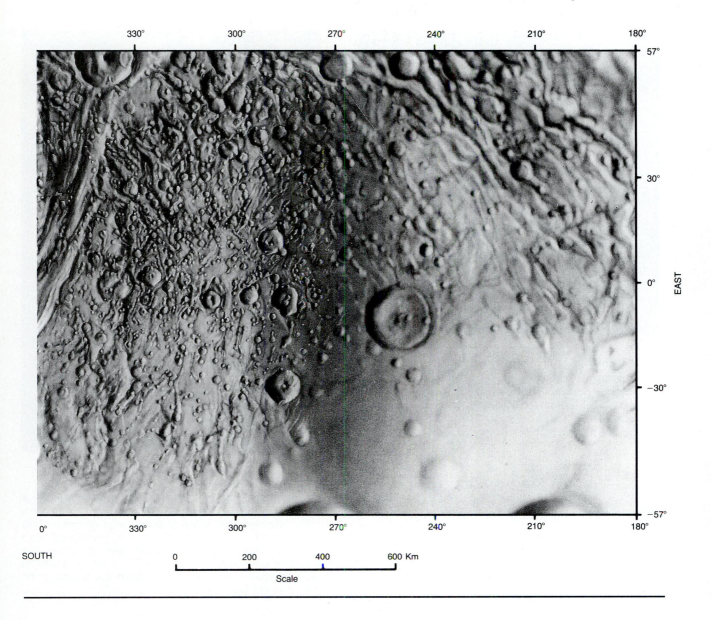

SOUTH

0   200   400   600 Km

Scale

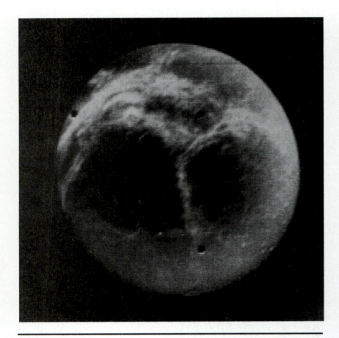

Figure 10.25

The surface of Dione shows a complex of strange bright markings or streaks that are distinctly different from the rays produced by impact. The streaks are thought to be deposits of ice and frost extruded along a system of fissures formed early in the history of Dione.

Figure 10.26

The densely cratered terrain of Dione is shown in sharp detail in this image, which has a resolution of about 3 km. Regions with different frequencies of craters indicate that there have been episodes of resurfacing since the early period of intense bombardment. Note the linear trough or valley near the top left of the image and also the white streaks extending around from the trailing hemisphere in the lower right.

Rhea

Rhea is the largest of Saturn's airless moons (1530 km in diameter). Its surface was photographed by Voyager 1 at a distance of only 59,000 km, providing our closest view of any of the icy satellites. Crisp details of surface features as small as 1.5 km were recorded. Figure 10.28 shows a typical view of the north polar region and shows that the surface is dominated by craters and resembles the cratered highlands of the Moon and Mercury, except that the surface material is brilliant white ice rather than dark silicate rock. There are, however, significant visible differences in the cratered surfaces that are of great geologic interest. The principal difference is that the large, fresh craters on the Moon and Mercury are surrounded by well-formed ejecta blankets, whereas the craters on Rhea are not. Presumably Rhea's weaker gravity field is responsible. Given the same ejecta velocities as on the Moon, crater ejecta on Rhea would spread out five times farther than lunar ejecta and would thus form a much thinner ejecta blanket. If Copernicus, a large, fresh, rayed crater on the lunar maria, had formed on Rhea, its ejecta would have spread around the entire globe.

There are also notable differences between the craters on Rhea and those on Callisto and Ganymede. As you will recall from Chapter 9, the craters on the Galilean satellites are typically flat-tened and some have lost nearly all of their topographic expression, presumably because of viscous flow of the once soft icy crust. Rhea's craters, in contrast, are commonly sharp and rugged and have significant topographic relief. Evidently, Rhea's lower gravity and small diameter allowed rapid cooling and enabled its crust to support rugged crater morphology relatively unmodified by viscous flow of the icy crust.

Many craters on Rhea have central peaks (Figure 10.29). Some craters are shallow, have subdued rims, and are generally degraded—showing their antiquity. Others have sharp, rugged rims, are unmodified by subsequent bombardment and are relatively young. Curious bright patches, which may be fresh ice exposed as the result of slumping or minor impacts, are visible on some craters. It is also apparent from Figure 10.29 that many of Rhea's large craters have irregular outlines, often polygonal, suggesting that a weak rubble layer with zones of weakness, like the Moon's megaregolith, makes up the upper crust.

Interpretation of the crater distribution on Rhea is the center of a controversy that bears not only on the impact history of the satellites of the outer planets, but also on the likelihood for their earlier disruption and reaccretion as well as on the relationship of moons to rings.

A few investigators suggest that there have been two distinct populations of impacting projectiles on the Saturnian satellites. One population is thought to include the original material from which the system was formed. The second population, which lacked large objects, may have been debris from later collisions within the Saturnian system, perhaps associated with ring formation, a hypothesis that is partly based on interpretations of the detailed images of Rhea. Figure 10.29 shows that the western area (Rhea's leading hemisphere) is marked by large craters, ranging in diameter from 30 to 100 km, in addition to a dense population of smaller craters. The smallest craters are only a few km across. The eastern area also has small craters, but larger, older craters are missing. A complicated impact history may also be indicated in Rhea's equatorial region. The central and upper right parts of Figure 10.28 show a smooth plain with more muted topography than the rugged area in the lower left. On the shaded relief map (Figure 10.29), several regions of relatively low crater density are apparent near the equator between 300 and 330 degrees longitude and between 20 and 60 degrees longitude. These smooth plains may have been produced in the midst of Rhea's bombardment, perhaps by volcanic resurfacing. Proponents of the two populations of impactors contend that part of the surface of Rhea lacks the large craters that are a normal part of the crater population on other planetary bodies of the solar system. This difference between the two types of cratered terrains on Rhea implies two periods of bombardment, with different sizes or populations of projectiles. The first period of bombardment involved projectiles (meteorites or comets) with a wide range of energies (size and or velocity). At some point, these large projectiles were all swept up, but bombardment by objects that formed the smaller craters continued. Meanwhile, part of Rhea's surface (on its trailing hemisphere) was resurfaced, perhaps by extrusion of volcanic fluids from the interior, thus obscuring the older cratered terrain. The new surface was then cratered, during the second period of bombardment, by projectiles that produced only small craters. Perhaps multiple periods of bombardment should be considered before the crater-

ing history based on observations of the Moon and Mercury is applied to all planetary bodies throughout the solar system.

On the other hand, many planetary scientists contend that the crater distribution on Rhea does not really vary from place to place in a statistically significant manner. They conclude that the variations in crater frequency are those expected of random cratering of a small body by one population of bodies. Moreover, they conclude that there is no evidence for the resurfacing of Rhea in the middle of its impact history. We will return to this controversy several times in the next few chapters.

The shaded relief map of Rhea shows that in addition to the densely packed craters that scar the surface there is a complex network of narrow linear grooves or troughs. One example, almost 200 km long, is located between 290 and 300 degrees longitude at 30 to 40 degrees north latitude. Another lies at about 110 degrees longitude and 15 degrees north latitude. The straight fracture systems are typically oriented at 45 degrees to the parallels of latitude, which suggests to some researchers that the fractures could be part of a global tensional stress pattern. An origin by cooling of the outer portion of Rhea, like that discussed for the fracture system on Tethys, might be called on here as well.

The distant global view of Rhea shows contrasting markings on the leading and trailing hemisphere (Figure 10.30). Like Dione, the leading hemisphere is bright and bland, whereas the trailing hemisphere has bright, wispy streaks superimposed on a darker background. It has been suggested that the streaks were originally formed by frost and ice extruded along fractures and fissures over much of the surface on both hemispheres. Subsequently, the leading hemisphere received many more impacts as it swept up debris in its orbital path. This pulverized and obliterated the striations on the leading hemisphere while the streaks were protected and preserved on the trailing side.

Iapetus

Iapetus circles Saturn in a lonely orbit nearly 3.5 million km from Saturn—almost three times as far away as Titan or Hyperion, the next closest satellites. In this far-distant reach of the early solar nebula, temperatures would have been low enough to freeze methane. Calculations based on the condensation model outlined in Chapter 2 suggest that materials condensed in this region of the early solar nebula would have yielded a mixture of water ice, silicate rock, and methane in a 55:35:10 ratio. The resulting planetary body accreting from such con-

NORTH

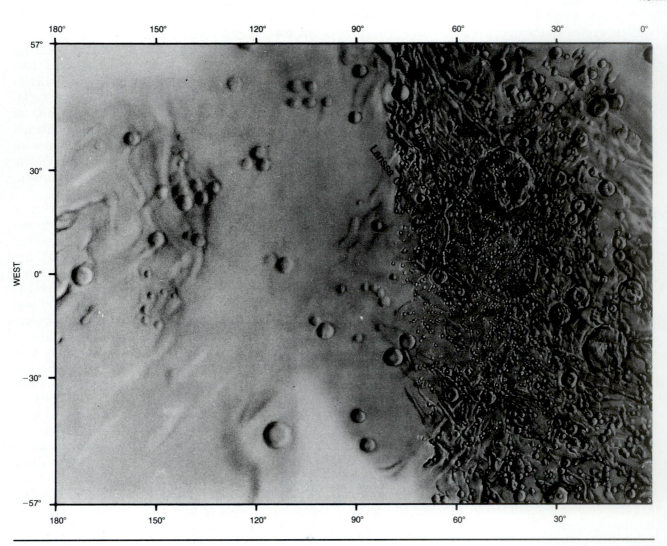

Figure 10.27

The shaded relief map of Dione shows the variety of contrasting terrain types on the satellite. Note (1) densely cratered terrain (0° to 90° longitude), (2) wispy terrain (210° to 0° longitude), and (3) the relatively smooth terrain, which lacks large impact structures.

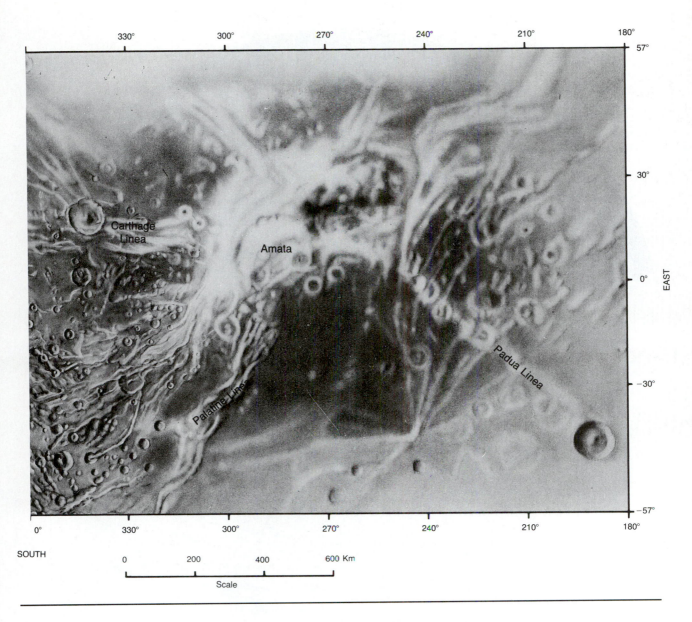

Figure 10.28

The surface of Rhea is dominated by impact structures. This shaded relief map of the north polar region shows many features in detail. Two populations of craters are apparent. One, dominated by large craters, is presumably older. The other, characterized by smaller craters, was formed later, after a period of resurfacing.

densates would have a density of about 1.1 g/cm³, very close to what has been determined for Iapetus (1.16 g/cm³). Thus, if the interpretation of the satellite's density is correct, Iapetus contains a significant fraction of methane, in one form or another, and could be unique among Saturn's satellites.

Only slightly smaller than Rhea, Iapetus has a diameter of 1460 km. However, it is not part of the regular system of Saturnian satellites in that its orbit is inclined 14.7 degrees from the equatorial plane. The other moons of Saturn all have nearly circular orbits and lie in the equatorial plane of the planet. Like other satellites of Saturn, Iapetus keeps the same face toward Saturn.

The best image of Iapetus was made by Voyager 2 (Figure 10.31) and shows the side of the moon that always faces away from Saturn. The north pole is in the area of the large crater near the terminator. The most spectacular surface feature on Iapetus is the irregular patch of dark material that covers the leading hemisphere. Indeed, the principal dark region is centered almost perfectly on this face (Figure 10.32). The bright material has an albedo comparable to that of slightly dirty snow, whereas the dark material has an albedo similar to coal or soot. The only known planetary substances with albedos as low are carbonaceous materials such as those found in some meteorites.

Two of the hypotheses that have evolved to explain the location and coloration of the dark patch on Iapetus are explained below. One is that the dark material originated somewhere other than on Iapetus, perhaps from Phoebe (the one known moon of Saturn further away from the planet than Iapetus), and had been transported inward and swept up by the leading side of Iapetus. The color of Phoebe and the dark side of Iapetus are not a perfect match, however, so there is no strong evidence that the material came from Phoebe.

An alternative view is that the material is an exotic type of volcanic ejecta that came from within Iapetus itself. The dark region would then be analogous to the maria on Earth's Moon. If you study Figure 10.31, you will note that the contacts between the dark region and the bright hemisphere are sharp, irregular, and complex, a pattern that would not likely develop if the dark material originated from space and dusted the leading edge of Iapetus. As can be seen in Figure 10.31, the dark material appears to be deposited selectively on the floors of many large craters near the boundary and even deep inside the bright hemisphere. These observations strongly suggest that the dark material is superposed upon the cratered terrain and is therefore younger. Although craters are clearly seen right at the boundaries of the dark region, there is no hint of bright, rimmed craters within the dark regions. This suggests that either the younger, dark material is very thick so that the brighter underlying material was not splashed out by subsequent impact or that the dark material is very young and covers all the bright features in the region. As shown on the map of Iapetus (Figure 10.32), the dark area appears to be free of bright features at the limits of resolution of the available images.

Taken together, the sharp definition and complex boundary between the dark and light terrain and the selective deposition of material on the floors of large craters imply a history of eruptions from the interior of Iapetus, evidence of another exotic type of volcanic eruption on a small icy moon in the outer solar system. This is not to say that the erupting material resembles any ordinary lava on Earth. One can speculate that the erupting material was a fluidized slurry of ammonia, water ice, and dark carbon-rich substances such as are found in some primitive meteorites.

Why then would the dark material be concentrated on one hemisphere? This may be a simple coincidence, and examples of such hemispheric distinctions are common in the solar system. The Moon has a distinct asymmetry in the distribution of its maria and highlands. Mars shows global asymmetry in its surface, with smooth plains located in the northern hemisphere and cratered highlands covering the southern half of the planet. Earth, too, is asymmetrical, with continents concentrated in the northern hemisphere and ocean basins in the southern.

The bright hemisphere of Iapetus appears to be more heavily cratered than any of the other satellites of Saturn and is therefore considered to have the oldest surface. Most craters are of small to intermediate size. Large impact striations and fracture systems like those on Tethys and Mimas are not seen.

The Small Satellites

In addition to the seven principal satellites, Saturn controls at least ten smaller satellites, ranging from 20 to 350 km in diameter (Figure 10.33). Many were discovered by the Voyager spacecraft, and presumably there are others yet undiscovered. Aside from some data obtained from imaging, little is known about these objects (Table 10.2). In particular, their masses are not all known, so their bulk densities cannot be calculated and their bulk compositions are uncertain. Perhaps the most interesting aspects about these small satellites are their orbits. Nearly all are remarkable in some way and provide new data on the complexities of the Saturnian system.

Phoebe. Phoebe is the outermost known satellite of Saturn and is perhaps the most anomalous. Voyager 2 images show that Phoebe is an approximately spherical object with a diameter of 220 km. This was surprising. Objects smaller than about 400 km in diameter lack the gravitational pull to mold themselves into spherical shapes and can maintain irregular shapes almost indefinitely. Phoebe also orbits Saturn in an eccentric, retrograde path. In addition, Phoebe has a very dark surface, with a reflectivity of less than 5 percent, and a gray color. The low albedo and color are similar to asteroids that are common in the outer solar system and are believed to have primitive compositions similar to carbonaceous meteorites. If

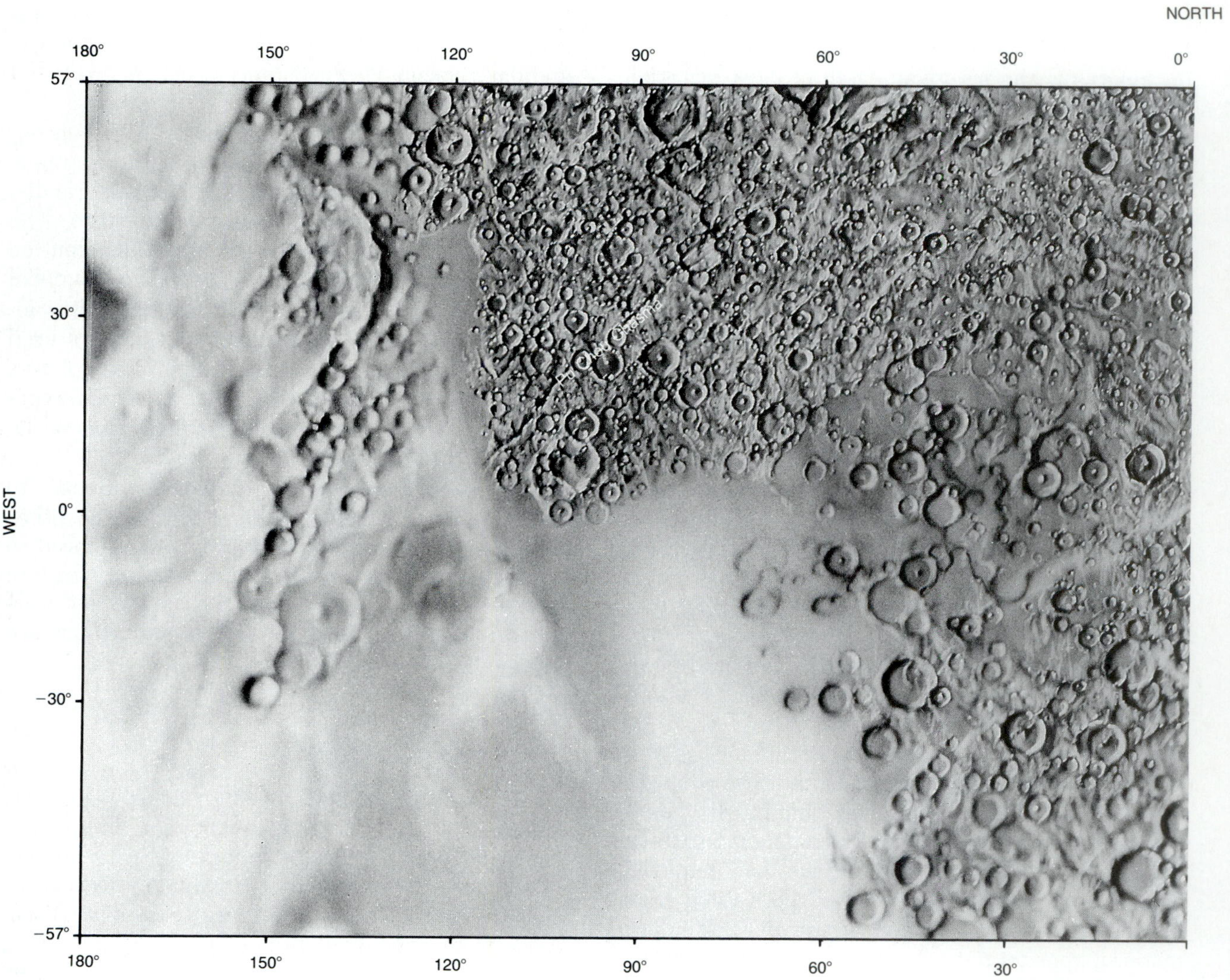

NORTH

Figure 10.29
The shaded relief map of Rhea was prepared from images obtained by Voyagers 1 and 2. In addition to the densely packed craters that scar Rhea's surface, the map shows several long, straight fractures typically oriented 45° to the east. The densely cratered surface of Rhea is much like the surfaces of the Moon and Mercury.

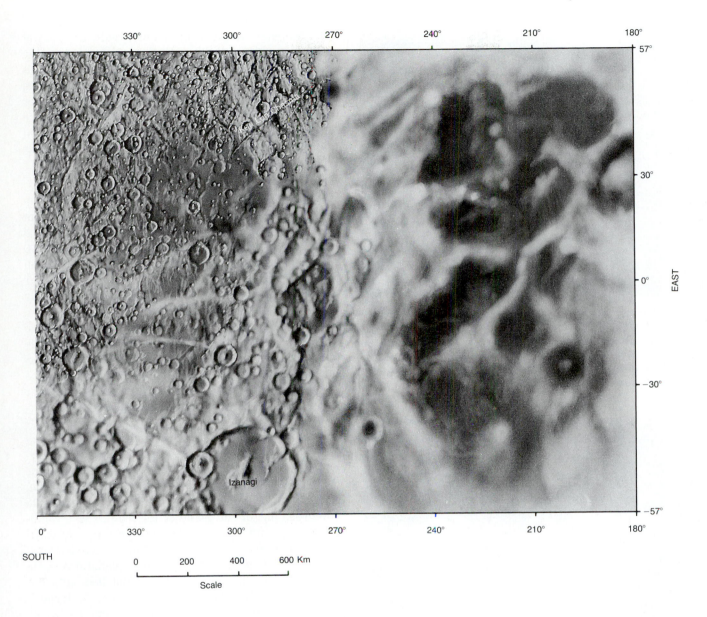

SOUTH

0   200   400   600 Km

Scale

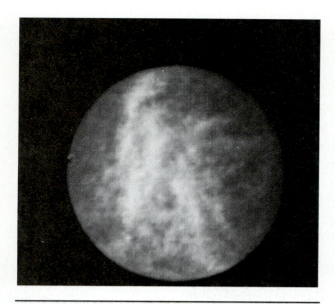

Figure 10.30

The wispy terrain of Rhea forms a pattern of bright markings superposed on the underlying cratered surface. The wispy streaks may have formed from frost extruded from fractures and fissures.

Figure 10.31

Iapetus is a strange world—quite unlike any other in the solar system. The leading hemisphere is fully ten times as dark as the trailing hemisphere. The bright icy terrain is heavily cratered and looks much like the surface of Rhea.

so, the images of Phoebe would be the first of such an asteroid. It is therefore believed to be very primitive and chemically unmodified (being so small it never heated up to begin differentiation).

Phoebe has little in common with the rest of the Saturnian satellites. It is dark and probably rocky; the others are light and icy. Phoebe's orbit is inclined; the others lie in the equatorial plane. Phoebe's orbit is retrograde; the orbits of the other satellites are prograde. These observations have been used to suggest that Phoebe is a captured asteroid and was not formed in the same area as the icy satellites of Saturn.

Hyperion. Hyperion orbits Saturn between Iapetus and Titan. It is an icy moon with an irregular shape. Larger than nearly spherical Phoebe, its major dimensions are about 380 by 290 by 230 km with angular features and facets as well as rounded edges (Figure 10.34). Its irregular shape suggests that it is a remnant of a larger body that was shaped by impact.

Several large impact structures ranging up to 120 km in diameter (with 10 km of relief) are visible on Hyperion. The surface is peppered with numerous craters 10 km in diameter and smaller. The irregular shape and cratered surface indicate that Hyperion may have the oldest surface in the Saturnian system.

The most interesting discoveries about Hyperion are its orbital attitude and peculiar spin. Hyperion appears to be tumbling on an axis that spins chaotically, but it lies nearly in the orbital plane. However, Hyperion's long axis is tilted about 45 degrees out of the plane of its orbit. These conditions are not the stable end products of tidal evolution, at least not in any simple gravitational relationship with Saturn. Two explanations seem possible. One is that Saturn has not had sufficient time to form a tidal lock on Hyperion. At Hyperion's distance from Saturn, tidal effects are weak, and the apparent dynamic imbalance could be ancient. A second possibility is that it is a remnant of a satellite disrupted in a recent massive collision. The impact could have jolted Hyperion's orientation askew and there has not been enough time for it to reestablish orbital equilibrium.

The Lagrangian Satellites. It has long been known that a small object could share an orbit with a larger one if it stayed 60 degrees in front or behind the major body. These are referred to as **Lagrangian points.** Theoretically, five points are possible, but only two are stable in real planetary systems. Jupiter has two such groups of asteroids (known as the Trojan asteroids because the first one discovered was named Hektor) 60 degrees ahead and 60 degrees behind in its solar orbit.

Even Tethys has two Lagrangian satellites (Calypso and Telesto), Tethys B and C, which share its orbit. Dione has one, Dione B or Helene; but, unlike the Tethys system, it has no satellite in its trailing position. A search for this satellite was made by Voyager 2 but none was found.

The Lagrangian satellites of Tethys and Dione are irregular objects of apparent icy composition. They are 30 to 40 km long and may be fragments of a larger parent body.

The Co-Orbital Satellites. Two satellites, Janus and Epimethus, were observed by Voyager 1 to be almost in the same orbit, and the motion of these two small moons is just as fascinating as their history. The orbit of the slower object is just 50 km larger than that of the inner and faster-moving body. Once every four years, the satellite in the inner and faster orbit catches up with its companion and they interact gravitationally. Since the space between the two orbits is less than the diameters of the satellites themselves, there is no room to pass, and, just short of collision, the two satellites attract each other gravitationally and exchange orbits. They then move slowly apart and the cycle starts over again. As far as we know, this type of orbital motion is unique in the solar system.

The co-orbital satellites, which measure 220 and 140 km in their longest dimensions, are the largest of the satellites inside the orbit of Mimas. Their densities are only about 0.7 g/cm^3, lower than water ice (0.92 g/cm^3). If they are made of ice, these moons must have about 30 percent empty space, suggesting that they are porous rubble heaps. They are irregular objects and appear to be battered with impact structures. All of these data suggest that these small moons are fragments of a larger parent that was fragmented by an impact more severe than the one that caused the large crater on Mimas.

The Shepherd Satellites. Three other small satellites orbit closer to Saturn and are called **shepherds** because of their role in confining the ring particles to their proper orbits. Prometheus and Pandora, the two larger moons, bracket the F ring and their gravitational perturbations of the orbital paths of the ring particles may be responsible for the ring's kinky braids. A particle escaping either inward or outward from the F ring will interact gravitationally with one of these "sheepdogs" and be "herded" back into line. Because of the gravitational focusing of these small satellites, the particles of the F ring are confined to a narrow band. Actually, the shepherd satellites of the F ring are nearly co-orbital, but their orbits are separated by about 2000 km, so there is ample room for them

to pass once each month with no difficulty. The inner F-ring shepherd, Prometheus, is oblong, being about 140 km in its longest dimension. Pandora is fairly spherical. Both show small craters on their surfaces.

The A-ring shepherd, Atlas, orbits just 800 km beyond the A ring and appears to keep the outer edge sharp and well defined. It is a small, oblong object just 40 km in its longest dimension. It has no visible markings, but its tiny gravitational force prevents material from the outer edge of the A ring from drifting farther out and seems to establish an outer boundary for the entire ring system.

Conclusions

Saturn and its entourage of moons and rings probably formed together, accreting from solids that condensed as a swirling Saturn-centered portion of the solar nebula cooled about 4.6 billion years ago. Saturn's massive atmosphere probably formed from the nebula when the planet's icy core grew large enough to trap the gaseous hydrogen and helium from the surrounding cloud. Three phases may characterize the gravitational contraction of Saturn: (1) an initial phase during which proto-Saturn was several hundred times its present size; (2) a collapse phase during which Saturn shrank to about five times its present size in about a year; and (3) a late phase during which Saturn slowly contracted to its present size over 4.6 billion years. Why isn't Saturn as large as Jupiter? Its icy core is almost as large as Jupiter's. Apparently, the nebula simply ran out of gas. During the T-Tauri phase of the Sun's development, huge volumes of gas were swept away. Apparently, this gas was escaping while Saturn was trying to grow.

The rings and satellites with regular orbits are thought to be made of solids that condensed from a disk of gas and dust that originated near the end of the second phase when the outermost part of the proto-Saturn was unable to contract with the rest of the mass because of the protoplanet's very rapid rotation. As Saturn formed, it must have become quite hot. Nonetheless, Saturn appears not to have been as warm as Jupiter at the stage when Saturn's satellites formed. All of Saturn's regular satellites are icy (some may even have ices of ammonia and methane), whereas the satellites that formed near Jupiter (Io and Europa) consist mostly of silicate rock, and those formed far from it have much more ice (Ganymede and Callisto)—evidence of a large thermal gradient centered on Jupiter.

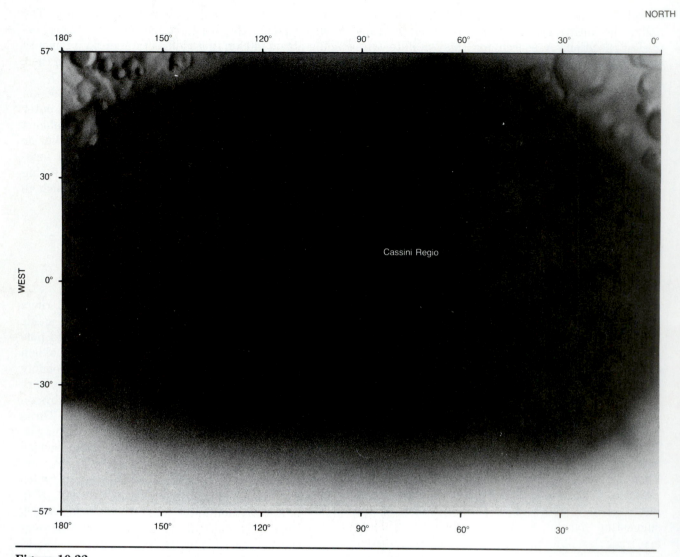

Figure 10.32

The shaded relief map of Iapetus shows the striking contrast between the dark and light terrains. This is the greatest contrast on any known object in the solar system. One idea is that the dark material represents an exotic type of volcanic ejecta; another holds that the material is shed onto the surface from space.

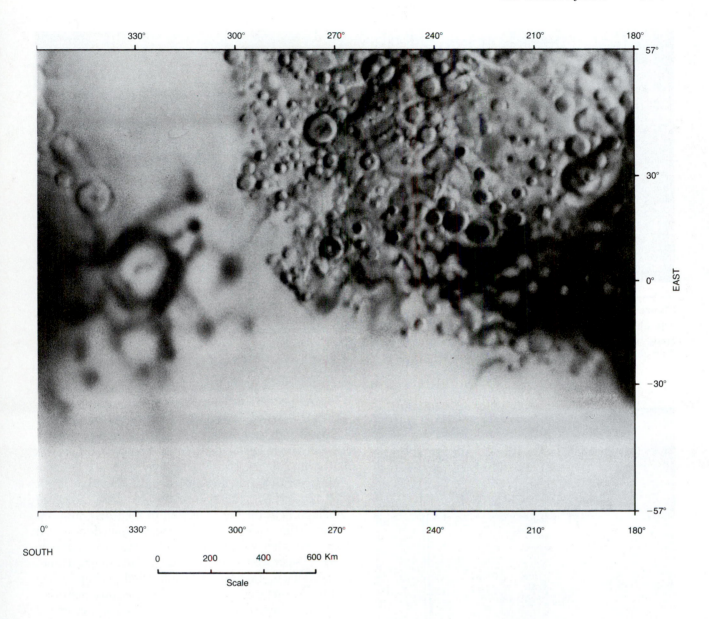

SOUTH

Scale

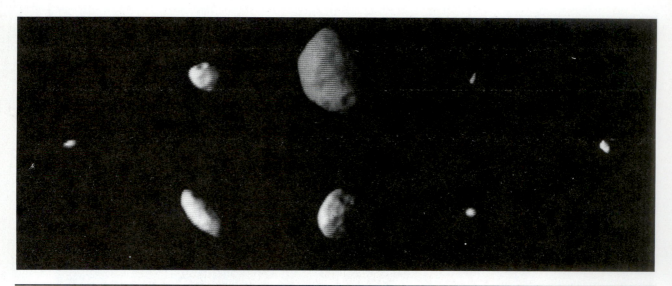

Figure 10.33

The small satellites of Saturn are irregularly shaped bodies composed mostly of ice. Many of them show scars from impact and closely resemble the small moons of Jupiter. This collage shows the eight inner satellites.

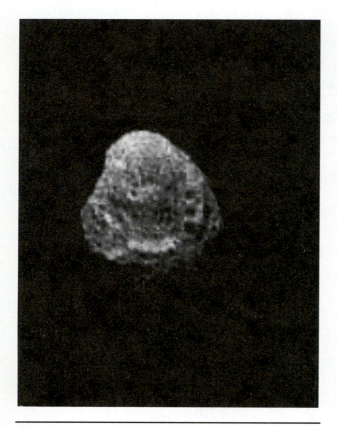

Figure 10.34

Hyperion is not much smaller than Mimas but is strikingly different. Its irregular outline suggests that it has been subject to impact fragmentation. It is composed of ice but has a low level of reflectivity, perhaps because it still retains its primordial dirty-ice crust.

After their collisional accretion and the formation of intensely cratered surfaces, the satellites of Saturn followed different histories. Titan, the largest of the Saturnian satellites, probably differentiated and degassed an atmosphere, which became nitrogen-rich and now obscures the surface. Most of the other satellites may not have acquired enough heat during accretion to melt and differentiate. Mimas and Iapetus preserve ancient cratered surfaces that were fractured by impact or tectonic processes (Mimas) or buried by dark debris (Iapetus). Dione, Tethys, Enceladus, and perhaps Rhea were resurfaced with (volcanic?) plains material, which covered and obliterated heavily cratered terrain. Subsequent collisions at reduced intensities produced less heavily cratered terrains that contrast with the older terrains on these bodies. These same satellites apparently fractured at some stage in their histories to produce large rifts or troughs—perhaps as a result of the cooling and contraction of their outer portions relative to their warmer interiors. Concurrently or subsequently to tectonism and resurfacing, material escaping from the interiors of Dione and Rhea created complex networks of bright streaks.

Among the satellites of Saturn, the thermal history of tiny Enceladus is unique. Tectonic disruption and volcanic resurfacing appears to have been continuous since the period of intense bombardment. The oldest terrain is only moderately cratered, and several generations of younger terrain types can be identified. Resurfacing processes by icy volcanism may be continuing today. The energy for extending the duration of the geologic development of Enceladus is probably the result of tidal interactions with Dione, with which it is in orbital resonance, and Saturn.

The meaning of the impact record on the satellites of Saturn is controversial. Some think the craters reveal a complex history of bombardment by two separate types of bodies, satellite disruption, and resurfacing during the early history of the Saturnian system. Others think it is the result of a single population of impacting bodies and discount the role of collisional disruption in the histories of the satellites and the rings.

Closely allied with the notion of two impactor populations is the idea that early collisions led to the collisional disruption of the satellites. It is important to remember that the concentrations and accelerations of impacting objects are controlled by Saturn's gravity field, so there may be a steep gradient in cratering rates from the outermost (low) to the innermost (high) satellites. Based on this notion, the cratering rate on Rhea could have been about twice that of Iapetus and the rate on Mimas about 20 times that on Iapetus. According to this interpretation, from Dione inward, the satellites were struck at least once by a projectile with sufficient energy to fragment these icy moons. When disruption of the inner satellites occurred, the fragments were probably reassembled by accretion. Dione and Tethys may have been disrupted and re-accreted once, Enceladus four times, and Mimas five times. The smaller satellites near the rings and the ring particles themselves could then be the products of a long and complex history of collisional disruption.

If we accept this hypothesis, we must consider the present Saturnian satellites to be re-accreted from the disruption of a system of larger satellites.

It is speculated that impact from large comets (perhaps 200 km in diameter) could disrupt and fragment these original satellites. Most of the denser material would be captured by Saturn itself, and re-accretion of the ice fragments would form the smaller satellites seen today. Titan was the only large moon to survive, probably because its orbit is far from Saturn and may have escaped the central focus of comet bombardment.

As an alternative to the notion of two populations of impacting bodies, many planetary scientists see the impact history of the Saturnian satellites as similar to that outlined for the other bodies of the solar system, including the Galilean satellites of Jupiter. A single group of impacting bodies, dominated by small objects, may have created the heavily cratered surfaces of the satellites of Saturn. Areas, such as those on Rhea, that appear to have fewer craters may just be flukes of the original random distribution of impacts on these small bodies. Following this idea, repeated disruption of the moons of Saturn is unlikely, and the satellites may date from the epoch of Saturn's formation. Moreover, based on the lower rate of impact derived from these studies, the rings of Saturn may have formed four billion years ago by impact fragmentation of a small moon, but it is unlikely that the rings of Saturn were formed within the last billion years by disruption of a single moon. Alternatively, the rings may represent particles that never accreted to form a large body.

Obviously, these very different interpretations of the history of Saturn's satellites have important consequences for understanding the early history of the solar system. New spacecraft and new techniques of study will likely be required before the final answers are in. In the meantime, the Saturn system will continue to be the focus of controversy.

Review Questions

1. Why might planetary rings be common but short-lived phenomena for the outer planets?
2. Compare the various possible modes of origins for planetary rings.
3. Describe the function of the shepherd satellites.
4. What are the principal differences between Titan and the rest of the satellites of Jupiter?
5. What is the evidence that the moons of Saturn consist mostly of water ice? Is this expected from our present understanding of planet formation from the solar nebula?
6. Why does Titan (like Earth) have an atmosphere rich in nitrogen when Saturn has an atmosphere rich in hydrogen and helium?
7. What is the significance of the differences between the small, icy moons Enceladus and Mimas?
8. How and why do craters on Mimas differ from craters of similar diameter found on the Moon?

9. Describe three ways in which large fracture systems can form on small icy satellites like those of Saturn.

10. Give an example of the difference between population 1 and population 2 craters as seen on the satellites of Saturn. What is the presumed difference in origin for these two classes of impacting bodies?

11. Why is the amount of radioactive heating thought to be related to the amount of rocky, silicate materials in a planetary body? Would a chunk of pure ice have any radioactive elements?

12. Why do most planetary satellites keep one side pointed in the direction of their orbital movement? In what ways might the trailing and leading hemispheres of a Saturnian satellite be expected to differ? Can you cite any examples of such differences between hemispheres?

13. What characteristics of Iapetus make it unique among the Saturnian satellites?

14. Does a generalized geologic history for a small, icy body in the outer solar system exist? What factors create variations on this theme?

15. What is the evidence that Phoebe and some of the other outer satellites of Saturn are captured rocky bodies?

Key Terms

Lagrangian Points

Regular Satellite

Retrograde Orbit

Roche Limit

Shepherd Satellite

Additional Reading

Gehrels, T., and M. S. Matthews. 1984. *Saturn*. Tucson: University of Arizona Press.

Lunine, J. I. 1994. Does Titan have oceans? *American Scientist*. Vol. 82, No. 2, pp. 134–143.

Morrison, D. C. 1982. *Voyagers to Saturn*. NASA SP-451, Washington, DC.

Owen, T. 1982. Titan. *Scientific American*. Vol. 246, No. 2, pp. 98–109.

Science. 1981. Vol. 212, No. 4491. (Issue devoted to Voyager 1 encounter with Saturn.)

Science. 1982. Vol. 215, No. 4532. (Issue devoted to Voyager 2 encounter with Saturn).

Sobel, D., 1994. Secrets of the rings. *Discover*. Vol. 15, No. 4, p. 86–91.

Soderblom, L. A., and T. V. Johnson. 1982. The Moons of Saturn. *Scientific American*. Vol. 246, No. 1, pp. 100–117.

Uranus Rising Over Miranda

TABLE 11.1

Physical and Orbital Characteristics of Uranus

Mean Distance from Sun (Earth = 1)	19.2
Period of Revolution	84.01 y
Period of Rotation	17.24 h
Inclination of Axis	98°
Equatorial Diameter	51,120 km
Mass (Earth = 1)	14.4
Volume (Earth = 1)	63
Density	1.28 g/cm^3
Atmosphere (main components)	H$_2$, He, CH$_4$
Temperature (at 1 bar)	78 K
Magnetic field (Earth = 1)	0.25 to 2.75
Gravity (at cloud tops; Earth = 1)	1.15
Known Satellites	15

CHAPTER 11

The Uranus System

 Miranda

 Ariel

 Umbriel

 Titania

 Oberon

Earth

Introduction

On January 24, 1986, the Voyager 2 spacecraft encountered the planet Uranus and its system of moons. At a distance of nearly 3 billion kilometers from Earth, Voyager sent back intriguing pictures of the atmosphere of the planet, measured its magnetic field, examined its system of rings, and photographed the surfaces of its major moons. The data obtained from this brief encounter provide the basis for our understanding of the Uranian system. Because of the expense and time involved in making and launching new spacecraft, it is unlikely that we will learn significantly more about the satellites of Uranus in our lifetimes.

The discoveries of Voyager show that the Uranus system is one of the strangest collections of planet, moons, and rings in the solar system. The giant blue-green planet (Figure 11.1) is tipped and lies on its side as it goes around the Sun; it possibly was shoved into that position eons ago by collision with a planet-sized body. To compound its oddness, the magnetic axis of Uranus does not pass through the center of the planet. Like Saturn and Jupiter, Uranus has rings, but they are thin, tenuous, and coal black, and fewer than half have circular orbits. Also, like Jupiter and Saturn, Uranus has a substantial system of icy moons. Five major satellites are visible with telescopes on Earth; ten smaller satellites were discovered by Voyager 2. The spectacular pictures sent back by Voyager show that several major satellites, most notably Ariel and Miranda, were geologically active early in their histories.

Major Concepts

1. Uranus, a large gas- and ice-rich outer planet, is the center of a system of at least 15 satellites and 10 dark rings.

2. Uranus is unique in that its spin axis is tipped almost 90 degrees from the normal solar system orientation. It seems that Uranus was tipped on its side when it was struck by a large body during its accretion. Moreover, its magnetic field is not centered on the spin axis; perhaps it is in the middle of a magnetic polarity shift.

3. Miranda (470 km in diameter), the smallest of the Uranian satellites, has a surface marked by large grabens and three large ovoidal terrains (coronae) shaped by tectonic events that postdate the era of heavy bombardment. These terrains may mark chunks of the planet redistributed by fragmentation and reaccretion.

4. The satellites Oberon (1520 km in diameter) and Umbriel (1170 km) both have heavily cratered surfaces that probably date to an early period of intense bombardment. Subsequent resurfacing by volcanism or tectonism was not significant.

5. The brighter, denser satellites, Titania (1580 km in diameter) and Ariel (1160 km), experienced resurfacing after the period of intense bombardment probably because of internal melting and icy volcanism. Both moons also have systems of grabens, which indicates that global expansion occurred late perhaps because of the more rapid cooling and contraction of the outer layers of these bodies compared to their warmer interiors. Tidal heating must have played a significant role in the tectonic histories of Miranda, Ariel, and Titania.

Uranus and Its Satellites

Many facts about Uranus are strange and unexpected. In contrast to all other planets in the solar system, Uranus spins on its side; that is, its axis of rotation lies nearly in the plane of its orbit (Figure 11.2). Thus, it rolls like a ball as it moves on its orbital path around the Sun, whereas other planets spin like tops. As Uranus revolves around the Sun, the north pole (over its spin axis) points directly toward the Sun at one time during its year, and the south pole does so at another. It is speculatively suggested that, during the final stages of accretion, a large, icy planetesimal the size of the Earth collided with a growing Uranus. An off-center impact could have knocked Uranus on its side, never to recover. Models for the development of the solar system cannot produce such an orientation without invoking a collision with another object.

The Uranian realm is a dark kingdom so remote from the Sun that it takes 84 Earth years to complete one revolution. Uranus is twice as far from the Sun as Saturn. Daylight on the moons of Uranus is roughly equivalent to a total solar eclipse on Earth. Uranus is also an extremely cold world. Since its time of origin, more than 4.5 billion years ago, temperatures in the Uranian system have never been higher than about 80 K. If the Earth were in the orbit of Uranus, our oceans would be frozen solid and our atmosphere of nitrogen would

liquify. Uranus is so far from the Sun that it was not discovered until after the invention of the telescope. William Herschel, a British astronomer, discovered Uranus in 1781.

The five major satellites of Uranus are compared in Table 11.2. Their relatively low densities and small diameters suggest that they are icy bodies that have much in common with the moons of Saturn.

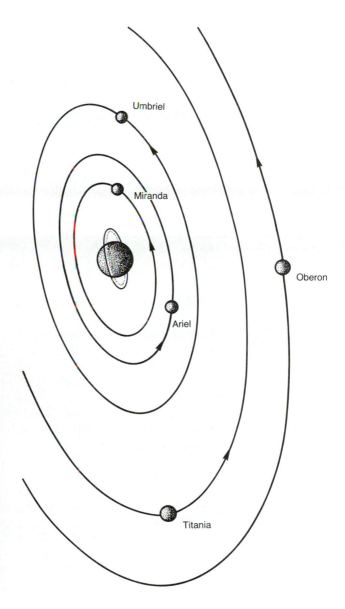

Figure 11.2

The spin axis for Uranus is tipped 98 degrees from the vertical orientation of the rest of the planets in the solar system. This anomalous tilt appears to be the result of an off-center collision with a large object late in its accretion history. Uranus has five major moons. Ten smaller moons are found within the orbit of Miranda.

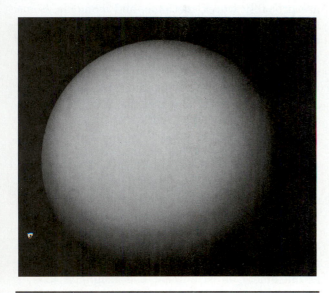

Figure 11.1

Uranus looks like a pale-blue billiard ball in this view produced by Voyager 2. Although computer-enhanced images reveal cloud structure and banding, these are invisible to the naked eye. The ring plane and the orbital plane of the satellites are also tipped.

TABLE 11.2
Characteristics of the Satellites of Uranus

Satellite	Diameter (km)	Density (g/cm³)	Comments
Cordelia	26	?	ring shepherd
Ophelia	32	?	ring shepherd
Bianca	44	?	
Cressida	66	?	
Desdemona	58	?	
Juliet	84	?	
Portia	110	?	
Rosalind	58	?	
Belinda	68	?	resonant with ring gap
Puck	154	?	
Miranda	470	1.15	coronae disrupt surface
Ariel	1160	1.56	fractures & volcanic flows
Umbriel	1170	1.52	dark, heavily cratered
Titania	1580	1.70	fractured
Oberon	1520	1.64	dark, heavily cratered
Moon	3500	3.30	

*Densities have uncertainties of 0.1 to 0.2 g/cm³.

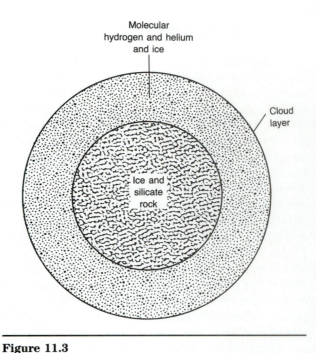

Figure 11.3

The interior of Uranus may consist of an inner core of dense silicate rocks about the size of Earth but with a much higher density, surrounded by a shell containing a mixture of ices of water, methane, and ammonia with the gases hydrogen and helium. Convection within the icy portion of the planet appears to have generated an irregular magnetic field.

The Interior and Atmosphere of Uranus

Because of its low density and hydrogen-rich composition, Uranus is often grouped with the gas giants Jupiter and Saturn. In fact, the density of Uranus is 1.28 g/cm³ (liquid water has a density of 1), significantly higher than either Jupiter or Saturn. Therefore, in contrast to gas- and liquid-dominated Jupiter and Saturn, Uranus must be richer in dense solids, like silicates or ice, than either of the larger planets. Apparently, Uranus is composed mostly of ices of water, ammonia, and methane, thought to be the most abundant ices in the solar system. Moreover, Uranus is substantially smaller than the larger planets, with only about 5 percent of the mass of Jupiter. The growth of Uranus may have been limited by the supply of nebular gas. Dissipation of the nebula occurred while Uranus was growing, stealing away some mass that could have become part of Uranus.

Various measurements made from Earth and Voyager 2 provide enough information so that a model of the interior of Uranus can be constructed (Figure 11.3). Uranus is believed to consist of three major components. A small, dense core of silicate rock may lie at its center. This core should be rich in iron and magnesium and is about 40 percent the size of Earth. It is probably surrounded by a thick mantle of ice or ice-rock mixture. The principal ices are probably water, methane, and ammonia. Some have speculated that inside Uranus this zone is molten. The outermost shell is gaseous, consisting mostly of hydrogen, helium, and traces of other gases. The data acquired by Voyager 2 about the rotation rate and tidal bulge of Uranus suggest that the core may consist of a mixture of dense silicates and ices. Probably about half of the mass of Uranus is ice (mainly water ice), less than half of it is rock, and the remaining 10 percent is hydrogen, helium, and other gases. There is a possibility that the ice and rock are separated into discrete layers, but some rock will dissolve in water at the high temperatures that must prevail inside Uranus.

Uranus has a small magnetic field, about ten times smaller than Earth's. However, it is not as simple as had been expected. To the complete astonishment of scientists, the magnetic axis is tilted approximately 60 degrees with respect to its axis of rotation. It is not known why. Perhaps Uranus is undergoing a magnetic reversal, in which the north and south poles switch places. Such reversals have occurred many times on Earth. The axis of the magnetic field is also offset from the planet's center. The magnetic field of Uranus, like those of other planets, appears to have developed in response to the convective movement of electrically conductive fluid (probably water) in the planet's icy layers. Convection may be driven by heat produced by radioactive decay of potassium, uranium, and thorium. These elements probably occur in the rocky part of the interior of Uranus.

Nonetheless, heat flowing from Uranus is also very low. The planet radiates only about 14 percent of the amount of heat that it receives from the Sun. This is about the amount of heat we expect to be generated by the decay of radioactive elements in its rock and ice core. There is no evidence that H and He are separating within the interior of Uranus. It appears that Uranus lacks the distinctive metallic hydrogen zone necessary for this type of differentiation to occur. Helium remains dissolved in molecular hydrogen and does not rain out. Uranus is too small and its internal pressure is too low for the transition from molecular to metallic hydrogen to be reached, so it retains the same proportions of hydrogen (about 74 percent) and helium (about 26 percent) in its atmosphere as does the Sun. Methane and ammonia have also been detected, but many other species are probably present in trace amounts.

In contrast to Jupiter, the atmosphere of Uranus does not dazzle the observer with details of its clouds, storms, and exotic spots. Visually, the most striking thing about Uranus may be just how bland it seems (see Figure 11.1). The planet is an almost featureless blue-green ball in which tumultuous clouds of ammonia or water ice, believed to occur at lower altitudes, are obscured. The color results from methane in the planet's atmosphere. Even with computer enhancement of the imagery sent back by Voyager 2, only a few deeply embedded methane clouds and faint smoggy haze were revealed. The visible clouds of Uranus are composed of methane. In sharp contrast, Jupiter's manifold atmospheric features are defined by clouds of ammonia ice particles. Voyager scientists discovered faint latitudinal bands in the atmosphere of Uranus, centered on the pole of rotation (Figure 11.4). The few small clouds reveal latitudinal (rather than poleward) circulation in the atmosphere. Uranus shares this characteristic with Jupiter, Saturn, Venus and, to a lesser extent, Earth and Mars, although solar radiation at Uranus is most intense at its poles rather than at its equator. Apparently the rotation of the planet is very important in determining atmospheric circulation patterns. Wind velocities are estimated to range up to 700 km/hr; they cause the upper cloudy part of the atmosphere to rotate more quickly than the interior.

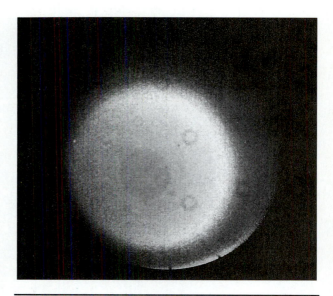

Figure 11.4
Color banding in the atmosphere of Uranus is only apparent after enhancement to bring out the subtle color variations in the methane clouds. In spite of significant solar heating of the polar regions, circulation in the atmosphere is dominated by flow parallel to the equator of the planet, just as it is on Earth and Jupiter. Apparently flow is dominated by the rotation of the planet and not by the distribution of solar heat. The spin axis is near the center of the light gray circle in the center of the disk. The small gray circles are flaws in the image.

The Uranian Ring System

Like Jupiter and Saturn, Uranus has a system of rings (Figure 11.5), but the particles in the rings are dark gray and were not detected by telescopes on Earth until 1977. They are some of the darkest objects ever studied in our solar system. Ten narrow rings have been identified, together with many dust bands. All lie within one planetary radius of the cloud tops. The rings are narrow, ranging from nearly 60 km to less than 2 km in width. For the most part, the rings of Uranus are made up of fairly large particles; in fact, boulder- to house-sized fragments seem to dominate. The most distant ring, the Epsilon ring, is made up almost entirely of black boulders; many are more than 10 m in diameter. Voyager 2 also discovered hundreds of dust bands within the ring system. If all of the ring particles were accumulated into one moon, it would have a diameter of only 150 km. What are the dark particles in the rings? Some think they consist of carbon-rich residues, made as methane ice in ring particles blackened and decomposed during exposure to solar radiation. Others suspect ring particles are made of exotic black rock, rich in organic compounds and ice. The rings lie in the equatorial

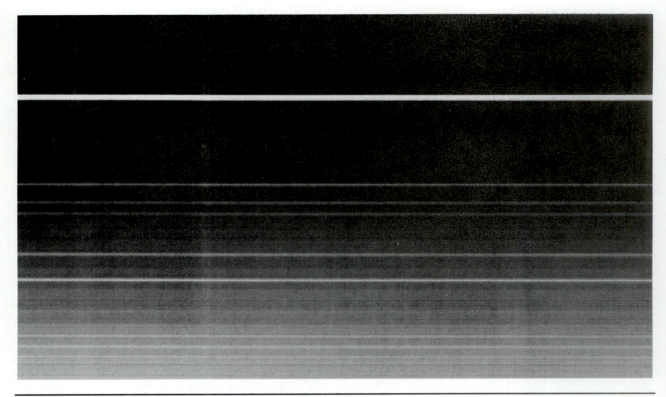

Figure 11.5

The rings of Uranus appear as delicate strands and are composed of charcoal-black particles, rather than the bright water-ice particles in the rings of Saturn. Two shepherd satellites were discovered on either side of the outermost (upper) ring.

plane of Uranus, tipped as it is. This suggests that the rings formed after Uranus fell on its side. If the rings existed before this planetary trauma, it is unlikely that they would reorient their orbits following the large impact.

Compared to Saturn's bright, complex rings, which are tens of thousands of kilometers across, the Uranian rings are but thin strands. Yet, they are important in that they provide further insight into the origin and evolution of ring systems and the origin of the solar system itself. The delicate rings are probably defined by shepherd satellites, although only two were found—one inside and one outside the most massive outer ring.

From studies of the narrow rings of Uranus, some have suggested that planetary rings may be short-lived phenomena that come and go during a planet's lifetime. For example, the dust rings discovered by Voyager 2 may have lifetimes of less than 1000 years because the extended atmosphere drags on the particles, slowing them down, until they spiral into Uranus. Based on the orbital characteristics of the shepherd satellites, the rings may have lifetimes of less than 600 million years. Although this is much longer than the duration of dust belts, it is much shorter than the age of the solar system. They must, therefore, have a continuing

source or they would have disappeared long ago. Uranian ring particles may be the flotsam produced by collisions of small satellites within the system or perhaps by collisions with cometary interlopers. Subsequently, the rings may in a sense erode themselves away as their large particles collide with each other, grinding down to dust that is swept away, ultimately showering down onto the planet below.

An important problem with the young ring hypothesis is that rings are present around all of the large outer planets. Would impacts have coincidentally produced rings about all of the planets in one brief moment of time? If rings are continually produced and eroded away, then a large amount of satellite material must have been consumed. Perhaps the rings are indeed long-lived relics of the disc of particles from which the satellites accreted. Other planetary scientists have hypothesized that the ring fragments condensed from the gas blown off Uranus by the giant collision that tilted the planet on its side.

The Moons of Uranus

Uranus has five major icy moons that were discovered using telescopes on Earth. In order outward from Uranus they are Miranda, Ariel, Umbriel, Titania, and Oberon. Like most other planetary satellites, all of the moons are in tidally locked orbits with the same face always pointing toward Uranus (Figure 11.2). Ten of the small satellites were identified by Voyager 2. All of the satellites occupy orbits lying in the plane of Uranus's equator. Thus, their orbits share the unusual inclination of the planet itself. They must have formed from a circum-Uranus nebula after the planet was bowled over on its spin axis. Oberon, Titania, Ariel, and Umbriel are quite similar in size (1100 to 1600 km in diameter) and are approximately the size of the intermediate moons of Saturn (Tethys, Dione, and Rhea). Miranda is considerably smaller, with a diameter of only 500 km, and is approximately the same size as Saturn's moon Mimas. Atmospheres have not been detected around any of the Uranian satellites.

Water ice, but not methane or ammonia, has been identified at the surface of all five major moons. (Methane and ammonia could be expected to be important constituents in the interiors or on the surfaces of these moons because temperatures at this great distance from the Sun may have been low enough for these volatile compounds to have condensed in the ancient solar nebula from which the planets formed. Moreover, methane has been detected on a moon of Neptune and on Pluto.) The surfaces of all the moons are dark or perhaps dirty mixtures of rock and ice, perhaps because of impact processes.

Like the moons of Saturn, the interiors of the Uranian moons are probably mixtures of low-density ice and denser silicate rock and iron. The Uranian moons are somewhat denser (1.3 to 1.6 g/cm^3) than those of Saturn, perhaps suggesting that they consist of a larger proportion of rocky material, perhaps about 60 percent silicate rock and 40 percent ice. At first glance this may seem to be opposite the expected trend for the cold, outer reaches of the solar system. In the ancient solar nebula, we might have expected the proportion of low-density methane ice (0.5 g/cm^3) to increase as we moved farther and farther from the Sun. The relatively high densities of these satellites tell us that, instead, much carbon resided in very volatile carbon monoxide instead of in methane. Apparently, carbon monoxide was a gas at the distance Uranus lies from the Sun. Consequently, the ratio of rock to ice was higher in the condensates that formed in the outer solar system.

We will discuss the satellites in the order of their occurrence outward from Uranus.

The Small Satellites

All of the ten newly discovered small moons orbit closer to Uranus than Miranda (Figure 11.2).

Two of these satellites, Cordelia and Ophelia, are ring shepherds, which gravitationally define the inner and outer edges of the outermost ring. The largest of the new satellites, Puck, is about 160 km in diameter and was photographed by Voyager 2. The photographs show that it is darker than any of the large Uranian moons and slightly irregular in shape. Like other small objects in the solar system, Puck is also cratered. The other small satellites, including one named Juliet, have diameters between 40 and 80 km and were not photographed in enough detail to show surface features, but they are all quite dark. This suggests that either icy materials at their surfaces darken by some sort of aging process involving methane ice or, as most investigators speculate, the dark material is an undifferentiated mixture of icy, silicate, and carbonaceous materials. If this latter explanation is correct, these small moons preserve some of the most chemically primitive material in the outer solar system.

Miranda

Voyager 2 flew within 29,000 km of the little moon, Miranda, and photographed remarkable detail of its surface features (Figure 11.6). Because of its small size, most scientists expected it to be a bland ball of water ice, little changed since its birth more than 4.5 billion years ago. What the pictures of Miranda revealed was one of the strangest planetary bodies in the solar system. Miranda is the smallest of the major Uranian moons (470 km in diameter) and occupies the innermost orbit. It has barely enough gravitational strength to pull itself into a sphere. In spite of its small size, its surface consists of complex and exotic terrains unlike anything seen on other satellites.

Two strikingly different terrain types have been delineated on Miranda (Figure 11.6). One is an old, heavily cratered surface that is similar to the lunar highlands and therefore is probably more than 4 billion years old. The other terrain type is completely different and consists of three roughly circular or oval areas of complex structures marked by parallel sets of alternating light and dark bands, scarps, and ridges. These terrains have been called *coronae* because of their ovoidal shape and the rings of distinctive terrain that surround them. Judging from the numbers of craters on the ovoids, it appears that the coronae are substantially younger than the heavily cratered plains. The most prominent of these exotic areas is located near the south pole (center, Figures 11.6 and 11.7). Its outer boundary is sharp and angular, and the internal pattern of ridges and bands displays many sharp corners. Inside this crustal fragment is a strange, angular V-shaped structure marked by bands of

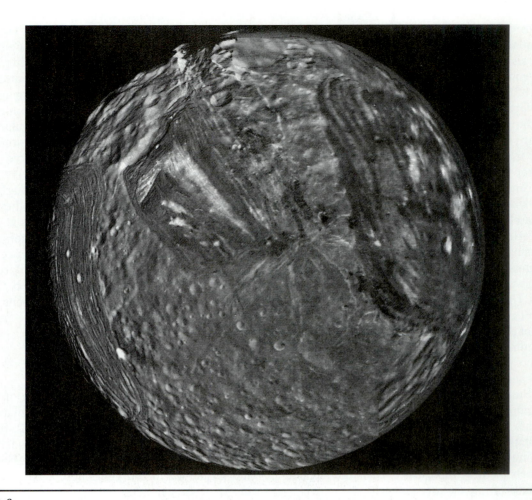

Figure 11.6

Miranda is a small moon 480 km in diameter, about the same size as Saturn's Mimas. Unlike Mimas, this Uranian moon shows evidence of significant crustal mobility and resurfacing after the period of heavy bombardment. Its surface consists of two very different types of terrains. One is heavily cratered and must be very old; the other consists of three different regions of banded terrain, or coronae, which are visible in this mosaic. The coronae have light and dark stripes, which truncate against the fault-defined margins of these areas. The trapezoidal corona near the top of the photo connects with a deep graben system. These disrupted regions may mark areas where ice rose toward the surface around dense masses of rock that sank.

light and dark material (Figure 11.7). The outstanding features in this area are that the outer boundary and the internal V have many sharp corners and that some ridges and grooves are truncated by sharp boundaries.

Another corona, shown on the right side of Figure 11.6, consists of light and dark discontinuous bands that curve smoothly around a core marked by light blotches on a darker terrain. The third corona is located to the left on Figure 11.6. It consists of closely spaced ridges surrounding an inner core of complex intersecting ridges and troughs. Both the ridged and banded terrains have an outer belt approximately 100 km wide that resembles a racetrack and wraps around an inner core (Figure 11.7). This outer band is actually a trough 5 to 6 km deep. Faults that cut across the

cores of the coronae truncate against the bounding beltway. The ridged and banded terrains somewhat resemble the grooved terrain on Ganymede. Faulting and extrusion of volcanic liquids or solid but plastic materials were undoubtedly involved in their formation. These three coronae that developed in the crust of Miranda are unique. Nothing like them has been seen anywhere else in the solar system.

Miranda also has a series of enormous faults that can be traced across the globe. Some are older than the complex coronae; others are younger. The most spectacular is a huge fault scarp 7 to 10 km high (Figure 11.8). It forms a rift valley near the top of Figure 11.6 and extends southward along the margin of the trapezoid, where it disappears in the unilluminated hemisphere. This steep rift valley is

Figure 11.7

Complex sets of V-shaped ridges, seen in this detailed view of the corona shown in Figure 11.6, must be the result of faulting of Miranda's icy lithosphere.

deeper than Valles Marineris on Mars. Other grabens can be seen as well; all show that Miranda experienced global expansion at some point in its history.

Why does Miranda have such strange and unusual terrains? How did such a surface develop on a tiny ball of ice? Several models have emerged to explain Miranda's coronae. One view holds that when Miranda accreted it was a uniform mixture of rock and ice. Because of internal heating, the small body began to differentiate. Masses of dense rock sank toward the center, and plumes or sheets of lighter ice rose past the margins of the rock and eventually breached the surface as volcanic liquids to create the coronae. The complex ovals are surface disturbances left by the sinking rock masses. The distribution of the coronae on the surface suggests some sort of organized flow of material within Miranda's interior—perhaps from a poorly developed convection pattern. According to this theory, the process of planetary differentiation was slow to start and then aborted, never to be completed. If it had been completed, more ice would have risen to the surface and the coronae would have been smoothed over. The relatively young age

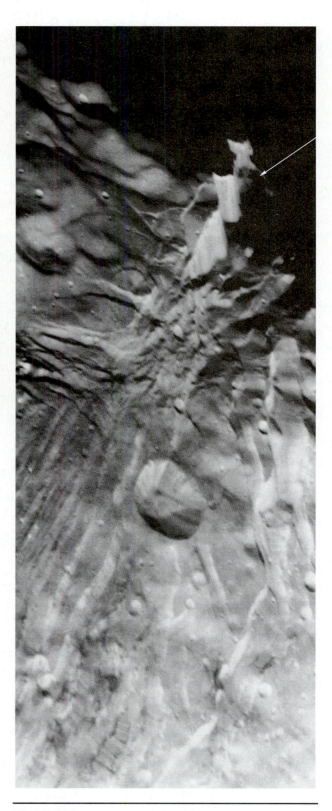

Figure 11.8

A brightly lit fault scarp 7 to 10 km high cuts the terminator of Miranda at the top of this photo. These faults are part of a graben system that connects with a corona.

of the coronae could be explained if Miranda was temporarily in a resonant obit with a forced eccentricity. Such a orbit could cause tidal heating and start internal differentiation.

Ariel

Beyond the orbit of tiny Miranda lie Ariel and Umbriel. These two satellites are roughly the same size (1160 and 1170 km in diameter), but they had strikingly different geological histories. Whereas Umbriel has the oldest and darkest surface of the major Uranian moons, Ariel has the youngest and brightest (Figure 11.9).

As for Saturn's moons, two types of cratered terrains have been identified in the Uranian system. One type is covered by closely spaced craters with diameters of 50 to over 100 km and is similar to the lunar highlands. The impactors responsible for the formation of these terrains and for the craters on the inner planets were probably the remnants of planetary accretion and moved in orbits around the Sun. These bodies were probably swept up rapidly during and after accretion of the moons. The other type of cratered terrain has far fewer craters in the size range of 50 to 100 km, and instead has many small craters. Some scientists think these craters were produced by the myriads of small objects created by satellite collisions in orbits around the planet—here Uranus.

The oldest surface on Ariel is a widespread cratered terrain consisting mostly of impact craters less than 60 km in diameter (Figure 11.9). As judged by crater frequencies, Ariel has the youngest surface of the Uranian moons. An older set of larger craters, seen for example on Umbriel, must have been covered by volcanic extrusions of ice and slush, modified by tectonic disruption or by isostatic adjustments while the lithosphere was still warm. Some small, highly flattened craters are still visible on Ariel, showing that viscous flow of the icy lithosphere has occurred locally.

The most striking feature on Ariel is the global system of fractures and faults that form spectacular, deep rift valleys (Figure 11.10). In some places, the rift valleys on Ariel are as deep as the Grand Canyon on Earth. Some of the older fault scarps are modified with young impact craters, whereas the younger scarps are remarkably fresh and free from superposed craters. These relationships show that the faulting occurred during the formation of the younger craters. The floors of most of the rift valleys are covered with smooth material, perhaps extrusions of slushy mixtures of water and ammonia ice that welled up through the fractures and spread out across the floors of the valleys like terrestrial glaciers. A sparsely cratered terrain occurs between the fracture system in the northern latitudes and was presumably formed by a sequence of flows that partly overlapped the older craters near the equatorial region. Occasionally, ghosts of underlying craters protrude from the plains. The youngest features on Ariel are bright-rimmed craters and their ejecta. Their abundance is roughly that expected for impacts during the last 3 or 4 billion years.

The geologic history of Ariel is complex—it is not a simple, undifferentiated sphere—and the major events are quite clear. After accretion, Ariel must have experienced the period of intense bombardment that scarred the surfaces of all other planetary bodies in the solar system and so once displayed a surface with relatively large impact craters. The craters on this surface, no longer apparent on Ariel, may have been obliterated by viscous relaxation together with watery volcanic extrusions of a low-melting-point liquid derived from Ariel's interior. Magma may be created by melting of an icy ammonia-water mixture in an undifferentiated mix of rock and ice. Thus, Ariel may be only partially or wholly differentiated into a core of dense silicates and an icy mantle like the Galilean satellites. A later period of bombardment by smaller objects created a new set of smaller impact craters that are still visible on the surface. As Ariel's surface layers, once heated by accretion, cooled and contracted around a still-warm interior, its lithosphere may have contracted and ruptured, creating a network of extensional faults and grabens. This extensional tectonism may have been accompanied by further extrusion of fluids along the fracture zones and could have produced the smoother plains. The few bright-rayed craters seen on Ariel were formed by subsequent impact over the past several billion years.

But where did the heat come from to keep Ariel active and warm for such a long time? As we have seen before, these small, icy moons rapidly lose their small amount of accretionary heat and have little from radioactive means. Is tidal heating the answer, as on Io? At present, no orbital resonances exist in the Uranian system of moons. This is critical because, to maintain significant tidal heating in a synchronously rotating satellite, a resonance is necessary since it forces the orbit to be eccentric (see Chapter 9). It is possible that Ariel and Umbriel, the next moon out, had a 2:1 resonance in the past. However, once stable resonance is achieved, it is difficult to disrupt. Thus, no entirely satisfactory solution is in hand. What is certain is that Ariel has been significantly modified by thermal processes after its accretion and heavy bombardment.

Figure 11.9

Ariel, 1150 km in diameter, is not as heavily cratered as Miranda but it also shows two types of terrains. The older terrain is covered with relatively small craters that may be scars of impacts from debris in orbit around Uranus, rather than around the Sun. Cutting across this terrain is a global system of extensional faults and grabens, which may be related to the global expansion and fracturing of the icy body. Some fault-bounded regions are covered by smooth, relatively young plains, which are probably the volcanic plains produced as watery volcanic liquids oozed out of the fissures produced by expansion.

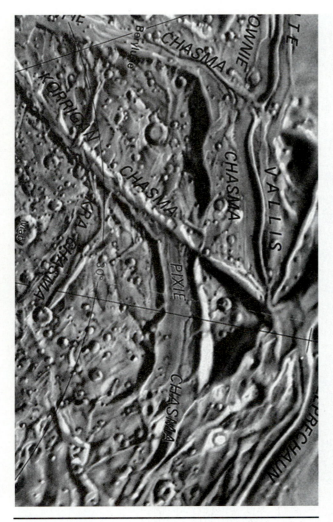

Figure 11.10

The smooth volcanic plains of Ariel are closely related to the global system of faults. Some faults are cut by impact craters, whereas others are fresh and have no superposed craters. The plains have relatively few craters and may be some of the youngest features in the Uranian satellite system.

Umbriel

Umbriel is a dull, gray world 1190 km in diameter, and is the darkest of the Uranian moons (Figure 11.11). The Voyager photographs of Umbriel are not as good as those of Miranda, but they show an ancient surface saturated with large, muted craters. This distribution of craters, especially those between 50 and 100 km in diameter, closely resembles that of the lunar highlands, as well as that of the most ancient, heavily cratered bodies in the solar system. Note how the crater population on similarly sized Ariel is strikingly different (Figures 11.9 and 11.11). There is a puzzling absence of the rayed craters visible on other Uranian satellites.

Little or no evidence of significant tectonic activity on Umbriel has been discovered. Thus, the surfaces of Umbriel, along with that of Oberon, appear to be the most ancient surfaces of the major satellites of Uranus, having remained relatively unaltered since the period of intense bombardment over 4 billion years ago. In this respect they are similar to Jupiter's Callisto.

Titania

Titania is the largest moon orbiting Uranus but is less than half the size of Earth's Moon. It is similar to Oberon in general global properties, such as size, density, and color, but the histories of the two satellites are quite different. In fact, the surface of Titania (1610 km in diameter) is more like the much smaller Ariel (1160 km in diameter).

Titania is dominated by impact craters and may at first glance look like Oberon. A comparison of Figures 11.12 and 11.13 shows that Titania has fewer craters than Oberon or Umbriel. Its surface must be younger. Moreover, most of the craters on Titania are small (less than 100 km across). This fact is important. Titania must have experienced the same period of intense bombardment as did Umbriel—indeed, as did all other planetary bodies in the solar system. Large, old craters were subsequently obliterated as the planet was resurfaced, perhaps by the extrusion of lava (slushy ice) or viscous relaxation of older, larger craters. The hypothetical new surface of low relief was subsequently modified by the impact of many smaller bodies forming terrains covered by smaller craters. Moreover, the crater density on Titania is not uniform. Several areas are distinctly smoother and less cratered than others, showing that the resurfacing processes on Titania extended over a significant period.

Like Ariel, Titania is crisscrossed by a near-global network of rift valleys and the extensional faults that bound them, suggesting late global expansion (Figure 11.12). The grabens are large structures with vertical relief on the fault scarps ranging from 2 to 5 km. Careful study of Figure 11.12 will show that the faults cut the larger craters and are not modified by the smaller ones. The rifts are, therefore, among the youngest features on Titania.

Thus, the surface features and geologic activity on Titania are reminiscent of those of Ariel, except that the tectonic activity and resurfacing of Ariel were more intense, more extensive, and more prolonged. These similarities and differences are evident in Figures 11.9 and 11.12. The global network of fault valleys is more fully developed on Ariel, and the plains are noticeably less modified by craters.

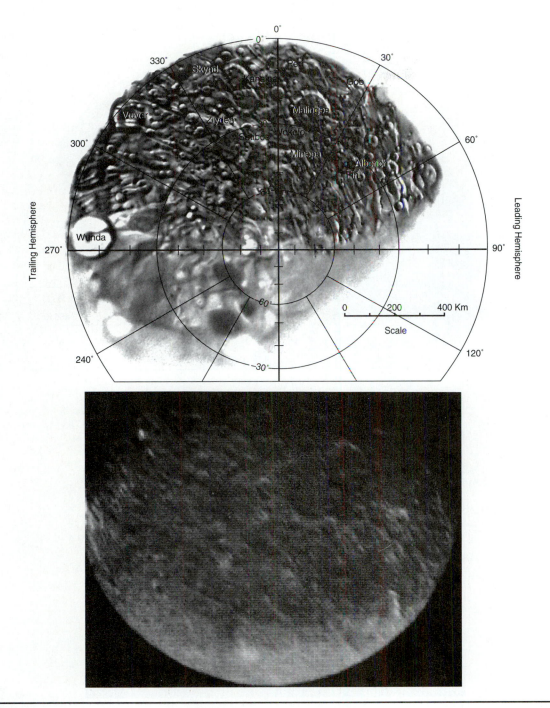

Figure 11.11

Umbriel's surface is nearly saturated with degraded craters. It is the least varied of the satellites of Uranus. Although it is about the same size as Ariel, it lacks the system of global faults and young terrains found on its neighbor. Something other than size apparently controlled the thermal, tectonic, and volcanic evolution of these moons.

Oberon

With a diameter of 1550 km, Oberon is only slightly smaller than Titania. Significant differences between these otherwise similar moons are obvious in Figures 11.12 and 11.13. The gray sur-face of Oberon is nearly saturated with large craters that range up to 100 km in diameter. Note the similarities in crater populations on Oberon and Umbriel (Figures 11.11 and 11.13) and how the crater population on Titania is strikingly different (Figure 11.12). The density of these craters is more

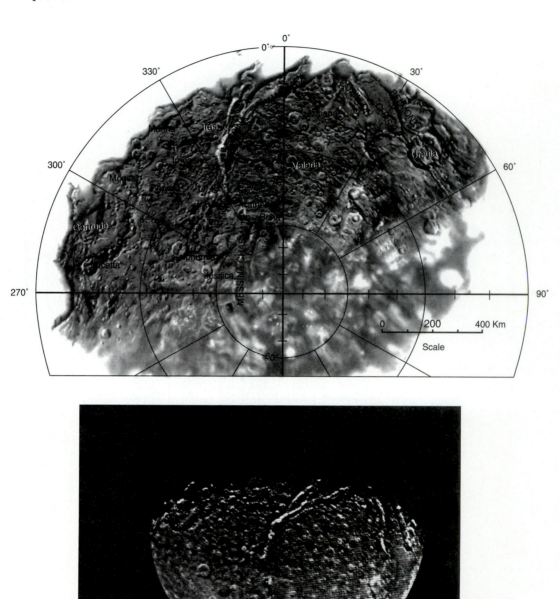

Figure 11.12

Titania's surface is marked by a large system of faults, probably produced as the icy moon expanded and cracked. This, the largest moon of Uranus, has a surface like Ariel's. However, its cratered terrain has more craters and is more extensive, but most of Titania's craters are relatively small. Large, old craters may have been obliterated by a global resurfacing event during early bombardment.

similar to that found on the lunar highlands, indicating that most of the surface was formed during intense bombardment. A high mountain is visible near the limb of Oberon. It may be a central peak of an otherwise invisible large impact structure several hundred kilometers in diameter. Bright rays are found around a few craters, but there is no global pattern of light and dark material. The floors of a few large craters are covered with dark material that could be lava composed of carbon-rich, dirty ice extruded after the period of intense bombardment, but there is no evidence of large, smooth plains. Several linear features that resemble faults are seen on Oberon, but there

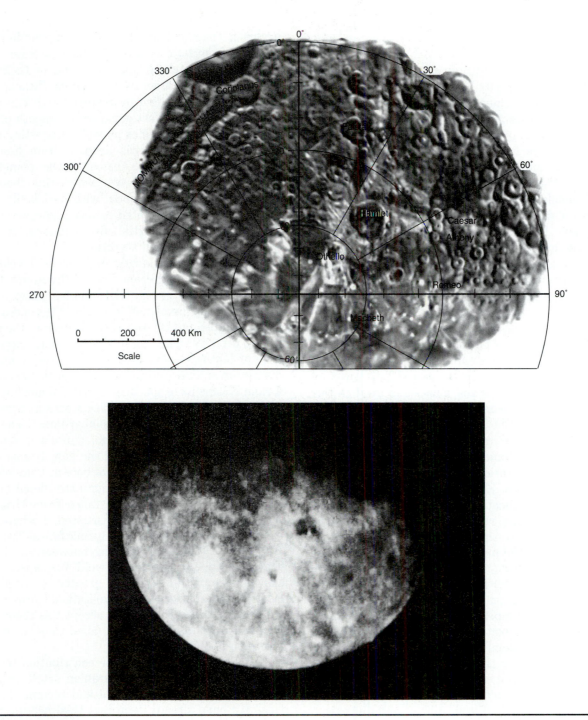

Figure 11.13
Oberon is also heavily cratered like smaller Umbriel. Its mottled gray surface shows only a few faults, suggesting that this satellite is relatively primitive.

is little other evidence of major tectonic activity.

These facts show that although Oberon is one of the larger Uranian satellites, it has been largely inactive since the period of intense bombardment some 4 billion years ago. It thus fits the classic picture of a small satellite made of ice in the outer solar system, too small and too cold to generate enough heat to produce significant geologic activity after the differentiation of ice from rock. It has acted as a passive recorder of each encounter with projectiles orbiting the Sun as well as those in orbit around Uranus itself.

Conclusions

Uranus with its strange orbital and magnetic field properties, and its host of rings and satellites, carries important clues about the nature and origin of the solar system. The Voyager data have provided us with an overview of these features. Nonetheless, we are left with many unanswered questions regarding the evolution of the bodies in the Uranian system. For example, why is Uranus so much smaller than Jupiter and Saturn? At the distance Uranus lies from the Sun, the temperature must have been about 60 K during condensation from the ancient solar nebula. The core of ice and rock that accreted was probably about the same size as the ones that formed at Saturn and Jupiter. Once it was large enough, nebular gases became gravitationally anchored to the icy protoplanet. Was it the dissipation of the gaseous nebula that halted the growth of Uranus?

Similarly, it is not clear that all of the moons of Uranus are internally differentiated into cores of hydrated silicate rocks and mantles of ice, like the moons of Jupiter. Some calculations suggest that during accretion the moons of Uranus may not have warmed sufficiently to allow the gravitational separation of ices and silicates. For small bodies like Miranda, accretionary heating probably was only a few degrees. Even if *all* the impact energy was converted to heat, the internal temperature of Miranda would have risen by only 30 K. Radiogenic heating might raise the temperature another 10 to 20 K. However, the probability of differentiation is increased if significant proportions of ammonia are mixed with the water ice. For example, pure water ice has a melting point almost 200 K above the present surface temperature (80 K), but if ammonia is mixed with the water ice, the mixture will melt at a much lower temperature, about 100 K above the temperature at the surface. This lowering of the ice melting point may increase the chance that even the small Uranian moons differentiated during their first few hundred million years of existence.

In the Uranian system we have discovered that two darker moons, Oberon and Umbriel—with relatively low densities (1.50 and 1.59 g/cm^3) but quite different sizes (1520 and 1180 km in diameter)—have relatively primitive surfaces. These objects have surfaces dominated by many craters and appear to have experienced no late expansion—at least none after the heavy cratering of their surfaces. Two of their orbital companions, Titania and Ariel, with significantly different sizes (1580 and 1160 km in diameter) have similar but slightly higher densities (1.68 and 1.66 g/cm^3) as well as surfaces disrupted by large extensional

grabens. These global rift systems appear to have resulted from late expansion of the satellites and consequent stretching and rupturing of their icy lithospheres. The fractured satellites, Titania and Ariel, are also less densely cratered and then only by smaller craters. Vast expanses of smooth plains are found on these bodies as well, suggesting that icy volcanic material was erupted from fissures associated with the expansion of the planetary bodies. We also outlined two contrasting theories for the evolution of Miranda, with its heavily cratered terrain and disrupted coronae; however, both ideas involved the partial differentiation of this tiny body after its final accretion.

Why are such dramatic differences developed on moons of such different sizes? Why aren't Titania and Oberon, moons with similar sizes, more alike? A radical explanation of the differences between Titania and Oberon harkens back to the reasoning used by some to explain the bizarre appearance of Miranda—that is, it relies on the complete disruption and re-accretion of some of the moons of Uranus. Some have speculated that Titania, but not Oberon, was shattered by collision with a large object late in the era of heavy bombardment. The debris from the collision reassembled into a new moon in a relatively short time, but the new Titania also had an entirely new surface that bore no trace of the early bombardment or previous extensional faulting. Oberon, at a greater distance from Uranus, would be less likely to be disrupted. Perhaps an early Ariel met the same cataclysmic fate as Titania. But how did Umbriel, which lies between Ariel and Titania, escape shattering as well? We expect that the number of collisions was greater on satellites closer to Uranus because of gravitational focusing of incoming objects. Even if we accept this theory, it does not explain why Oberon failed to develop extensional features.

Was tidal heating another contribution to the heat budget of any of the Uranian satellites and consequently to their thermal and tectonic histories? We have already seen that tidal heating was significant for the evolution of Jupiter's Io and Europa and also Saturn's Enceladus. The Uranian moons are not now in orbital resonance and thus cannot be receiving a thermal input from tidal flexing. However, it is possible that Miranda, Ariel, and Umbriel have passed through a resonance in the past. In fact, the noncircularity of their present orbits may have been produced when former resonant configurations were destroyed. Ariel might have been significantly heated if it resided in resonance with Umbriel for a long time. However, heating of Umbriel would have been negligible, because of its greater distance from Uranus and smaller tidal distortions. This idea is consistent with the heavily cratered surface of Umbriel. The temporary trapping of Miranda in an orbital position where tidal flexing or chaotic rotation was important might have caused a thermal event that resulted in its partial resurfacing and development of the coronae. Unfortunately, this list of possible resonances and potentials for tidal heating does not include tectonically modified Titania.

Perhaps part of the explanation for the differences among the satellites lies in the compositions of the satellites as reflected in their densities. Both Titania and Ariel, which appear to have longer tectonic histories with late lithospheric extension, also have higher densities, perhaps implying that they have more of a dense silicate component and less ice in their interiors than the less dense Oberon and Umbriel. Incorporated in the silicate materials are small but significant quantities of heat-producing radioactive elements. Therefore, it is conceivable that Titania and Ariel, with their larger silicate fractions, contained just enough radioactive elements to augment their differentiation and tectonic processes started by accretion. Thus, cooling and consequent fracturing of this pair might have occurred later during the period of reduced cratering rates. Oberon and Umbriel, the darker, less rocky, bodies less well endowed with radioactive elements, may have cooled more quickly and perhaps expanded during the intense bombardment associated with the end stages of accretion. Little evidence of ancient geologic events such as volcanism and faulting has survived the early resurfacing of Oberon and Umbriel by impact processes.

Review Questions

1. Among the planets of the solar system, Uranus has unique orbital properties. What are they, and what is a possible explanation for them? What suggests that the orbit was changed before the satellites formed?
2. What might account for the compositional differences between Uranus and Saturn?
3. Why is the atmosphere ice ratio of Uranus smaller than that of Saturn?
4. Which is more important for the movement of gases in the atmosphere of Uranus, the rotation of the planet or the input of solar energy? Would the answer be the same for Earth?
5. In what way do the ring particles of Saturn and Uranus differ?
6. Prepare a table that compares the diameters, densities, surface compositions, ages, and features of the five major satellites of Uranus.
7. How do the densities of the satellites of Uranus and Saturn compare? Is that predicted by simple cooling of the solar nebula? How might the difference in density be explained?
8. Contrast the resurfacing histories of Oberon-Umbriel and Titania-Ariel. What geological differences can account for their different histories?
9. How does Miranda differ from its icy neighbors?
10. Do any common themes about the evolution of small, icy bodies emerge from a comparison of the satellites of Jupiter, Saturn, and Uranus?

Additional Reading

Bergstrah, J. T., E. D. Miner, and M. S. Matthews. 1991. *Uranus*. Tucson:University of Arizona Press, 1076 p. (Technical reports, photos, and maps derived from Voyager data.)

Cuzzi, J. N., and L. W. Esposito. 1987. The Rings of Uranus. *Scientific American*. Vol. 257, No. 1, pp. 52–67.

Johnson, T. V., R. H. Brown, and L. A. Soderblom. 1987. The Moons of Uranus. *Scientific American*. Vol. 256, No. 4, pp. 48–60.

Ingersoll, A. P. 1987. Uranus. *Scientific American*. Vol. 256, No. 1, pp. 38–45.

Science. 1986. Vol. 233, pp. 1–132. (Reports on Voyager 2 encounter with the Uranian system.)

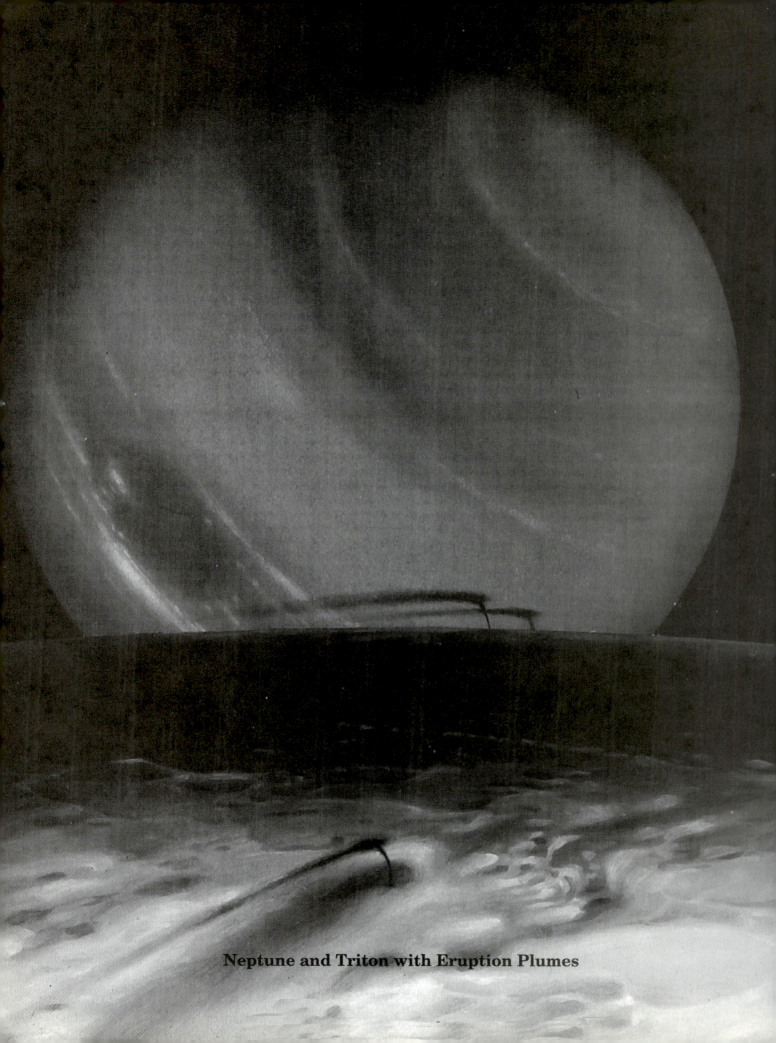

Neptune and Triton with Eruption Plumes

TABLE 12.1

Physical and Orbital Characteristics of Neptune

Mean Distance from Sun (Earth = 1)	30.1
Period of Revolution	164.8 y
Period of Rotation	16.05 hrs
Inclination of Axis	28.8°
Equatorial Diameter (km)	49,500
Mass (Earth = 1)	17.2
Volume (Earth = 1)	58
Density	1.64 g/cm^3
Atmosphere (main components)	H_2, He, CH_4
Temperature (at 1 bar)	69 K
Gravity (at cloud tops; Earth = 1)	1.12
Magnetic Field (Earth = 1)	0.24
Known Satellites	8

CHAPTER 12

The Neptune System

1989N1

Triton

Nereid

Earth

Introduction

Neptune, a planet that rivals Earth for the designation as the blue planet, orbits the Sun in the outer solar system far beyond Uranus. Until August 1989, little was known about Neptune or its satellites, as no spacecraft had yet visited them. All of this changed when Voyager 2 hurtled past Neptune on its way out of the solar system. Voyager 2 flew closer to Neptune than any other planet on its 12-year-long tour of the planets. True to its history, the Voyager spacecraft returned remarkable images and information about Neptune that refined our views of planetary evolution in this frigid part of the outer solar system.

Neptune is a giant planet with a banded, blue atmosphere, decorated with brilliant white clouds of methane ice, unlike its neighbor Uranus. The rings of Neptune, surmised from Earth, were confirmed. Neptune has at least eight moons, six of which were discovered by Voyager. One newly discovered moon is as large as Saturn's Mimas. The largest of Neptune's moons, Triton, moves in a highly anomalous, retrograde orbit. Moreover, Triton joins Titan as one of the few moons in the solar system to have an atmosphere, albeit a tenuous one. Triton is so far from the Sun and temperatures (37 K) are so low on its surface that nitrogen, a gas even on Titan, is frozen solid to form a large ice cap. The surface of this icy moon is much younger than might be expected for such a small body; large regions have been shaped by volcanic and tectonic processes that must have been active over much of the history of the planet. In fact, at least one form of volcanic eruption is active.

Major Concepts

1. Neptune is a large gas- and ice-rich planet of the outer solar system. It has at least eight moons and a distinctive system of rings.
2. Neptune is similar in size and density to Uranus but has a normal rotation direction. Neptune probably has an internal structure like that of Uranus, with a relatively undifferentiated core of rocky and icy materials cloaked by a thick hydrogen- and helium-rich atmosphere.
3. Neptune's three principal rings are narrow, and one is marked by discontinuous groups or clusters of particles. Like the rings of Uranus, they are made of dark particles.
4. Neptune has three satellites greater than 300 km in diameter. Little is known about Nereid other than its eccentric orbit. Proteus, a newly discovered moon, is larger than Nereid and nearly as large as Saturn's Mimas, but it has an irregular shape and a cratered surface.
5. Triton, the largest moon of Neptune (2700 km in diameter) has a retrograde orbit and a thin atmosphere of nitrogen and methane. Presently, an ice cap of nitrogen covers much of the southern hemisphere. No heavily cratered terrain dating from the period of intense bombardment is preserved on Triton. Highly deformed surfaces were probably formed by repeated fracturing and subsequent filling with viscous volcanic ice or flooding by mixtures of molten ices of water, methane, or nitrogen. The relative youth of its features may be the result of tidal heating after Triton was gravitationally captured by Neptune.

Neptune and Its Satellites

Photos returned by the Voyager spacecraft show that Neptune is a beautiful blue planet with a banded, cloudy atmosphere (Figure 12.1). Neptune's bluish color is imparted by methane (which preferentially absorbs red light) in its atmosphere. Nonetheless, hydrogen and helium are the most abundant elements in the planet. Thus, along with Uranus, Neptune is commonly grouped with Jupiter and Saturn as a low-density giant with a hydrogen-rich atmosphere. Yet these smaller planets differ significantly from their larger neighbors in space. Neptune is the smallest of the outer gas giants. It is only about 6 percent as massive as Jupiter, but its volume is still greater than 60 Earths. Moreover, Neptune's blue coloration is radically different from any of the other outer planets, and its magnetic field is distinctive, as we shall see. Uranus and Neptune are said to be the twins of the outer solar system. They have similar sizes, rotation rates, atmospheric compositions, and magnetic fields. Nonetheless, compared to Uranus, Neptune is significantly more dense; it has a banded atmosphere with bright white clouds, and it emits a significant amount of heat. Indeed, no two planets are really very much alike; and as with Venus and Earth, "twins" of the inner solar system, more important differences are bound to become apparent as the study of Neptune and Uranus intensifies.

Neptune is also a bit less exotic than its neighbor Uranus in that its dynamic properties are normal. Its axis is inclined at an Earthlike 28.8 degrees. It is 49,500 km in diameter; that makes it 3000 km, or 6 percent, smaller in diameter than Uranus. Neptune is so far from the Sun (nearly 4.5 billion kilometers away) that it takes the planet 165 years to complete one orbit. Strange as it may seem, Neptune has yet to complete one orbit since it was discovered in 1846. The Voyager measurements refined the length of a Neptunian day, showing that it rotates once every 16 hours. The temperature near the cloud tops is very low, only 69 K, but the temperatures at the surfaces of the moons are even lower. Daytime temperatures hover near 38 K on these satellites. The magnetic field of Neptune is distinctive but relatively small, only a fraction of that developed in Earth. Like the field at Uranus, the pole of Neptune's magnetic field is tilted compared to the spin axis. The center of the field is also displaced from the center of the planet.

Neptune has a system of at least eight moons; their sizes are compared in Table 12.2 on page 429. Triton is quite large; with a diameter of about 2700 km, it is somewhat smaller than Jupiter's Europa. The other moons all have diameters of less than 500 km. The surfaces of the moons are very cold; temperatures on Triton hover around 37 degrees above absolute zero—the temperature at which molecular motion stops. Ices, not stable on the warmer moons of the solar system, have thus formed on these satellites. It is the geology of these ices, not silicate rocks, that we must attempt to understand. The new data from Voyager also showed the presence of several narrow rings around Neptune, resolving the decade-long debate over the possible presence of rings.

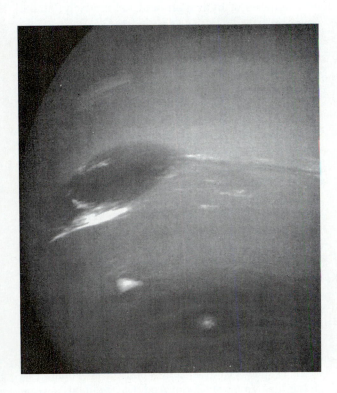

Figure 12.1

Neptune's banded, cloudy atmosphere is apparent in this image obtained by Voyager 2 in August of 1989. The Great Dark Spot is visible at 20 degrees south of the equator. Another dark spot that looks like an eye is located farther south. These spots are great, turbulent storms in the atmosphere of Neptune. The bright clouds are composed of methane ice crystals far above the main deck of haze and clouds.

The Interior and Atmosphere

Because of their broadly similar diameters, densities, and locations in the solar system, Uranus and Neptune are thought to have similar internal structures. A model for the interior structure of Neptune is shown in Figure 12.2. The model has

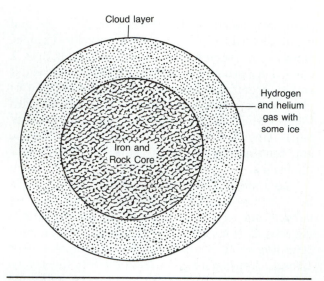

Cloud layer

Hydrogen and helium gas with some ice

Iron and Rock Core

Figure 12.2

The internal structure of Neptune is probably similar to that of Uranus. Neptune has a slightly higher density than Uranus and is thought to possess a slightly larger core of rock and ice (composed of water, methane, and ammonia) and a thick atmosphere of hydrogen, helium, and methane gas.

similarities to those discussed for both Uranus and Saturn. Neptune is thought to possess a core of rock and water ice and a thick atmosphere of hydrogen and helium. Although smaller in diameter than Uranus, Neptune is more massive (approximately 17 times greater than Earth) and consequently has a density of 1.64 g/cm³, roughly one-third denser than Uranus. These differences suggest that Neptune possesses a larger proportion of dense rocky materials than Uranus. Otherwise, the interior models for the two planets are similar. It appears that the rocky portion is about 45 percent of the mass of Neptune, or about seven times the mass of Earth. Because of the enormous pressure, the rocky materials would have much higher densities than Earth itself. Slightly less than half of the mass of Neptune may be icy materials. The principal ice is probably water, but ices of methane (CH_4) and ammonia (NH_3) are probably important constituents as well. Such volatile compounds, not present in any abundance at Jupiter, probably precipitated from the ancient nebula in the cool fringes of the outer solar system. From measurements of Neptune's density and rotation rate, it has been deduced that the planet's internal constituents may not be completely differentiated into discrete shells of rock, ice, and gas. Rather, the rocky and icy constituents appear to be rather well mixed, as they are for Uranus.

From the information already in hand, it can be deduced that the bulk compositions of both Uranus

and Neptune differ markedly from the nearly solar compositions of Jupiter and Saturn. The compositions of Jupiter and Saturn more nearly reflect the composition of the nebula from which the planets formed. For example, Neptune and Uranus are enriched in methane, compared to the composition of these other planets or, for that matter, of the Sun itself. This enrichment probably exists because a larger proportion of carbon was contributed to Uranus and Neptune by the solids that condensed in the cool fringes of the nebula. Carbon may have been in the form of carbon monoxide or carbon. Subsequently, carbon and hydrogen may have reacted to form the methane now seen. Another important difference between Jupiter-Saturn and Uranus-Neptune is reflected by the much thinner, gaseous envelopes of Uranus and Neptune. When compared to Jupiter or Saturn, these two planets are substantially depleted in hydrogen and helium. This is probably because during the accretion process, the rock and ice cores of Uranus and Neptune did not grow as large as those parental to Jupiter and Saturn. Therefore, they did not retain massive atmospheres of hydrogen- and helium-rich nebular gases.

Neptune has a small, but significant, magnetic field. The strength of Neptune's magnetic field is the weakest of the giant planets and is only about one-fourth that of Earth's. Neptune's magnetic field has a surprising orientation. The magnetic axis is tipped relative to the spin axis by about 50 degrees. When the tilted magnetic field of Uranus was discovered, scientists suggested that Voyager had caught the field in the middle of a reversal (when the magnetic north and south poles switch places). Magnetic polarity reversals are known to occur rapidly on Earth; subsequently, the field keeps the same orientation for millions of years. Thus it seems that the possibility of finding two planets both experiencing magnetic polarity reversals simultaneously is small. Another explanation may be necessary. Moreover, Neptune's magnetic field is not centered in the middle of the planet but is offset about 10,000 km toward the south rotational pole. Thus the magnetic field surrounding the planet is extremely complex, and its orientation with respect to Neptune's orbit is constantly changing. Voyager 2 entered the field at a time when the southern magnetic pole was pointed directly toward the Sun. These significant differences suggest that the mode of generation of magnetic fields in Neptune and Uranus may differ substantially from that which generates the fields that envelop Saturn and Jupiter. As we have explored the outer solar system, we have found increasingly complicated magnetic fields surrounding the planets. The implication is that the dynamo-generating region is not as deeply seated as in Jupiter (or Earth, for that matter) but is closer to the surface. Could it be that the electrically conductive medium that generates the magnetic field is not in the center of Neptune, but resides in its watery shell (Figure 12.2) where electrically charged ions can exist? Could convection in such a shell produce a magnetic dynamo?

The gaseous outer layer of Neptune, like those found around Jupiter and Saturn, consists principally of hydrogen (H_2) and helium (He), which are the most abundant elements in the solar system. Methane (CH_4) exists in an abundance of several percent. Traces of acetylene have also been detected. Acetylene and other hydrocarbon gases appear to form when methane is ripped apart by energy from sunlight, and the fragments recombine in new ways. These new molecules are relatively heavy and sink through the atmosphere. Eventually, they break down and react to form methane again. As in the atmospheres of Jupiter and Saturn, carbon dioxide and other oxidized gases are present in minor quantities. The contribution of the gas layer to the mass of Neptune probably amounts to about 10 percent.

The cloudy, banded atmosphere of Neptune is the most striking visual difference between Uranus and Neptune (Figure 12.1). Bands of varying brightness are apparent on all of the images of Neptune's atmosphere returned by Voyager. In contrast to the case for Uranus, no special processing of the images was needed to make the circulation of the atmosphere apparent. The apparent surface of Neptune is created by a deck of thick clouds and haze in its upper atmosphere. Two dramatic dark spots are visible in the southern hemisphere. These deep blue, ovoidal spots seem to change shape as they churn around in a counterclockwise direction. The dark spots are probably great, hurricanelike storms in the atmosphere, like those found in the atmospheres of Jupiter and Saturn and best exemplified by Jupiter's Great Red Spot. The Great Dark Spot is the larger and more northerly of the two dark spots (Figure 12.3). It lies at a latitude of about 20 degrees south, the same position as Jupiter's Great Red Spot. The more southerly dark spot lies at about 50 degrees south and looks like a large cyclopean eye peering out into blackness of space (Figure 12.4). These vast, rotating storms are relatively permanent features and probably result from shear between streams of winds moving in different directions; winds with velocities of over 100 km/hr race through the atmosphere. Towering above the dark spots are brilliant white clouds of methane ice. Constantly changing,

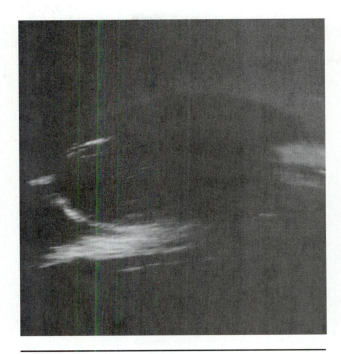

Figure 12.3

The Great Dark Spot is similar in origin to Jupiter's Great Red Spot. Both are storms created by shear between oppositely moving wind streams. Neptune's Great Dark Spot has a length about the same as the diameter of Earth. The bright clouds on the southern margin of the dark spot are methane clouds that lie much higher in the atmosphere and do not rotate with the lower storm.

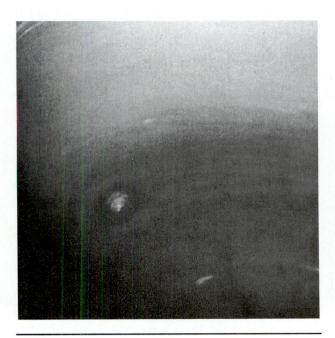

Figure 12.4

The southern dark spot is smaller and contains a bright center of high methane ice clouds.

wispy clouds form on the southern margin of the Great Dark Spot, but they do not rotate with the spot because they are much higher in the atmosphere. Bright clouds also develop in the core of the smaller, more southerly storm system. These clouds extend at least 50 km above the bluish deck of haze below them.

What could cause the appearance of the atmospheres of Uranus and Neptune to be so different? The circulation of the atmospheres of the inner planets is driven in large part by solar energy and the rotation of the planet. If these were the source of energy driving the motions in the atmospheres of Neptune and Uranus, we would expect that Uranus, closer to the Sun, would have a more vigorous atmosphere. Perhaps the answer lies in another important difference between these giant planets. Neptune radiates a comparatively large amount of energy from its interior; Uranus emits almost no internal heat. Some investigators have suggested that this is a consequence not of any difference in heat production mechanisms, but rather of a difference in the way the two planets lose heat. In our discussions of the inner planets, we noted that convection is a much more efficient means of moving heat than conduction. Perhaps Neptune can rid itself of heat by convection because of a more homogeneous internal structure. Uranus, on the other hand, with a more differentiated structure, may have several internal barriers to efficient convection. Another important difference between the planets that could account for some atmospheric differences lies in the tilt of the spin axes of the two planets. Much of the heat that reaches Uranus from the Sun falls on its rotation poles, whereas solar energy is most intense at the equator of Neptune, as it is on most planetary bodies.

The Rings of Neptune

Although planetary rings are common in the outer solar system, the various ring systems are quite dissimilar. Jupiter possesses a narrow diffuse ring; Saturn is encompassed by a spectacular, bright system of broad rings over 70,000 km across with thousands of individual rings; Uranus has nine narrow rings made of dark particles; and Neptune's recently described rings contain clumpy arcs (Figure 12.5). The rings of Neptune were first found by patient observers using telescopes on Earth who suggested that the rings were incomplete arcs that extended only part of the way around the planet. The true character of the rings was revealed by the Voyager images.

Figure 12.5

The three principal rings of Neptune are shown in this Voyager image. Between the distinct rings lie broad diffuse zones that are sparsely populated with particles but are still detectable in this image. Like those of Uranus, the orbiting ring particles are very dark and may contain radiation-darkened methane.

The rings of Neptune are a family of three principal rings and several diffuse sheets with different concentrations of small dark particles, all in orbit about the planet (Figure 12.5). The rings lie between 42,000 and 63,000 km from the center of Neptune. Four moons lie within the ring system. If they should become torn apart by impact or tidal disruption, a spectacular ring system like Saturn's could be created. Like other planetary ring systems, the rings lie within the Roche limit, implying that tidal disruption or halted accretion are somehow important in their generation or in keeping the small ring particles from recombining to form a larger moon. The brighter streaks that appear in images of the outer, or main, ring (Figure 12.6) have been called ring arcs. They range in width up to about 15 km and extend up to 50,000 km along a

Figure 12.6

Ring arcs found in the outer ring of Neptune consist of clumps of ring particles in an otherwise continuous ring.

given orbit. The leading edge of each arc seems to fade away, while the trailing edge of the arcs looks more sharply defined. The clumps appear to consist of higher concentrations of dark particles. Features similar to the ring arcs are found in the F-ring and are embedded in the gaps found in Saturn's rings; these features must be produced by the same processes. The continuous parts of the rings are so tenuous (Figure 12.6) that they could not be detected from Earth. Between the rings, extending toward the planet from the diffuse inner ring, are particles widely distributed in zones or sheets reminiscent of the broad rings of Saturn, but populated by much smaller concentrations of dark particles. The innermost dust sheet is similar to Saturn's D-ring.

The dark ring-particles are probably produced by the same mechanism that produced the dark ring-particles of Uranus. The surfaces of the smaller moons of these two planets are also relatively dark, and some scientists think that the dark moons and dark ring-particles have a common origin. They suggest that both consist of radiation-darkened methane ice. Compared to the Uranian rings, however, Neptune's rings are populated by smaller particles. Dust-sized particles constitute from 30 to 60 percent of the main rings. Only the diffuse ring associated with the middle ring (Figure 12.5) appears to lack large proportions of dust.

How can a ring system with ring arcs be maintained? One explanation might be that the arcs consist of particles collected at the Lagrangian points of a small moon. These stable points lie in the same orbital path as another larger object, but

precede and follow the larger object by 60 degrees; for example, two of Saturn's moons have Lagrangian satellites. Others have suggested that each ring arc is defined by a small shepherd satellite that pulls the ring particles into clusters. Four of the six newly discovered moons of Neptune (Despina, Galatea, Thalassa, and Naiad) lie within the ring system. Proteus and Larissa are too far away to affect ring properties. If shepherd satellites exist, they have not yet been discovered. However, newly discovered Galatea has an inclined orbit and might be pulling some particles into these arcuate clumps. Another alternative being explored by some scientists is that the arcs are composed of fragments of a relatively recent fragmentation of a small satellite. With the passage of time, the fragments could spread out along the orbit to form a continuous, smooth ring. Such impact fragmentation may be an important ring-forming process for all of the outer planets.

The Satellites

Neptune's system of large satellites is smaller than that which surrounds Uranus. Only three satellites larger than 300 km in diameter have been discovered, and one of these was discovered by Voyager 2, which also discovered five other small moons (Figure 12.7). Triton and Nereid were discovered by telescopic observations and are the classical moons of Neptune. In all the solar system, these two moons have the strangest orbital properties (Figure 12.8). Triton revolves about Neptune in a retrograde direction, and Nereid has an extremely eccentric orbit that takes it millions of kilometers from Neptune. In fact, it takes Nereid almost as long to orbit Neptune as it takes Earth to orbit the Sun. The largest of the recently discovered satellites is larger than Nereid, but because it orbits so close to Neptune, it could not be seen from Earth because of the glare from the planet. The density of Triton and brightnesses, or albedos, of the satellites suggest that, like those of Uranus, they consist largely of mixtures of water ice and rock. Spectral observations show that ices of nitrogen and methane are important at the surfaces as well. These are the only bodies we are likely to visit in our lifetimes that will give us an idea of what Pluto may be like. Below we discuss the satellites in their order outward from Neptune.

The Small Satellites

Voyager discovered six satellites embedded within and near the ring system of Neptune (Fig-

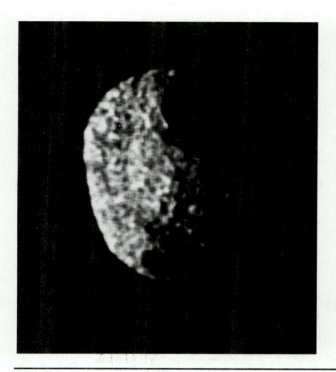 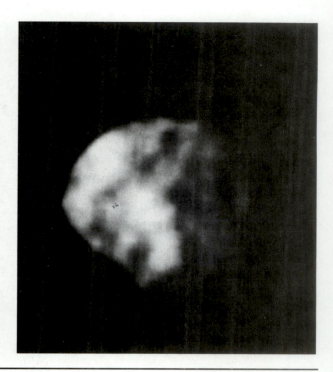

Figure 12.7

Two of the small moons of Neptune are compared at the same scale. Proteus is shown at the left and Larissa at the right. Most of the satellites are irregular, even the largest that is over 400 km in diameter, suggesting that they are collisional fragments. These small satellites are also relatively dark bodies like the moons of Uranus.

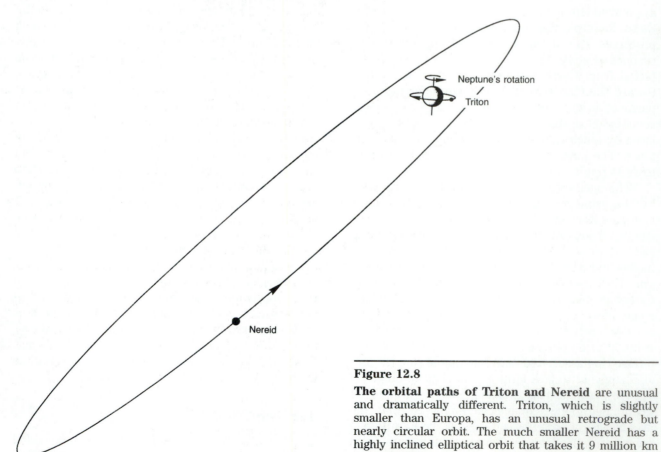

Figure 12.8

The orbital paths of Triton and Nereid are unusual and dramatically different. Triton, which is slightly smaller than Europa, has an unusual retrograde but nearly circular orbit. The much smaller Nereid has a highly inclined elliptical orbit that takes it 9 million km from Neptune.

TABLE 12.2
Characteristics of Major Satellites

Satellite	Diameter (km)	Density (g/cm³)	Mean Distance from Neptune (km)
Nereid	340		5,567,000
Triton	2,700	2.07	355,000
Proteus	400		117,600
Larissa	190		73,600
Galatea	90		62,000
Despina	150		52,500
Thalassa	80		50,000
Naiad	54		48,200
Moon	3,500	3.3	

ure 12.7). From this information, their approximate sizes and orbits have been calculated. All these satellites are dark, like the moons of Uranus. Five of the satellites are quite small and, as noted above, four orbit within the ring system but do not appear to be ring shepherds. None were photographed closely enough to resolve any of the details of their surface features, but they appear to have irregular shapes.

The outermost of the newly discovered satellites is also the largest. Proteus orbits just outside the ring system. It completes an orbit every 1.1 days. This satellite is almost as large as Mimas, and yet it is not spherical (Figure 12.7). The outline of the body is trapezoidal, with a shape like a rounded cobble. We pointed out earlier that Mimas barely had enough gravitational energy to pull itself into a sphere; apparently Proteus did not have quite enough mass to deform its surface into a spherical shape after the last events that shaped it. Vague, nearly circular features are discernible on its surface, and a large trough appears to extend from north to south across the face of the satellite. Most likely, Proteus was shaped by impact. Like the ring particles, Proteus is very dark and reflects about as much light as does soot. The dark coloration of Proteus and the other small moons is probably the result of radiation damage of methane ice at the surface.

Triton

Triton, Neptune's inner moon, is a relatively large satellite (2700 km in diameter), comparable in size to Jupiter's Europa and Io or to Earth's Moon (Figure 12.9). Some of the most spectacular images taken by Voyager were the high resolution views it provided of Triton. The various geological processes and features found on Triton remind us of

bodies as complex as Mars, Europa, Ganymede, Enceladus, and Ariel. Before we discuss the geological terrains of Triton, we need to discuss several of its unique properties, including its orbital characteristics and the composition of the ices on its surface.

Like Nereid, Triton has remarkable orbital properties. Although Triton is like most other moons in that it has a nearly circular, synchronous orbit and always keeps the same face pointed toward its primary, Triton revolves about Neptune in a retrograde fashion (that is, it revolves about Neptune in a direction opposite to the planet's rotation). Triton is the only large satellite in the solar system to do so. In addition, the orbital plane of Triton is tipped at an unusually steep 21 degrees away from the planet's equator. Because of these peculiarities, its orbit is very slowly decaying. It has been calculated that within about 10 billion years Triton will pass within Neptune's Roche limit; however, this is longer than the remaining life of the Sun. A retrograde orbit cannot originate by simple accretion of a moon from debris in orbit around Neptune. Triton's strange orbit suggests instead that it was gravitationally captured by Neptune after the satellite formed elsewhere, perhaps destroying an originally larger system of satellites in the capture process. Others have suggested that Triton's unique orbit is the result not of capture but of a large collision with another planetary body sometime in its past. Both hypotheses require some sort of cosmic catastrophe. Moreover, in either case, Triton's early orbit would have been very eccentric. The fact that Triton's orbit is now nearly circular almost demands that, during its orbital evolution toward a circular orbit, large tidal forces were exerted on it as Triton alternately drew near to and far from Neptune. Calculations of the amount of energy produced by this tidal flexing suggest that Triton may have been heated dramatically while the orbit became circular. Thus, although the orbit may have become circular rapidly, the heating may have kept Triton's interior warm for a billion or more years after it was captured. What evidence would we look for to test this hypothesis? The surfaces of the Galilean satellites of Jupiter should give us a clue.

Triton is remarkable, too, in that ices of methane (CH_4) and nitrogen (N_2) have been identified on its surface. In fact, methane may be more common at the surface than water ice. There is no spectral evidence of water ice yet. The photographs of Triton may show us the only detailed view of such an exotic surface. Triton's surface temperature is extremely low at 37 K and is the coldest object we have yet visited. Nonetheless, sublima-

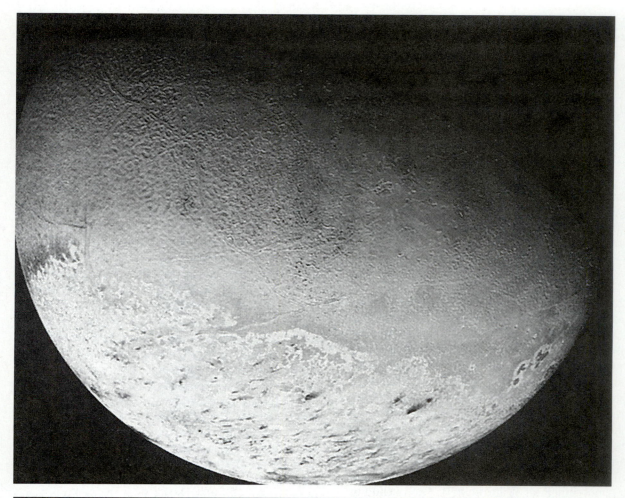

Figure 12.9

Triton is the largest of Neptune's satellites. It has two fundamentally distinct hemispheres as shown in this Voyager image. The southern hemisphere is covered by a large cap of what appears to be nitrogen ice. This cap may cycle from pole to pole during seasonal changes on Triton. Exposed in the northern hemisphere are several lightly cratered terrains created by volcanic and tectonic processes. The equator runs parallel to the top of the ice cap.

tion of methane and nitrogen ices contributes to a transparent and very tenuous atmosphere with a current surface pressure of 0.01 mb, just a tiny fraction of that present on Mars (6 mb on average). Nitrogen is more abundant than methane in the atmosphere because it is more volatile than methane and has a higher vapor pressure. Titan, Saturn's largest satellite, and Earth are the only other planetary bodies in the solar system known to have nitrogen-rich atmospheres. If by some cosmic juggling process Earth could be moved to the position of Triton, it would possess a shell of nitrogen ice nearly 15 m thick.

However, a simple, globe-encircling shell of nitrogen ice has not developed on Triton, for good reason. As noted above, Triton's orbit about Neptune is inclined; in addition, Neptune's spin axis is also inclined. These tilts cause Triton's polar re-

gions to point toward the Sun occasionally during a complicated cycle of seasons within seasons that lasts about 600 years. At one extreme, the polar axis of Triton tilts 40 degrees toward the Sun, basking in constant sunlight while the other is in constant shadow. These seasonal changes are enough to produce significant variations in the amount of solar energy that reaches the surface of Triton over the course of a season. Consequently, there may be large seasonal variations in the position and extent of a polar cap of nitrogen ice. Voyager photos show that the present nitrogen ice cap extends across most of the southern hemisphere nearly to the equator (Figure 12.9). The southern hemisphere of Triton is tilted now toward the Sun; summer in the southern hemisphere will peak about the year 2000. As summer sets in, nitrogen in the ice cap should be vaporized, and the

The Neptune System **431**

cap of ice will shrink and may disappear completely. At the same time, winter holds sway in the northern hemisphere and nitrogen should freeze out of the atmosphere, forming a cap of ice that gradually grows southward. The atmosphere functions as the transport medium as nitrogen flows back and forth between the poles. Thin hazes and clouds seen in the atmosphere of Triton may be graphic evidence of the cycling of volatile materials from one pole to the other. The intricate patterns along the margins of the southern ice cap (Figure 12.10) appear to result from sublimation. The southern hemisphere is now moving into its summer season. We predict that a north polar cap is developing, which was out of view of Voyager's cameras at the time it flew by. Triton joins Mars and Earth as planetary objects that possess polar ice caps that change with the seasons.

Now let us focus on the surface geology of Triton, for it is unlike anything else in the solar system. Based on superposition and surface features, three distinctive types of terrain were identified on Triton. From oldest to youngest they are

(1) highly fractured plains, (2) flooded volcanic plains, and (3) the polar ice cap. Significantly, no heavily cratered terrains are present on Triton.

Fractured Plains (Cantaloupe Terrain).

As revealed in the images taken by Voyager 2, much of the surface of Triton may consist of relatively dark plains crisscrossed with many linear fracture systems (Figure 12.11). Most of the fractures appear to be the result of rifting of the lithosphere to form fault-bounded grabens. Some grabens are over 1000 km long, and even the youngest intersect one another at high angles (Figure 12.11). Many of these fractures have been filled by linear protrusions, or ridges, that rise above the surrounding terrains (Figure 12.11). The most straightforward interpretation of these fracture fillings is that they were fashioned by a form of icy volcanism, similar to that seen on other satellites such as Enceladus and Ariel. Nonetheless, the well-defined character of these linear strands shows that they were not formed by flooding from a fluid like liquid water. Instead, they seem to be extrusions of a viscous, perhaps nearly solid, material. The composition of these volcanic flows could range from water to watery mixtures with ammonia or methane. Either of these latter two compounds, when added to water, will reduce the melting temperature of water and make the production of watery magmas more likely below the frigid surface of Triton. The presence of ammonia may have increased the viscosity of these flows.

The terrain across which the young fractures cut appears to consist of similar but older and more subdued fractures and icy volcanic flows. Perhaps this terrain has a long history of repeated fracturing and volcanic activity. The fractures have allowed molten, or partially molten, material from inside Triton to well up onto the surface. The multiple generations of fracturing have created a very complex topography of irregular depressions and narrow ridges that together create a surface that looks something like the pitted and ridged skin of a cantaloupe (Figure 12.11). Locally, it may overprint older cratered terrains.

The fractured regions of Triton are its oldest terrains, but they are not heavily cratered like many of the moons of Uranus and Saturn; nor are they as lightly cratered as the volcanoes of Io or the plains of Enceladus. The impact crater frequency on these surfaces is similar to that found on the lunar maria. If the flux of impacting bodies was about the same on Triton as it was on the Moon, then these terrains must be less than 3 billion years old.

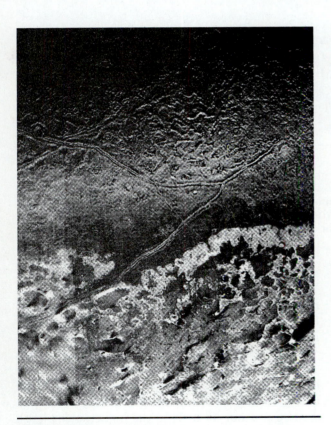

Figure 12.10

The northern margin of Triton's ice cap is defined by polygonal lines and marked by intricate dark patches. The patches may be underlying terrains exposed when the ice sublimes. The cap is obviously superimposed on the older fractured terrain shown in this photo. A halo of brighter frost appears around the edge of the ice cap.

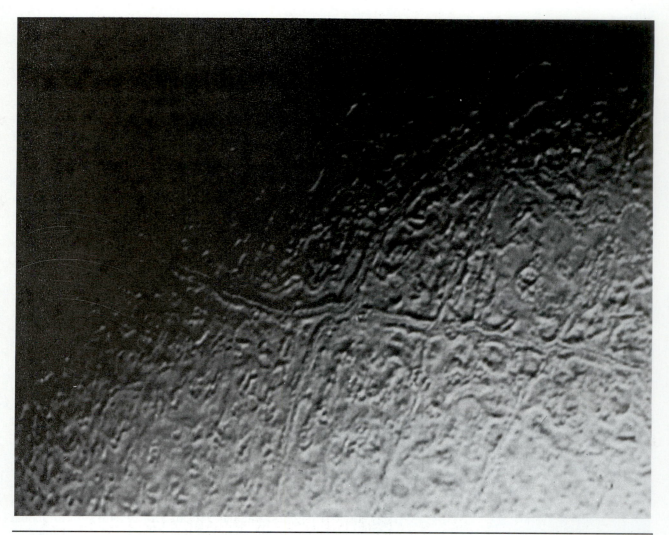

Figure 12.11

The fractured terrains of Triton are exposed in the northern hemisphere. They consist of large regions cut by a multitude of grabens of varying ages. The youngest grabens stand out as long, linear traces, some hundreds of kilometers long. Filling these fractures are ribbons of high-standing material that may represent ices that extruded through the fractures. Repeated rifting and associated icy volcanism appear to have shaped this terrain.

Flooded Volcanic Plains. A fascinating terrain unique to Triton cuts across or covers portions of the fractured plains. It consists of relatively smooth plains cut by arcuate scarps and large shallow depressions that are elliptical to sub-circular (Figure 12.12). The irregular depressions have many similarities to the floors of the large collapse calderas found atop the volcanoes of Earth, Mars, and Io. Some flat areas several hundred kilometers across are bounded by low scarps 200 to 400 m high. This region appears to have been shaped by floods of fluid volcanic liquids forming large lava lakes—a different kind of icy volcanism than that in the cantaloupe terrain. Floods of softer, hotter, and more fluid lavas seem to have been involved. The composition of these lavas is unknown but could consist of a mixture of water,

nitrogen, and methane, with a low melting point. Low ammonia contents would make these liquids more fluid. The floods appear to have accumulated in lakes or to have ponded in calderas. Multiple levels of cooling and stagnation are apparent. In many places, it appears as if a volcanic liquid cooled to form a solid lid and then drained away beneath part of the lid causing that part to founder and melt. Cooling of the new lower level of liquid created another lid next to the higher, older level. These depressions can be interpreted as volcanic calderas themselves or as the result of the accumulation of fluid lava flows in pre-existing depressions. These alternatives can be thought of as the caldera versus the lake hypotheses. The fundamental distinction between them lies in the location of the vent and the magma body it taps. A caldera has a

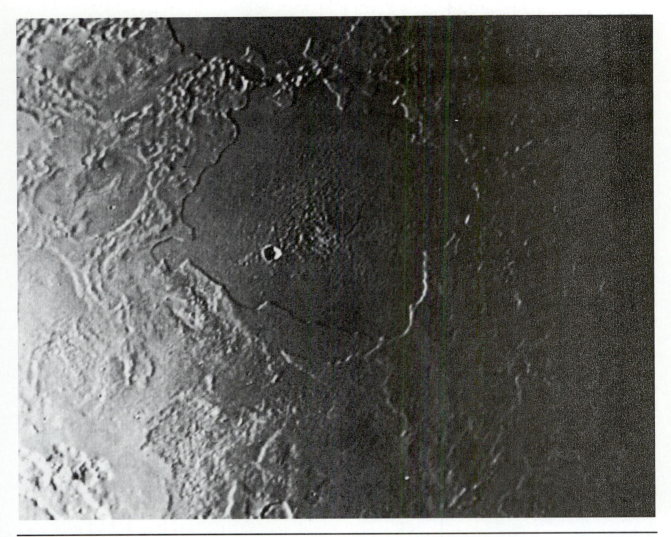

Figure 12.12

The flooded volcanic plains of Triton are dotted with several large caldera or lava-lake complexes like that shown here. Multiple levels of icy lava lakes are obvious. Repeated draining, collapsing, melting, and renewed cooling created these features. An impact crater formed on this frozen lake.

magma body beneath it, into which the roof collapses to create a depression; a lava lake can accumulate in any kind of depression and need not lie above the chamber that fed its lava flows. Complex superposition relationships between many generations of these floods and their partially collapsed surfaces have created the multitude of scarps in this region of Triton.

The flooded plains are superimposed on and are less cratered than the fractured, cantaloupe terrains, but many relatively small impact craters are present and imply that the plains formed within the last several billions of years.

Polar Ice Cap.

The youngest and brightest terrain on Triton is the large bright ice cap that covers the south polar region and extends nearly to the equator. Triton is one of the few planetary objects that has an ice cap. Both Earth and Mars have frozen ice caps, water on the one hand, and water and carbon dioxide on the other. Ice caps serve as important, but relatively temporary, sinks for atmospheric volatiles.

Triton's ice cap reflects nearly 90 percent of the sunlight that falls on it and is one of the brightest surfaces in the solar system. Clean, freshly fallen snow reflects about 75 percent of the light that falls on it. Measurements made from Earth suggest that the cap is composed of nitrogen ice. After the identification of nitrogen on Triton, speculation about Triton included the possibility of seas of liquid nitrogen. Perhaps it would have been appropriate that Triton, mythical son of the sea god Neptune, should have had a sea, even one of liquid nitrogen. As it turns out, Triton is too cold for a nitrogen ocean to exist as a liquid, and an ice cap has formed instead.

The ice cap is probably a constantly changing feature of the landscape of Triton. Presently, the outline of part of the cap is roughly polygonal (Figure 12.9), rather than smoothly lobate as are other planetary ice caps. The northern edge of the cap nearly reaches the equator of Triton. The northern limit is extremely irregular, with many patches of darker terrain apparent behind the front. These darker patches probably are exposures of the underlying terrains. The irregularities of the cap appear to be the result of sublimation of the nitrogen along low scarps or by the deposition of bright ice in irregular low spots. The polar cap must change shape and thickness with the complicated seasonal cycles on Triton. Around the rim of the ice cap, but on other, older terrains, a bright halo exists (Figure 12.10). The halo may consist of frosts of ices that condensed onto the surface where the temperature was lower near the edge of the cap.

Volcanic Eruptions.

Enigmatic dark streaks are apparent atop the ice cap near its northern margin (Figure 12.13). These streaks are perhaps the most intriguing features identified on Triton. In actuality, the streaks are not as dark as

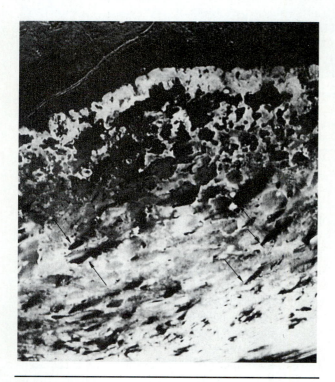

Figure 12.13

The dark streaks that cross the bright ice of Triton's ice caps may represent young volcanic ejecta thrown onto the ice cap by small volcanoes. The transition from liquid to gaseous nitrogen may power these eruptions. The dark particles may be darkened, methane-ice fragments torn from the walls of the vent.

they seem in this specially processed photograph and are really quite bright (about like snow). They appear dark here because they are significantly darker than the very bright background. Most of the streaks in the photographed area are aligned parallel to one another in a northeasterly direction. Each streak has a roughly triangular outline, with a well-defined southern apex and an irregular, diffuse northern termination. The apexes of some streaks arise from what appear to be small circular patches on the ice cap. After studying these streaks with stereoscopes so that their three-dimensional character could be seen, scientists discovered that they are eruption plumes of active volcanoes. Triton thus joins Earth and Jupiter's moon Io as the only known planetary bodies with active volcanoes. Several eruptions were identified. The pictures show that dark vertical plumes rise to heights of nearly 8 km before Triton's winds sweep the volcanic ejecta across the sky. But Triton's volcanism is unlike that on Earth or Io. It is apparently powered by the vaporization of liquid nitrogen or methane. Indeed, the volcanism on Triton may be more like geysers on Earth. The "magmas" in these small volcanoes may form as slightly higher temperatures are reached below the surface of the cap or as ices melt at the surface to form small pools. No special heating process is required; rather, the melting point of nitrogen or methane ice is reached because of the natural temperature gradient that probably exists on Triton. Higher spring and summer temperatures may cause cyclical volcanic seasons. If liquid nitrogen is formed at the base of the ice cap, it can rise because of its lower density, but before it can extrude onto the surface as a quiet lava flow, it reaches its boiling point and explodes violently because of the tremendous volume change that occurs when liquids convert to gases. The darker color of the streaks could be caused by radiation-darkened methane particles entrained in the eruption blast and scattered in a small eruptive plume across the surface of the ice cap. The preferred alignment of the streaks must be the result of prevailing winds sweeping across the cold cap toward the warmer equator. All of the dark streaks are very young. They have developed on top of a polar ice cap that buries all other terrains on the planet; if our theories of seasonal variation are correct, the cap may be only a hundred years or so old, and, thus, the volcanoes could be this young, too.

Interior and Composition.

The density of Triton is about 2 g/cm³; apparently it consists of about one-third ice and two-thirds silicate rock. When we compare the densities of the satellites of Saturn (1.2 to 1.4 g/cm³), Uranus (1.4 to 1.7 g/cm³), and Neptune (2.0 g/cm^3), a distinct trend of increasing density with increasing distance from the Sun emerges. This systematic change in density has given rise to speculation about the proportions of ice and rock inside these small bodies. Perhaps the proportion of low-density ice in the outer part of the nebula was reduced, compared to dense silicates. This could have happened if volatile carbon monoxide (CO) existed in preference to water or methane ice because of the lower pressure and temperature in the nebula near Neptune and Uranus than around Saturn. The resultant mixture of solids could contain more dense rocky solids and less ice.

Another aspect of the geology of Triton is uncertain—is its interior differentiated? Presently we don't know, but if Triton is indeed a captured planetesimal, then the exchange of energy that occurred during capture may have led to extensive internal melting and differentiation of the satellite.

Geologic History. Apparently, Triton accreted from icy materials enriched in rocky components compared to ice (water, methane, and ammonia) in the outer part of the solar system. Accretion probably formed a heavily cratered surface. At some point early in its history as a Sun-orbiting planetesimal, it was captured by Neptune's gravitational net and placed in a retrograde orbit. The ensuing circularization of its orbit produced great tidal interactions between Neptune and Triton, heating and probably differentiating Triton. The heat drove tectonic and volcanic processes on Triton for perhaps as long as 500 million years afterward. Volatile gases (methane and nitrogen, produced by the decomposition of ammonia) were released from the ices that once contained them to accumulate at the surface. A thin atmosphere was created, but much of it condensed to form a sheet of nitrogen ice at the coldest spot on the planet—the pole experiencing winter. The location and details of the polar cap have changed constantly throughout the millennia of seasonal changes that followed.

Again, Voyager has revealed the characteristics of an enigmatic satellite. Triton, with a diameter of less than 3000 km, has a young surface. Tectonic rupturing of its surface obliterated all of the heavily cratered terrain that probably developed following its accretion. Associated with the tectonism, repeated volcanic outpourings were important parts of its relatively recent geologic history (the last several billion years). If icy Triton had been left to its own devices—its own radioactive and accretionary heat sources—geologic activity would not have persisted beyond the period of intense bombardment. Thus the evidence from its orbital properties and the youthfulness of its surface features are both consistent with a geologic scenario calling for the gravitational capture of Triton after its accretion elsewhere (but not too far away) in the solar system. The resulting modification of captured Triton's orbit raised large tides in its lithosphere, heated its interior, and remolded its surface. Enough heat was apparently created to keep Triton warm for some billion or so years after capture.

The presence of icy materials on its surface, and presumably in its interior as well, with very low melting points may also have extended the duration and nature of geologic activity on its surface. Nitrogen forms a liquid at temperatures that may be encountered only a short distance beneath the present surface of Triton. When materials are liquid or at least are soft enough to flow and deform, internally driven geologic processes can create new surfaces and modify old ones through tectonic and/or volcanic processes.

Nereid

Nereid, the outermost satellite of Neptune, has a diameter (340 km) in the same class as, but significantly smaller than, Uranus's Miranda or Saturn's Mimas. Nereid has a highly inclined and eccentric orbit (Figure 12.8). At closest approach, it is within 1.3 million km of Neptune; its most distant point lies 9.7 million km away. For Nereid to complete one revolution around Neptune requires 359 Earth days. Because of this strange orbit, it has been speculated that Nereid may be a captured, outer solar system asteroid or that it formed as a satellite nearer Neptune but was sent into its present orbit when Triton was captured. The capture of Triton profoundly affected the satellite system. The present satellites may have accreted from the debris of an earlier satellite system destroyed by Triton's capture.

Pictures of the surface of Nereid obtained by the Voyager spacecraft were taken at a great distance and therefore do not show any details about its surface geology. However, the images do show that Nereid is not round and has the irregular appearance of many small, solar-system objects shaped by collisional fragmentation. As it rotates, Nereid shows marked brightness variations, either as the result of its irregular shape or because of bright and dark surface markings like Iapetus.

Conclusions

The characteristics of Neptune continue the trends established for the planets of the outer solar system. Composed mostly of ice and rock and with less hydrogen, Neptune is slightly more dense than Uranus—apparently the result of having slightly more rock in its interior. Like Uranus, Neptune is circumscribed by narrow rings of dark particles. Clumps of high concentrations of small particles are especially obvious in the ring system, creating *ring arcs*. Also like Uranus, Neptune has a strangely oriented magnetic field and a thick atmosphere colored by methane. But unlike its neighbor, Neptune emits a significant amount of heat, apparently enough to create visible banding and turbulent storms in its atmosphere unlike any seen at Uranus.

Seven of the eight known moons of Neptune are small, icy bodies about which little is known. Six of the moons were discovered by Voyager 2 and circle the planet in orbits in and near the ring system. These bodies probably formed or are remnants of bodies that formed from a disk of material that rotated around Neptune. On the other hand, the outermost satellite, Nereid, may have originated elsewhere as an asteroid or satellite of another planet. We know most about Triton, the largest of Neptune's moons. It promises to be an exciting object for continued study, with an extended geologic history marked by a multitude of tectonic and volcanic events—some of which are presently active. Its constantly changing ice caps and thin atmosphere also mark it as a distinctive satellite. Exotic ices of nitrogen and methane are found at its surface. Geologic activity was driven after the period of heavy bombardment by the tides raised on Triton when it was captured by Neptune and placed in its retrograde orbit. The importance of distinctive energy sources and materials for the geological history of a planetary body is nowhere more obvious than on Triton.

Review Questions

1. Outline the major differences and similarities between Neptune and Uranus. Consider their orbital properties, sizes, atmospheric compositions, internal structures, magnetic fields, rings, and satellite systems.
2. Neptune has a banded atmosphere with prominent clouds and storm systems; Uranus has a nearly featureless atmosphere. Is there an explanation for this obvious difference between the planets?
3. Prepare a table that compares the diameters, surface compositions, ages, and features of the major satellites of Neptune.
4. Make a graph showing the densities of the satellites of Jupiter, Saturn, Uranus, and Neptune versus their relative distance from the Sun. Explain any apparent trend and anomalies to the trend.

5. Why is the retrograde orbit of Triton important to its geological history? Would the same hold true for the eccentric orbit of Nereid?
6. Compare the origins of the atmospheres of Earth, Titan, and Triton—all bodies with atmospheres dominated by nitrogen.
7. Contrast the apparently volcanic plains of Triton with those of the Moon. What factors cause the differences?
8. If you could design a spacecraft that could land on only one moon of Neptune *or* Uranus, which would you choose as the most scientifically revealing? What information would you gather?
9. Compare the volcanic activity of Earth, Io, and Triton. Consider major geologic factors such as type of material, melting temperatures, and source of heat.

Additional Reading

Kinoshita, J. 1989. Neptune. *Scientific American*, No. 11, pp. 82–91.

Science. 1989. Vol. 246, pp. 1422–1478. (Reports on Voyager 2 encounter with Neptune.)

Charon and the Sun Seen from Pluto

CHAPTER 13

The Pluto System

TABLE 13.1

Physical and Orbital Characteristics of Pluto and Charon

	Pluto	Charon
Mean Distance from Sun (Earth = 1)	39.4	
Period of Revolution	247.7 y	6.4 d
Period of Rotation	6.4 d	6.4 d
Inclination of Axis	88° to 122°	
Equatorial Diameter	2,284 km	1,192 km
Mass (Earth = 1)	0.0026	
Density (of both)	2.06 g/cm^3	
Atmosphere (main components)	N_2	
Surface Pressure	1 to 3 microbars	
Surface Temperature	37 K	
Known Satellites	1	

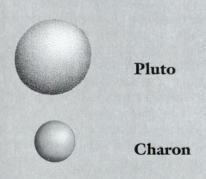

Pluto

Charon

Earth

Introduction

Beyond the orbit of Neptune lies a genuinely unexplored planetary system centered on Pluto. There are no detailed images of the surfaces of these small bodies because no spacecraft have yet visited their realm—this planetary incognita. The Voyager spacecraft did not fly by Pluto and its satellite Charon before leaving the solar system. Our best views of the planet have come from the Hubble Space Telescope. The surface features and many characteristics of Pluto will remain mysterious unless a flyby mission becomes reality in the mid-1990s. A fortunate circumstance has nonetheless resulted in a tremendous increase in our knowledge of this remote planetary system. In 1985, Pluto and Charon became an eclipsing binary system, with Charon passing before and behind Pluto as seen from Earth. This special geometry has allowed planetary scientists to calculate accurate orbital and physical parameters for the pair and to deduce something of the nature of these distant bodies. Detailed spectroscopic studies, conducted with telescopes on Earth, are also helping us probe the composition of the ices at its surface.

Pluto is a planet of extremes. It is the farthest, the smallest, the coldest, and the darkest. Pluto is distinctive among the outer planets in that it lacks a thick, hydrogen-rich atmosphere. It has a surface layer of frozen nitrogen and a tenuous atmosphere. Indeed, it is much more similar to the moons of Neptune than to any of the major planets. What follows is a brief summary of what we know about the nature of these distant objects, as well as some speculations. Pluto and Charon promise to be just as fascinating as the recently revealed system of bodies at Neptune. Their compositions and orbital evolution will prove to be key tracers of events in the ancient evolution of the outermost solar nebula.

Major Concepts

1. Pluto, the smallest and outermost of the planets, is icy and has a satellite companion and a tenuous atmosphere. No images of this planet have been made by spacecraft. Our best images come from the Hubble Space Telescope.

2. Pluto and Charon form a double-planet system with an elliptical orbit about the Sun. Pluto has a surface dominated by nitrogen-ice and an atmosphere formed by vaporization of this same ice.

3. Pluto has a density significantly higher than those of the moons of Uranus and Saturn and about the same as Triton, suggesting that it contains a large proportion of rocky materials. In many ways, it must be similar to Triton. This may suggest that Pluto and Triton formed as Sun-orbiting, icy planetesimals where the solids that condensed from the ancient nebula were enriched in rock compared to ice.

Pluto

Since it was discovered in 1930, Pluto has been known as the most distant of the planets of our solar system. However, Pluto has a highly elliptical orbit that requires 250 years to complete (Figure 13.1). The distance between Pluto and the Sun varies from a minimum of 4.5 billion km to a maximum of 7.4 billion km. So strongly does Pluto's orbit deviate from circular that periodically Pluto is the eighth planet from the Sun. Presently, Pluto is closer to the Sun than Neptune. This situation will prevail until the end of the century. Besides being highly elliptical, Pluto's orbit is inclined nearly 20 degrees out of the plane of the solar system. Only asteroids and comets have similarly inclined orbits. Moreover, Pluto's orbit is in a 2:3 resonance with Neptune. Other orbital oddities ensure that the two never collide.

Controversy has surrounded the size of Pluto since its discovery. Pluto is so small and distant that accurate measurements of its diameter are difficult to make from Earth. In the 1950s, Pluto was thought to be larger than Mercury (4800 km diameter) and about the same size as Saturn's largest satellite, Titan, with a diameter of about 5800 km. If this had been true, Pluto would have been more like Earth because the high calculated density suggested a rocky or even metallic compo-sition. Two decades of new observations show that Pluto is neither rocky nor Earthlike.

Late in 1978, a natural satellite of Pluto was discovered. This discovery has substantially changed our notions of Pluto's size, density, and probable origin. The discovery photograph (Figure 13.2) shows Charon, the newly found moon, as extending or smearing the disk of Pluto. Charon is so close to Pluto and so far from Earth that its disk has not been resolved, or separated, from that of Pluto; it appears only as a bump on this speckled image constructed using an Arizona telescope. Much better images of Pluto and Charon have been obtained from the Hubble Space Telescope (Figure 13.3). This large telescope, placed in orbit by the Space Shuttle, is not hampered by the obscuring effects of Earth's atmosphere. Evidence from these photographs, and from careful observations of the pair as they pass in front of one another, shows conclusively that Pluto is the smallest planet in the solar system, with a diameter of only about 2300 km. Thus, Pluto is smaller than seven planetary satellites, including Earth's Moon. Its closest ana-log is probably Triton, a satellite of Neptune.

Judging from telescopic observations, Charon has a diameter of almost 1200 km. These measure-ments show that Charon is more than half as large as Pluto, about which it orbits. Although by no means the largest planetary satellite in the solar

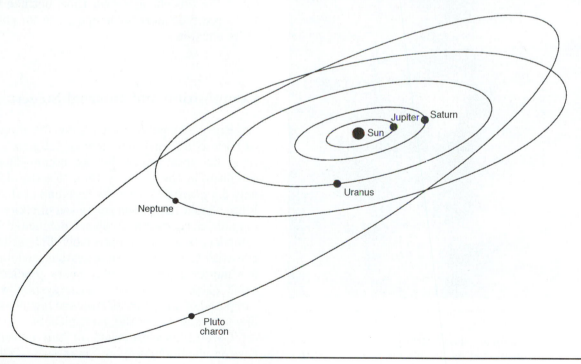

Figure 13.1

Pluto's orbit around the Sun is very elliptical; for about 20 years, Pluto is actually closer to the Sun than Neptune.

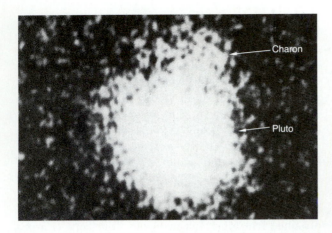

Figure 13.2

This splotchy telescopic image of Pluto showed the existence of a moon orbiting this tiny, icy planet. The slight bulge at the top of the image is caused by Charon, Pluto's moon. The two bodies are so close that they cannot be resolved as separate objects by telescopes on Earth.

Figure 13.3

Pluto and Charon are shown clearly as separate disks in this image constructed by the Hubble Space Telescope, although no surface features are visible.

system, Charon has the distinction of being the largest satellite as compared to its primary. It may be appropriate to think of Pluto and Charon as a double planet, unique in the solar system.

Charon orbits Pluto at a distance of only about 20,000 km (Earth's Moon orbits at a distance nearly 20 times as great). As seen from the surface of Pluto, Charon would be a dim, faintly glowing globe, but its apparent diameter would be nearly six times as large as the Moon appears from Earth. (In contrast, the Sun, 40 times more distant than from Earth, is merely a bright star in Pluto's perennial twilight.) Like many other planetary satellites, Charon revolves once about Pluto for each rotation on its axis (6.4 days) and therefore keeps the same face pointed toward Pluto at all times. Unlike other systems, however, Pluto also keeps the same face pointed toward Charon, which therefore appears to remain locked in a fixed position in the sky, like a geosynchronous communication satellite. A viewer on the opposite side of Pluto would never see Charon. Both bodies exert substantial gravitational forces upon one another because of the similarity in their sizes and their relative proximity. The revolution of Charon about Pluto is from north to south, out of the plane of the solar system, suggesting that Pluto, like Uranus, has an axis of rotation tilted by nearly 90 degrees so that its equator is perpendicular to its orbital plane. The tilt of the axis changes over time, but like Uranus, Pluto receives more solar energy at its poles than at its equator.

Composition and Internal Structure

Early estimates of Pluto's density ranged from as high as 8 g/cm^3 to as low as 1.2 g/cm^3, greater than the total range for all other planets and satellites in the solar system. (Earth's density is only 5.5 g/cm^3.) If near the high end of that range, Pluto should have been composed of rocky or even metallic components. A planet dominated by such components would be extremely difficult to reconcile with the nebular condensation model outlined in Chapter 2. Instead, this theory predicts that a small planet in the outer solar system would be composed of a mixture of rock and low-density ices, like the satellites of the giant planets.

Careful measurements of the movement of Charon about Pluto allow the masses of Pluto and Charon to be calculated with greater accuracy. Pluto and Charon together have a mass that is only about 0.26 percent of Earth's mass and less than 20 percent the mass of the Moon. New measurements of the diameter of Pluto vary from 2300 to 2400 km. This may seem like a small difference but it translates into important differences in the calculated density and internal structure for Pluto. The mean density of the Pluto Charon system has been determined to be about 2.06 g/cm^3. Data from the Hubble Space Telescope suggest that Charon's density may be as low as 1.3 g/cm^3.

The new value for Pluto's density partly supports the prediction that it is composed of ice and rock. Moreover, Pluto is more dense than the icy satellites of Saturn and Uranus, but very similar to Triton, which has a density of 2.02 g/cm^3. A reasonable model of Pluto's internal structure that satisfies the density requirements calls for 70 to 80 percent rock with a density of about 3 g/cm^3, with the remainder being low-density ices of water, methane, nitrogen, and carbon monoxide. If this model is correct, Pluto contains a higher proportion of rocky material than the moons of Saturn or Uranus, which can be modeled as about 50–50 mixtures of ice and rock. Pluto apparently has more rock than even Ganymede and Callisto, the large moons of Jupiter. Condensation models for the solar nebula suggest that the outermost regions of the ancient nebula may have been water-poor and rich in carbon monoxide. Carbon monoxide should not condense as readily as methane ice, thereby leaving the solids richer in denser silicates and poorer in low-density ices. The rather high densities of both Pluto and Triton support this idea.

Figure 13.4 shows a possible internal structure for Pluto. This model assumes that Pluto is differ-

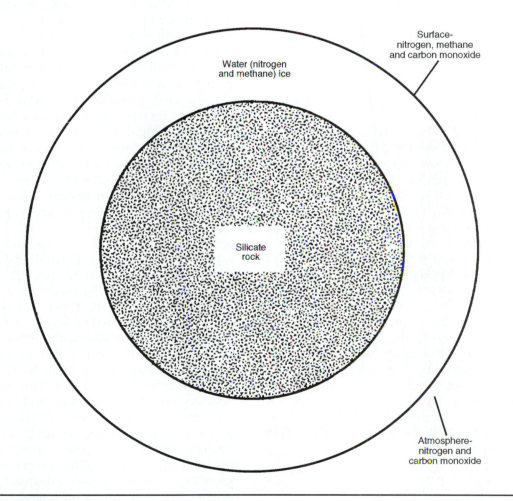

Figure 13.4

Pluto's interior may consist of up to 70 to 80 percent rock, with the remainder being water and methane ice. This configuration, reminiscent of Triton and Europa, is suggested by its relatively high density. The model assumes that Pluto is differentiated into a rocky core and an icy mantle, but that is uncertain. The ices in Pluto are probably dominated by water ice, but spectroscopic studies show that bright methane ice is present. Sublimation of methane ice from the surface may give Pluto a thin atmosphere.

entiated into a rocky core and an icy mantle, but even that assumption is uncertain. Accretion of this small body may not have produced enough heat to melt the ice-and-rock mixture sufficiently to allow gravitational separation and planetary differentiation. However, the high rock content probably led to melting of water ice because of heat released by radioactive decay. The ices in Pluto probably consist largely of water ice, but spectroscopic studies, completed using a sensitive telescope on top of Hawaii's Mauna Kea volcano, show nitrogen (N_2) ice is the most common ice at the surface. Water ice has been detected only on Charon thus far. Carbon monoxide (CO) ice and methane (CH_4) ice are also present at the surface. The nitrogen ice may be concentrated in a discrete icy shell on top of a mantle of mostly water ice. Temperatures at the surfaces of Pluto and Charon may hover just below 40 K, reminding us that the image evoked by Pluto's hellish namesake (Hades of Greek mythology) has little to do with reality.

Pluto's Atmosphere

A tenuous atmosphere, probably of nitrogen, envelops Pluto. The pressure exerted by this atmosphere may only be a few microbars (1×10^{-6} bars). Sublimation of nitrogen ice at the surface must be the source of this wispy atmosphere. Because of Pluto's eccentric orbit, temperature increases are expected to occur when it comes closer to the Sun, and temperature decreases as it moves farther from the Sun. Because of these small temperature changes, repetitive cycling of ice and gas between the surface and atmosphere probably occurs. During orbit-related climate changes, the atmosphere should periodically freeze and fall to the surface during an exotic snowstorm. Pluto's atmosphere may also contain carbon monoxide. There may be very little methane in the atmosphere of Pluto, in spite of the presence of methane ice on the surface. Pluto may simply be too cold for significant amounts of methane to sublime. Our best estimates of its temperature lie between 35 and 37 degrees above absolute zero.

Origin of Pluto and Charon

Because Pluto is such an oddity—an icy planet among the gas-rich outer planets—there has been much conjecture about its origin. The chaotic nature of Pluto's orbit even makes it difficult to know the original location of Pluto. It might have formed in its current orbit; or it might have accreted in a less eccentric orbit, and then evolved into the present situation. Since its discovery, the highly inclined and elliptical orbit of Pluto has been cited as evidence that Pluto originated as a satellite of Neptune. Perhaps a chance, close encounter between Triton (either as a satellite of Neptune or as an outer solar system renegade) and proto-Pluto ejected Pluto from Neptune's satellite system and pushed Triton into its retrograde orbit around Neptune. Other scientists speculate that a large impact on a primitive Neptune-orbiting Pluto fragmented the satellite, ejecting the material from Neptune's gravitational grasp and creating the present Pluto-Charon double planet. These Neptune origin scenarios are extremely unlikely because they require two chance events to occur nearly simultaneously. First, some body is required to move Pluto out of Neptune's grasp, and then a quick means of shifting it into resonance before it collided with Neptune is needed. Another problem is that if Pluto had been thrown from the Neptune system, Pluto and Charon should have fused together again in less than a million years.

The most probable explanation is that Pluto and Triton accreted from materials that condensed in the same frigid part of the outer solar nebula as sun-orbiting planets. In this part of the nebula, the ratio of ice/rock that condensed was low, explaining why Pluto has so much rocky material inside it. In sub-nebulas that form around major planets, the ice/rock ratio is much higher, like that found in the moon systems around Uranus and Saturn. After accretion, Triton, with its retrograde orbit, was captured by Neptune's gravitational field. Pluto was subsequently captured in a 3:2 orbital resonance with Neptune. In this view, Pluto is little more than a big comet. Triton and Pluto may be outer solar system "asteroids," the last survivors of the multitude of icy planetesimals that accreted to form the outermost planets. But where did Charon come from? Did Charon form by fission from a rapidly spinning molten Pluto? This seems unlikely because it is difficult to understand how such rapid spin rates could be achieved. Did Charon accrete as a separate planetesimal in orbit around Pluto? Probably not. This hypothesis would be unable to explain Charon's odd orbital properties. The simplest solution seems to be a catastrophic collisional origin of Charon from a once larger Pluto. Fragments blasted away from Pluto may have re-accreted in orbit to form Charon. The low density of Charon may indicate that it is little more than a porous mass of rubble. A massive collision might also explain why Pluto's spin axis is tipped at an odd angle. We have invoked this same sort of chance event to explain the origin of our own Moon.

Conclusions

Many questions about the nature and origin of Pluto and Charon remain unanswered or even unformulated. One day a prolonged visit to Pluto could be extremely important to our perceptions of the outer solar system. Unlike Triton, Pluto probably was never heated by tidal massaging, and therefore it may come closer than any other planet to preserving primordial abundances of the ices that condensed from the nebula billions of years ago. The enigma of Pluto is likely to persist until a spacecraft, with cameras, spectrometers, and other instruments, can penetrate this remote part of the solar system to scrutinize Pluto. The goals for such a mission have already been formulated and a late 1990s launch date has been discussed by NASA. A small spacecraft could reach Pluto in six or seven years after launch. If this mission to the last unexplored planet is successfully funded, the quality of the new images and other data will exceed that obtained by Voyager at Triton.

Important questions that require answers include the following:

- What is the composition of Pluto?
- What are the sizes and densities of Pluto and Charon?
- What is the internal structure of Pluto? Is it differentiated?
- Was Pluto ever geologically active like Triton?
- What was the impact history of Pluto?
- Are the features of Charon and Pluto different as a result of different surface compositions?

Though answers to these and other questions may come from a flyby mission, the only certain thing, according to one of Pluto's investigators, is that we will be surprised by what we find. Until then, the results of Voyager investigations of Triton provide insights useful for interpreting the natures of Pluto and Charon. Continued Earth-based investigations will yield most data for improving our understanding of these planets on the edge of the solar system.

Review Questions

1. Why is it appropriate to consider Pluto-Charon as a double planet?
2. How did the discovery of Charon change our view of the nature of Pluto?
3. In terms of its origin (not composition), is the atmosphere of Pluto more like that of Mars or Jupiter?
4. What other bodies in the Solar System is Pluto most like? What does this imply about its origin?

5. Early estimates of the density of Pluto were greater than the density of the Earth. If these estimates had been supported by later measurements, would they conflict with the theory for planet formation from a solar nebula?
6. NASA has proposed sending a spacecraft to Pluto. What type of studies would you recommend to understand Pluto better?

Additional Reading

Binzel, R. P. 1990. Pluto. *Scientific American.* Vol. 262, p. 50.

Mulholland, D. 1982. The Ice Planet. *Science.* Vol. 82, pp. 64–68. (Discusses Pluto and the discovery of Charon.)

Sobel, D., 1993, The last world. *Discover.* No. 5, pp. 68–76.

Tombaugh, C. W., and P. Moore. 1980. *Out of the Darkness: The Planet Pluto.* Harrisburg, PA: Stackpole Books. (An account of Pluto co-authored by its discoverer.)

Comet, Earth, and Moon

CHAPTER 14

Comets and Small Bodies of the Outer Solar System

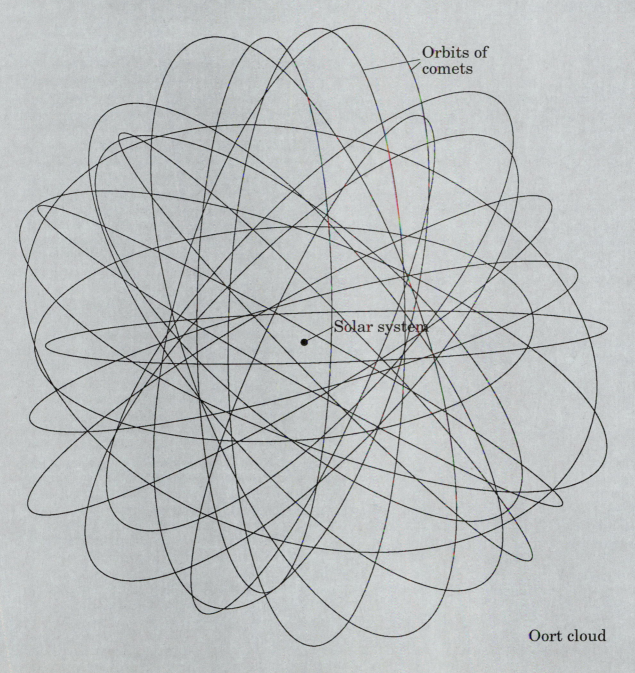

Orbits of comets

Solar system

Oort cloud

Introduction

Although among the smallest members of the solar family, comets hold answers to some of the biggest questions regarding the origin of the solar system. The birth of comets is intimately tied to the birth of the solar system itself. They appear to have formed in the outer solar system as direct condensates from the ancient solar nebula. By virtue of their icy compositions and small sizes, they may have preserved evidence of their conception and birth 4.6 billion years ago. Comets may also have played an important role in the subsequent history of the solar system. Many of the impactors that created the heavily cratered surfaces of the planets and dominated the early history of the solar system—from Mercury to the satellites of Neptune—were probably comets. When they struck, they may have given each planet some of its volatile constituents. Some say comets may have also implanted organic molecules that were the building blocks of life on Earth. Thus, the ultimate goal of studying a comet is to understand its birth and death and thereby relate it to the formation, evolution, and present state of the solar system.

Major Concepts

1. Comets are small bodies composed basically of ice and dust. Their icy nuclei partially vaporize when they come close to the Sun, forming large diffuse comas and spectacular tails of gas and dust.
2. Comets may have formed near Uranus and Neptune by condensation of gas in the ancient solar nebula. Subsequent gravitational perturbations from Jupiter probably ejected them to a distant cloud of comets that presently envelops the solar system. Periodically, some comets are gravitationally forced into shorter elliptical orbits that take them into the inner solar system.
3. Several small Sun-orbiting bodies, only recently discovered, and comets are remnants of the planetesimals that accreted to form the outer planets and their satellites. Unaffected by the processes of planetary differentiation, comets may tell us about the nature of the solids that condensed in the outer reaches of the ancient solar nebula.

TABLE 14.1

Orbital Characteristics of Selected Comets

Comet	Distance from Sun		Last Appearance (Year)	Orbital Period (y)
	Closest (AU)	Farthest (AU)		
Long Period Comets				
Donati	0.58	313	1858	2000
Humason	2.13	400	1962	2900
Morehouse	0.95	Large	1908	Large
Burnham	0.50	Large	1960	Large
Kohoutek	0.14	Large	1973	Large
Short Period Comets				
Encke	0.34	4.2	1987	3.3
Tempel II	1.4	4.8	1988	5.3
Giacobini-Zinner	1.0	6.1	1985	6.6
Halley	0.59	28.4	1986	76
Swift-Tuttle	0.96	47	1992	120

Comets

Long regarded as omens of ill luck and catastrophe, comets may ultimately provide us with critical data necessary to explain many mysteries associated with the formation of the planets. Individual comets represent a very small portion of the total mass of the solar system, and most revolve about the Sun at such great distances that they are almost lost from our solar system. As a result, we tend to ignore them, but the recent return of Halley's comet to the inner solar system brought comets back into the spotlight.

Most comets are thought to be concentrated in a loosely defined group called the **Oort Cloud** in the very outer reaches of the solar system. There is no direct evidence for the existence of the Oort Cloud, but its presence is inferred from the highly elliptical nature of comet orbits, which take many of them very far from the Sun (Figure 14.1). Moreover, the orbits of comets that have long orbital periods are not confined to the same plane as the planets. Thus, a loose cloud of comets speculated to surround the solar system extends to 200,000 times

as far from the Sun as Earth—one tenth of the distance to the closest star. Some estimates claim that this reservoir of comets contains as many as 7 trillion comets with a total mass 50 times that of Earth's.

Another speculative group of comets may lie closer to Pluto, in what is called the **Kuiper belt**. This group may produce comets with short orbital periods that commonly have orbits lying in the same plane as the rest of the planets. The Kuiper belt may be populated by a remnant of the group of comets ejected to form the Oort Cloud. The total number of comets in this Sun-orbiting belt is probably small compared to that estimated for the Oort cloud. There may be fewer than 1000 bodies in this group.

Because individual comets are so small, they are not visible until they come near the orbit of Mars. Once inside our viewing range, most comets appear as the one in Figure 14.2. The spherical, diffuse region at the front of the comet is called the **coma**; it may be 100,000 to over 1,000,000 km in diameter. Inside the coma is a small, solid body called the **nucleus**, which is probably no more than

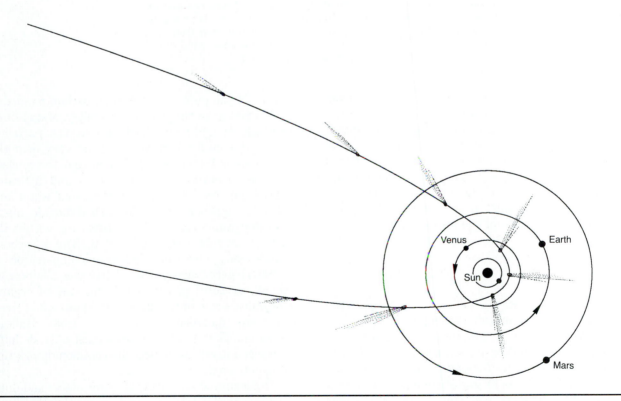

Figure 14.1

The orbital paths of the comets we see in the inner solar system take them in highly elliptical routes around the Sun. Some comet orbits take thousands of years to complete. Many orbits are also highly inclined, even more than Pluto's. A loose group of comets, called the Oort Cloud, is postulated to extend out to beyond 50,000 AU. (One AU is the distance from the Earth to the Sun; Pluto lies only 39 AU from the Sun.) These cosmic icebergs are the farthest-traveled members of the solar system.

Figure 14.2

Comets develop this classical structure once they reach the inner solar system. The head of the comet consists of a very small solid nucleus of ices (water, ammonia, methane, and carbon dioxide) plus some rocky material, surrounded by a diffuse gaseous coma of volatilized ices. The head may be over 100,000 km in diameter but the nucleus is probably only a few kilometers in diameter. Comets glow only because of light reflected from the Sun. A comet's long tail is formed as the force of the almost imperceptible solar wind drags material away from the coma. The tail always points away from the Sun.

1 to perhaps 100 km across. Streaming away from the coma, in a direction that is always away from the Sun, is a long bright **tail**. These magnificent tails may extend for several tens of millions of kilometers. A comet does not develop a coma or tail until it is near the Sun. The coma and tail are produced when ices in the nucleus sublimate or vaporize, carrying dust particles with them, and stream away because of the **solar wind**.

Comet Halley

The 1986 passage of Halley's comet through the inner solar system provided an opportunity for five spacecraft to study it, adding to the information gained from ground-based studies of comets. The chemical composition of cometary nuclei is critical to understanding the origin of comets and of low-temperature nebular condensates. Spectroscopic studies of Halley's comet indicate that the gas and dust cloud consisted of a variety of molecules composed of volatile elements—hydrogen, nitrogen, carbon, and sodium. Eighty percent of the gas consists of water. Significantly for our models of the

nature of the outer solar nebula, carbon monoxide ice is present in the nucleus of Halley. Magnesium, iron, silicon, and nickel, probably in dust particles, can be detected when comets come very near the Sun. This is interpreted to mean that the nucleus consists of carbonaceous materials and hydrated silicate minerals mixed in a matrix of water and other ices (carbon monoxide, carbon dioxide, methane, and ammonia). The temperature within the nucleus may be only 25 to 40 K. The spacecraft found many particles similar in composition to primitive carbonaceous chondrites (see Chapter 3). About one-third of the coma consisted of organic molecules, some of the same sort identified in these primitive meteorites. Evidence from Halley's comet shows that silicates may exist only as fluffy, dust-sized particles, hence, the common description of comets as dirty snowballs.

Although obscured by streams of gas and dust, photographs of Halley's nucleus by the European Space Agency probe Giotto and the Soviet Vega showed it to be a tumbling irregular dark mass 15 km long and about 7 km wide (Figure 14.3). Estimates of the density of the nucleus range from 0.1 to 0.4 g/cm^3. The nucleus is apparently quite fluffy

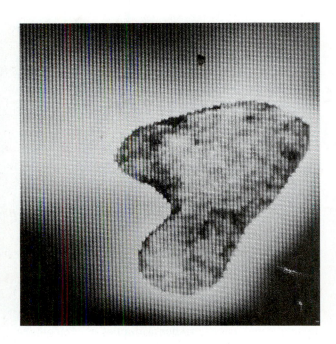

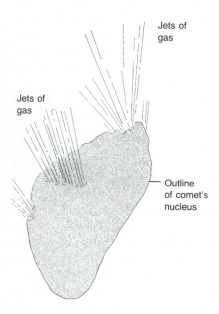

Jets of gas

Jets of gas

Outline of comet's nucleus

Figure 14.3

Halley's comet contains a dark, irregularly shaped nucleus. Surprisingly, the nucleus is one of the darkest objects yet seen in the solar system. Dark carbonaceous residues appear to coat bright ices buried inside the comet. Jets of gas burst through the dark crust and bring material from the icy interior when the comet is close enough to the Sun to vaporize. Halley's nucleus measures about 16 km across in its longest direction.

and loosely held together. Most of the comet is empty space, a cosmic sponge cake. Spacecraft photos also showed narrow jets of gas erupting from several areas on the day side of Halley's nucleus. These gas jets appear to form along cracks and allow the gas formed by sublimation of internal ice to erupt. The jets are formed by a process very much like that which forms the geysers on Triton. Halley's jets erupt dust at a rate of as much as 10 tons per second. The ejected material forms the coma and is swept away to form the tail. The comet's nucleus is about as dark as the dark material in the splotch on Saturn's Iapetus. Apparently, when near the Sun, a thick residue of dark carbon-

aceous and silicate dust is left behind and builds up on the surface of the nucleus as the more volatile ices beneath the crust are vaporized by the Sun's warmth and are erupted through the jets (Figure 14.4). The dark dusty lag deposit may thicken and create an armor that coats and protects the icy interior. Nonetheless, Halley loses 1/1000 of its mass every time it enters the inner solar system by these processes.

If a comet could be seen when in the portion of its orbit far from the Sun, it would probably look like a dark asteroid. The gases are frozen solid under the surface. There would be no coma or tail in this part of its orbit. The distinction between asteroids and comets is, in fact, fuzzy. Asteroids are bodies that do not degas and develop bright comas, but what happens to a comet nuclei buried in the silicate debris accumulated at its surface after millennia of sublimation? Vaporization of volatiles from the interior might be choked off and the bodies would look like asteroids.

Outer Solar System Planetesimals

Although not as abundant as the small bodies of the solar system, a growing number of small icy bodies have been discovered in the outer solar system. Increasingly sensitive telescopes have been used to track down these icy equivalents of the asteroids.

Discovered in 1977, *Chiron* is a small dark asteroid-sized object that occupies a lonely position in the outer solar system between the orbits of Saturn and Uranus. It has a diameter somewhere between 200 and 340 km, but its surface has never been photographed by a spacecraft and even its density is not known. Chiron is the only small Sun-orbiting body discovered in this vast area of the solar system. A diffuse cloud of carbon dioxide gas envelops it, suggesting that Chiron is really a comet. Chiron's orbit never takes it close enough to the Sun to vaporize water ice and develop a large coma or tail. If this is true, Chiron is the largest comet known.

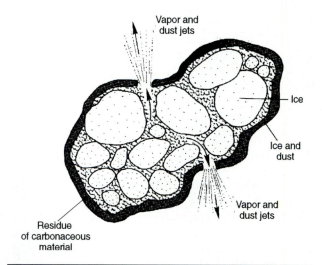

Figure 14.4

The internal structure of a comet is shown in this sketch based on observations of Halley's comet. Comet nuclei consist of ice and smaller amounts of carbonaceous and silicate dust. As the comet nears the Sun the ices are warmed up and vaporize, with the gas jetting through cracks to the surface. A residue of dark carbonaceous materials forms a black insulating crust over the ice.

Pholus is another small outer solar system body discovered even more recently, in 1992. It has a diameter of about 200 km and a very eccentric orbit that carries it across the orbits of Saturn, Uranus, and Neptune.

Six other small bodies have been discovered with orbits beyond that of Neptune. Most do not yet have formal names. These may show that the Kuiper belt of comets is more than a hypothetical model. Four of the bodies lie from 32 to 35 times as far from the Sun as Earth. (Neptune orbits the Sun at about 30 times the Earth-Sun distance.) They may be about half as large as Chiron, with diameters of about 100 km. Two other bodies orbit the Sun at a distance beyond that of Pluto's. At least one of them is estimated to have a diameter as much as 200 km and a reddish color. The color may be caused by a hydrocarbon-rich surface deposit. It is difficult to tell if any of these bodies have stable long-lived orbits and are thus legitimate members of the Kuiper belt. They may have remained in the same orbits over the entire history of the solar system. Alternatively, they could have been recently kicked into their present orbits. These mavericks may be on their way to becoming short-period comets if they receive other gravitational bumps from the giant planets.

Such recent discoveries suggest that the outer solar system is littered with chunks of material left over from the formation of the gas giants, just like the remnants of accretion still found in the inner solar system that we call asteroids. These bodies may have formed in the same part of the ancient solar nebula that many comets are thought to have originated. Moreover, they have many similar characteristics, including ice-rich compositions. Phoebe, one of the outer captured satellites of Saturn, is similar in size and color to Chiron. It joins the ranks of Pluto, Charon, and Triton as representatives of a once much larger population of outer solar system planetesimals. The major differences between these bodies and comets may be that their orbits never bring them near to the warm hearth of the inner solar system. Therefore, they lack comas and tails and are not decreasing in size with each passing orbit.

Origin and History of Comets

The orbits of comets and the large content of volatile materials show that they formed in the cool (as low as 20 K) part of the solar nebula. Some planetary scientists have postulated that comets formed as small planetesimals between Uranus and Neptune or even farther out near Pluto. There is no reason to think that the ancient solar nebula ended at Pluto. Studies of other stars show that nebulas around young stars extend as far as 20 times farther than the Pluto-Sun distance away from the central star. If condensation of solids occurs there, the solids are most likely to contain a large proportion of ice, just like comets in our solar system. Such an extended nebular disk may not form large planets for several reasons. First, the density of the material in the solar nebula probably dropped off with increasing distance from the Sun. There was simply less material from which to assemble planetesimals and planets. Second, because of the long orbital periods in the outer solar system, the time between collisional encounters between planetesimals would be long. Consequently, accretion would be slow.

After an icy body formed near Uranus or Neptune, Jupiter's gravitational energy may have altered its orbit and catapulted it into the distant Oort Cloud, a shell of this cosmic debris left over from planet formation. Later gravitational perturbations, perhaps from other stars, further altered the orbits of some comets, fueling a steady stream of "new" comets with shorter orbital periods that pass through the inner solar system. Anciently, those that entered the inner solar system may have produced some of the impact features we see today on the planets and their satellites. Indeed, it may have been a small comet that exploded over Siberia in 1908, knocking down trees for kilometers but creating no crater. If a short-period comet does not collide with a planet or moon, its ices eventually vaporize, scattering its dusty carbonaceous and silicate fractions through the solar system.

The final days of comet Shoemaker-Levy 9 illustrate what must have been the fate of thousands of comets before it (Figure 14.5). This comet was discovered in 1993 shortly after it swept close to Jupiter. As the comet passed by Jupiter, the massive gravitational force of the giant planet ripped the icy body into a series of at least 21 perfectly aligned fragments, strung out like pearls on a cosmic string. The fragments became satellites of Jupiter, swinging around the planet in a highly eccentric orbit. Each newly formed nucleus was surrounded by a coma and tail of gas and dust, making it difficult to measure the exact size of the icy fragments. The best estimates are that the largest was little more than 1 km in diameter. Even then, each comet nucleus may itself have been made of many separate smaller bodies. Moreover, the composition of the comet is still debated. Was it an ice-rich cometary snowball, or was it a rocky remnant of a once icy comet?

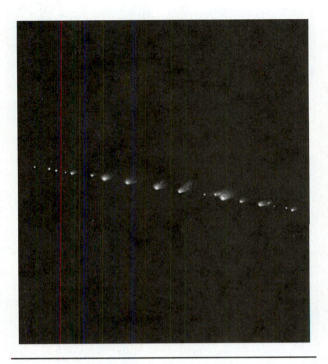

Figure 14.5
Comet Shoemaker-Levy 9 broke up into a string of at least 21 objects as a result of a gravitational encounter with Jupiter. Each of the 21 bodies had its own nucleus and coma.

From 16 to 22 July 1994, the jewels on this necklace of comets collided one by one with Jupiter. Spectacular fireballs in the hydrogen-rich atmosphere marked each exploding comet fragment. The largest fragment released the equivalent of several million megatons of TNT (Figure 14.6). Before exploding, the fragments penetrated through the clouds on the nighttime side of Jupiter. Rapid rotation of the planet quickly brought the sites of impact into view. Bright expansive plumes were visible from Earth, even with the telescopes available to amateur astronomers. The best images were acquired by NASA's Hubble Space telescope.

Computer models and careful observations from telescopes and satellites have helped us to understand this cometary impact. As each fragment entered the atmosphere, it flashed like a shooting star from the friction generated by its rapid flight. Each chunk probably tunneled about 75 to 150 km below the cloud layer, leaving behind it a trail of hot, pressurized gas and cometary debris. Temperatures reached almost 2000 K. The heat released hurled an upward expanding fireball back through the tunnel created by the falling fragment (Figure 14.6). A powerful shock wave closely followed the fireball creating the spherical plume. These huge plumes, larger in diameter than Earth itself, rose hundreds of kilometers, punching

15 sec after impact   300 Km

Cloud top

50 sec

70 sec

(A) Computer simulations of the collision show the sequence of events associated with each impact. First, a fireball forms along the trail of the falling comet. Second, the heated gases rise back along the tunnel. Third, a large fireball punches through the cloud tops.

(B) The string of comets collided one by one with Jupiter as seen in this artist's conception.

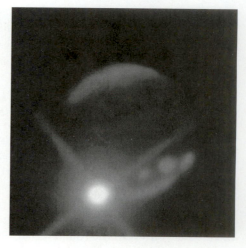

(C) A flash of light marks the rise of the plume from the largest impact through Jupiter's atmosphere. This photo was taken through an Australian telescope.

(D) The impacts left a string of dark scars on Jupiter aligned in a ring around its southern hemisphere. The Great Red Spot is a large storm, not an impact feature. The Hubble Space Telescope took this picture.

Figure 14.6

The impact of comet Shoemaker-Levy 9 into the atmosphere of Jupiter was one of the most spectacular events in the recent history of the solar system.

454

through the cloud tops within minutes of the explosion. Momentarily, the light from Jupiter increased fiftyfold. The bright plumes disappeared in a matter of hours and were replaced by outward expanding ripples or waves in the atmosphere and by enigmatic dark scars. Each impact site had a complicated structure made of a ring inside a broad dark arc (Figure 14.6). These blemishes are not impact craters; the atmosphere is far too fluid for such a structure to form, but their exact origin is not known. The inner dark ring is probably a ripple in Jupiter's atmosphere set up in response to the comet explosion. Its outward expansion was clocked at 800 meters per second. Perhaps the outer arc consisted of dark sootlike particles formed in the heat of the collision that are suspended temporarily in the atmosphere. Ammonia and sulfur compounds were identified in the hot material in the plumes.

The importance of this kamikaze comet is underscored by understanding just how rare such impacts are. Today, 4.5 billion years after the solar system was cleared of most orbiting debris, it is rare for a comet to go into orbit around Jupiter. Shoemaker-Levy 9 is the first to do so in recorded history. It is even less likely that the comet would also break up, and it must be a once-in-a-millennium event for the broken comet to collide with a planet and afford us the opportunity to actually see the impacts.

Ultimately, those comets that do not collide directly with a planet break apart in orbit because of repeated solar heating. The breakup of these cosmic icebergs is probably caused by heating, gas jetting, and consequent chaotic spinning of the comet nucleus. In effect, the comet becomes "unglued" when its icy matrix vaporizes and jets to the surface. Several comets have been known to fragment, leaving in their wakes streams of particles. Debris from disruption as well as debris jetted from the comet eventually string out along the orbital path of the comet. If this littered path intersects Earth's orbit, meteor showers in the atmosphere of Earth are produced on an annual basis. Examples of comet-related meteor showers include the Leonid shower that recurs every November and the Perseid shower in August. The latter is dust and debris shed from comet Swift-Tuttle that reappeared in the inner solar system in 1992; its last visit was during the American Civil War. Some volatile-rich micrometeorites may have reached Earth via comet disruption.

Conclusions

Although there are probably more theories concerning the nature and origin of comets than there are facts, it seems that these small bodies are the remnants of the icy planetesimals that formed the outer planets and their satellites. In this way they are analogous to the rocky asteroids that are the remnants of the planetesimals that formed the inner planets. The cold storage of these icy planetesimals for so long may mean that they preserve some ancient icy materials deep inside them. This is made difficult by the fact that comets have experienced some chemical differentiation since their presumed origin 4.6 billion years ago. Nonetheless, some of the most important geologic information to be gained from comets will be the composition and structure of their nuclei. Studies of these vagabonds of the solar system may help us answer such questions about the nebula as these: What were its constituents? How, in what sequence, and at what temperature did ices, silicates, and carbonaceous materials condense? How long ago did condensation occur?

Because of the information that could be gained from comet studies, several additional flyby and perhaps sample-return space missions are planned by NASA and the space agencies of other countries for a short-period comet with a well-established orbit during the next few decades. Since comets are composed of primitive volatile-rich material, such probes with imaging, chemical, and magnetic devices could reveal much more about the physical nature of comets, their structure, and composition and could yield more pieces to the puzzle that is the early history of the solar system. A sample of this material may be as close as we will ever get to examining the icy condensates of the solar nebula.

Review Questions

1. List the principal constituents of comets.
2. In terms of physical characteristics and origin, how is a comet different from an asteroid?
3. Describe the probable changes that have occurred in the orbit of a short-period comet that presently passes through the inner solar system.
4. Why do some comets have two tails? Do these tails shows the direction of movement?
5. What is the ultimate fate of comets that enter the inner solar system?
6. What do you think the next investigations of comets should concentrate on? Why?

Key Terms

Coma
Kuiper Belt
Nucleus
Oort Cloud
Solar Wind
Tail

Additional Reading

Balsiger, H., H. Fechtig, and J. Geiss. 1988. A close look at Halley's comet. *Scientific American*. Vol. 259, No. 3, pp. 96–103.

Sagan, C., and A. Druyan. 1985. *Comet*. New York: Random House.

Whipple, F. L. 1985. *The Mystery of Comets*. Washington, DC: Smithsonian Institution.

SATURN

EARTH

JUPITER

CHAPTER 15

Comparison of the Planets

MARS

MERCURY

MOON

EARTH

IO

EUROPA

GANYMEDE

CALLISTO

VENUS

TITAN

Introduction

This has been an extraordinary period of exploration. We have landed on the Moon and sampled its rocks. We have mapped the surface of Mars, tested its soil for evidence of life, and explored its huge canyons and grand volcanoes. We have surveyed the diverse landscapes of the moons of Jupiter and Saturn and have discovered that small, icy satellites can have complex histories of impact, volcanism, and tectonic deformation. We have photographed the exotic surfaces of the moons of Uranus and Neptune, which present some new and exciting insights into the origin of planetary bodies. We have begun to acquire detailed information about rocky asteroids and icy comets. In short, we discovered new worlds and looked back at Earth from space with greater understanding of how our planet functions and why it is unique. Every object in the solar system contains part of a record of planetary origin and evolution. These new worlds are important to us on Earth because they tell a great deal about the phenomena and forces that shape and control planetary environments and vividly show how things on our own planet might have been different. They also show how things might develop in the future. By understanding the history of the planets and the reasons for their diversity, we have greatly increased our understanding of Earth, our home.

Major Concepts

1. The solar system consists of the Sun and its family of planetary bodies. They were formed at the same time and from the same nebula, and although each body is unique, they all have much in common.

2. The great variety of surface features, compositions, and internal structures of the planets was produced and modified by geologic processes that can be understood as the results of the interaction of planetary matter and energy.

3. Geologic processes that acted on almost all planetary bodies of the solar system include impact cratering, internal differentiation, volcanism, and tectonism. Physical and chemical interactions with atmospheres and other surface fluids were important on some planets and moons.

4. The terrestrial planets (Mercury, Venus, Earth, the Moon, and Mars) are composed mostly of rocky materials surrounding metallic cores. Although they probably started out much the same with impact-dominated surfaces, they evolved along different paths because of differences in size, composition, internal thermal energy, and distance from the Sun.

5. The satellites of the outer planets and Pluto are composed mostly of ice surrounding cores of rock or mixtures of rock and ice. Most have changed very little since they originated, but a few have had sufficient internal energy to develop crustal deformation and volcanism after the period of heavy bombardment. Except for Pluto, the giant outer planets are large bodies of gas, ice, and melted ice.

The Diversity of the Planets

The exploration of the solar system reveals a family of planets of great diversity, which we attempt to summarize graphically in Figures 15.1, 15.2, and 15.3. The Moon and Mercury are small, rocky bodies with ancient surfaces marked by an abundance of impact craters and smooth plains only slightly wrinkled by tectonic processes. Mars has a variety of landforms including giant volcanoes, huge canyons, and large terrains eroded by catastrophic floods of water. Earth has continents and ocean basins, and its very young surface is continually washed with an ocean of water, modified by volcanic activity, and deformed by a mobile lithosphere. Venus is also a dynamic planet with rifts, volcanoes, "continental" highlands, and eolian processes, but it has no liquid water. Even the small, mostly icy moons of the outer solar system present evidence for a vast array of geologic phenomena. Io has volcanoes erupting sulfur. Titan has a thin secondary atmosphere of nitrogen that may hide a hydrocarbon sea. There are the complex, grooved terrains of Ganymede, the surprising rifts and plains of Enceladus, the dark terrain of Iapetus, the coronae of Miranda, and the ancient cratered surfaces of Callisto, Tethys, Oberon, and others. The giant outer planets are radically different from the other bodies of the solar system in that they lack solid surfaces and consist mostly of hydrogen and helium inherited from the ancient solar nebula—the cocoon from which all of the planets emerged 4.6 billion years ago. Delicate rings of ice particles have formed around the giant planets. The outermost planet, tiny methane-frosted Pluto, has more in common with the moons of the outer solar system than with the other outer planets.

Aside from the differences apparent on their surfaces, there are also dramatic variations in density, internal structure, and albedo (see Figures 15.2 and 15.3) among the planets and their satellites. This diversity reflects systematic variations in their compositions. The albedo, or brightness, is caused by the composition of the surface materials; water ice is bright, and silicates and methane ices are dark. The density reflects the proportions of dense silicates, moderate-density ices, and light gases in the interiors of the planets.

Here, then, is a system of planets that at first seems bewildering in its variety. Indeed, a grand theme revealed by the exploration of the planets is their great diversity. Yet all of the planetary bodies were formed in the same little corner of the universe, at the same time, and from the same materials. Why are they so different? Why have some planets remained geologically active for 4.6 billion years, whereas others have remained essentially unchanged since they formed? Why do some have atmospheres, oceans, and continents, whereas others are giant spheres of gas or tiny balls of ice? Why is the balance of matter and energy just right for life to evolve on Earth and so hostile on others? Are there a few fundamental laws that govern planetary development? Our purpose in this closing chapter is to re-emphasize a few basic processes that can explain the diversity of the various planetary bodies as we review some differences and similarities among them.

The Fundamentals of Geology— Matter and Energy

Answers to some questions posed above may be found by analyzing the type and amount of material that make up a planet and the nature and magnitude of energy available at its surface or in its interior. One of the general principles to emerge from this venture into comparative planetology is that the science of geology is an exploration of the interaction of *planetary matter* (in its various forms) and *energy* (in its various forms).

Planetary Matter

Planetary matter can be described by its chemical composition, mass, and size. These properties of matter help determine at which temperature the material melts or freezes, how strong it is, how it transfers heat, and how quickly it cools. Moreover, a planet's composition determines its abundance of long- and short-lived radioactive elements, which affects how much internal energy it possesses.

The chemical compositions and sizes of the planets are not random but are strongly controlled by their distances from the Sun. As we saw in Chapter 2, our solar system was spawned 4.6 billion years ago in a cold, diffuse cloud of gas and dust deep within a spiral arm of the Milky Way galaxy. The huge cloud was made up largely of the two lightest elements, hydrogen and helium, and only small concentrations of the other elements. Under the force of gravity, the giant cloud collapsed and assumed the shape of a rotating disk. Solid material within the disk was segregated according to composition. In the center of the disk, near the hot proto-Sun, only relatively refractory compounds, such as silicates and iron, condensed from the gaseous nebula to form solid particles. The outer part of the cloud was naturally colder so that other, more voluminous substances such as water, ammo-

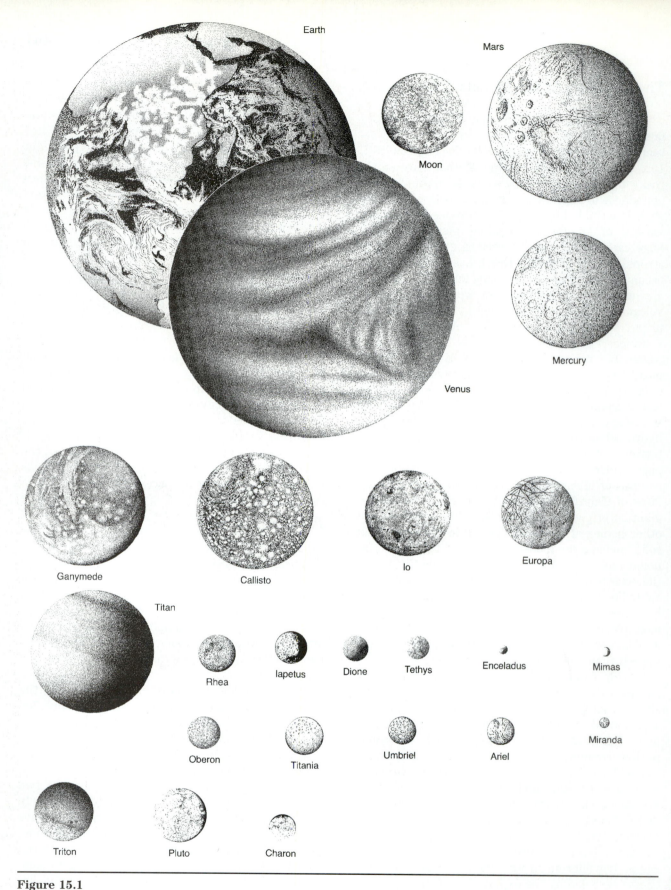

Figure 15.1

The surface features of the planets and moons in the solar system are shown here at the same scale. The dominant landform of the planets is the impact crater. Many of these craters are ancient and date from a period of heavy bombardment. Impact-related fractures cross the surfaces of some of the smallest bodies. Volcanic plains and shield volcanoes are apparent on some planets, even at this small scale. Tectonic modifications represent the response of the lithosphere to stresses derived from the thermal or tidal evolution of the planet. Large domes and rifts are apparent on several bodies.

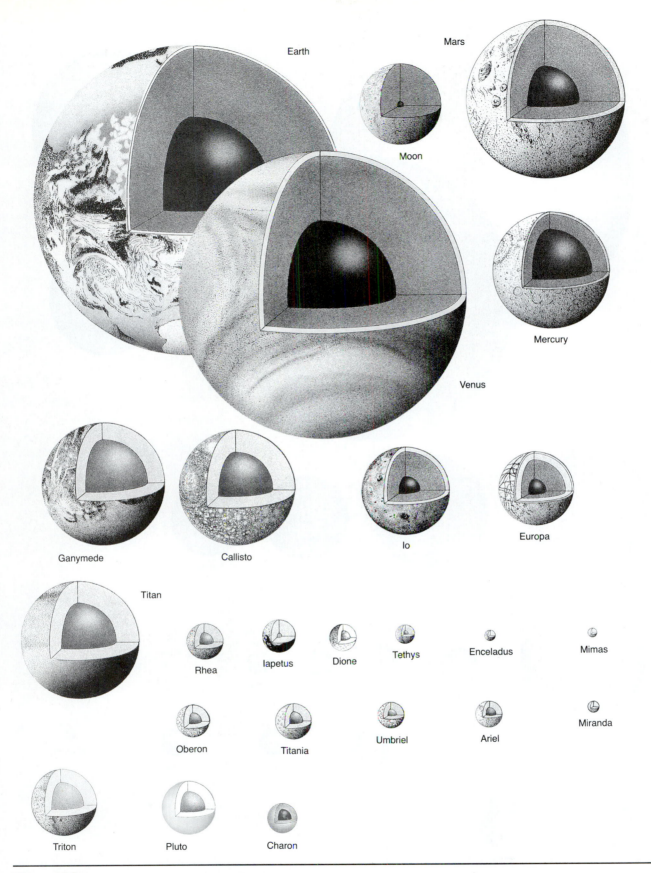

Figure 15.2

The internal structures of the planets and moons are dominated by concentric layers of diverse compositions and mechanical properties. The inner planets and Io probably have dense cores of iron metal and thick mantles and crusts of silicates. In contrast, the other moons of the outer planets and Pluto may have cores of silicates surrounded by mantles of water ice. Although internal differentiation was an important result of accretionary heating in many planets, moons, and asteroids, some small icy satellites of Saturn, Uranus, and Neptune may not be differentiated. The interiors of these small objects may consist of more-or-less homogeneous mixtures of ice and silicate rock.

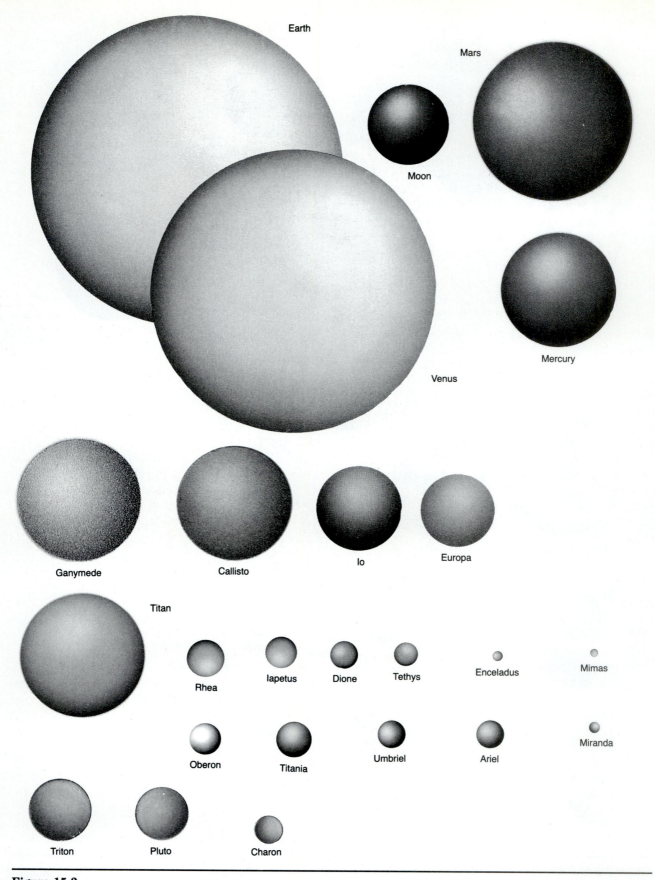

Figure 15.3

The compositions of the surfaces of the moons and planets are emphasized in this diagram showing their relative brightness or albedo. The inner planets, their moons, and the asteroids are composed of dark, rocky materials. The satellites of the outer planets have surfaces that are either bright and dominated by water ice or dark and dominated by methane ice. Ices of nitrogen, and liquid and solid hydrocarbons may be important on a few of these bodies.

nia, methane, and so forth also solidified. Thus, early in the history of the solar system, there was a separation and differentiation of solid matter that exerted a profound control on the nature of the future planets. Denser particles that condensed at high temperatures were the only solids in the central region near the Sun, whereas low-temperature, low-density ices dominated the condensates near the fringes of the nebular disk. The planets formed by collisional accretion of the particles that were present at a particular distance from the Sun. The silicates and iron in the inner solar system accreted to form the inner planets. The ices in the outer part of the nebula eventually formed the cores of the giant worlds—Jupiter, Saturn, Uranus, and Neptune. Because of the large size of their icy cores, huge masses of nebular gas—mostly hydrogen—were gravitationally attracted to these planets. In contrast to the planets of the inner solar system, subsequent T-Tauri winds were unable to strip this gas away from the giant planets; the features we see on these planets are all in their primary atmospheres. Therefore, the outer planets are larger but less dense compared to the rocky inner planets. In addition, most satellites of the outer planets have large proportions of water ice with smaller amounts of silicate and metal. By contrast, the inner planets have, at best, discontinuous caps of polar ice or shallow oceans of molten water on spheres dominated by silicates and metals. This same sequence of icy low-density to rocky high-density bodies is seen in the four Galilean satellites (Figure 15.2), which are thought to have formed in a temperature gradient focused on Jupiter. Even the rings of icy particles that encircle these outer planets are apparently remnants of icy fragments that did not accrete to form sizable moons or are the result of recent impact fragmentation of small satellites.

Thus, although all of the planets in the solar system formed from the same solar nebula, the differentiation of the nebula caused the composition of the inner and outer planetary bodies to be distinctly different. The trend in the compositions of the planets supports the fundamental idea that condensation in a thermally zoned nebula resulted in the dichotomy between the rocky inner planets and the icy outer planets and their satellites.

Planetary Energy

After the planets were formed, their subsequent histories were controlled largely by the amount and type of energy operating on the planet. Without energy there can be no change. Three major forms of energy are important in this con-

text: (1) internal heat (including that derived from radioactivity, differentiation, and tidal friction), (2) impact of meteorites and comets, and (3) thermal energy from the Sun. In a planet, as energy interacts with matter, a new rock body is formed and a landform or structure is produced. For example, when internal heat causes melting and volcanic eruption of lava or ash, new rock bodies (lava flows and plutons) and new surface features (volcanoes) are created. Similarly, when a meteorite collides with a planet, its kinetic energy produces an ejecta blanket (a new rock body) and a crater (a new landform). Tectonic deformation of the crust also produces new rock bodies and terrain types (rifts and mountain belts). In addition, solar energy drives the hydrologic systems found on several planets, allowing water to erode new landscapes and deposit sediments to form new layers of rock.

The amount and type of energy that drive geologic processes on and within planets varies greatly from planet to planet and has varied throughout the history of each planet. For example, the thermal energy of a planet and the geologic activity it produces are very sensitive to the planet's size. Most planets formed relatively hot because of accretion. From this initial state, small planetary bodies tended to cool rapidly because of their large surface areas compared to their masses. Cooling may be moderated by the production of heat from differentiation and from radioactivity. Nonetheless, small planets, such as the rocky Moon or icy Mimas, cooled rapidly and had short thermal histories. As a result, ancient surface features are well preserved on these bodies because, soon after formation, their internal heat was rapidly radiated to space. Larger bodies retain their internal heat longer and, as a result, have prolonged periods of volcanism and crustal deformation that obliterated ancient cratered terrains.

But other factors, recently discovered by the space program, also affect a planet's thermal evolution. If the Voyager missions to the outer planets had not revealed the tumultuous volcanic eruptions of Io or the smooth plains of Enceladus, we might not have considered tidal heating, caused by the effects of a distinctive orbital environment, to be important to the thermal history of a planet. We have also learned that a planet's accretion history (slow and cool versus fast and hot) is important in a planet's thermal history.

The relationship of composition, size, and energy in a planet's evolutionary history can be shown in a three-dimensional diagram (Figure 15.4). The vertical axis represents the *composition* of a planetary body and ranges from silicates at the base to icy materials at the top. The present *ther-*

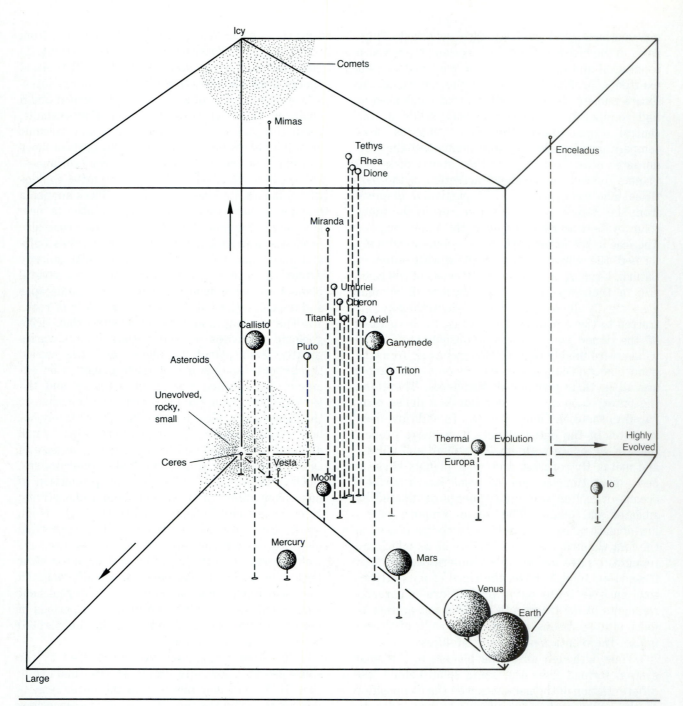

Figure 15.4

The size, composition, and thermal evolution of a planetary object are its fundamental characteristics and determine the nature of its surface and interior. The three-dimensional chart shown here uses the vertical axis to represent the composition of the planetary body and ranges from rocky silicates at the bottom of the cube to icy objects at the top. The present thermal state, shown along the back of the cube, ranges from cold and inactive bodies to moons and planets that are still hot and experiencing volcanism. The size of the planet is shown on the remaining axis with small bodies at the back of the cube and large ones at the front. In terms of these three factors, various planets and moons are shown within the cube. In general, larger bodies are more thermally evolved and show signs of recent volcanic activity (Earth and Venus), whereas smaller bodies have ancient surfaces unmodified by recent volcanism (Mimas, Oberon). The correlation between size and evolution implies that these bodies derive their internal heat energy from the decay of radioactive elements and that their rates of heat loss are controlled by the surface area to mass ratio. However, several moons of the outer solar system diverge strongly from this trend, showing signs of a high degree of thermal evolution in spite of their small sizes (Io and Enceladus). These bodies appear to have a size-independent source of energy related to tidal heating.

mal state, shown along the back of the diagram, ranges from cold and inactive planetary bodies to those that are still hot and experiencing tectonic deformation and volcanism. The *size* of the planet is depicted along the remaining axis with small bodies at the back of the diagram and large ones at the front. Within the diagram are the positions of various planetary bodies that have solid surfaces. These three parameters describe the most important aspects of a planet's available energy and the matter acted upon.

In general, larger bodies are more thermally evolved and show signs of recent tectonic and volcanic activity (Earth and Venus), whereas smaller bodies have ancient surfaces unmodified by recent tectonism or volcanism (Mimas and Oberon). For example, contrast the history of Mars with the other inner planets. Mars has a diameter about half that of Earth or Venus, but it is almost half again larger than Mercury and about twice as large as the Moon. It has thus retained its internal heat longer and has been more tectonically and volcanically active than the Moon and Mercury but less than Earth and Venus. It is believed that Mars started much like Venus and Earth, with a cratered surface, an atmosphere, and active volcanism. However, because of its smaller size and more rapid loss of heat, the volcanic and tectonic activity of Mars was limited by comparison with Earth. Large lithospheric domes and rifts developed over mantle plumes, but lithospheric plates did not develop, shift, and recycle as on Earth.

The correlation between size and evolution of the planets implies that the rates of heat loss on many planets and satellites are controlled by their surface area to mass ratios. Consequently, an important source of heat must be the decay of radioactive elements—a quantity related to the mass and total composition of the planet. However, several moons of the outer solar system diverge strongly from this trend of increasing tectonic evolution and size. Io, Europa, and Enceladus show signs of a high degree of thermal evolution in spite of their small sizes. These bodies appear to have a source of energy independent of size—a source of energy related to tidal interaction in their respective satellite systems and the creation of heat by friction as they are gravitationally flexed along their slightly elliptical orbital paths.

In brief, size, composition, and energy content control a planet's thermal evolution, including the rate of volcanism, the extent of atmospheric degassing, the intensity of internal differentiation, and the rate at which the lithosphere thickens. Lithospheric thickness, density, and strength control volcanic and tectonic processes on a planet, how and when rifting may occur, if mountain belts or volcanoes can form, and if recycling of the lithosphere back into the mantle is possible. As a result, many geologic differences among the planets can be explained by (1) different amounts of internal energy, as expressed by the rate of heating or cooling, and (2) different kinds of planetary materials, including silicates, ices, liquid water, and atmospheric gases. These differences drove the fascinating histories of the moons and planets of the solar system along unique paths.

Geologic Processes on the Planets

With these basic ideas about planetary matter and energy in mind, let us consider the important geologic processes that have shaped the planets— impact cratering, internal differentiation, volcanism, tectonism, and the flow of surface fluids. The characteristics of each geologic process and the landforms and rocks they create depend on the nature of the energy and of the material that this energy acts upon.

Impact Cratering

Cratering by meteorite or comet impact has been one of the most pervasive geologic process in the solar system. Indeed, the planets grew to their present sizes by collisional accretion from smaller bodies. Most of the planetary bodies we have studied retain an imprint of an early episode of intense bombardment, which declined rapidly during the first several hundred million years of the solar system's history (4.6 to 3.9 billion years ago). Mercury and the Moon are excellent examples and remain as fossils of this early stage in planetary development. They provide a valuable record of this first chapter in the history of the solar system. Most of the small, icy bodies in the outer solar system are also dominated by impact craters. Besides this similarity, a tremendous diversity in crater shapes and sizes is apparent.

Impact cratering is the result of the transfer of a projectile's kinetic energy to a planet's surface. As energy is transferred, a crater is formed whose features depend not only on the amount of energy the projectile had but also on the composition and nature of the surface materials. As a result, impact craters do not all look the same; large craters are distinctly different from small craters in their general geometry. Small craters are bowllike with smooth walls and floors; larger, complex craters develop terraced walls and central peaks; and still larger basins develop multiple concentric rings.

The morphologic changes that accompany changes in crater size are reflections of the amount of energy expended by a projectile as it collides with a planet's surface—energy that is proportional to the mass and velocity of the falling projectile.

Changes in crater features are also affected by the manner in which the target materials respond to the passage of the shock wave and subsequent relaxation to normal temperatures and pressures. Thus, a projectile with a small size or low velocity excavates a simple depression, scattering the ejected debris around the crater rim. At higher energies, a deeper crater is excavated, which enhances the likelihood for crustal rebound to form central peaks or rings. Simultaneously, failure occurs at the crater rim to form terraces and outer rings. The diameter of the crater at which these features appear on different planets varies with the size or strength of the gravitational field of the body (Figure 15.5). Craters up to 100 km in diameter on Amalthea and 30 to 40 km on Mimas do not have central peaks; on Earth, in contrast, the appearance of central peaks may occur at diameters as small as 3 to 5 km. Apparently, the process of rebound is accentuated on planets with large sizes and gravitational forces.

In other details as well, the response to impact cratering stems from the physical and chemical characteristics of a planet's surface. For example, the production of large volumes of impact melt is facilitated on icy bodies with their low temperatures of melting and is hindered on silicate crusts because of their higher melting temperatures.

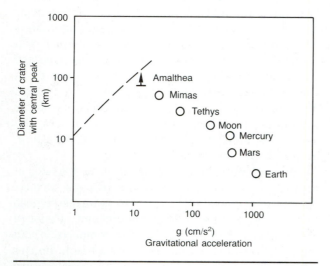

Figure 15.5

The occurrence of central peaks in impact craters depends on the size of the planet or moon and also the energy of the impact. Even small craters on the relatively large Earth develop central peaks, whereas the largest craters seen on Amalthea lack central peaks.

Groundwater also greatly affects crater morphology. As we saw in Chapter 6, the water-rich regolith on Mars creates ejecta that flows like mud, rather than arching away in ballistic trajectories. In addition, the longer-term response of a surface is affected by the planet's chemical and physical properties, especially its strength or viscosity, which are measures of a material's ability to flow. Simple bowl-shaped craters on icy satellites of the outer planets are systematically 20 to 40 percent shallower than on the rocky terrestrial planets. Apparently the rebound of crater floors is easier in ice than silicate rock. The low-strength icy lithosphere of Ganymede, for example, allowed crater excavations to become almost totally erased by viscous flow of the ice. Stronger silicate lithospheres will support larger craters without relaxing, but (as we suggested for the early history of Mercury or present-day Venus) when silicate lithospheres are warm and weak, they too may respond in a similar viscous fashion.

The major role of *large impacts* in the evolution of the planets has become appreciated only recently. Several important planetary features believed to be the result of giant impacts serve to illustrate this idea. For example, the high density of Mercury may be the result of a giant impact that stripped away parts of the outer silicate layers of the already differentiated planet, leaving it enriched in the dense iron that had drained to form the core. A late impact on Venus with a Mars-sized object may have slowed its spin and reversed its rotation direction compared to that of all other planets. Moreover, without large collisions, all of the planets should have spin axes that are perpendicular to the ecliptic. A large collision may, therefore, be required to explain the tilt of both Earth and Mars. The origin of the Moon may be rooted in the collision of a Mars-sized object with Earth. Fragments of this collision later re-accreted in Earth orbit to form the Moon. The global dichotomy on Mars can be traced back to a giant impact basin in the northern hemisphere. The shapes of the asteroids are the result of fragmentation during large impacts. Perhaps the small, icy satellites of Saturn and Uranus were fragmented several times only to re-accrete later. Moreover, the rings that encircle the outer planets may be created by the collisional fragmentation of small, icy moons. At the very least, several satellites sustained massive impacts that created global fracture systems and large craters. In addition, planetary collisions with large bodies in the outer solar system may have tipped Uranus on its side and fragmented Pluto to form a double-planet system. Collisions with large bodies late in the accretion histories of the planets can be understood by remembering that just as the planets grew larger by accretion, so did the impacting planetesimals. Therefore, as the planets approached their final sizes, they were probably struck by larger and larger planetesimals.

Impact cratering has been important over the entire 4.6-billion-year history of the solar system. But has cratering influenced life on Earth? Many scientists think so. An impact that occurred 65 million years ago, at the end of the Cretaceous Period, may have caused a mass extinction of many forms of Earth life, including dinosaurs. Others suggest that throughout much of geologic time, periodic extinctions have been caused by impacts on Earth. Even today the human species is not isolated from its cosmic environment and may be vulnerable to the catastrophic effects of large impact events.

Internal Differentiation

Planetary differentiation is another fundamental geologic process. The interiors of all of the planets and most of their satellites are layered as the result of thermally and gravitationally driven separations of elements with distinctive chemical affinities. The magnitude and significance of this geologic process are dramatically displayed in the many variations of composition and internal structure found in the planets. Figure 15.2 shows this diversity and emphasizes other basic themes as well.

The planets, larger satellites, and even many asteroids became differentiated during their very early histories. The most important source of thermal energy was probably accretionary heating, caused by the conversion of kinetic energy of a meteorite or comet to thermal energy stored in the planet. The extent of differentiation as a result of accretionary heating is critically dependent on how quickly a planet accretes. Planets that accrete quickly store the heat in the growing mass; those that accrete slowly radiate the impact-created heat before it can be buried by subsequent additions of matter. Other important heat sources may have been the decay of short-lived radioactive isotopes and energy released by differentiation itself. Nonetheless, small bodies, such as the icy moons of Saturn, Uranus, and Neptune, may not have acquired enough energy during accretion for complete internal differentiation. Most of the moons in the outer solar system are made largely of water ice with smaller amounts of silicate and metal; the interiors of these bodies may consist of icy rinds surrounding undifferentiated mixtures of rock and ice. In Figure 15.2, the cores shown at the center of the moons of the outer planets represent only one possible configuration and are shown to emphasize the proportions, not exact distributions, of rock and ice in the smaller bodies.

Moreover, heat, derived from accretion and from the decay of radioactive elements in the rocky fractions of the small icy satellites, was rapidly transmitted to their surfaces and radiated to space. Consequently, most of these moons lacked sufficient internal energy to generate substantial tectonic systems that could change their surfaces. Instead, they retain densely cratered surfaces, like those of Mercury and the Moon, and record only ancient events because of their rapid thermal evolutions. Callisto, the outermost Galilean moon of Jupiter, may be considered typical of this class of satellites. Although each of the moons of the outer solar system has a distinctive geologic history and surface features, many can be grouped with Callisto because their rapid thermal evolutions soon terminated their internal dynamics. Their surfaces have changed little since the period of intense bombardment. This class of satellites includes Mimas, Tethys, Dione, Rhea, and Iapetus (moons of Saturn); Oberon, Titania, Umbriel, and Ariel (moons of Uranus); and Proteus and Nereid (moons of Neptune).

Volcanism

Most of the solid planets and their moons experienced volcanism during their histories. Volcanic landforms can be as dramatic as the giant volcanoes on Mars and the large ash-flow calderas of Earth and Io or as subtle as the smooth plains of Enceladus that apparently lack features related to vents. In addition, the composition of the volcanic materials varies from planet to planet, depending on the planet's composition. Eruptions may consist of lavas of molten silicates, such as those on Earth, Venus, Mars, the Moon, some asteroids, and probably Mercury and Io; or they may be volcanic flows of molten water mixed with other volatiles, such as those on Ganymede, Europa, Ariel, Triton, and many others; or they may even be exotic lavas made of molten sulfur as on Io. In spite of these variations, an important generalization is that volcanic activity is a sensitive indicator of the mechanism of heat loss and the thermal state of the interior of a planet. Thus, when the rate of volcanism is compared with the age of a planet, it usually shows the progressive cooling of the planet. Moreover, most small bodies have short histories of volcanism.

Mercury and the Moon, for example, were too small to retain much internal energy. They heated initially and became differentiated with metal cores, and rocky mantles and crusts. Oceans of magma may have entirely enveloped these bodies for a short time. Internal heat also produced short episodes of volcanic activity, during which lava was extruded and covered some early formed, densely cratered terrain. The Moon's lithosphere thickened rapidly during its early history; presently it is 1000 km thick—ten times thicker than Earth's. This thick rigid shell makes it nearly impossible for molten lava to reach the surface and prohibits lateral movements like those that produce continental drift on Earth. Mercury is similar to the Moon in many ways. A period of extensive volcanism is recorded on Mercury's surface, but like the Moon, volcanic activity stopped early in its history. A rigid lithosphere must have developed well before the end of the period of intense bombardment. Thus, on Mercury and the Moon there has been little or no geologic activity after the volcanic event that produced floods of lava several billion years ago.

In the outer solar system, Io and Europa are notable exceptions to these generalities about small planetary bodies in that each has a prolonged volcanic history resulting from a unique source of internal energy. Io, the innermost moon of Jupiter, is a small body that is tremendously hot on the inside but frigid on the outside. Ordinarily, because of its small size, it would be expected to be a cold, dead world like the Moon, but internal heat is generated in Io because of the tidal actions of Jupiter and its large satellites. As a result, Io is the most volcanically active body in the solar system and has a very young surface that is continually being resurfaced by volcanism. Water and other volatiles lighter than sulfur, which were extruded during its prolonged volcanic history, have escaped into space because Io's small gravitational field could not hold them. Sulfur, on the other hand, is too heavy to escape Io's gravitational field and has been concentrated on the surface. Many scientists believe that sulfur functions on Io much like water does on Earth—melting and erupting like geysers. Deeper sources may produce molten silicate lava flows. The volcanic history of Io, created by geologic processes unknown before the space program, is dominated by its distinctive energy source and the distinctive composition of its surface materials.

Europa is smaller and farther from Jupiter than Io. Europa's almost perfectly smooth icy surface is marred with sets of tan streaks, which are similar to fracture systems in sea ice in the polar regions of Earth. Europa's internal heat produced magmas of slushy water that extruded through fissures in the crust and coated the surface with fresh ice. The process must be similar to fissure eruptions of silicate lava on the Moon, Mercury, Mars, and Earth; but on Europa the lava is water. The near absence of impact craters on Europa indicates that the surface is very young, and the resurfacing processes of water eruptions have continued up to quite recent times. As with Io, much of Europa's internal energy is derived by tidal flexing of the planet.

Not only are volcanic histories different from planet to planet, but the volcanoes themselves are strikingly different. The various types of volcanoes reflect the style of eruption, volume of magma erupted, composition of the magma, and the characteristics of the path taken by the magma to the surface. The style of eruption varies from quiet eruptions of lava (the small eruptions of martian low shields) to violent gas-driven explosions that devastate large areas (the ash-flow calderas of Earth's continents). The volume of magma in a single eruption or in a group of related eruptions also helps to shape a volcano's features and, just as important, reflects the amount of energy inside a planet. Individual eruptions range from fractions of a cubic kilometer to huge eruptions that may pour thousands of cubic kilometers of magma across a planet's surface in just a few days. Small cinder cones, low lava shields, great shield volcanoes with collapse calderas, sheets of flood lavas, and huge ash-flow shields result partly from variations in the supply of magma. Magma composition controls the amount of volatiles and the way the magma flows or fragments. Eruptions of water, like those that occurred on the icy outer satellites, produce vast thin sheets; silicate magmas may form similar large sheets but also may pile up around their vents to produce cinder cones, shield volcanoes, or other types of edifices. Finally, the way that magma rises to the surface also affects the final appearance of a volcano. A central, long-lived conduit may produce a large shield volcano, like Olympus Mons on Mars, or a stratovolcano, like Mt. St. Helens on Earth, whereas magma sources that are spread over a large area create fields of small isolated volcanoes with short lives, like those on Earth's Snake River Plain.

Fundamental controls on volcanic activity then are (1) a source of heat to partially melt the interior of a planet, (2) the composition of the magma source, (3) a pathway to the surface, and (4) the nature of the vent and its environment. In the simplest cases, volcanic processes are driven by thermal energy that, if it increases locally, causes partial melting of rock or ice. The amount of liquid depends on the amount of heat and the composition of the rock. If the liquid produced is less dense than its surroundings, then it can rise to the surface and erupt. If it is more dense, as happens with liquid iron in a silicate planet, it may sink toward the core of the planet. Thus, volcanic processes are part of the overall differentiation of a planet.

Tectonism

The large-scale deformation that produces the structure or architecture of a planet's lithosphere is the result of tectonism. Here again, internal heat is the driving force, but the response (e.g., flowing, buckling, folding, fracturing) of the crust depends on its strength. The strength of crustal rocks, in turn, depends mostly on their composition and temperature. Gravitational potential energy is also important in shaping some tectonic features (e.g., folded mountains of Venus and crater palimpsests on Ganymede).

As we have emphasized, the tectonic history of a planet is strongly related to its thermal history. Planetary lithospheres that are cool, rigid, and thick do not yield to tectonic stresses because of their strength. In contrast, warm-and-ductile or thin-and-brittle lithospheres are weak, and they fold or fracture when stresses are sufficient. Perhaps the simplest forms of tectonics are those of global extent that result from planetary heating and expansion or cooling and contraction. The global pattern of thrust-fault systems on Mercury is an example. The Moon is another; it cooled rapidly to develop a thick lithosphere. As it cooled, it contracted and its surface was deformed as a result. Subsequently, the lithosphere has not deformed for almost 3 billion years. In this way, the Moon is similar to many other small planets composed of ices or silicates. The only apparent tectonism, not related to impact, is the result of planetary or local expansion or contraction that occurred long ago. In contrast, the surfaces of Earth, Venus, Io, Europa, Ganymede, Enceladus, and Triton have been dramatically reshaped by tectonic processes.

Ganymede is an example of the tectonically more-complex bodies; it has a thick outer shell of water ice, but its surface is unique in that it consists of a baffling array of structural features unlike any in the solar system. Many features appear to result from breaking and lateral movement of crustal fragments, a system similar in some respects to the tectonic plates on Earth. This has produced two distinct terrain types. The older is dark and is nearly saturated with craters, but it has been fractured and split apart and many fragments have shifted about. Found between these fragments is the younger terrain, brighter, smoother,

and probably created by eruptions of water. It is crisscrossed by a series of complex grooves and stripes, features that result from deformation and cracking of an icy crust.

Enceladus, a tiny, icy moon of Saturn, also has a complex, grooved terrain similar in some ways to the terrain on Ganymede, showing unexpected tectonism on such a small icy body. It is clear from the large tracts of very smooth, crater-free plains that tectonism on Enceladus must be very young. Some source of heat is necessary to soften the interior of Enceladus and cause the volcanic activity that resurfaced the satellite to create the smooth plains. What type of energy activates the surface of this tiny ball of ice? Can tidal heating be sufficient to warm it and produce crustal deformation and volcanic activity?

A convenient measure of the duration and intensity of active tectonism on a planet's surface is the proportion of the surface that is not heavily cratered. All of the planets experienced an early bombardment by meteoritic projectiles, but only those with sufficient internal energy have been resurfaced by destruction and burial of the cratered terrains (Figure 15.6). For example, some small, icy moons of Jupiter, Saturn, Uranus, and Neptune (e.g., Callisto, Rhea, Mimas, and Umbriel) and, probably, the asteroids have surfaces un-

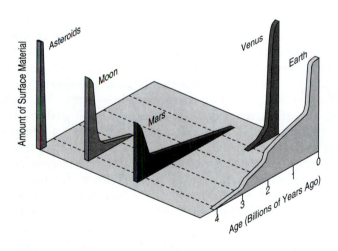

Figure 15.6

The ages of rocks and surfaces on the inner planets show variations that reflect the amount of internal energy available to drive geologic activity such as volcanism and tectonism. Most small bodies, such as asteroids, have very old ages because they lost their heat shortly after they formed. Their geologic features are dominated by impact craters formed during the intense bombardment that tailed off about 3.8 billion years ago. Larger planets, such as Venus and Earth, have rocks and surface features that formed after the period of intense bombardment.

marked by tectonic or volcanic processes younger than the early intense bombardment. Other planetary bodies, both silicate and icy, such as Mercury, the Moon, Tethys, and Dione, have surfaces dominated by heavily cratered terrain with only small areas buried by younger volcanic flows, cut by rifts or folded to form ridges. Mars, Ganymede, and Enceladus have lost about 50 percent of their ancient cratered terrain to tectonic processes and have an intermediate status. The planets with the most sustained thermal evolution—Earth, Venus, Io, and Europa—have been completely resurfaced many times and retain no heavily cratered regions.

The Role of Surface Fluids

The geologic processes related to the movement of fluids on the surface of a planet can completely resurface a planet many times. These processes derive their energy from the Sun and the gravitational forces of the planet itself. As these fluids interact with surface materials, they move particles about or react chemically with them to modify or produce new materials. The energy, the nature of the fluid, and the composition and environment (temperature and pressure) of the surface materials determine the patterns of modification.

On solid planets with atmospheres and hydrospheres only tiny fractions of their masses flow as surface fluids. Yet the movements of these fluids have dramatically altered their appearances. To emphasize the importance of surface fluids, let us consider Venus, Earth, and Mars—the terrestrial planets that have atmospheres.

Venus and Earth are commonly considered twins but not identical twins. They are about the same size, composed of roughly the same mix of materials, and may have been comparably endowed with carbon dioxide and water. However, the twins evolved differently, largely because of their distance from the Sun and the resultant differences in solar energy that they receive. With a significant amount of internal heat, Venus may continue to be geologically active with volcanoes, rifting, and folding. However, it lacks any sign of a hydrologic system: there are no streams, lakes, oceans, or glaciers. Space probes suggest that Venus may have started with as much water as Earth but was unable to keep its water in liquid form. Receiving more heat from the Sun, water, outgassed from the interior, evaporated and rose to the upper atmosphere where the Sun's ultraviolet rays broke the molecules apart. Much of the freed hydrogen escaped into space and Venus lost its water. Without water, Venus became less and less like Earth and kept an atmosphere filled with carbon dioxide. The carbon dioxide acts as a blanket creating an intense greenhouse effect and driving surface temperatures high enough to melt lead and to prohibit the formation of carbonate minerals. Volcanoes continually vented more carbon dioxide into the atmosphere. On Earth, liquid water removes carbon dioxide from the atmosphere and combines it with calcium, from rock weathering, to form carbonate sedimentary rocks. Without liquid water to remove carbon from the atmosphere, the level of carbon dioxide in the atmosphere of Venus remains high. During the early years of the solar system, temperatures at the surface of Venus were only slightly warmer than those on Earth. Can a few tens of degrees make such a difference? We must understand the answer to this question in order not to stress the environment on Earth beyond its limits. Moreover, what role does water play in the internal dynamics of a planet? Is the apparent lack of continents and plate tectonics on Venus a result of its water-poor interior and high surface temperatures?

Like Venus, Earth is large enough to be tectonically active and for its gravitational field to hold an atmosphere. Unlike Venus, it is just the right distance from the Sun so that temperature ranges allow water to exist as a liquid, a solid, and a gas. Water is thus extremely mobile and moves rapidly over the planet in a continuous hydrologic cycle. Heated by the Sun, the water moves in great cycles from the oceans to the atmosphere, over the landscape in river systems, and ultimately back to the oceans. As a result, Earth's surface has been continually changed and eroded into delicate systems of river valleys—a remarkable contrast to other planetary bodies where impact craters dominate. Few areas on Earth have been untouched by flowing water. As a result, river valleys are the dominant feature of its landscape. Similarly, wind action has scoured fine particles away from large areas, depositing them elsewhere in sheets of loess or as vast sand seas dominated by dunes. These fluid movements are caused by gravity flow systems energized by heat from the Sun. Other geologic changes occur when the gases in the atmosphere or water react with rocks at the surface to form new chemical compounds with different properties. An important example of this process was the removal of most of Earth's carbon dioxide from its atmosphere to form waterlaid carbonate rocks that now form extensive layers covering the stable platforms or folded into mountain belts. However, if Earth were a little closer to the Sun, our oceans would evaporate; if it were farther from the Sun, the oceans would freeze solid. Because liquid water was present, self-replicating molecules of carbon, hydrogen, and oxygen developed life early in Earth's history and have radically modified its surface, blanketing huge parts of the continents with greenery. Life thrives on this planet and helped create its oxygen- and nitrogen-rich atmosphere and moderate temperatures. Indeed, if alien scientists were to study Earth with a passing spacecraft, the composition of its unique atmosphere would be one of the most important evidences of the existence of life.

Although Mars has only a thin atmosphere, the effects of moving surface fluids are abundant. Distinctive wind-formed streaks and plumes cross its surface. A vast sea of sand dunes encircles its north polar cap. Sheets of loess may be common. The surface of Mars reveals river channels, large and small, produced by flowing water. These channels seem to require that water participated in some type of hydrologic cycle over a long period. Mars may have developed a dense atmosphere very early in its history and possessed liquid water and a moderate climate warmed by a carbon dioxide greenhouse. Rainfall and flooding may have produced significant erosion at this time. But Mars is smaller than Earth, has less internal heat, and may not have outgassed as much water or carbon dioxide. In any case, eventually atmospheric pressure and temperature dropped and the water froze; Mars was left cold and dry. Today, great dust storms rage on Mars altering its surface and atmospheric temperatures. Earth's atmosphere acts not only as a window for sunlight but also as a blanket for heat. Unlike carbon dioxide, which thickens the blanket and creates a warming greenhouse, dust thrown into the atmosphere closes the window, blocking solar energy from reaching the surface below. Mars, therefore, remains cold, with all its water locked up in its ice caps or frozen in the pore spaces of its rocks and soil. Important questions persist about the evolution of the fluid envelope around Mars. Under what conditions did liquid water flow in small rivulets or great floods across the surface? Did life evolve when surface water was more abundant? What role do dust storms play in atmospheric conditions and surface temperatures of the planet?

A comparison of the terrestrial planets has awakened our appreciation for the susceptibility of a planetary surface to large environmental changes. Perhaps, if nothing more, our studies of the diversity of compositions and conditions of solar system bodies should remind us of the delicate balance of energy and evolution that allows us to exist at all.

Extreme examples of fluid- or gas-rich planets are the giant outer planets—Jupiter, Saturn, Ura-

nus, and Neptune—which present only the tops of their thick colorful atmospheres to the view of telescopic observers and passing spacecraft. At the other end of the spectrum lie most of the solar system's planetary bodies that lack atmospheres and the consequent movement of fluids. None of the planetary satellites of the inner solar system, the asteroids, or Mercury have significant atmospheres. There are at least three reasons why these bodies lack atmospheres. Some were poor in volatile elements to begin with (Mercury and the Moon); others were too small to hang on to released gas (some asteroids); and still others were too small to differentiate and release volatiles from their interiors (Phobos, Deimos, and some asteroids). Without atmospheres and hydrospheres, their surfaces are modified only by impact today.

Small planetary bodies in the outer solar system also lack atmospheres and surface features resulting from the flow of fluids. Here we specifically exclude water magmas because of their high temperatures compared with the surface temperature. (Exceptions are Titan with its nitrogen atmosphere, which holds the possibility of supporting a liquid-gas methane cycle, and the extremely tenuous atmospheres of Triton and Pluto.) The absence of fluids on the surfaces of these planets is not because they are poor in volatiles or that they are too small to have become differentiated. They are quite simply too cold. Water is an important constituent on these bodies but was released to their surfaces and frozen to form icy shells. The energy from the Sun that reaches these bodies is insufficient to maintain water in its molten, erosive state; in a solid form, water ice behaves like rock. Another complicating factor is the generally small size of these bodies and their resultant inability to gravitationally retain light volatile substances such as hydrogen, helium, or nitrogen gas.

Concluding Remarks

By studying other planetary bodies, we gain a greater understanding of our own Earth. Its size and composition are just right for the development of a tectonic system that even today, 4.6 billion years after Earth formed, recycles the lithosphere, creates continents and ocean basins, and concentrates ores and minerals. Earth appears to have accreted some solids that condensed at low enough temperatures to contain significant amounts of water. During its differentiation, some of this water was released from the solids and accumulated at the surface. Earth's gravitational field was strong enough to hold this water and an atmosphere, which was also extruded from its interior. Earth is just the right distance from the Sun so that water at the surface can exist as solid, liquid, and vapor and can move in a hydrologic cycle. If Earth were a little closer to the Sun, the oceans would evaporate; if farther from the Sun, they would freeze solid. Because liquid water is present, huge volumes of carbon dioxide were removed from Earth's atmosphere and concentrated in layers of rock. As a result, greenhouse warming of Earth is only moderate compared to the inferno present on Venus.

Another result of Earth's composition and distance from the Sun was the evolution of life. Life, and especially human life, is not a passive presence on the planet; life modifies and drives a variety of geologic processes on Earth. For example, the evolution of plants enriched the atmosphere in free oxygen and allowed a radiation-filtering ozone layer to form in the upper atmosphere. Billions of years later, humans are upsetting this delicate balance by infusing the atmosphere with carbon dioxide created by burning fossil fuels and risking greenhouse warming. Moreover, we have created complex compounds that are destroying the protective ozone layer. The studies of other planets have taught us that Earth is a small place, an oasis in space, a home that we are still trying to understand. By exploring the worlds in our solar neighborhood, we are beginning to understand how Earth works and why it is unique, why we have an atmosphere, moderate climates, continents, and ocean basins, and why we alone have life. Will the intelligence that allowed us to explore the planets and begin to understand them give us the ability to live within the limits of our own natural system?

Additional Reading

Allegre, C. J. 1992. *From Stone to Star*. Cambridge, MA: Harvard University Press.

Beatty, J. K., B. O'Leary, and A. Chaikin (eds). 1990. *The New Solar System*. London: Cambridge University Press.

Broecker, W. S. 1987. *How to Build a Habitable Planet*. Palisades, NY: Eldigo Press.

Consolmagno, G. H., and M. W. Schaefer. 1994. *Worlds Apart: A Textbook in Planetary Science*. Englewood Cliffs, NJ: Prentice-Hall.

Greeley, R. 1994. *Planetary Landscapes*. 2nd edition. London: Allen and Unwin.

Hartmann, W. K. 1993. *Moons and Planets*. 3rd edition. Belmont, CA: Wadsworth Publishing Company.

Morrison, D. 1993. *Exploring Planetary Worlds*. New York: Scientific American Library.

Taylor, S. R. 1992. *Solar System Evolution: A New Perspective*. Cambridge, England: Cambridge University Press.

GLOSSARY

A

abrasion The mechanical wearing away of a rock by friction, rubbing, scraping, or grinding.

absolute age Age of a rock or surface measured in Earth years in contrast to relative time, which involves only the sequence of events.

accretion The growth of planets from smaller bodies by collisions. In our solar system this occurred over 4.5 billion years ago.

achondrite A stony meteorite that lacks chondrules. Some have igneous textures and compositions and are similar to basalts: they may be derived from lava flows on the surface of asteroids.

albedo The brightness of an object or planetary surface. Bright objects have high albedos, and dark ones have low albedos. The albedo is a measure of the reflectance of an object.

alluvial fan A fan-shaped deposit of sediment built by a stream, where it emerges from an upland into a broad valley or plain. Alluvial fans are common in arid and semiarid climates on Earth and on Mars.

alpha particle The nucleus of a helium atom, consisting of two neutrons and two protons.

amphibole An important group of iron- and magnesium-rich silicates. Amphibole crystals are constructed from double chains of silicon–oxygen tetrahedra.

andesite A fine-grained igneous rock composed mostly of plagioclase feldspar with mafic minerals. Andesite has an intermediate silica content and is commonly erupted above subduction zones on Earth but is not restricted to these settings. Stratovolcanoes commonly erupt andesite.

anomaly A deviation from the norm or average.

anorthosite A coarse-grained intrusive igneous rock composed primarily of calcium-rich plagio-clase. Anorthosite is an important rock type in the lunar highlands.

anticline A fold in which the limbs dip away from the axis.

antipode A point on the surface of a planet, or sphere, exactly opposite another point.

aphanitic texture A rock texture in which individual crystals are too small to be identified without the aid of a microscope.

aquifer A permeable horizon or zone below the surface of a planet through which groundwater moves.

ash fall An accumulation of volcanic fragments erupted explosively that fall to the ground without subsequent flow.

ash flow A turbulent blend of unsorted pyroclastic fragments mixed with hot gases erupted explosively from a fissure or crater.

ash-flow shield A large shieldlike volcano composed of ash-flow tuff that usually surrounds a central collapse caldera. The caldera may be up to 100 km across. Highly silicic magmas, such as rhyolites, commonly erupt from such volcanoes on Earth. Such volcanoes are found on Earth, Io, and perhaps Mars.

asteroid A small, rocky planetary body orbiting the Sun. Asteroids are numbered in the tens of thousands. Most are located between the orbit of Mars and the orbit of Jupiter. Their diameters range downward from 1000 km.

asteroid belt The region between the orbits of Mars and Jupiter where most asteroids are found.

asthenosphere The zone in a planet directly below its lithosphere that is fluid enough to convect. This material may be partially molten.

astronomical unit (AU) The mean distance from the Earth to the Sun—approximately 150 million km.

atmosphere   The mixture of gases surrounding a planet or moon. Atmospheres may be outgassed from the interior of a planet, or they may have been trapped gravitationally from the nebula out of which the planets formed.

atom   The smallest unit of an element. Atoms are composed of protons, neutrons, and electrons.

axis   An imaginary straight line about which a body rotates.

B

ballistic trajectory   The path followed by an unpowered projectile; typically a high arch from the launching or ejection point to the point where it collides with the surface of the planet.

barchan dune   A crescent-shaped dune where the tips or horns point downwind. Barchan dunes form in deserts where sand is scarce.

basalt   A dark-colored, fine-grained igneous rock composed of plagioclase feldspar and pyroxene. Olivine may or may not be present. Basalt is the most common volcanic rock on the inner planets.

basaltic plains   Volcanic regions with many small low-shields and fissure eruptions of basaltic lavas; they are common on the inner planets.

basin, impact   A large impact crater, usually taken to be larger than 200 km in diameter. Most impact basins have multiple rings forming their perimeters.

basin, structural   A circular or elliptical down-warp in a planet's lithosphere.

batholith   A large body of intrusive igneous rock exposed over an area of at least 100 km^2.

bed   A layer of rock 1 cm or more in thickness.

biosphere   The totality of life on or near Earth's surface. No other planet in the solar system is known to have a biosphere.

block faulting   A type of normal faulting in which segments of the crust of a planet are broken and displaced to different elevations and orientations.

blowout   A small depression excavated by wind erosion.

braided stream   A stream with a complex of converging and diverging channels separated by bars or islands. Braided streams form where more sediment is available than can be removed by the discharge of the stream.

breccia   A general term for rock consisting of angular fragments in a matrix of finer particles.

Examples include sedimentary breccias, volcanic breccias, and impact breccias.

butte   A somewhat isolated hill, usually capped with a resistant layer of rocks and bordered by fragments. A butte is an erosional remnant of a formerly more-extensive slope.

C

calcite   A mineral composed of calcium carbonate ($CaCO_3$).

caldera   A large, more-or-less circular depression or basin associated with a volcanic vent. Its diameter is many times greater than that of the included vents. Calderas are believed to result from subsidence or collapse and are associated with many types of volcanoes, including shield volcanoes and ash-flow shields.

carbonaceous   Containing carbon other than as carbonate.

carbonate mineral   A mineral formed by the bonding of carbonate ions (CO_3) with ions such as calcium (Ca) or magnesium (Mg).

carbonate rock   A rock composed mostly of carbonate minerals; limestone is an example.

cement   Minerals precipitated from groundwater in the pore spaces of a sedimentary rock and binding the rock's fragments together.

central peak   A hill or group of mounds located in the center of moderately sized impact craters.

chondritic meteorite   A meteorite with chondrules, or a meteorite that is chemically similar to those that do.

chondrule   Spherical or nearly spherical bodies found in some stony meteorites. They range up to 10 mm in diameter and consist of silicate minerals.

cinder   A fragment of volcanic ejecta from 0.5 to 2.5 cm in diameter.

cinder cone   A cone-shaped hill composed of loose volcanic fragments surrounding a central vent. Most are composed of basalt.

clastic   Pertaining to fragments (such as mud, sand, and gravel) produced by the mechanical breakdown of rocks, or describing the rocks formed chiefly of consolidated fragments.

coma   The diffuse, fuzzy head of a comet that consists of gaseous material and surrounds the dense nucleus. The coma is created as a comet nears the Sun and its ice vaporizes from the heat.

comet   A relatively small, icy body that orbits the Sun. The orbits of most comets are highly eccentric. When a comet is in the part of its orbit near the Sun, it develops a coma and a long tail. Comets are thought to have formed in the outer solar system.

composite volcano   A moderate-size volcanic cone built by extrusion of ash and lava. Synonymous with **stratovolcano.**

compound   A substance composed of two or more chemical elements.

compression   A system of stresses that tends to reduce the volume of or shorten a substance.

condensation   A chemical reaction in which a solid crystallizes or a liquid precipitates from a gas.

conduction   The transfer of heat energy through solids by molecular impact without movement of the material itself.

continent   A part of Earth's crust, from 20 to 60 km thick, composed mostly of granitic rock. Continents rise abruptly above the oceanic crust because of their relatively low densities.

convection   The transfer of heat energy by movement of material. Fluids convect as a result of density differences produced by heating and consequent expansion, cooling, and contraction.

core   The central part of a differentiated planet, usually composed of denser material. The inner planets have cores of iron in a liquid or solid state. The outer planets have cores of silicates and metal. The icy satellites may also have cores of silicates.

core refrigeration   A process that takes place in the central parts of large stars that leads to collapse and nova or supernova explosions.

corona   Elliptical, strongly deformed terrains found on Venus and Miranda.

crater frequency   The number of impact craters of a certain size per unit area; high crater frequencies indicate old surfaces, and low crater frequencies indicate young surfaces.

crater, impact   A circular depression created by impact of a meteorite or comet; they are usually surrounded by an ejecta blanket.

crater, volcanic   An abrupt more-or-less circular depression formed by extrusion of volcanic material or by collapse of part of a volcano.

cross-cutting relations, principle of   The principle that a rock body is younger than any rock across which it cuts.

crust   The outermost layer, or shell, of a differentiated planet. The crust is defined on the basis of its chemical composition, not its mechanical properties.

crustal warping   Gentle bending (upwarping or downwarping) of a planet's crust.

crystal   A solid, polyhedral form bounded by naturally formed plane surfaces.

crystallization   The process of crystal growth. It occurs as a result of condensation from a gaseous state, precipitation from a solution, or cooling of a melt.

cyclone   A storm in the atmosphere of a planet with counterclockwise (in the northern hemisphere) circulation of winds.

_____ **D** _____

deflation   Erosion of loose rock particles by the wind.

deflation basin   A shallow depression formed by wind erosion of loose particles.

degradation   The general lowering of the surface of the land by processes of erosion or impact.

delta   A large, roughly triangular body of sediment deposited at the mouth of a river.

dendritic drainage pattern   A branching stream pattern resembling the branching of certain trees.

density   The measure of concentration of matter in a substance: mass per unit volume, expressed in grams per cubic centimeter (g/cm^3).

differentiation, planetary   The process by which the materials in a planetary body are separated according to density and chemical affinity so that an originally homogeneous body is converted into a zoned, or layered, body with a dense core, a mantle, and a crust.

dike   A tabular, intrusive igneous rock that cuts nearly vertically across strata or other surrounding rocks.

distributary   Any of the numerous stream branches into which a river divides where it reaches its delta.

divergent plate boundary   A lithospheric plate boundary formed where the lithosphere splits into plates that drift apart from one another as new lithosphere is created between them. See also **mid-ocean ridge.**

dome, structural   An uplift that is circular or elliptical in map view, with beds dipping away in all directions from a central area.

dome, volcanic   A volcano formed from extremely viscous lavas that takes the shape of a steep-sided dome. Rhyolite domes are usually less than 1 km across on Earth.

drumlin   A smooth, glacially streamlined hill that is elongate in the direction of ice movement. Drumlins are usually composed of unsorted glacial sediments.

dunes   A low mound of fine-grained material that accumulates as a result of sediment transport in a current system on the surface of a planet. Dunes have characteristic geometric forms that are maintained as they migrate.

dust   Very fine-grained material, smaller than sand grains, usually carried by the wind.

————————   **E**   ————————

earthquake   A series of elastic waves propagated in a planet, initiated where stress along a fault exceeds the strength of a rock so that sudden movement occurs along the fault.

eccentric orbit   A noncircular orbit.

ecliptic   The planet Earth and most of the other planets orbit about the Sun.

ejecta   Rock fragments, glass, and other material thrown out of an impact crater or a volcano.

ejecta blanket   Rock material (crushed rock, large blocks, breccia, and dust) ejected from an impact crater and deposited over the surrounding area.

elastic deformation   Temporary deformation of a substance, after which the material returns to its original size and shape.

electron   A very small elementary particle with a negative electrical charge. An electron normally moves about the nucleus of an atom.

energy   The capacity for doing work.

eolian   Pertaining to wind.

equilibrium   A state of balance between opposing chemical or physical forces.

erosion   The processes that loosen sediment and move it from one place to another on a planet's surface. Agents of erosion include water, ice, wind, and gravity.

eruption   The ejection of material from a volcano; material may be erupted as lava, as pyroclastic fragments, or as gas.

escape velocity   The velocity required for an object to escape the gravitational control of another planetary object.

excavation   The stage of impact crater development where material is thrown out from the growing cavity.

extension   A system of stresses such that a body is pulled apart by tensional forces.

extrusive rock   A rock formed from a mass of magma that flowed out on the surface of a planetary body.

————————   **F**   ————————

fault   A surface along which a rock body has broken and been displaced.

feldspar   A mineral group consisting of silicates of aluminum and one or more of the metals potassium, sodium, or calcium. Feldspars are the most common minerals in the crust of Earth, the Moon, and probably other inner planets.

fission, nuclear   The breaking or splitting apart of a nucleus; energy is usually released as a result. Some of the heat-producing decay of uranium occurs by fission.

fissure eruption   Extrusion of lava along an open crack.

flood lava   An extensive flow of lava erupted chiefly from open fissures. Relatively smooth, featureless plains are produced. On the inner planets, these lavas are commonly basaltic; on the moons of the outer planets, they are composed mostly of water.

fluvial   Pertaining to a river or stream.

flux   The number of objects passing through a given area per unit time; for example, the meteorite flux corresponds to the number of meteorites striking a surface in a given time.

fold   A bend, or flexure, in a rock body.

folded mountain belt   A long, linear zone of a planet's crust where rocks have been deformed by horizontal stresses. On Earth, they form at convergent plate boundaries.

footwall   The block beneath a dipping fault surface.

fossil   Naturally preserved remains or evidence of past life, such as bones, shells, casts, impressions, and trails.

frost wedging   The forcing apart of rocks by the expansion of water as it freezes in fractures and pore spaces.

fusion, nuclear   The combination of two or more nuclei to form a different, usually heavier, element; energy is commonly given off as a result. The Sun produces its energy by nuclear fusion.

G

galaxy A group of associated stars, nebulas, star clusters, and planets. Our galaxy contains billions of stars and is shaped like a spiral with a massive central condensation.

geologic column A diagram representing divisions of geologic time and the rock units formed during each major period on a given planetary body.

geologic time scale The time scale determined by the geologic column and by absolute dating of rocks and surfaces on planetary objects.

geothermal gradient The rate at which temperature increases with depth in a planetary body.

geyser A thermal spring that intermittently erupts steam and boiling water.

giant planets The large planets in the outer solar system—Jupiter, Saturn, Uranus, and Neptune.

glacier A mass of ice that is thick enough to flow plastically.

gneiss A coarse-grained metamorphic rock with a characteristic layering resulting from alternating layers of mafic minerals.

graben An elongate fault block that has been lowered in relation to the blocks on either side.

granite A coarse-grained igneous rock composed of alkali feldspar, plagioclase feldspar, and quartz, with smaller amounts of mafic minerals. Granite is rich in silica and is common on Earth's continents.

gravity The tendency for matter to attract itself together.

ground ice Frozen water below the surface of a planetary body; it generally occurs in pore spaces of rocks and soil.

groundwater Liquid water below the surface of a planetary body; it generally occurs in pore spaces of rocks and soil.

H

half-life The time required for half of the radioactive atoms in an object to decay.

hanging wall The block above a dipping fault surface.

head, cometary The nucleus and coma of a comet.

horst An elongate fault block that has been uplifted in relation to the adjacent rocks.

hot spot The expression at the surface of a planet of a mantle plume, or column, of hot, buoyant rock rising in the mantle beneath the lithosphere.

hydrologic system The system of moving water at the surface of a planet.

hydrosphere The waters of a planet, as distinguished from the rocks, the atmosphere, and the biosphere.

hydrostatic pressure The pressure within a fluid (such as water) at rest, exerted on a given point within the body of the fluid.

I

ice Solids formed of volatile materials, particularly water, methane, ammonia, and nitrogen.

ice sheet A thick, extensive body of ice that is not confined to valley. If an ice sheet is thick enough, it may move by glacial flow. Ice sheets or caps form at the poles of several planets and moons.

igneous rock Rock formed by cooling and solidification of molten minerals (magma). Igneous rocks can be intrusive or extrusive and can consist of a variety of materials, such as silicates on the inner planets or water on the icy satellites.

interstitial Pertaining to material in the pore spaces of a rock. Groundwater and ground ice are interstitial materials.

intrusive rock Igneous rock that, while it was fluid, penetrated into or between other rocks and solidified.

ion An atom or combination of atoms that has gained or lost one or more electrons and thus has a net electrical charge.

iron meteorite Meteorite composed mostly of metallic iron. They are thought to have formed as the cores of small differentiated asteroids.

island arc A chain of volcanic islands on Earth formed above a subduction zone.

isostasy A state of equilibrium, resembling flotation, in which segments of a planet's crust stand at levels determined by their thickness and density. Isostatic equilibrium is attained by flow of material in a viscous fashion.

isotope One of the several forms of a chemical element that have the same number of protons in the nucleus but differ in the number of neutrons and thus differ in atomic weight.

J

jovian planet   The giant planets Jupiter, Saturn, Uranus, and Neptune.

K

karst   A landscape characterized by sinks, solution valleys, and other features produced by groundwater activity.

Kelvin   A temperature scale that begins at absolute zero, equivalent to –273 degrees Centigrade or Celsius.

kinetic energy   Energy of motion; the kinetic energy of a moving object is equal to one-half the product of its mass multiplied by the square of its velocity.

KREEP basalt   A type of basaltic rock found on the Moon that has high concentrations of potassium, rare earth elements, and phosphorous thought to be formed from the residue of the Moon's magma ocean.

L

lag deposit   A residual accumulation of coarse fragments that remains on the surface after finer material has been removed by wind.

landform   Any feature of a planet's surface having a distinct shape and origin. Collectively, the landforms of a planet constitute the entire surface configuration.

landslide   A general term for relatively rapid types of mass movement.

latitude   A north–south coordinate on the surface of a planet; it ranges from 0 degrees at the equator to 90 degrees at the pole.

lava   Magma that reaches the surface of a planet and erupts to form a stream of molten rock.

lava tube   A cylindrical opening in the central part of a lava flow, formed as liquid magma drains out from beneath a solid crust.

light   Electromagnetic radiation that is visible to the eye.

light-year   The distance light travels in a vacuum in one year, about 9.46 trillion km.

limestone   A sedimentary rock composed mostly of calcium carbonate.

liquid   The state of matter in which a substance flows freely and lacks crystal structure. Unlike a gas, a liquid retains the same volume independent of the shape of its container.

lithosphere   The relatively rigid outer zone of a planetary body that includes the crust and part of the mantle above the softer asthenosphere.

loess   Unconsolidated, wind-deposited dust.

longitude   An east–west coordinate on the surface of a planet ranging from 0 to 360 degrees, or from 0 to 180 degrees east and west of a central meridian.

longitudinal dune   An elongate sand dune oriented in the direction of the prevailing wind.

M

mafic rock   An igneous rock containing more than 50 percent minerals that are rich in iron and magnesium silicates.

magma   A mobile melt that can contain suspended crystals and dissolved gases as well as liquid. A magma must have high temperature relative to the surface temperature of a planet. Thus water is not a magma on Earth, but on the cold, icy moons of the outer solar system, liquid water is much hotter than the surface and behaves like a magma.

magnetic field   The volume affected by the magnetism of an object. Planets generate magnetic fields by internal dynamos.

magnetic pole   One of several (usually two for a dipole) points on a planet at which the density of magnetic lines of force is the highest. The needle of a compass aligns itself along these lines of force.

magnetic reversal   A complete 180-degree reversal of the polarity of a planet's magnetic field.

mantle   The zone of a planetary body's interior between the base of the crust and the core.

mare *(pl. maria)*   Any of the relatively smooth, low, dark areas of the Moon. The lunar maria were formed by the extrusion of floods of basaltic lava.

mass   A measure of the total amount of material in an object.

mass movement   The transfer of rock and soil downslope by direct action of gravity without a flowing medium (such as a river or glacial ice).

mechanical weathering   The breakdown of rock into smaller fragments by physical processes such as frost wedging.

melt   A substance altered from the solid state to the liquid state.

mesa   A flat-topped, steep-sided highland capped with a resistant rock formation. A mesa is smaller than a plateau but larger than a butte.

metamorphic   Pertaining to the processes or products of metamorphism.

metamorphism   Alteration of the minerals and textures of a rock by changes in temperature and pressure and by a gain or loss of chemical components.

meteor   A meteorite in transit through a planet's atmosphere before it strikes the surface; a shooting star.

meteorite   Any particle of solid matter that has fallen to Earth, the Moon, or another planet from space.

meteoroid   A meteorite before it reaches a planet.

midocean ridge   A divergent plate boundary formed in one of Earth's ocean basins where new lithosphere is formed.

molecule   Two or more atoms bound together; the smallest particle of a chemical compound.

moon   A natural satellite of one of the nine planets in the solar system.

mountain   A general term for any landform that stands above its surroundings. In the stricter geologic sense, a mountain belt is a highly deformed part of planet lithosphere created by tectonic processes rather than volcanic or impact processes.

multiring basin   Large impact crater surrounded by several concentric rings.

----------- **N** -----------

nebula   A body of gas and dust residing within a galaxy. Some nebulas are the birthplaces of stars and planets.

neutron   An elementary particle that resides in the nucleus of an atom. A neutron has about the same mass as a proton but has no electrical charge.

normal fault   A steeply inclined fault in which the hanging wall has moved downward in relation to the footwall.

nova   A star that suddenly increases in brightness by hundreds to thousands of times and then fades away gradually. Novas develop as relatively large stars explode from internal nuclear reactions.

nuclear reaction   Reaction between or in atomic nuclei; contrasts with chemical reactions that involve the electrons of an atom. Such reactions include fission, fusion, and radioactive decay.

nucleus, atomic   The dense central part of an atom composed of neutrons and protons. The nucleus is positively charged and contains most of the mass of an atom.

nucleus, comet   The central portion of a comet composed of icy solids that vaporize to form a head when the comet is in the part of its orbit near the Sun.

----------- **O** -----------

occultation   An eclipse of a star or a planet by another planetary object.

olivine   A silicate mineral with magnesium and iron but no aluminum. Olivine is common in the mantles of the inner planets.

Oort cloud   A hypothetical group of comets that orbit the Sun at great distances. Chance encounters with stars or other comets change the orbits of these icy bodies to make them enter the inner solar system.

orbit   The path followed by one body in its revolution about another. Determined by gravitational interactions, planetary orbits range from circular to extremely elliptical. The planets follow orbits around the Sun, and satellites have orbits around a planet.

orogenic   Pertaining to deformation of a planet's lithosphere to the extent that a folded mountain belt is formed.

outer planets   The planets beyond the orbit of Mars—Jupiter, Saturn, Uranus, Neptune, and Pluto.

outflow channels   Large channels on Mars thought to be created by catastrophic floods of water released from below the surface of the planet.

oxidation   Chemical combination of oxygen with another substance.

----------- **P** -----------

p-wave   A compressional seismic wave or "primary" wave.

palimpsest, crater   An impact crater that has been nearly destroyed by the rebound of the crater floor by isostatic adjustment. Palimpsests are common on the icy surfaces of the moons of Jupiter.

parabolic dunes   A dune shaped like a parabola with the concave side toward the wind.

partial melting   The process by which minerals with low melting points liquefy within a rock body as a result of an increase in temperature or a decrease in pressure (or both) while other minerals in the rock are still solid.

patera   Pertains to low-profile volcanoes found on Mars and Io.

patterned ground   Distinctive geometric patterns created by alternate freeze-thaw processes. Polygonal patterns of cracks are one type of patterned ground found in Earth's polar regions and on Mars.

period   A time interval; for example, a geological period or the time required to complete one orbit.

permafrost   Permanently frozen ground.

physiographic   Pertains to the assemblage of landforms or topography in a region on the surface of a planet.

plagioclase feldspar   A group of feldspar minerals that range from calcium to sodium aluminum silicates. Plagioclase is common at the surface of the inner planets.

planet   A relatively large body that orbits a star. Planets are not luminous. Smaller bodies include asteroids and comets.

planetary nebula   A shell of gas expanding away from the surface of a small star. For example, the Sun will die when nuclear burning reactions reach its surface, ripping a layer of gas away to create a planetary nebula.

planetesimal   Small solid bodies from which the planets accreted.

plate   A broad segment of the lithosphere (including the rigid upper mantle plus the crust) that floats on the underlying asthenosphere and moves independently of other plates.

plate tectonics   The theory of planetary dynamics in which the lithosphere is broken into individual plates that participate in convection of the upper mantle of a planet. The lithosphere on such a planet is created and destroyed by recycling back into the mantle.

plateau   An extensive upland region.

pluton   A body of intrusive igneous rock.

pore fluid   A fluid, such as groundwater, that occupies pore spaces of a rock.

pore space   The spaces within a rock body that are unoccupied by solid material. Pore spaces include spaces between grains, fractures, gas bubbles, and voids formed by dissolution.

potential energy   Energy stored in an object that can be converted to another form; for example, the gravitational potential energy of an object increases with height above the surface of a planet and can be converted to kinetic energy if the object is allowed to fall back to the surface.

proton   A relatively large elementary particle with a positive charge. Together with neutrons, protons reside in the nuclei of atoms.

protostar   A forming star, before nuclear fusion starts.

pyroclastic   Pertaining to fragmental rock material formed by volcanic explosions.

pyroxene   A group of silicate minerals composed of single chains of silicon–oxygen tetrahedra. Most are rich in iron and magnesium.

Q

quartz   A silicate mineral composed of silicon–oxygen tetrahedra joined in a three-dimensional network.

R

radar   A technique for observing distant objects that uses reflected radio waves.

radiation   The process of energy transfer by electromagnetic waves.

radio telescope   A telescope designed to examine radio waves emitted or reflected from distant objects.

radioactive decay   The spontaneous disintegration of an atomic nucleus with the emission of energy.

radiogenic heat   Heat generated by radioactivity.

radiometric dating   Determination of the age in years of a rock or mineral by measuring the proportions of an original radioactive material and its decay product.

rampart craters   Impact craters surrounded by ejecta that has fluid-flow features. The ejecta is thought to have contained water, helping it flow across the surface of a planet.

rarefaction   The process by which the floor of an impact crater returns to the normal pressure of a planet's surface after the passage of a compressive shock wave.

ray crater   An impact crater that has a system of rays extending away from the crater rim. Only young craters have rays.

recrystallization   Reorganization of elements of the original minerals in a rock resulting from changes in temperature, pressure, or the activity of pore fluids.

refractory   A material that melts or vaporizes only at very high temperatures. Silicate minerals are refractory compared to ices.

regolith   The blanket of soil and loose rock fragments overlying solid rock on a planet's surface.

relative age   The age of a rock or an event as compared with some other rock or event.

relative dating   Determination of the order of a series of events in relation to one another without reference to their ages measured in years. Relative geologic dating is based primarily on superposition, faunal succession, and cross-cutting relations.

relative time   Geologic time as determined by relative dating, that is, by placing events in chronologic order without reference to their ages measured in years.

relief   The difference in altitude between the high and the low parts of an area.

remnant magnetism   Permanent magnetism of rocks in a planet's lithosphere as opposed to that caused by the planet's magnetic field.

remote sensing   Any means of determining the physical or chemical properties of an object from a distance. Telescopes are remote-sensing devices, but so too are the cameras on spacecraft.

resolution   The degree to which small objects can be seen on an image or photograph of the surface of a planet. High-resolution images reveal very small objects, whereas low-resolution images allow only large features to be detected.

resonance, orbital   Pertaining to the orbital periods of satellites that are simple multiples of one another. If one satellite orbits its primary twice for every orbit of a satellite closer to the planet, it displays a 2:1 resonance. Tidal interactions are accentuated by resonance in satellite systems.

retrograde motion   An orbital or spin direction opposite to most bodies in the solar system. As viewed from the north, most planets and satellites move in a counterclockwise direction; retrograde motion would thus be clockwise rotation, or revolution.

reverse fault   A fault in which the hanging wall has moved upward in relation to the footwall—a high-angle thrust fault.

revolution   The orbital motion of one body around another.

rhyolite   A fine-grained volcanic rock composed of quartz, alkali feldspar, and plagioclase. It is the extrusive equivalent of a granite.

rift system   A system of faults resulting from extension.

rift valley   A valley of regional extent formed by block faulting in which tensional stresses tend to pull the crust apart. Synonymous with **graben.**

rille   An elongate trench, or cracklike valley on the Moon's surface. Rilles can be sinuous (lava channels or collapsed lava tubes) or relatively linear structural depressions (grabens).

ring   A stream of small particles that orbits one of the outer planets. The particles are usually made of ice.

river system   A river with all of its tributaries.

roche limit   The orbit closest to a planet where a satellite can withstand the tidal forces exerted by its primary. Inside the roche limit, a moon will be disrupted. The limit varies for bodies of different sizes and different strengths, but it is usually calculated for a satellite with very low strength, like liquid water.

rock   An aggregate of minerals that forms the solid part of a planetary body.

rotation   Spinning of a body about an axis running through it.

S

sand   Sedimentary material composed of fragments ranging in diameter from 0.0625 to 2 mm. Sand particles are larger than silt but smaller than pebbles. Much sand on Earth is composed of quartz grains, but other materials can also form sand.

sandstone   A sedimentary rock composed mostly of sand-size particles, usually cemented by calcite, silica, or iron oxide.

sapping   The process by which valleys are created by the erosion and undercutting caused by spring water.

satellite   A planetary body that orbits about a larger one; for example, a moon of a planet.

scarp   A cliff produced by faulting or erosion.

sea ice   Relatively thin, floating masses of ice that form on terrestrial oceans by freezing of sea water, in contrast to icebergs that form when glaciers enter the sea.

seamount   An isolated, conical mound rising more than 100 m above Earth's ocean floor. Seamounts are probably submerged shield volcanoes formed above hot spots.

secondary crater   A crater formed by the impact of material ejected from another crater.

sediment   Material (such as gravel, sand, and dust) that is transported and deposited by wind, water, ice, or gravity; material that is precipitated from solutions that exist at surface temperatures.

sedimentary rocks   Rock formed by the accumulation and consolidation of sediment.

seismic   Pertaining to earthquakes or to waves produced by natural or artificial earthquakes.

seismograph   An instrument used to detect seismic waves.

shale   A fine-grained, clastic sedimentary rock formed by compaction of clay and mud.

shield   An extensive area of Earth's continents where igneous and metamorphic rocks are exposed and have relatively low relief and flat surfaces. Rocks of the shield are usually old and many formed as the roots of ancient mountain belts that have now been eroded away.

shield volcano   A large volcano shaped like a flattened dome and built up almost entirely of numerous flows of fluid (usually basaltic) lava. The slopes of shield volcanoes seldom exceed 10 degrees, so that in profile they resemble a shield or broad dome.

shock wave   A strong compressional wave that lasts for a very short period of time, as, for example, when a meteorite strikes the surface of a planet.

silicate   A variety of rock-forming minerals composed of variously arranged tetrahedra of silicon and oxygen plus other elements. Silicate minerals are especially important in the crusts and mantles of the inner planets and the cores of the icy satellites of the outer planets.

slope retreat   Progressive recession of a scarp or the side of a hill or mountain by mass movement or stream erosion.

slump   A type of mass movement in which material moves along a curved surface of rupture.

solar system   The system that includes the Sun, planets, moons, asteroids, comets, and other objects that orbit the Sun.

solar wind   The stream of ions flowing away from the Sun.

soil   The surface material of the planets, produced by the disintegration of rocks.

solid   The state of matter in which substance has a definite shape and volume and some fundamental strength.

spatter cone   A low, steep-sided volcanic cone built by accumulation of splashes and spatters of lava (usually basaltic) around a fissure or other vent.

spectral   Pertaining to the various wavelengths of electromagnetic radiation; for example, a spectral class of asteroids is defined based on the character of the light reflected from it.

spectrum   The variety of wavelengths (colors for visible light) in electromagnetic radiation; a prism separates these colors to show the spectrum.

spin-orbit coupling   Pertains to the orbital and rotational movements of a planet. Coupling occurs when the rotation rate of a planet or a moon is a simple multiple (one, two, or three times) of its revolution rate around the Sun or a planet. The spin and the orbit of the Moon are coupled in such a fashion.

spreading center   A plate boundary formed by tensional stress along a terrestrial oceanic ridge. Synonymous with **midocean ridge, divergent plate boundary.**

sputtering   The process of ejection of small amounts of material from a planet's surface as the result of an impact of ions from the solar wind. This process is only effective on bodies that lack atmospheres.

stable platform   The part of a terrestrial continent that is covered with flat-lying or gently tilted strata and underlain by a complex of igneous and metamorphic rocks. The stable platform has not been extensively affected by crustal deformation.

star   A large self-luminous object that creates energy by nuclear fusion. The Sun is the star at the center of our solar system.

star dune   A mound of sand with a high central point and arms radiating in various directions.

stony meteorite   A meteorite composed mostly of silicate minerals, in contrast to metallic iron. Achondrites and chondrites are stony meteorites.

stony-iron meteorite   A meteorite composed mostly of an intimate mixture of silicates and iron metal.

stratigraphy   The study of rock strata on or near the surface of a planet.

stratovolcano   A volcano built up of alternating layers of ash and lava flows. Synonymous with **composite volcano.**

stratum (*pl. strata*)   A layer of rock.

stream valley   A valley produced by fluvial erosion.

strike-slip fault   A fault in which movement has occurred parallel to the trend of the fault.

subaerial   Occurring beneath the atmosphere or in the open air, with reference to conditions or processes (such as erosion) that occur there.

subduction zone   An elongate zone in which one lithospheric plate descends beneath another, a fundamental feature of Earth's plate tectonic system.

sublimation   The process by which a material changes state from a solid directly to a vapor without passing to a liquid.

subsidence   A sinking or settling of a planet's lithosphere with respect to the surrounding parts.

supernova   A very large nova, or stellar explosion.

superposition, principle of   The principle that, in a series of strata that has not been overturned, the oldest rocks are at the base and the youngest are at the top.

s-wave   A seismic "shear" wave.

synchronous rotation   The rotation of a satellite or planet that has equal orbital and rotational periods; 1:1 spin–orbit coupling.

---------------   **T**   ---------------

T Tauri star   A variable star showing rapid and erratic changes in the electromagnetic energy it emits. T Tauri stars are thought to be young stars from which large amounts of matter are swept away during this stage of their development.

tail, comet   Gases and particles of solid that stream away from a comet in the inner solar system.

tectonics   Regional or global structures and deformational features of a planetary object.

tension   Stress that tends to pull materials apart.

terminator   The line separating the sunlit from the dark hemisphere of a planetary object.

terra (*pl. terrae*)   A densely cratered highland on the Moon.

terrace   A nearly level surface forming narrow shelves on the inside of a crater.

terrestrial planets   The planets most like Earth, with lithospheres of silicate minerals—Mercury, Venus, the Moon, Earth, and Mars.

tessera   Terrains on Venus that have been intensely modified by tectonic processes. They consist of interlacing ridges and valleys.

thermal energy   Heat energy associated with the motion of the particles in a solid, liquid, or gas.

thermo-karst   A karstlike terrain created by collapse as volatile cements or rocks (like water ice) sublime, rather than by solution of rock in water.

thermonuclear reaction   Fusion under conditions of high temperature such as in stars.

thrust fault   A low-angle fault (45 degrees or less) in which the hanging wall has moved upward in relation to the footwall.

tidal heating   The process of frictional heating of a planetary object by the alternate growth and decay of a tide in its lithosphere.

tide   Periodic deformation of the lithosphere or hydrosphere of a planetary body that is caused by the gravitational attraction of another object such as the Sun, a satellite, or a planet.

topography   The shape and form of a planetary body's surface.

trajectory   The path followed by a projectile.

transform fault   A special type of strike-slip fault forming the boundary between two moving lithospheric plates, usually along an offset of the oceanic ridge. A fundamental feature of Earth's plate tectonic system.

transverse dune   An asymmetrical dune ridge that forms at right angles to the direction of prevailing winds.

trench   A narrow elongate depression of Earth's oceanic lithosphere oriented parallel to the trend of a continent or island arc. Trenches mark the sites of subduction zones.

tributary   A stream flowing into or joining a larger stream.

turbulent flow   A type of flow in which the path of motion is very irregular, with eddies and swirls.

---------------   **U**   ---------------

undifferentiated   Pertains to a planet that has not become internally differentiated and is thus relatively homogeneous.

upwarp   An arched or uplifted segment of a planetary lithosphere.

V

vent The point of extrusion for volcanic materials. Vents may be fissures, craters, cinder cones, etc.

ventifact A pebble or cobble shaped and polished by wind abrasion.

vesicle A small hole formed in a volcanic rock by a gas bubble that became trapped as the lava solidified.

viscosity The tendency within a body to resist flow. An increase in viscosity implies a decrease in fluidity, or ability to flow.

volatile A substance that can be vaporized at a relatively low temperature. Water and carbon dioxide are volatile materials.

volcanic ash Dust-size particles ejected from a volcano.

volcano The accumulation of extrusive igneous rocks around a volcanic vent.

volume A measure of the amount of space occupied by an object.

W

weathering The process by which rocks are chemically altered or physically broken into fragments as a result of exposure to atmospheric agents and the pressures and temperatures at or near the surface of a planet, with little or no transportation of the loosened or altered materials.

wind shadow The area behind an obstacle where air movement is not capable of moving material.

wrinkle ridge A sinuous, irregular segmented ridge on the surface of the lunar maria, Mars, and Mercury, believed to be the result of deformation of the lava.

Y

yardang An elongate ridge carved by wind erosion.

ILLUSTRATION CREDITS

Cover Paintings Teryl Bodily.

Figures 1.2, 1.6, 1.7, 1.8, 1.9 1.10, 1.11, 1.12, 1.13, 1.15 Jet Propulsion Laboratory.

Figures 2.4, 2.5 Lick Observatory, University of California, Santa Cruz.

Figure 2.12 Modified after W. K. Hartmann, 1983, *Moons and Planets*, 2nd ed.

Figure 2.22 National Air Photo Library, Dept. of Energy, Mines, and Resources, Canada.

Chapter 3 Opener Jet Propulsion Laboratory.

Figures 3.1, 3.3, 3.4, 3.5 Center for Meteorite Studies, Arizona State University.

Figure 3.7 Modified after A. Chaikin, 1981, in *The New Solar System*.

Figures 3.8, 3.9, 3.12 Jet Propulsion Laboratory.

Figure 3.10, 3.11 National Space Science Data Center, NASA.

Figure 3.13 Modified after H. V. McSween, *Meteorites and Their Parent Planets*.

Figure 3.14 Modified after C. R. Chapman, 1981, in *The New Solar System*.

Figure 3.15 Modified after H. Y. McSween, *Meteorites and Their Parent Planets*.

Chapter 4 Opener Teryl Bodily.

Figure 4.1 Kit Peak Observatory.

Figure 4.3 After E. M. Shoemaker, 1960.

Figure 4.4 Center for Meteorite Studies, Arizona State University.

Figure 4.6 U.S. Geological Survey, Astrogeology Branch.

Figures 4.2, 4.7, 4.8, 4.9, 4.10, 4.11, 4.12, 4.13, 4.14, 4.15, 4.17, 4.18, 4.19, 4.20, 4.21, 4.23, 4.26, 4.27, 4.28, 4.29, 4.30, 4.31, 4.32, 4.34, 4.35, 4.36, 4.37 National Space Science Data Center, NASA.

Figure 4.22 Modified after J. E. Guest and R. Greeley, 1977, *Geology on the Moon*.

Figure 4.24, 4.46 Modified after W. K. Hartmann, 1983, *Moons and Planets*, 2nd ed.

Figures 4.39, 4.40, 4.41, 4.42, 4.43 G. H. Ladel, NASA.

Chapter 5 Opener National Space Science Data Center, NASA.

Figure 5.1, 5.3, 5.4, 5.5, 5.6, 5.7(A), 5.8, 5.9, 5.10, 5.11, 5.12, 5.13, 5.14, 5.15 National Space Science Data Center, NASA.

Figure 5.2 After B. C. Murray and others, 1975, *Journal of Geophysical Research*.

Figure 5.7(B), 5.15, 5.18 Modified after R. G. Strom, 1987, *Mercury: The Elusive Planet*.

Figure 5.17 After W. K. Hartmann, 1983, *Moons and Planets*, 2nd ed.

Chapter 6 Opener A. McEwen, U.S. Geological Survey Astrogeology Branch.

Figure 6.2(A), 6.3(A), 6.13, 6.17, 6.36, 6.43, 6.50, 6.51 U.S. Geological Survey Astrogeology Branch.

Figure 6.2(B), 6.3(B), 6.5, 6.6, 6.7, 6.8, 6.9, 6.10, 6.11, 6.15, 6.16, 6.18, 6.20, 6.21, 6.22, 6.23, 6.24(A), 6.24(B), 6.25, 6.26, 6.27, 6.29, 6.30, 6.31, 6.32, 6.33, 6.35, 6.37, 6.38, 6.39, 6.40, 6.41, 6.42, 6.44, 6.45, 6.46 National Space Science Data Center, NASA.

Figure 6.14 National Air Photo Library, Dept. of Energy, Mines, and Resources, Canada.

Figure 6.19 EROS Data Center. Sioux Falls, SD.

Figure 6.34 A. McEwen, U.S. Geological Survey Astrogeology Branch.

Figure 6.47 Modified after M. H. Carr, 1981, *The Surface of Mars*.

Figures 6.48, 6.49 J. Inge, U.S. Geological Survey Astrogeology Branch.

Figure 6.52 Modified after W. K. Hartmann, 1983, *Moons and Planets*, 2nd ed.

Figure 6.53 Modified after T. A. Mutch and others, 1976, *The Geology of Mars*.

Chapter 7 Opener Jet Propulsion Laboratory.

Figure 7.1, 7.2 Jet Propulsion Laboratory.

Figure 7.3, 7.5   U.S. Geological Survey Astrogeology Branch.

Figure 7.4, 7.6, 7.14, 7.15, 7.17, 7.19, 7.20, 7.21, 7.22, 7.25, 7.26, 7.28, 7.30, 7.32, 7.34, 7.35, 7.36, 7.37, 7.38, 7.40, 7.41, 7.42, 7.43   G. H. Pettengill, Magellan Project, Planetary Data System, and National Space Science Data Center, NASA.

Figures 7.7, 7.8, 7.9, 7.10   Jet Propulsion Laboratory.

Figure 7.12, 7.47   Modified after R. J. Phillips and M. C. Malin, 1983, in *Venus*.

Figure 7.13, 7.18, 7.46   Modified after W. K. Hartmann, 1983, *Moons and Planets*, 2nd ed.

Figure 7.23, 7.27   Modified after J. W. Head and others, 1992, *Journal of Geophysical Research*.

Figure 7.24   D. L. Bindschadler and G. Schubert and W. M. Kaula, University of California, Los Angeles.

Figure 7.29, 7.31, 7.45   Modified after J. W. Head and L. Crumpler, 1992, *Science*.

Figure 7.38(B)   S. W. Squyres, Cornell University.

Figure 7.39   Modified after S. W. Squyres and others, 1992, *Journal of Geophysical Research*.

Figure 7.44   Modified after E. R. Stofan and others, 1992, *Journal of Geophysical Research*.

Chapter 8   Opener EOSAT Corporation.

Figure 8.1, 8.27   National Space Science Data Center, NASA.

Figure 8.4   Modified after P. C. Cloud, 1979, *Scientific American*.

Figure 8.8(B), 8.39, 8.51, 8.54   National Air Photo Library, Dept. of Energy, Mines, and Resources, Canada.

Figure 8.10, 8.13(B), 8.46, 8.55   EROS Data Center.

Figure 8.11   U.S. Geological Survey.

Figure 8.12   U.S. Geological Survey Astrogeology Branch.

Figure 8.14   Woods Hole Oceanographic Institute.

Figure 8.15   K. C. Macdonald.

Figure 8.19, 8.24, 8.31, 8.42, 8.48   Earth Satellite Corp.

Figure 8.23(B), 8.55   Modified after W. K. Hartmann, 1983, *Moons and Planets*, 2nd ed.

Figure 8.25   V. Sharpton, Lunar and Planetary Institute.

Figure 8.28, 8.29, 8.32, 8.40, 8.44, 8.52   U.S. Department of Agriculture Air Photo Lab.

Figure 8.30   Modified after W. Hildreth, 1981, *Journal of Geophysical Research*.

Chapter 9 Opener A. McEwen, U.S. Geological Survey Astrogeology Branch, and Teryl Bodily.

Figures 9.1, 9.3, 9.8, 9.10, 9.11, 9.12, 9.15, 9.24, 9.25, 9.26, 9.27, 9.28, 9.29, 9.31, 9.33, 9.34, 9.35   Jet Propulsion Laboratory.

Figure 9.5, 9.6, 9.7, 9.14, 9.16, 9.17, 9.18, 9.20, 9.22, 9.23, 9.32   U.S. Geological Survey Astrogeology Branch.

Figure 9.9   A. McEwen, U.S. Geological Survey Astrogeology Branch.

Chapter 10   Opener Jet Propulsion Laboratory.

Figures 10.1, 10.3, 10.4, 10.5, 10.6, 10.7, 10.8, 10.10, 10.11, 10.12, 10.15, 10.18, 10.20, 10.22, 10.25, 10.26, 10.30, 10.31, 10.33, 10.34   Jet Propulsion Laboratory.

Figure 10.16   Modified after W. K. Hartmann, 1983, *Moons and Planets*, 2nd ed.

Figure 10.17, 10.19, 10.21, 10.23, 10.24, 10.27, 10.28, 10.29, 10.32   U.S. Geological Survey Astrogeology Branch.

Chapter 11   Opener Jet Propulsion Laboratory.

Figures 11.1, 11.4, 11.5, 11.6, 11.7, 11.8, 11.9, 11.11 (bottom), 11.12 (bottom), 11.13 (bottom)   Jet Propulsion Laboratory.

Figure 11.10, 11.11 (top), 11.12 (top), 11.13 (bottom)   U.S. Geological Survey Astrogeology Branch.

Chapter 12   Opener Teryl Bodily.

Figures 12.1, 12.3, 12.4, 12.5, 12.6, 12.7, 12.10, 12.11, 12.12, 12.13   Jet Propulsion Laboratory.

Figure 12.9   A. McEwen, U.S. Geological Survey Astrogeology Branch.

Chapter 13   Opener Teryl Bodily.

Figure 13.2   U.S. Naval Observatory.

Figure 13.3   Hubble Space Telescope Science Institute, NASA.

Chapter 14   Opener Teryl Bodily.

Figure 14.2, 14.3   Jet Propulsion Laboratory.

Figure 14.5, 14.6(D)   H. Hammel, MIT, and Hubble Space Telescope Science Institute, NASA.

Figure 14.6(A)   Modified after Sandia National Laboratory, 1994, *EOS*.

Figure 14.6(B)   D. Seal, Jet Propulsion Laboratory.

Figure 14.6(C)   P. McGregor, Siding Spring Observatory, Australian National University.

Chapter 15   Opener National Space Science Data Center, NASA.

Figure 15.4   Modified after L. A. Soderblom and T. V. Johnson, 1983, *Scientific American*.

Figure 15.6   Modified after C. J. Allegre, 1992, *From Stone to Star*.

INDEX